Transgenic Herbicide
Resistance in Plants

Transgenic Herbicide Resistance in Plants

V.S. Rao
International Weed Scientist
and Affiliate Member
Department of Plant Sciences
University of California
Davis, CA
USA

CRC Press
Taylor & Francis Group
Boca Raton London New York

CRC Press is an imprint of the
Taylor & Francis Group, an **informa** business
A SCIENCE PUBLISHERS BOOK

Cover illustrations provided by the author, Dr. V.S. Rao.

CRC Press
Taylor & Francis Group
6000 Broken Sound Parkway NW, Suite 300
Boca Raton, FL 33487-2742

First issued in paperback 2020

© 2015 by Taylor & Francis Group, LLC
CRC Press is an imprint of Taylor & Francis Group, an Informa business

No claim to original U.S. Government works

ISBN-13: 978-1-4665-8737-3 (hbk)
ISBN-13: 978-0-367-73860-0 (pbk)

Visit the Taylor & Francis Web site at
http://www.taylorandfrancis.com

and the CRC Press Web site at
http://www.crcpress.com

Dedication

This book is dedicated to Sri Rāmachandra

and

To my loving wife Nirmala Devi

Preface

Currently, the field of weed science is facing a growing challenge in the form of rapid evolution of weeds resistant to a wide spectrum of herbicides whose use has tremendously increased ever since the commercialization of phenoxyacids, beginning with 2,4-D, in 1946. Initially, plant biologists and weed scientists linked the chemistry of the herbicide to the biology of plants and selected suitable newer herbicides to combat global weed problems both in cropped and non-cropped situations. Over the past 45 years, these herbicides have also become the cause and effect of about 240 weed species showing resistance to 155 of them belonging to 22 of the 25 known herbicide sites of action around the world. This trend is not going to abate any time soon. Instead, it is expected to exacerbate.

An altogether different, but not unrelated, development took place in the late 1980s which culminated in the use of genetic engineering to produce crop varieties resistant to herbicides, particularly the nonselective ones whose field use is precluded because of crop toxicity. In this fast-developing area, termed transgenic engineering, two approaches are usually followed to insert herbicide resistance into a crop plant. In one approach, either the plant enzyme or other sensitive biochemical target of herbicide action was made to be insensitive to the herbicide, or the unmodified target protein was induced to overproduce, thus permitting normal plant metabolism to occur. The other approach was the introduction of an enzyme or enzyme system that degrades or detoxifies the compound in the plant before the herbicide reaches the site of action. Plants modified by both approaches were obtained by transferring non-plant (and even from within the plant kingdom) genes that encode herbicide resistance traits into the target plant genome. In doing so, plant scientists linked the biology of plants to the chemistry of the herbicide.

In this process of genetic engineering, scores of herbicide-resistant transgenic varieties, beginning with the bromoxynil-resistant 'BXN' cotton line in 1994, have been developed in several crops. This technology was taken further by applying it to engineer resistance in crops to other biotic stresses like insects, plant pathogens, etc. as well abiotic stresses, besides improving qualitative and quantitative traits of crops. Currently, two transgenic traits—herbicide-resistance and insect-resistance—dominate the crop biotech technology by accounting for about 95 percent of the global area under biotech varieties in maize, soya bean, cotton, and rapeseed (canola).

Despite the obvious pecuniary and non-pecuniary benefits derived by farmers of more than two dozen countries which adopted these transgenic crop varieties over the past two decades, transgenic technology has raised more risks and issues so far than could be resolved both at the farmer level and consumer level. After all, farmers and consumers are the vital components in the success of any agricultural technology. Consequently, development of transgenic crops has been viewed more as a profit-driven rather than need-driven process.

At the present time, few topics in global agriculture are more polarizing and controversial than transgenic engineering.

However, transgenic engineering has the arsenal to address various problems plaguing agriculture in a more effective way than ever and to meet the future food and fiber needs of the rising global population. As in the case of the atom, electronics, computers, and communication, the more we understand the intricacies of this plant-related technology, the better we will be able to utilize it for the betterment of mankind.

This work has been designed to bring out a comprehensive reference-cum-textbook on the basic principles of herbicide resistance and transgenic engineering besides a detailed discussion on the current status of transgenes used to engineer herbicide (also insect) resistance in crops, and development of herbicide-resistant transgenic crop events and stacked varieties and hybrids. The book also deals with the role of transgenic technology in phytoremediation of soils and environment from organic contaminants such as herbicides, insecticides, fungicides, oil spills, explosives, etc. and inorganic pollutants that include natural elements. Additionally, it deals with other chapters that discuss the global adoption and regulation of transgenic crops, while bringing to the fore the benefits transgenic crops were reported to have derived to the global community and discussing the various risks and issues that concern and affect both farmers and consumers. The presentation is based on facts, tinged with impartiality.

As it is impossible to review the entire research done in each of the areas under discussion, I chose to include only the more useful material to bring relevance and objectivity to the subject under discussion. I also chose to exclude some topics because of space constraints. I accept full responsibility for choice of information, presentation, interpretation, and discussion. However, I cannot assume responsibility for the contents of the original references as well as the success or otherwise of a gene expression, gene-sequencing, genotyping, transformation methods/protocols, etc. The reader is advised to refer to the original source for more information. If trade and commercial names are mentioned, it was only as a matter of convenience to the reader but not as an endorsement of a particular product or variety. I generally followed American orthography while writing this book.

During the course of working on this edition, I have been the recipient of constant encouragement given by my loving wife, Nirmala Devi, as well as my beloved daughter, Madhavi Lata Rajavasireddy, and beloved son Srinivas R. Vallurupalli. Not to be outdone by them were my son-in-law Rajiev and daughter-in-law Neelima, my dear grandsons Nikhil and Milind, and granddaughters Ria, Reva and Rayna who have always surrounded me and given me immense joy.

Acknowledgment

I have reproduced in this book certain copyrighted material, including some figures and the accompanying discussions and material published in scientific journals, books, and the Internet. I have received help in other ways too to help me prepare this book. I gratefully acknowledge the help extended and permissions granted to the following highly regarded scientists and individuals.

1. Dr. Joseph Di Tomaso, Professor of Weed Science and Director of the Weed Research and Information Center, Department of Plant Sciences, University of California, Davis, California, USA for facilitating access to the library facility by having me nominated as an affiliate member of the University.

2. Dr. Chris van Kessel, Chair and Professor of Department of Plant Sciences, University of California, Davis, California, USA for approving my name for affiliate membership of the University to facilitate access to its library facility.

3. Dr. Ian Heap, the Australian weed scientist and Director of the International Survey of Herbicide-Resistant Weeds for granting me permission to use a significant portion of the material available at his website www.weedscience.org. on "International Survey of Herbicide Resistant Weeds."

4. Dr. Govindjee, Professor Emeritus, Biochemistry, Biophysics, and Plant Biology, University of Illinois, Urbana, Illinois, USA for readily providing me the "Z-Scheme of electron transport in photosynthesis" published by Govindjee and Wilbert Veit in 2010 and granting permission to insert it in Chapter 2.

5. Dr. Patrick F. Byrne, Professor of Plant Breeding and Genetics, Department of Soil and Crop Sciences, Colorado State University, Fort Collins, USA for providing the "Simplified representation of a gene construct that contains a transgene, promoter, and a terminator inserted into plant's genome" and granting permission to insert it in Chapter 4.

This page will be incomplete without my acknowledging the inestimable and invaluable help provided by my loving wife Nirmala Devi for ungrudgingly enduring the loss of precious personal life while I was obsessed with the sole objective of completing the manuscript by the due date. My heartfelt gratitude also goes out to my beloved daughter Madhavi Lata Rajavasireddy and son Srinivas R. Vallurupalli for providing me their computer expertise and taking care of computer related problems at crucial stages of preparing the manuscript and publication of the book.

Contents

Acronyms

Genes, Gene Elements

*Aac*C3; *aac*C4 **genes**	3-*N*-aminoglycoside acetyl transferases 3 and 4
ACCase	Acetyl-CoA carboxylase enzyme
*aad*1 **gene**	Aryloxyalkanoate dihydrogenase gene 1 (from *Sphingobium herbicidovorans*)
addA	Aminoglycoside-3-adenyltransferase gene
adh1	Alcohol dehydrogenase gene 1 (class I)
AHAS	acetohydroxyacid synthase gene
alcA	Alcohol dehydogenease I gene promoter
AlcR	alcohol dehydogenease transactivator protein
als	Acetolactate synthase gene
m*ALS*	mutated Acetolactate synthase gene
*amp*R	Ampicillin restriction gene
*aph*IV	Aminoglycoside phosphotransferase type IV gene
APX	Ascorbate peroxidase promoter
Arab-SSUlA/CTPl	**A. thaliana* small subunit 1A ribulose-1,5-bisphosphate carboxylase
ARG7	Arginosuccinate lyase locus 7
AroA	Aaromatic amino acid gene (glyphosate-resistance)
aroA-M1	Aaromatic amino acid gene mutant 1 gene
AtuORF1 3'UTR	**A. tumefaciens* 3′ untranslated region (UTR) comprising the transcriptional terminator and polyadenylation site of open reading frame 1
AtuORF23 3'UTR	*A. tumefaciens* 3′ untranslated region (UTR) comprising the transcriptional terminator and polyadenylation site of open reading frame 23.
AtUbi10	*A. thaliana* polyubiquitin UBQ10 comprising the promoter 5′ untranslated region and intron
avhppd-03	**A. fatua* p-hydroxyphenylpyruvate dioxygenase
BADH	Betaine aldehyde dehydrogenase gene
bar	Bialaphos resistance gene
bla	β-lactamase gene, codes for antibiotic ampicillin
ble	Bleomycin resistance gene
bxn	Bromoxynil resistance gene from *Klebsiella anthropic* subsp. *ozaenae*
CaMV 35S	Cauliflower mosaic virus of the 35S RNA promoter
CAT	Chloramphenicol acetyltransferase
*Cla*I	Endonuclease restriction enzyme (cleaves double-stranded DNA) in a site-specific manner: recognition site is: ATCGAT
CMP	Cestrum yellow leaf curling virus

CO1	mitochondrial gene *cytochrome-C oxidase-1*
colE1	Carries a gene for colcin E1 (the *cea* gene), a bacteriocin toxic to bacteria including *E. coli*
cp4 epsps	*epsps* gene from *Agrobacterium* spp. Strain CP4
Cre-*lox*	Cyclization recombination locus X (site-specific recombination system of bacteriophage P1)
cry1A.105	Gene from *B. thuringiensis* subsp. *kumamotoensis*, encodes Cry1A.105 protein comprising of the Cry1Ab, Cry1F and Cry1Ac proteins
cry1Ab	Gene from *B. thuringiensis* subsp. *kurstaki* (*B.t.k.*) strain HD-1, encodes Cry1Ab δ-endotoxin
cry1Ac	Gene from *B. thuringiensis* var. *kurstaki* HD73 encodes Cry1Ac δ-endotoxin
cry1F	Gene from *B. thuringiensis* var. *aizawai* strain PS811, encodes Cry1F protoxin
cry1Fa2	Gene from *B. thuringiensis* var. *aizawai*, encodes modified Cry1F δ-endotoxin
cry2Ab	Gene from *B. thuringiensis* strain 14-1, encodes Cry2Ab δ-endotoxin
cry2Ae	Gene from *B. thuringiensis* subsp. *Dakota*, encodes Cry2Ae δ-endotoxin
cry3Bb1	Gene from *B. thuringiensis* subsp. *kumamotoensis* strain EG4691, encodes Cry3Bb1 δ-endotoxin
cry34Ab1	Gene from *B. thuringiensis* strain PS149B1, encodes Cry34Ab1 δ-endotoxin
cry35Ab1	Gene from B. *thuringiensis* strain PS149B1, encodes Cry35Ab1 δ-endotoxin
CS-dmo	Coding sequence of dicamba monooxygenase
csrt-1	Mutated form of acetolactate synthase (contains a single nucleotide change, resulting in a single amino acid substitution in the ALS protein)
CsVMV	Cassava vein mosaic virus promoter
CTP	Chloroplast transit peptide
ctp2	DNA sequence for the N-terminal of chloroplast transit peptide from *A. thaliana* EPSPS gene
dhfr	Dehydropholate reductase gene
dmo	Dicamba monooxygenase gene from *Stenotrophomonas maltophilia*
dms	Dicamba O-demethylase (dicamba) gene
***Eco*RI**	Endonuclease enzyme isolated from strains of *E. coli*. Recognition site: GAATTC
***Eco*RI site**	*E. coli* recognition site of endonuclease: cleaves the phosphodiester bonds of DNA at specific nucleotide sequences
ecry3.1Ab	Synthetic form of *cry3A* and *cry1Ab* genes from *B. thuringiensis* encodes Cry3A-Cry1Ab δ-endotoxin
5'e1	Tapetum specific E1 gene (GE1) of *Oryza sativa*
epsps	5-enolpyruvylshikimate-3-phosphate synthase gene
epsps **ace5**	Modified *epsps* gene derived from *A. globiformis
epsps **grg23ace5**	Modified *epsps*-glyphosate resistance gene 23 from *A. globiformis
2m-*epsps*	Double mutant *epsps* gene

ept	*Estrogen-induced pituitary tumor 1* gene
E9 3'	3′ non-translated region of the pea RbcS gene E9
EcR	Ecdysone receptors from moths
ER	Human estrogen receptor
Fad	Fatty acid desaturase enzyme
Flp/*frt*	Flp: enzyme flippase; *frt*: flippage recognition target from *Saccharomyces cerevisiae*
FMV 35S	Figwort mosaic virus 35S RNA
FMV/Tsf1	Figwort mosaic virus 35S RNA from the Transferrin 1 gene of *A. thaliana*
gat	Glyphosate *N*-acetyltransferase gene
gat4621	Modified *gat* gene based on the sequences of the three *gat* genes from *Bacillus licheniformis*
GFP	Green fluorescent protein
Gin/*gix*	Gin: G inversion; *gix*: the recombination sites (*Gin-gix* system from bacteriophage Mu)
gm-hra	*Glycine max*-herbicide resistant acetolactate synthase gene
gox	Glyphosate oxidoreductase gene
GT	Glucocorticoid receptor from rat
Gumbi	*Glycine max* ubiquitin
GUS	β-D-glucuoronidase
H4A748	Promoter that codes for the S-phase specific Histone 4
His3	Histidine 3
3′histonAt	The 3′ untranslated region of the histone H4 gene of *A. thaliana*
hpp	Hydroxyphenylpyruvate dioxygenase gene
hppdPfW336	p-hydroxyphenylpyruvate dioxygenase gene of **P. fluorescens* strain A32 modified by replacing amino acid Gly with Trp at position 336
hsp70	Heat shock protein 70 gene
intron1 h3At	First intron of Gene II of the histone H3.III variant of *A. thaliana*
IPT	Isopentenyltransferase gene
*lox*P	Locus X-over P1
manA	Phosphomannose isomerase gene
matK	Chloroplast gene: maturaseK
mcry3A	Synthetic form of *cry3A* gene from *B. thuringiensis* subsp. *tenebrionis*, encodes mcry3A δ-endotoxin
neo	Neomycin phosphotransferase gene
nos	Napoline synthase gene *A. tumefaciens*
nos 3'	Napoline synthase 3′-polyadenylation signal
3′nos	The 3′ polyadenylation signal from non(un)-translated region of napoline synthase from *A. tumefaciens*
*Not*I	DNA restriction enzyme from **N. otitids-caviarum*
*npt*II/*neo*	Neomycin phosphotransferase II from *E. coli*
*npt*III	Neomycin phosphotransferase III from *Streptococcus faecalis R* plasmid
OCS	Octopine synthase from *A. tumefaciens*
ORF	Open reading frame
ORF25 polyA	3′ polyadenylation signal from ORF25 (*A. tumefaciens*)
ORI ColE1	Origin of replication for colcin E1
Ori-pBR322	Origin of replication from pBR322 for maintenance of plasmid in *E. coli*

ori-V	Origin of replication for *Agrobacterium*-derived from the broad host range of plasmid in RK2
ORIpVS1	Origin of replication from the *Pseudomonas* plasmid pVS1
Ori-322	Origin of replication in *E. coli* plasmid pBR322
OsAct2	*Oryza sativa actin2* gene
OsCc1	*Oryza sativa cytochrome c* gene1
OsDMC1	*Oryza sativa* disrupted meiotic complementary DNA1 gene
OsTubA1	*Oryza sativa* α-tubulin gene which consists of four exons and three introns
OTP	Optimized transit peptide which directs translocation of proteins to chloroplasts. It is derived from plant sequences obtained from maize and sunflower ribulose 1,5-bisphosphate carboxylase oxygenase (RuBisCo)
pat	Phosphinothricin acetyl transferase gene *S. viridochromogenes* strain Tu494
pcd	Phenmedipham hydrolase gene derived from *Arthrobacter oxydans* strain P52
PCLSV	Peanut chlorotic streak caulimovirus promoter
P-*e35S*	Cauliflower mosaic virus 35S RNA containing the enhancer region that directs transcription in cells
PG	Polygalacturonase gene
PGD1	Phosphogluconate dehydrogenase 1 promoter
Ph4a748	The promoter region from the gene for histone H4 from *A. thaliana*
Ph4a748At	A sequence including the promoter region of the histone H4 gene derived from *A. thaliana*
PhL	Phosphorylase-L gene
pmi	Phophomannose isomerase (Mannose-6-phospho isomerase) from *E. coli*
*pin*II	Proteinase inhibitor II from potato (**S. tuberosum*)
pol	Polymerase gene
Ppo	Polyphenol oxidase gene
PqrA	Paraquat resistance gene
Ps7s7	Duplicated promoter region derived from subterranean clover stunt virus genome segment 7
psbA	Q_B protein (D_1 protein: PS II reaction center) from *Amaranthus hybridus*
P35S3	Promoter region of the CaMV 35S transcript
PSsuAra	A rubisco small subunit gene from *Arabidopsis thaliana*
*pvu*I	*Proteus vulgaris* restriction enzyme I
rbcl	Chloroplast gene: *ribulose-bisphosphate carboxylase*
rbcS	Ribulose-1,5-bisphosphate carboxylase small subunit
RuBisCo	Ribulose bisphosphate carboxylase oxygenase enzyme
SAG21	Senescence-associated gene 21
SCP1	Super core promoter 1
Sul	Sulfonamide resistance encoding gene
T-35S	CaMV 35S 3′ polyadenylation signal
Tahsp17.3′	**T. aestivum* 3′ untranslated region heat shock protein
tCUP	**T**obacco **c**onstitutive **p**romoter
Tdc	Tryptophan decarboxylase gene
TetR	A tetracycline repressor protein

tetO	Tetracycline operator sequence
TEV	Tobacco etch virus
5'tev	5' end of tobacco etch virus
tfdA	Gene which encodes ferrous iron-dependent dioxygenase that uses α-ketoglutarate as a co-substrate
TpotpC	Coding sequence of an optimized transit peptide, containing sequence of the RuBisCo small subunit genes of *Z. mays* and *H. annuus*
TPotp Y	Optimized transit peptide derivative (position 55 changed to tyrosine) containing sequence of the RuBisCo small subunit genes of *Z. mays* and *H. annuus*
tRA	A tetracycline transactivator fusion protein
TS- CTP2	Targeting sequence from the *ShkG* gene encoding the chloroplast transit peptide region of *A.thaliana* EPSPS
Ubi	*Ubiquitin*
ubi4	Ubiquitin gene 4
ubi7	Ubiquitin gene 7
ubi9	Ubiquitin gene 9
Ubi Zm1	Ubiquitin *Z. mays* promoter, and the first exon and intron
UBQ10	Ubiquitin-10 gene promoter poly (UBQ10) of *A. thaliana*
uidA	Gene which encodes β-D-glucuoronidase (*E. coli* K12)
3'-UTR	3' untranslated region
3'-UTR of wheat HSP 17.3	3'-UTR of wheat heat shock protein 17.3
vip3Aa20	*Gene from B. thuringiensis* strain AB88, encodes modified Cry3A δ-endotoxin
XhoI	*X. holcicola* type II restriction endonuclease which cleaves DNA to give specific double-stranded fragments with terminal 5'-phosphates
xylA	Xylulose isomerase encoding gene
zm-epsps	*Z. mays*-5-enolpyruvylshikimate-3-phosphate synthase
ubi Zm1	Ubiquitin *Z. mays* promoter, and the first exon and intron
ZFNs	Zinc finger nucleases
Zm-hra	Modified acetolactate synthase (als) from *Zea mays*
ZmPer5 3' UTR	*Zea mays peroxidase 3' untranslated region*
***Zm ubi1* (*ubi Zm1*)**	*Zea mays ubiquitin and the first exon and intron*

Agrobacterium tumefaciens; *Arabidopsis thaliana*; *Arthrobacter globiformis*;
Avena fatua; *Bacillus thuriginensis*; *Escherichia coli*; *Helianthus annuus*;
Nocordia otitids-caviarum; *Pseudomonas fluorescens*; *Solanum tuberosum*;
Streptomyces viridochromogenes; *Triticum aestivum*; *Xanthomonas holcicola*;
Zea mays

The Author

Dr. Vallurupalli Sivaji Rao, native of Gudlavalleru, Krishna District, Andhra Pradesh, India, is an eminent weed scientist with a Bachelor's degree in Agriculture from Karnataka University (1961) and Master's degree in Agronomy from Osmania University (1963), both in India. He followed them up with a Ph.D. degree in weed science from Cornell University, USA, in 1973. He has worked in several universities (including Cornell), governmental research institutions, and international and non-governmental organizations in India and USA for over 50 years, teaching and researching in the field of weed science in crops like rice, tea, sorghum, maize, wheat, alfalfa, and groundnut. He has vast experience of conducting lab research on herbicide action mechanisms as well as field research to develop farmer-oriented weed management programs. He has held the positions of senior scientist, professor, director, advisor, and consultant. Currently, he is an Affiliate Member of Department of Plant Sciences, University of California, Davis, California, USA.

Dr. Rao is credited with scores of research publications in addition to authoring the widely popular reference-cum-textbook "Principles of Weed Science", with its first edition published in 1983 and second edition in 2000. He has served the Indian Society of Weed Science as Executive Vice President and was Editor-in-Chief of Indian Journal of Weed Science during 1982–90. He has also served on several national and state committees in India as a member and chairman between 1980 and 1994. Dr. Rao is the recipient of the Lifetime Achievement Award of Indian Society of Weed Science for 2010. He is a life member (emeritus) of some professional societies in India and USA including the Weed Science Society of America. He and his wife have been living in California, USA since 1995.

CHAPTER 1

Introduction

Knowledge is both bliss and power. The pursuit of knowledge is a characteristic of the human mind which constantly raises questions only to answer them. This act quietens the mind, but only briefly, before embarking on another quest to find answers for more questions. Thus, the human mind constantly empowers itself. Actually, it is science that truly empowers the human mind to solve problems that plague us. This was what the Greek mathematician, engineer, and astronomer Archimedes experienced 2,200 years ago after discovering the principle of buoyancy while taking a bath, and then getting out of the public bath-tub and running naked through the town of Syracuse, Sicily shouting the Greek words "Eureka! Eureka!" (I have found it! I have found it!).

Today, science faces new challenges, perhaps some of the greatest ever. Prominent among them is the need to feed the world's growing population against the backdrop of shrinking arable land and crop-related resources, constraints that limit the potential of crop yields, and imposition of a greater burden on the planet Earth than any other human activity in the history of mankind. After all, crop plants are the first level of the human food chain.

Mankind is currently tasked with increasing agricultural productivity in a manner that embraces the principles of agricultural sustainability without compromising the ability of future generations to meet their own needs. The challenges to productivity include slower rate of growth due to increased competition from weeds, insects, diseases, etc.; decelerating productive and cultivable agricultural land; rapidly growing demand for food and feed; increasing undernourishment; growing carbon dioxide emissions leading to global warming; depleting water/irrigation availability leading to more frequent droughts than ever; escalating salinity in irrigated regions; shrinking labor force; non-remunerative price for crop produce, etc.

The current global population of 7.3 billion is expected to reach 9.3 billion by 2050, with the Asian continent (5.142 billion), powered by India (1.62 billion) and China (1.47 billion), accounting for 55 percent [United Nations 2012a]. In order to meet the increased demand for food both by the rising population and the expected increase in per capita food consumption in developing and underdeveloped nations, agricultural production needs to increase by 50 percent by 2030 and 70–100 percent by 2050 [Tomlinson 2011]. This is equal to an additional one billion tons of cereals and 200 million tons of livestock products each year. This can be met by an increase in cultivable area and enhanced production level per unit of land. Both, however, are beset by several constraints.

Some 1.6 billion ha of the world's best, most productive lands are currently used to grow crops. Parts of these lands are being degraded through farming practices that result in water and wind erosion, the loss of organic matter, topsoil compaction, salinization and soil pollution, and nutrient loss. FAO reports that 70 percent of the world population will live in urban areas by 2050, up from 49 percent now. This suggests that there will be fewer people to depend on agriculture for their livelihood. Salinization (caused by high levels of salts in water, greater mobility of salts from ground water, changes in climate favoring salt accumulation, and human activities such as land clearing, etc.) of world's arable land is poised to go up from the present 25–30 percent by 2020 and as much as 50 percent by 2050. Of the current area under irrigation, salinity affects 40 percent of it in various degrees. Crop irrigation increases salinity owing to trace elements in irrigation water [GM Science Update 2014]. Besides, production in over 70 percent of dryland agriculture is also limited by salinity stress worldwide [GM Science Update 2014].

Drought and desertification, which determine water availability and quality as well as biodiversity, cause a loss of 12 million ha each year (23 ha min^{-1}) where 20 million tons of grain could have been produced. Environmental degradation is caused by depletion and changes in the availability and quality of resources such as air, water, and soil. According to the Global Land Assessment of Degradation published by FAO, nearly two billion ha worldwide have been degraded since the 1950s, representing 22 percent of the world's cropland, pastures, forests, and woodlands. In particular, some of the Asian, African, and Latin American countries have the highest proportion of degraded agricultural land as revenue-poor national governments pursue seemingly lucrative policies of deforestation for industrial expansion, urbanization, and population growth. This degradation is likely to increase in the coming decades because of changes in the availability and quality of resources such as air, water, and soil.

Of the water that is available for use, about 70 percent has already been in use for agriculture [Vorosmarty et al. 2000]. Many rivers no longer flow all the way to the sea, while 50 percent of the world's wetlands have disappeared, and major groundwater aquifers are being mined unsustainably, with water tables in parts of India, China, Mexico, and North Africa declining by as much as 1 m yr^{-1} [Somerville and Briscoe 2001].

Compounding the challenges mentioned earlier are the predicted effects of climate change [Lobell et al. 2008]. As the sea level rises and glaciers melt, low-lying croplands will be submerged and river systems will experience shorter and more intense seasonal flows, as well as more flooding [Intergovernmental Panel on Climate Change 2007]. As yields of our most important food, feed, and fiber crops decline precipitously at temperatures greater than 30°C, heat and drought will increasingly limit crop production [Schlenker and Roberts 2009].

Considering that there is little possibility of future increase in cultivable land and the prospect of adverse effects due to climate changes, all of the required increase in agricultural production must largely come from the same land while using less water. However, growth of cereal production has been decelerating over the past 50 years. It came down from 3.2 percent in 1960 to 1.8 percent in 2000 and 1.2 percent in 2010. Currently, the major four crops of wheat, rice, maize, and soya bean, which together provide two-thirds of global agricultural calories, are increasing at the non-compounding rates of 0.9 percent, 1.0 percent, 1.6 percent, and 1.3 percent per annum respectively [Ray et al. 2013]. At these rates, production of these crops are likely to increase at 38 percent, 42 percent, 67 percent, and 55 percent respectively. These rates of increase significantly fall short of the projected demands of 76 percent, 59 percent, 101 percent, and 84 percent for wheat, rice, maize, and soya bean respectively by 2050.

Production of rice, the staple food crop of 55 percent of world population in 2050, must be doubled from the current level. At today's levels of increase of wheat (0.9 percent) and rice (1.0 percent) production, there will be no change in the *per capita* wheat and rice harvests to 2050 [Ray et al. 2013]. A similar scenario will prevail in the case of maize and soya bean as well. Currently, the top three countries that produce rice and wheat, China, India, and USA, have very low rates of crop yield increase.

Yields are no longer improving on 24–39 percent of world's most important croplands areas [Lin and Huybers 2012]. Many of these areas are in top crop-producing nations, having rising population, increasing affluence, or a combination of these factors [Tilman et al. 2011; United Nations 2012b; Finger 2010; Brisson et al. 2010; Lin and Huybers 2012; Ray et al. 2012]. This may increase the difficulty of meeting future crop production goals.

If the world is required to boost the production in these top four global crops that are now responsible for directly providing ~43 percent of the global dietary energy and ~40 percent of its daily protein supply [FAO 2013] from yield increases alone, it has to necessarily increase the production per unit of land. This can only be achieved by raising yield threshold levels and removing the various crop-production constraints such as biotic and abiotic stresses via biotechnological approaches like conventional breeding, genetic engineering, etc.

The losses caused by these biotic and abiotic stresses, which already result in 30-60 percent yield reductions globally each year, occur after the plants are fully grown: a point at which most or all of the land, water, and funds required to grow a crop has been invested [Dhlamini et al. 2005]. For this reason, a reduction in losses to weeds, insects, and pathogens as also to environmental stresses is equivalent to creating more land and more water. Also, crops and other plants are routinely subjected to a combination of different abiotic [Miller 2006] and biotic stresses. In drought areas, for example, many crops encounter a combination of drought, heat, and salinity stresses, besides infestation by weeds, insects, and pathogens.

Agricultural research, like any field of science, continues to evolve in order to meet the current and future challenges. British physicist and ecologist Lord Robert May had said in 2002 that "we couldn't feed today's world with yesterday's agriculture and we won't be able to feed tomorrow's world with today's... but we can try to do it in a way that is more environmentally sensitive by producing crops that are water tolerant, salt tolerant, and resistant to particular insects without putting chemicals on them that are potentially hazardous to wildlife" [IFR 2002].

Transgenic technology in which desired genes and traits are inserted into the plant genome offers a better chance to raise crop yields from the present near-stagnant levels and mitigate some of the biotic and abiotic stresses the crops are subjected to. It also offers new avenues of plant improvement in a shorter period compared to conventional breeding and the fresh possibility of incorporating new genes with low problem of incompatibility.

Crop Yield

Yields can be intrinsic and operational. Intrinsic yield, the maximum potential that can be achieved, is obtained when crops are grown under ideal conditions. By contrast, operational yield is obtained under field conditions, when biotic and abiotic stresses are considerably less than ideal. Genes that improve operational yield reduce losses caused by these stress factors.

The priority for plant biotechnologists is to raise the intrinsic yields of crops. The major barrier in increasing the yield potential of a crop, however, is the complexity of genes involved in this quantitative trait, requiring several genes to be manipulated. Many of the genes now being considered for increasing yield involve greater genetic, biochemical, and phenotypic complexity than current genes for herbicide tolerance and insect tolerance, and this complexity will sometimes exacerbate the tendency of transgenic crops to produce side effects, some of which may be unacceptable [Gurian-Sherman 2009]. Yield potential may be raised by improving, for example, by enhancing of efficiency of photosynthesis, protein and lipid metabolism, etc. Although several attempts have been made in the past in this area of biotechnology, one general difficulty is that many improvements have been aimed at aspects of plant physiology that are several steps removed from grain yield [Gurian-Sherman 2009].

Biotic Stress

When a plant is affected by stress, it is subjected to sub- and supra-optimal physiological conditions. Stress results in the formation of a reduction in water potential. Biotic stress on plants is caused by other living organisms, known by a general term 'pest', such as weeds, insects, bacteria, viruses, fungi, nematodes, and parasites. Of these, weeds constitute a major and continuing biotic constraint affecting cropping systems worldwide. Although agricultural crops are damaged by thousands of pest species, less than 10 percent of them are generally considered to cause major problems. It is generally known that yield is progressively reduced with increasing number of these pests. The recognition that pest infestations are a major reason for crop varieties not realizing their yield potential fully had prompted early farmers to select and breed plants that survived the infestation. Overall, weeds cause the highest potential loss (34 percent) followed by insect pests (18 percent), and pathogens (16 percent) [Oerke 2006].

Removal of biotic stress factors do not lead to yield increase *per se*. However, the losses caused by them can be minimized by using pesticides. The global market for different pesticides and agrochemicals which grew at 9.8 percent annually between 2007 and 2013 [World Pesticide and Agrochemical Market 2014] is expected to increase approximately by 8.7 percent from the current $50 billion market. Of these, herbicides account for over 40 percent of the market followed by insecticides and fungicides with 27 percent and 21 percent respectively. Although these agrochemicals protect crops from losses, some 20 to 40 percent of the world's potential crop production is still lost annually. Excessive use and misuse of herbicides, insecticides, and fungicides have their own attendant secondary problems such as weeds developing resistance to herbicides, insects to insecticides, and pathogens to fungicides, aside from the adverse impact pesticide products have on the environment and ecosystem.

It is increasingly difficult to discover a new herbicide, and even more difficult to find one with a novel mode of action. Today, approximately 500,000 compounds must be screened to discover a potential herbicide compared with one per 500 compounds screened in the 1940s. Given the difficulty of discovering new herbicides, as it has been particularly over the past 20 yr, expanding the utility of existing ones that have a broad weed-control spectrum and good environmental profile through genetically enhanced resistance is a useful strategy from the agro-ecological point of view.

Development of resistance in weeds due to the continuous use of same herbicides or the ones that have similar molecules is not a new phenomenon ever since their first commercial introduction, beginning with 2,4-D in 1946. It is, in fact, similar to that exhibited by insects to insecticides since 1914 and plant pathogens to fungicides from 1940. Initially, development of herbicide-resistant weeds was slow, but picked up pace from the early 1980s (vide Chapter 2). At the present time, after nearly six decades of the first reports of biotypes of two weed species becoming resistant to 2,4-D in Hawaii, USA and

Ontario, Canada in 1957, 235 species (138 dicots and 97 monocots) have been identified to develop resistance to scores of herbicides worldwide thus far (vide Chapter 2) [Heap 2014]. A great majority of cases of herbicide resistant weeds have so far been found in developed countries. About one-third of these resistant species have been from the U.S. which uses herbicides most intensively. These numbers do not remain static. They keep increasing as and when more weed species are found resistant in virtually every corner of the globe where herbicides have been in continuous use.

The direct consequences of evolution of herbicide-resistant weeds include reduced crop production, higher weed management costs, and tremendous increase in seed banks leading to greater weed pressure on subsequent crop(s). Furthermore, some herbicides which are effective against a wide range of weed species are either nonselective to crops or selective to only specific crops, and so they have little or limited practical utility in most of the cropping situations. Continuous application of the herbicide(s) in question will only lead to a faster evolution of weed species. Generally, a combination of herbicides with cultural and mechanical methods may only delay but not prevent evolution of herbicide resistant weeds [Green and Owen 2011]. At this juncture, inserting a gene from a plant or non-plant source encoding a protein that confers the herbicide-resistant trait was considered to be an indispensable tool for solving problems associated with development of resistance to herbicides. Currently, several genes derived from non-plant sources (vide Chapter 5) are employed to confer resistance to several crops to some of the herbicides used (vide Chapter 6).

Unlike weeds, insects not only cause direct yield losses by damaging and consuming plants, but also act as vectors for many viral diseases. The damage they inflict often facilitates secondary microbial infections on crops. Herbivorous insects and mites are a major threat to global food and feed production. Larval forms of lepidopteran insects are considered the most destructive pests, with about 40 percent of all insecticides directed against heliothine species [Brooks and Hines 1999]. Many other species within the orders Acrina (ticks and mites), Coleoptera (beetles and weevils), Diptera (flies), Hemiptera (aphids, hoppers, cicadas, and shield bugs), and Thysanoptera (thrips) are also considered agricultural pests with significant economic impact. There are innumerable insect pests that can devastate agricultural production. Of these, the ones most notable for their destructive capacity include the Migratory locust (*Locusta migratoria*), Colorado potato beetle (*Leptinotarsa decemlineata*), Boll weevil (*Anthonomus grandis*), Japanese beetle (*Popillia japonica*), and aphids (of family Aphidoidea) which serve as vectors of plant viruses [Ferry and Gatehouse 2010]. Another destructive insect pest is the Western corn rootworm (*Diabrotica virgifera virgifera*) which is called the billion dollar bug due to its economic impact in USA alone [Ferry and Gatehouse 2010].

Insect pest control is heavily dependent on insecticides which are more expensive and damaging to the environment than herbicides. These chemicals are nonselective, killing harmless and beneficial insect species along with target insects, and eventually accumulating in water and soil. They are also hazardous to human health if overused or misused. In the 1960s, the American biologist and conservationist Rachel Carson brought to the attention of the world, the detrimental environmental and human impacts resulting from overuse or misuse of some insecticides. Constitutive exposure to insecticides can lead to the evolution of resistance in insect populations, leading to reduced insect control. Generally, insecticides are too expensive for farmers in the developing world and in any case are often ineffective against sap-sucking insects including the rice brown plant hopper (*Nilaparvata lugens*) [Christou and Chapel 2009].

Therefore, insect pests became the second most important target for transgenic technology. The transgenic engineering of plants to express insect-resistance genes offers the potential to overcome the shortcomings of continued heavy use of insecticides. In doing so, genes that are specific towards a particular insect species are isolated from bacteria and other non-plant sources. Furthermore, the proteins encoded by these transgenes within plants allow effective control of insects that feed or shelter within the plant. Several of these genes have been sourced from bacterial toxins, lectins, and protease inhibitors. The foremost of the toxins have been derived from the spore-forming bacterium *Bacillus thuringiensis* (*Bt*). *Bt* toxins are known as crystal (Cry) proteins or δ-endotoxins (vide Chapters 5 and 6).

Crop plants have also evolved resistance mechanisms that protect them from pathogen species. One of the most prevalent and problematic bacterial diseases in food crops is bacterial blight of rice, causing losses in excess of US$250 million every year in Asia alone. The gene, *Xa21*, isolated from rice-related wild species *Oryza longistaminata* was shown to confer resistance to all known isolates of the blight pathogen *Xanthomonas oryzae* pv. *oryzae* in India and the Philippines [Khush et al. 1990]. This gene encodes a receptor tyrosine kinase [Song et al. 1995]. The transfer of this plant gene to susceptible rice varieties resulted in plants showing strong resistance to a range of isolates of the pathogen [Wang et al. 1996; Tu et al. 1998; Zhang et al. 1998].

Viruses, which also cause significant crop losses, are another major target of transgenic technology. Virus resistance can be achieved by introducing into the crop plant one or more genes from the virus itself. One way to achieve pathogen-derived virus resistance is to express a coat protein gene which can block virus replication. This strategy has been demonstrated in rice, the host to more than 10 disease-causing viruses. Tungro virus is the most damaging viral disease in rice in South Asia and Southeast Asia. This is caused by a combination of two viruses: rice tungro bacillus virus (RTBV) and rice tungro spherical virus (RTSV). Rice hoja blanca virus (RHBV) causes 100 percent yield losses in rice in Central and South America. Pathogen-derived resistance to these diseases has been achieved in experimental plants by exposing coat protein genes from RTBV, RTSV, and RHBV [Kloti et al. 1996; Lentini et al. 1996; Sivamani et al. 1999].

Another virus-resistant transgenic crop is papaya which is seriously affected by papaya ringspot virus (PRSV) in Hawaiian Islands. PRSV is a potyvirus with single-stranded RNA. The inserted transgene was designed with a premature stop codon in the PRSV coat protein sequence to prevent expression of a functional coat protein because, at the time of engineering, it was thought that the protein itself was an important factor in resistance. RNA analysis revealed that the plants with the best resistance exhibited the least detectable message, suggesting the involvement of an RNA silencing mechanism [Tripathi et al. 2006].

Abiotic Stress

Abiotic stress, a natural part of every ecosystem, causes changes in soil-plant-atmosphere continuum and metabolic activities within the plant, leading to reduced crop production. Abiotic stress is an integral part of "climate change," a complex phenomenon with a wide range of unpredictable impacts on the environment. Prolonged exposure to abiotic stress factors such as temperature (heat, chilling, freezing) drought, cold, ozone, salinity, flooding, intense light, and nutrient imbalance (mineral toxicity and deficiency) leads to an altered metabolism and damage to biomolecules. These responses cause deterioration and destruction of crop plants, resulting in low productivity.

In the case of mitigation of salinity, two major genes Na(+) exclusion in durum wheat, *Nax1* and *Nax2*, that were previously identified as the Na(+) transporters TmHKT1;4-A2 and TmHKT1;5-A, have been transferred into bread wheat in order to increase its capacity to restrict the accumulation of Na(+) in leaves [James et al. 2011]. The recent introgression of an ancestral form of the *HKT1;5* gene from the more Na$^+$-tolerant wheat relative *Triticum monococcum* into a commercial durum wheat species (*Triticum turgidum* ssp. *durum*), which is susceptible to salinity, has increased grain yields on saline soil by 25 percent [Schroeder et al. 2013]. These results indicate that both *Nax* genes have the potential to improve the salt tolerance of bread wheat and even other crops. Besides, these genes have the potential to confer an extra advantage under a combination of waterlogged and saline conditions [James et al. 2011]. Combining HKT transporter traits with vacuolar Na$^+$ sequestration mechanisms provides a potentially powerful approach to improve the salinity tolerance of crops [Schroeder et al. 2013]. In rice, the most aluminum tolerant of the cereal crops, *NRAT1* gene, which encodes a protein that confers further tolerance to this metal, has been identified by Cornell scientists [ISAAA 2014].

In the short and medium term, as more genes are identified that confer salinity-tolerant traits, their introduction by transgenic methods, alone or in combination, should elevate salinity tolerance in other crop species in future.

The presence of abiotic stress can also have an effect of reducing or enhancing susceptibility to a biotic stress like weed competition, insect infestation, pathogen (fungi, bacteria, viruses, etc.) infection, etc. This interaction between biotic and abiotic stresses is orchestrated by hormone (e.g., abscisic acid) signaling pathways that may induce or antagonize one another. Specificity in multiple stress responses is further controlled by a range of molecular mechanisms that act together in a complex regulatory network. Therefore, the subject of abiotic stress is gaining considerable significance in genetic engineering.

Gene Manipulation

One branch of science that humans have been pursuing from pre-historic times is agriculture. It began when man struggled to survive. Initially, he grew crops in quantities sufficient to support his family. Later, he needed to support others not engaged in agriculture and the growing population. In this pursuit, he had to grow more crops and reap more yields from the same land year after year. For this, he devised and followed best production practices to increase crop yields and feed others. In order to increase yields, he resorted to domestication of wild crop species followed by manipulation of plant genes through selection and much later, by crossing to evolve more useful and productive cultivars. In the process, he altered the genomes of plant species for thousands of years, by choice or otherwise, to derive cultivars with improved quantitative and qualitative traits. This became the forerunner of the birth of a new branch of science called 'genetics' following the path-breaking findings of the Austrian monk, Gregor Mendel, on heredity and segregation of heritable traits after crossing many generations of garden pea between 1856 and 1863.

A flurry of research activities followed on both sides of the Atlantic over the next several decades to give birth to the field of modern or conventional plant breeding. This resulted in an enormous number of improved varieties in economic crops, including the two vital food crops wheat and rice, thus ushering in 'Green Revolution' in the last century. Spectacular advances developed in molecular biology, biotechnology, and genetic engineering (vide Chapters 3 and 4). The speed at which this progress has been achieved in these fields is probably unparalleled in the history of science, barring inventions of the atom, computer, and communication.

Both conventional breeding and genetic engineering are used to improve the genetic traits of plants for human use. Their main goal is to develop crop varieties that express good agronomic characters as well as to enable crop plants withstand biotic and abiotic stresses.

All plant traits are encoded by genes. A plant has 10,000–50,000 genes, depending on the species. Each of the genes is associated with specific traits. Many genes encode enzymes that catalyze specific biochemical reactions. Before the advent of genetic engineering, plant breeders used the genes of a plant to select specific desirable traits. In conventional plant breeding, the genetic composition of plants is modified by making crosses and selecting new superior genotype combinations. However, conventional breeding has limitations. First, crop improvement depends solely on the desirable genes available naturally, created by induced mutations, or their shuffling for desired recombinations. Second, breeding can only be done between plants that can sexually mate with each other. This limits the new traits that can be added to those that already exist in that species. Third, when plants are crossed, many traits are transferred along with the trait(s) of interest. These include those with undesirable effects on yield potential. Fourth, there is no guarantee of obtaining a particular gene combination from the millions of crosses generated. Undesirable genes can be transferred along with desirable genes, or while one desirable gene is gained, another is lost because the genes of both parents are mixed together and re-assorted more or less randomly in the offspring [ISAAA 2012]. These problems limit the improvements that plant breeders can achieve. It was here that genetic engineering found its niche and utility over three decades ago.

Three key elements have essentially transformed biotechnology into genetic engineering. These are: identification of DNA as the carrier of genetic information by Avery Colin McLeod and MacLyn McCarty in 1944, discovery of the structure of DNA by James Watson and Francis Crick in 1953, and discovery of a recombinant technique by which a section of DNA is cut from the plasmid of an *E. coli* bacterium for transfer into the DNA of another by Stanley Cohen and Herbert Boyer in 1973.

The first genetically engineered plant was tobacco when Fraley et al. [1983] produced in 1982, an antibiotic-resistant tobacco plant by using the soil bacterium *Agrobacterium tumefaciens* to insert a small segment of DNA (T-DNA), from a Ti plasmid, into the plant cell. This was followed by the herbicide-tolerant tobacco in 1986, the frost-resistant strawberry and potato in 1987, and the virus-resistant tobacco in 1992 (vide Chapter 4). The year 1994 saw the first transgenically engineered, but short-lived, whole food tomato line 'Flavr Savr' developed by the Davis, California-based Calgene and the first herbicide-resistant crop tobacco which was engineered to become resistant to bromoxynil developed by the European Union. These path-breaking developments laid the groundwork for all transgenic crops developed over the next two decades (vide Chapters 4 and 6). Transgenic plants have been developed in scores of species to be useful for agriculture and forestry. Some of these include maize, soya bean, cotton, canola (rapeseed), sugar beet, rice, wheat, potato, tobacco, papaya, lucerne (alfalfa), linseed (flax), pea, tomato, squash (zucchini: *Cucurbita pepo*), sugarcane, sorghum, sweet pepper (*Capsicum annuum* var. *annuum*), brinjal (eggplant: *Solanum melongena*), banana and plantain, creeping bentgrass (*Agrostis stolonifera*), chicory (*Cichorium intybus* var. *foliosum*), pine (*Pinus* spp.), poplar (*Populus* spp.), Jatropha (*Jatropha curcas*), petunia, etc.

The traits inserted in these crops include herbicide resistance; insect resistance; virus resistance; resistance against fungal and bacterial infections; abiotic stress tolerance;

delayed fruit ripening; male sterility; production and quality of biofuels (e.g., Jatropha, high starch-to-sugar maize, low-lignin poplar, etc.); pharmaceutical products of therapeutic value; phytoremediation of the soil contaminated by explosives (TNT, RDX, etc.), toxic elements (mercury, selenium, etc.) and organic pollutants (polychlorinated biphenyls); production of drugs (e.g., Elelyso: taliglucerase alfa from carrot cells for the treatment of Gaucher's disease, a rare genetic disorder); production of bioplastics, detergents, substitute fuels, and petrochemicals; etc. As weeds and insect pests are the primary targets for transgenic technology, a vast majority of commercially grown transgenic plants are modified for herbicide resistance, insect resistance, and both [James 2006].

Initially, genetic engineering was done to carry genes that deliver single traits. Later, plants have been transformed to carry two or more genes that code for proteins having different modes of action and enzymes. As multi-trait stacks are tightly linked, they exhibit an extremely low rate of segregation, essentially behaving as a single gene. Biotech stacks provide better chances of overcoming the myriad of problems in the field such as weeds, insect pests, diseases, and environmental stresses, low yields and nutritional quality, etc. simultaneously.

In genetic engineering, a genetic material is inserted followed by selection. Insertion is done by a vector-mediated transformation or one of the vector-less direct methods into the host plant cell and then, with the help of genetic elements in the construct, the genetic material inserts itself into the chromosomes of the host plant. Genetic engineers must also insert a 'promoter' gene from a virus as a part of the package, to make the inserted gene express itself. This process is profoundly different from conventional breeding even if the primary goal is only to insert genetic material from the same species.

However, the technique of genetic engineering offers a new type of genetic modification. It enables direct and purposeful transfer of one or just a few genes of interest from species, families, and even kingdoms which could not previously be sources of genetic material for a particular species, and even to insert custom-designed genes that do not exist in nature. Thus genetic engineering allows movement of genetic material from any organism to any other organism. It also offers the ability of creating a new genetic material and expression of products like never before.

The plant genome is a complex entity made up, in part, of genes and genetic elements that interact in complex regulatory pathways to create and maintain the organism. The new genetic material that enters the genome of the host plant must fit into this total complex or it may end up destabilizing it. Genome is like a complex computer program or an ecological community. When a new sub-program is introduced within the larger complex computer program, it may fit in well or can create unpredictable effects and may ultimately cause the whole program to crash. Similarly, in a complex ecological community, the introduction of a new species may survive or cause a catastrophic effect on the ecosystem. Unlike in a computer program, the changes that a new genetic material or species may bring about cannot be predicted or be evident in a short time span.

The gene that is transferred, called transgene, holds information that will give the host plant a trait. However, it cannot control the location where the trait is inserted into the genome with any precision or with a guarantee of stable expression. Regardless of the method of transformation, the site of insertion of the transgene is fairly random. As the effect of a gene on the host plant is governed by its location, the lack of control over location is the cause of unexpected effects. This is not altogether an unexpected phenomenon because transgenic engineering, unlike conventional breeding, involves organisms with desperate evolutionary backgrounds. Thus transgenic engineering is more a random process than conventional breeding.

Transgenic engineering, however, is not bound by the limitations of traditional plant breeding. It physically removes the DNA from one organism and transfers the gene(s) for one or a few traits into another. Since crossing is not necessary initially, the 'sexual' barrier between species is overcome. Therefore, traits from any living organism can be transferred into a plant.

Although there are many diverse and complex techniques involved in transgenic engineering, its basic principles are reasonably simple [ISAAA 2012]. There are five major steps in the development of a genetically engineered crop. For every step, it is very important to know the biochemical and physiological mechanisms of action, regulation of gene expression, safety of the gene, and the gene product to be utilized. Even before a genetically engineered crop is made available for commercial use, it has to pass through rigorous safety and risk assessment procedures before being approved for commercialization.

The length of time in developing a transgenic plant depends upon the gene, crop species, available resources, and regulatory approval. It may take 6–15 yr before a transgenic line is ready for commercial release [ISAAA 2012]. This transgenic technology has been used over the past two decades to develop scores of transgenic crop lines incorporated with various qualitative and quantitative traits in several crops around the world.

As transgenic technology in the West was driven predominantly by the potential commercial gain, research has so far focused mainly on the weed (and insect) problems farmers of industrialized nations faced. There has been little interest and attempt in producing crops with resistance to the weed species that plague subsistence farmers in the developing world, even though this would have an immediate impact on global food security [Christou and Chapel 2009].

Currently, close to 180 million ha are under transgenic crops globally, and this area may rise to 400–500 million ha by 2030 at a time when over 120 crops are expected to be transgenically engineered with desired traits and adopted worldwide. This, however, is dependent on the extent of adoption by developing nations, particularly the growth engines China, India, Brazil, Argentina, and South Africa. Further expansion is also dependent on global regulatory procedures and how best the risks and issues (vide Chapter 9) associated with them are answered to the satisfaction of farmers and consumers.

Once a transgenic crop line is developed, it needs to go through regulatory system before it is commercialized. However, there is no uniform global regulatory framework in place. Each country has its own regulatory framework (vide Chapter 8). Even within a country, there is a wide variation in review and assessment because the biotech variety intended for food use undergoes through a different perspective from the one used for non-food or feed purpose. Many a time, assessment, approval, and regulation are based not entirely from technology standpoint.

Although genetic engineering provided a significant breakthrough in terms of substituting land scarcity for agriculture and enhancing the production efficiency of certain edible crops, few topics in agriculture are more polarizing and controversial than this growing field of biotechnology. This is because the proponents vehemently praise the virtues of the technology and the progress made thus far in offering farmers alternatives to herbicides and insecticides while the opponents zealously point out the perils it has brought upon the farmers, agro-ecology, and soil ecosystem aside from its potential impact on health of consumers. However, what has been found thus far is that the benefits derived from biotech crops by farmers varied with the crop, the traits it carried, farm size, and the country that adopted them. The success achieved by one farmer and one country is no guarantee that other farmers and countries will also taste them. In reality, every technology has its benefits and risks. Transgenic engineering is no exception.

Future of Transgenic Engineering

The transgenically engineered crop varieties developed over the past 20 yr have certainly offered a means to enhance global agricultural sustainability. This technology holds promise to meet the agricultural needs of the 21st century, particularly in regard to increasing plant productivity, both directly and indirectly, while enhancing quality of the produce. It also aids in reducing the footprint of pesticide chemicals (herbicides, insecticides, fungicides, etc.) and carbon on the environment. Transgenic engineering has the required arsenal to address various problems plaguing agriculture in a more effective way than ever. As in the case of atom, electronics, computer, and communication, the more we understand the intricacies of this plant-related technology, the better we will be able to utilize it for the betterment of mankind.

Currently, there are scores of useful genetically engineered traits in the pipeline. Future transgenic technology (second and third generations) is expected in the following:

1) Enhancing the photosynthetic efficiency of crop plants as well as protein and lipid metabolism;
2) Increased nitrogen use efficiency and reducing the detrimental environmental impacts such as water eutrophication caused by nitrogen compounds in fertilizers and greenhouse gas emissions emanating from their synthesis;
3) Improved phosphorus efficiency and availability;
4) Enhanced nutritional quality in staple food crops such as rice, wheat, maize, and sorghum (β-carotene, iron, protein, etc.) besides other nutritious legume crops chick pea, pigeonpea, groundnut (peanut), etc.;
5) Better drought tolerance in most of the crops in the light of global warming and shrinking water resources;
6) Greater tolerance of crops to frost, salinity, and flooding;
7) Longer shelf life in tomato and major fruit crops like banana, mango, apple, etc.;
8) Higher levels of health-promoting antioxidants like flavonols and flavonoids in fruits and juice;
9) Silencing of polyphenol oxidase to avoid bruising and browning of potatoes and apples;
10) Greater tolerance of plants to arsenic;
11) Restoration of fertility;
12) Enhancing plant characteristics (panicle size, seed quantity per panicle, etc.) for higher yields;
13) Lowering of seed-shattering habit of crops, particularly food crops;
14) Apomixis in fruit crops;
15) Male sterility and self-incompatibility;
16) Lower lignin content in tree crops for paper making;
17) Providing renewable alternatives to fossil fuels, such as feedstocks for biofuels, e.g., sugar (sugar beet, sugarcane), starch (maize, wheat), oil (rapeseed), and woody ligno-cellulose (poplar, willow, Miscanthus, etc.);
18) Developing oilseed crops that accumulate omega-3-long-chain polyunsaturated fatty acids (LC-PUFAs) such as docosahexaenoic acid (DHA) and eicosapentaenoic acid

(EPA) representing a potential sustainable terrestrial source of fish oils which can reduce the risk of cardiovascular disease;

19) Biofortification (nutritional enhancement) of staple crops, fruit crops, vegetable crops, and animal products with higher vitamin A, iron, zinc, etc.;

20) Phytoremediation of organic contaminants (herbicides, insecticides, oil spills, explosives, industrial chemicals, etc.) and inorganic pollutants (natural elements: cadmium, cobalt, iron, lead, mercury, selenium, tungsten, etc.); and

21) Biopharming which includes production of vaccines, therapeutics, antibodies, and enzymes from crops like rice, tobacco, potato, maize, lucerne (alfalfa), barley, etc.

With regard to herbicide-resistant crops, transgenic research may be directed more towards developing crop varieties resistant to herbicides other than glyphosate and glufosinate besides multi-herbicide tolerant stacks. This helps in controlling weeds resistant to these nonselective herbicides to which several crops have become or are becoming tolerant in the recent past. Furthermore, herbicide-resistant traits may be combined with traits other than those with insect resistance.

Any new plant variety, developed through transgenic engineering, carries the risk of unintended consequences. This is because transgenic plants contain desirable traits which offer a range of benefits above and beyond those that emerged from innovations in traditional agricultural biotechnology. However, this technology is complex because it deals with genes, the actions of genes, and interactions with other genes, more often derived from non-plant sources.

Although transgenic engineering has proved its utility in successfully transforming many crops with desired traits, agro-ecological and human safety concerns remain a contentious issue. This calls for using **intragenic** engineering technology, which involves insertion of DNA fragments from the same plant species or cross-compatible species in a sense or antisense orientation (vide Chapter 4), as an alternative. In this method, the undesirable genes are silenced and desirable genes are enhanced by linking with beneficial genes by using tissue-specific or near-constitutive promoters. Attempts have been made to successfully use this technology (vide Chapter 4) in tomato to redesign and improve the quality of Calgene's Flavr Savr tomato; to eliminate discoloration of a high-yielding potato variety; and to enhance oleic acid and oil content and shattering resistance in rapeseed (canola). This suggests that intragenic engineering can be used to insert traits such as resistance to herbicides, diseases, salinity, drought, frost, flooding, etc.

Another alternative to transgenic engineering with regard to herbicides is employing non-transgenic or partial-transgenic technology. The former was used to develop imidazolinone herbicide-resistant lines in maize, rapeseed (canola), rice, wheat, sunflower, and lentil, while the latter was used to produce β-carotene-rich 'Golden' rice variety by inserting two genes, one derived from a plant (daffodil) and the other from a soil bacterium.

One promising area of future genetic engineering is the artificial chromosome technology involving mini-chromosomes. The current technology is beset with certain limitations. One of them is that stacking of multiple genes in one germplasm takes many years, but still faces the possibility of segregation of transgenes in later generations [Yu and Birchler 2007; Yu et al. 2007; Xu et al. 2012]. Another limitation is linkage drag, which refers to the reduction in fitness of a cultivar due to deleterious genes introduced along with the beneficial gene during backcrossing [Xu et al. 2012; Yu and Birchler 2007]. A mini-chromosome is an extremely small version of a chromosome, the threadlike linear strand of DNA and associated proteins that carry genes and functions in the transmission of hereditary information. A normal chromosome is made of both centromeres and telomeres with much intervening DNA while a mini-chromosome contains only centromeres and telomeres, the end section of a chromosome, with little else. Mini-chromosome technology enables circumventing the conventional problems of genetic engineering. It opens up new possibilities for the development of crops carrying multiple genes that confer resistance to herbicides, insects, viruses, fungi, and bacteria as also for the development of proteins and metabolites that can be used to treat human illnesses.

One of the most important bio-technological developments over the last 10 years, together with advances in bioinformatics, is the availability of high throughput DNA sequencing methods at very affordable prices. These techniques allow the production of Gigabases of DNA sequences for around US$1000, and as a consequence, the amount of information present in DNA databases has doubled every 18 months. As a result of these high throughput methods, the genome sequences of the main plant species are now known. The resulting datasets include the genomes of model species, such as *Arabidopsis thaliana*, those of the main crops, rice, maize, and soya bean, and other important species such as poplar, cotton, grapevine, apple, cassava, and sorghum [GM Science Update 2014].

At the same time, the genomic variation within a species has become accessible due to resequencing of different breeding lines (cultivars). In the case of *Arabidopsis* and rice, more than 1000 sequences from different cultivars have been obtained and published [1001 Genomes Project n.d.; Huang et al. 2012]. This genome sequence data is helping to identify the genetic basis of domestication of the main crop species, and the many major genes affecting the performance of crops, including yield and disease resistance. Progress is also being made towards the identification of minor genes affecting quantitative traits that would have been more difficult to identify using classical molecular biology and genetics.

Over the past 5 years, newer methods to produce genetically modified plants are being increasingly developed. These enable mutations at very specific locations or to target gene sequences at specific sites. Classical methods of genetic engineering cannot predict the location of the new gene in the plant genome, and the insertion may cause the new gene to have low levels of expression or to insert into another useful gene. In contrast, the targeted transgenic methods enable the new gene in a specific location to avoid unforeseen effects.

These new methods, based on the use of site-directed nucleases (SDN) [Podevin et al. 2013; Goldstein et al. 2012; EFSA GMO Panel 2012], prepare the target site DNA for modification or insertion of a new sequence. Two examples of SDNs are Zinc-finger nucleases and TAL-nucleases in which a hybrid protein comprises a nuclease domain from a bacterium to cleave DNA and a sequence-specific DNA binding domain with a motif from a plant pathogen Xanthomonas. The SDN is expressed in cells of the crop to be engineered by a break at a defined location in the genome, where a mutation or foreign DNA can be introduced. Methods using SDNs are likely to become the normal method in transgenic engineering in future [GMO Science Update 2014].

Missing Links

Despite the significant advances made thus far, there are several missing links in today's transgenic technology. One of them is the country-specific and region-specific technology. Not all countries are alike when it comes to agriculture. Farmers of developing countries, with smaller farm size, poor capital, expensive inputs, and inadequate techniques stand to lose more than their counterparts in developed countries. Many of them do not even have access to good seeds and the available technology at affordable prices. If they had all of them, they would have achieved much better yields as has been proved in certain parts of the developing world. Any new farming technology will be successful only when it is oriented to the needs of farmers. They will then use it voluntarily without the need for it to be forced upon them.

Another missing link is the relatively weaker participation of public research organizations in developing of biotech crops. In fact, many of the early discoveries have emanated from universities and public research institutions. Despite the fact that several of them are also involved in this field, their research agenda is being increasingly influenced and taken over by the private sector in ways never seen in the past [Altieri 1998]. The public sector, aided by government funding, will be able to serve farming communities better if they could ensure the availability of ecologically sound aspects of biotechnology and making such knowledge available in the public domain for the benefit of society [Altieri 1998].

Yet another missing link—the vital one—is the consumer. The success of any technology is dependent on consumer acceptance. Consumers always exercise choice when they buy any product, be it agricultural or non-agricultural. Denial of this basic right will eventually lead to failure of technology as has been the case in the past. Transgenic technology will be no exception if they (and their animals) are denied the choice of what they consume. Consumers are not yet fully aware of the potential risks—both short term and long term—foods derived from transgenic crops have on human and animal health. In order to make these food products acceptable to consumers all over the world, biotech industry, food companies, and farmer markets are required to answer and allay their concerns satisfactorily by resorting to strict product traceability and labeling standards, besides printing of nutritional facts (including genetic information) on the labels.

This means that transgenic engineering has a long way to go before being fully accepted by both farmers and consumers. The American poet Robert Frost wrote: "The woods are lovely, dark and deep. But I have promises to keep, and miles to go before I sleep, and miles to go before I sleep." This is most relevant in the case of the fast growing field of plant biotechnology which still has a long way to go before the global agricultural needs of the next four decades are met, free of risks and issues to farmers and consumers.

References

Altieri, M.A. 1998. The Environment Risks of Transgenic Crops: An Agroecological Assessment: Is the failed pesticide paradigm being genetically engineered? Pesticides and You. Spring/Summer 1998.

Brisson, N., P. Gate, D. Gouache, G. Charmet, O. Francois-Xavier et al. 2010. Why are wheat yields stagnating in Europe? A comprehensive data analysis for France. Field Crops Res. 119: 201–212.

Brooks, E.M. and E.R. Hines. 1999. Viral biopesticides for heliothine control-fact or fiction. Today's Life Sci. Jan/Feb 38–44.

Christou, P. and T. Chapel. 2009. Transgenic crops and their applications for sustainable agriculture and food security. pp. 1–22. *In*: N. Perry and A.M.R. Gatehouse (eds.). Environmental Impact of Genetically Modified Crops. CAB International, Wallingford, Oxon, U.K.

Dhlamini, Z., C. Spillane, J.P. Moss, J. Ruane, N. Urquia and A. Sonnino. 2005. Status of Research and Application of Crop Biotechnologies in Developing Countries: Preliminary Assessment. Food and Agriculture Organization of the United Nations Natural Resources Management and Environment Department; Rome, Italy.

EFSA GMO Panel. 2012. Scientific opinion addressing the safety assessment of plants developed using Zinc Finger Nuclease 3 and other Site-Directed Nucleases with similar function. EFSA J. 10(10): 2943.

FAO. 2013. Food balance sheets. Part of FAOSTAT—FAO database for food and agriculture. Rome: Food and agriculture Organization of United Nations.

FAO. 2012. World Agriculture Towards 2030/2050. The 2012 Revision.ESA E Working Paper No.12-03.

Ferry, N. and A.M.R. Gatehouse. 2010. Transgenic crop plants for resistance to biotic stress. pp. 1–65. *In*: C. Kole, C.H. Michler, A.G. Abbott and T.C. Hall (eds.). Transgenic Crop Plants. Springer, Berlin/ Heidelberg, Germany.

Finger, R. 2010. Evidence of slowing yield growth—The example of Swiss cereal yields. Food Policy 35: 175–182.

Fraley, R.T., S.G. Rogers, R.B. Horsch, P.R. Sanders, J.S. Flick et al. 1983. Expression of bacterial genes in plant cells. Proc. Natl. Acad. Sci. USA 80: 4803–4807.

GM Science Update. 2014. A report to the Council for Science and Technology. March 2014. https://www. gov.uk /.../cst-14-634a-gm-science-update.pdf.

Goodstein, D.M., S. Shu, R. Howson, R. Neupane, R.D. Hayes, J. Fazo, T. Mitros et al. 2012. Phytozome: a comparative platform for green plant genomics. Nucleic Acids Res. 40(D1): D1178–D1186.

Green, J.M. and M.D.K. Owen. 2011. Herbicide-resistant crops: utilities and limitations for herbicide-resistant weed management. J. Agri. Food Chem. 59(11): 5819–5829.

Gurian-Sherman, D. 2009. Failure to Yield: Evaluating the Performance of Genetically Engineered Crops. Union of Concerned Scientists. UCS Publications, Cambridge, MA, USA, 43 pp.

Heap, I. 2014. International Survey of Herbicide Resistant Weeds. Online. Internet.21 September. Available www.weedscience.com.

IFR (Institute of Food Research). 2002. All About Wheat: Meeting the World's Food Challenges. John Innes Centre & Institute of Food Research.

Intergovernmental Panel on Climate Change. 2007. Climate Change 2007: Impacts, Adaptation and Vulnerability. Cambridge University Press, Cambridge, UK.

ISAAA. 2014. A fix to Aluminum Tolerance to open Arable land. Crop Biotech Update. May 7.

ISAAA in Brief. 2012. Pocket K No. 17: Genetic Engineering and GM Crops. International Service for Acquisition of Agri-Biotech Applications.

James, C. 2006. Global Status of Commercialized Transgenic Crops: 2006. ISAAA Briefs No. 35. ISAAA, Ithaca, NY.

James, R.A., C. Blake, C.S. Byrt and R. Munns. 2011. Major genes for Na$^+$ exclusion, Nax1 and Nax2 (wheat HKT1;4 and HKT1;5), decrease Na$^+$ accumulation in bread wheat leaves under saline and waterlogged conditions. J. Exp. Bot. 62(8): 2939–2947.

Kloti, A., J. Futterer, R. Terada, D. Bieris, J. Wunn, P.K. Burkhardt, G. Chen et al. 1996. Towards genetically engineered resistance to tungro virus. pp. 763–767. *In*: G.S. Khush (ed.). Rice Genetics III. IRRI, Los Banos, The Philippines.

Khush, G.S., E. Bacalangco and T. Ogawa. 1990. A new gene resistance to bacterial blight from *O. longistaminata*. Rice Genetics Newsletter 7: 121–122.

Lentini, Z., L. Calvert, E. Tabares, I. Lozano, B.C. Ramirez and W. Roca. 1996. Genetic transformation of rice with viral genes for novel resistance to rice hoja blanca virus. pp. 780–783. *In*: G.S. Khush (ed.). Rice Genetics III. IRRI, Los Banos, The Philippines.

Lin, M. and P. Huybers. 2012. Reckoning wheat yield trends. Environ. Res. Lett. 7. doi:10.1088/1748-9326/7/2/ 024016.

Mittler, R. 2006. Abiotic stress, the field environment and stress combination. Trends Plant Sci. 11: 15–19.

Oerke, E.-C. 2006. Crop losses to pests. J. Agri. Sci. 144: 31–43.

Podevin, N., H.V. Davies, H.V. Hartung, F. Nogué and J.M. Casacuberta. 2013. Site-directed nucleases: a paradigm shift in predictable, knowledge-based plant breeding. Trends in Biotechnol. 31(6): 375–383.

Ray, D.K., N.D. Mueller, P.C. West and J.A. Foley. 2013. Yield trends are insufficient to double global crop production by 2050. PloS One 8(6): e66428. doi: 10.1371/journal.pone.0066428.

Ray, D.K., N. Ramankutty, N.D. Mueller, P.C. West and J.A. Foley. 2012. Recent patterns of crop yield growth and stagnation. Nature Communications 3: 1293. doi:10.1038/ncomms2296.

Schlenker, W. and M.J. Roberts. 2009. Nonlinear temperature effects indicate severe damages to U.S. crop yields under climate change. Proc. Natl. Acad. Sci. USA 106: 15594–15598.

Schroeder, J.I., E. Delhaize, W.B. Frommer, M.L. Guerinot, M.J. Harrison, L. Herrera-Estrella, T. Horie et al. 2013. Using membrane transporters to improve crops for sustainable food production. Nature 497(7447): 60–66.

Sivamani, E., H. Huet, P. Shen, C.A. Ong, A. de Kochko, C. Fauquet and R.N. Beachy. 1999. Rice plant (*Oryzasativa*) containing rice tungro spherical virus (RTSV) coat protein transgenes are resistant to virus infection. Mol. Breeding 5: 177–185.

Somerville, C. and J. Briscoe. 2001. Genetic engineering and water. Science 292: 2217.

Song, W.Y., G. Wang, L. Chen, H. Kim, L.Y. Pi, T. Holsten, J. Gardner et al. 1995. A receptor kinase-like protein encoded by the rice disease resistance gene, *Xa-21*. Science 270: 1804–1806.

Tomlinson, I. 2011. Doubling food production to feed 9 billion: a critical perspective on a key discourse of food security in the UK. J. Rural Studies xxx: 1–10. doi:10.1016/j.jrurstud.2011.09.001.

Tripathi, S., J. Suzuki and D. Gonsalves. 2006. Development of genetically engineered resistant papaya for papaya ringspot virus in a timely manner: a comprehensive and successful approach. pp. 197–240. *In*: P.C. Ronald (ed.). Plant-Pathogen Interactions: Methods and Protocols. Methods Mol. Biol. 354. Humana Press, Totowa, NJ.

Tu, J., I. Ona, O. Zhang, T.W. Mew, G.S. Khush and S.K. Dutta. 1998. Transgenic rice variety 'IR72' with Xa21 is resistant to bacterial blight. Theor. Appl. Genetics 97: 31–36.

United Nations. 2012a. Urbanization Prospects. The 2011 Revision: Highlights. Population Division. Economics and Social Affairs. United Nations. ESA/P/WP/224, March 2012.

United Nations. 2012b. World Population Prospects: The 2012 Revision. Department of Economic and Social Affairs/Population Division, Population Estimates and Projections Section. Available: http:// esa.un. org/unpd/ wpp/ index.htm.

Vorosmarty, C.J., P. Green, J. Salisbury and R.B. Lammers. 2000. Global water resources: vulnerability from climate change and population growth. Science 289: 284–288.

Wang, G.L., W.Y. Song, D.L. Ruan, S. Sideris and P.C. Ronald. 1996. The cloned gene, Xa-21, confers resistance to multiple *Xanthomonas oryzae* pv. *oryzae* isolates in transgenic plant. Molecular Plant-Microbe Interactions 9: 850–855.

Xu, C., Z. Cheng and W. Yu. 2012. Construction of rice mini-chromosomes by telomere-mediated chromosomal truncation. The Plant J. 70(6): 1070–1079.

Yu, W. and J.A. Birchler. 2007. Minichromosomes: The next generation technology for plant engineering. Univ. Missouri, Division of Biological Sciences.

Yu, W., F. Han, Z. Gao, J.M. Vega and J.A. Bircher. 2007. Construction and behavior of engineered minichromosomes in maize. Proc. Natl. Acad. Sci. USA 104(21): 8924–8929.

Zhang, S., W.Y. Song, L. Chen, D. Taylor, P. Ronald, R. Bleachy and C. Fauquet. 1998. Transgenic elite indica rice varieties, resistant to *Xanthomonas* pv. *oryzae*. Mol. Breeding 4: 551–558.

CHAPTER 2

Herbicide Resistance

Ever since the widespread usage of phenoxy herbicides beginning 1946, weed scientists began pondering over the possibility of development of herbicide-resistant weed populations similar to those exhibited by insects against an inorganic insecticide in 1908 but documented in 1914, and plant pathogens against fungicides since 1940. In 1950, Blackman [1950] warned "… repeated spraying with one type of herbicide will sort out resistant strains within the weed population." In 1954, McCall [1954] wondered whether weeds were becoming more resistant to herbicides. The same year saw a report from the U.K. suggesting that continuous application of 2,4-D has led to resistance of weed species normally susceptible to it. This was followed by two other reports against 2,4-D in 1957, one from Hawaii where biotypes of *Commelina diffusa* (spreading dayflower) in sugarcane fields [Hilton 1957], and another from Ontario, Canada, where biotypes of *Daucus carota* (wild carrot) in sections of highway weeds [Switzer 1957] exhibited resistance.

These and a few other warnings were largely ignored until the first confirmed report of herbicide resistance against simazine and atrazine which failed to control *Senicio vulgaris* in 1968 in a Washington nursery, where they had been used since 1958 [Ryan 1970]. Since then, herbicide resistance problems have been accelerating. Consequently, management of weeds have become increasingly more difficult and complex.

At the present time, 238 weed species (138 dicots and 100 monocots) infest 84 crops and non-cropping areas in 65 countries have been identified to develop resistance to 155 different herbicides belonging to 22 of the 25 herbicide families with as many sites of action [Heap 2014]. As many species showed resistance to herbicides of multiple sites of action, the number of unique resistant cases are much higher (Fig. 2.1; Appendix: Table 1). For example, *Lolium rigidum* is resistant to herbicides of 11 sites of action.

The most wide-spread herbicide-resistant weed species in the world, with four and more sites of action other than *Lolium rigidum* include *Echinochloa crus-galli* var. *crus-galli* (10), *Poa annua* (9), *Eleusine indica* (7), *Alopecurus myosuroides* (6), *Amaranthus tuberculatus* (=*A. rudis*) (6), *Echinocloa colona* (6), *Lolium perenne* ssp. *multiflorum* (6), *Amaranthus palmeri* (5), *Ambrosia artemisiifolia* (5), *Avena fatua* (5), *Conyza canadensis* (5), *Raphanus raphanistrum* (5), *Amaranthus retroflexus* (4), *Bromus tectorum* (4), *Chenopodium album* (4), *Conyza bonariensis* (4), *Ischaemum rugosum* (4), *Kochia scoparia* (4), *Setaria viridis* (4), *Sisymbrium orientale* (4), and *Sorghum halepense* (4) [Heap 2014]. 27 showed resistance to three (3) sites of action and 41 to two (2) sites of action, while the remaining species exhibiting action at no more than one site (vide Appendix: Table 1) [Heap 2014].

Among weed families, Poaceae contributed the most number of resistance species of 75 followed by Asteraceae (37), Brassicaceae (21), Amaranthaceae (12), Cyperaceae

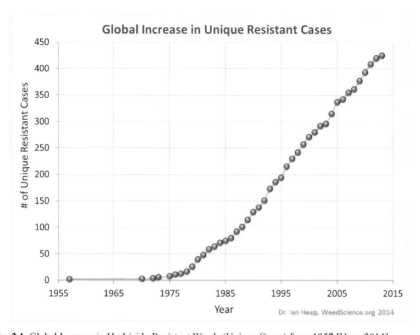

Fig. 2.1. Global Increase in Herbicide-Resistant Weeds (Unique Cases) from 1957 [Heap 2014].

(10), Scorphulaceae (9), and Chenopodiaceae (8), Alismataceae (6), Polygonaceae (6), and Caryophyllaceae (5) [Heap 2014].

Of the various crops in which herbicide-resistant weed species were found so far, wheat tops with 65 followed by maize/corn (58), rice (50), and soya bean/soybean (46), rapeseed/canola (20), and cotton (17) [Heap 2014].

Generally, '**resistance**' is defined as the **inherited ability** of a plant species/biotype to survive and reproduce following exposure to a dose of herbicide normally lethal to the wild type [WSSA 1998]. This is the dose normally used for satisfactory weed control. In both crop plants and weeds, herbicide resistance may be naturally occurring or induced by genetic engineering or selection of variants produced by tissue culture or mutagenesis [WSSA 1998]. Generally, this heritable resistance trait is found in crop plants, as against weeds, thus forming the basis for herbicide selectivity. Resistant weeds are those plant species that express the genetic variation required to evolve mechanisms to escape control. Like other crop pests, herbicide-resistant weeds are the result of intensive selection pressure in weed populations. When a herbicide causes selection pressure, susceptible plants are killed while the resistant plants survive to reproduce without confronting any competition from susceptible plants.

The term **herbicide-resistance** is normally used while referring to a) evolution of resistance to herbicides over a period of time and b) the resistance trait introduced in a plant transgenically. This is also the term used in this book in both instances. However, the term herbicide-tolerance is also used intermittently, particularly when the source of published information cited in the book has used it.

On the other hand, '**tolerance**' is the **inherent ability** of a species to survive and reproduce after herbicide treatment [WSSA 1998]. This implies that there is no selection or genetic manipulation to make the plant tolerant; it is naturally tolerant. Tolerance may also be considered the natural or normal variability of response to herbicides that exists within a species and can easily and quickly evolve.

Evolution, Spread, and Types of Herbicide Resistance

Herbicides do not induce resistance. Instead, they select for resistant individuals that naturally occur within the weed population. The more a herbicide is used, the greater the likelihood of encountering a resistant individual in a field. Once a resistant plant is selected, repeated use of a herbicide over multiple generations allows the resistant plants to proliferate as susceptible plants are eliminated. Once a resistance gene has occurred within a population, failure of the herbicide can be rapid.

There are two pre-requisites for the evolution of herbicide resistance in plant populations: a) the occurrence of heritable variation in genetic composition for herbicide resistance and b) natural selection for increased resistance to herbicides.

In response to repeated treatment with a particular herbicide or class (family) of herbicides, weed populations change in genetic composition such that the frequency of resistance alleles and resistant individuals increase [Jasieniuk et al. 1996]. In this way, weed populations become adapted to the intense selection pressure imposed by herbicides. The evolution of resistance under continuous application of a herbicide may be considered as an example of recurrent selection in which there is a progressive and, sometimes, rapid shift in average fitness of populations of weeds exposed to it. This shift in fitness, a genetic trait, is directly related to an increase in frequency of the resistance trait (phenotype) in the population. The selection pressure for herbicide resistance is contributed by: a) efficiency of the herbicide, b) frequency of herbicide use, c) duration of herbicide effect, d) method of herbicide use, e) selection pressure that is characteristic of the herbicide, and f) resistance mechanism in weed species.

The intensity of selection in response to herbicide application is a measure of the relative mortality in target weed populations and/or the relative reduction in seed production of survivors; this will be proportional, in some manner, to herbicide dose [Maxwell et al. 1990]. The duration of selection is a measure of the period of time over which phytotoxicity is imposed by herbicide. Both intensity and duration will interact to give seasonal variation in the process of selection which will, in turn, depend upon the phenology and growth of a weed species. For example, in the case of a preemergence herbicide that inhibits seedling emergence over a time period, the intensity of selection may be much higher on weed seedlings emerging early in the life of a crop than those emerging later. The occurrence and speed of evolution of herbicide resistance are determined by: a) number of alleles involved in the expression of functional resistance, b) frequency of resistance alleles in natural (unselected) populations of weed species, c) mode of inheritance of the resistant alleles, d) reproductive and breeding characters of the weed species, e) longevity of weed seeds in the soil, f) intensity of selection which differentiates resistant biotypes from susceptible ones, and g) absolute fitness of resistance and susceptible genotypes.

Factors Leading to the Evolution of Herbicide Resistance

Factors that lead to, or stimulate and accelerate, the evolution of herbicide resistance are manifold. These include biological characteristics of the weed species, characteristics and application of the herbicide, and cultural practices adopted for weed control.

Weed Characteristics

The most likely weed characteristics that favor increase in resistance against a particular herbicide are: a) annual growth habit, b) high seed production, c) relatively rapid turnover of the seed-bank due to high percentage of seed germination each year, d) several reproductive generations per growing season, e) extreme susceptibility to a particular herbicide, and f) frequency of resistant gene(s) among weeds.

Weed species less likely to develop resistance generally have a) a slower generation time, b) incomplete selection pressure for most herbicides, c) ability to adapt to changing environment, d) lower fitness for resistant biotypes, and e) extended seed dormancy in the soil. These factors increase the number of susceptible biotypes in the population.

Herbicide Characteristics

The following properties of herbicide molecule build resistance in weeds: a) mechanism of action, b) a single site of action, c) frequency of herbicide use, d) broad spectrum control, and e) longer residual activity in the soil.

Cultural Practices

Complete reliance on herbicides for weed control can greatly enhance the occurrence of herbicide resistant weeds. Cultural practices can also increase the selective pressure (discussed latter in the Chapter) for the development of herbicide resistant biotypes. These are: a) shift away from multi-crop rotations towards mono-cropping, b) little or no cultivation or tillage for weed control or no elimination of weeds that escape herbicide control, c) continuous or repeated use of a single herbicide or several herbicides that have the same mode of action, d) higher herbicide use rate relative to the amount needed for weed control, and e) weeds in orchard and vineyard systems as well as roadsides.

The level of herbicide resistance in weeds varies with weed biology and resistance mechanism. In some cases, resistance occurs when the species survives a labeled rate of application, while in other cases the species can survive up to 1,000 times the labeled rate. There are two levels of herbicide resistance characteristics: low-level and high-level. Low-level resistance includes, a) continuum of plant responses from slightly injured to nearly dead, b) display of an immediate response by majority of plants, and c) presence of susceptible plants in the population, especially when herbicide resistance is determined early. In high-level resistance a) plants are slightly injured to uninjured, b) few plants have intermediate responses, and c) susceptible plants can be present in the population.

Genetic Variation and Mutation

Evolution of herbicide resistance is dependent on the extent of genetic variation and frequency of occurrence of mutation. In a susceptible population, genetic variation must be present for occurrence of the evolution of herbicide resistance. Most weed species contain adequate genetic variations that allow them to survive under a variety of environmental stresses. Genetic variation, which has a direct relationship with natural selection, may be measured by such quantitative measures as plant height, time to flowering, and total biomass.

Genetic variation can be preexisting or arise *de novo* by mutation (or recombination) following herbicide application. We can thus distinguish two situations as regards genetic variation for herbicide resistance in non-selected populations: a) factors affecting the acquisition of resistance by novel mutation and b) factors affecting the probability of preexisting variation for resistance.

Given the existence of genetic variation, the rate of evolution will be determined by the mode of inheritance of resistance traits and intensity of selection. The evolution of resistance under persistent applications of herbicides may be considered as an example of recurrent selection in which there is a progressive, and sometimes rapid shift in average fitness of weed populations exposed to herbicide. Once established, gene flow via seed and propagule distribution contributes to the spread of resistant weeds. A major determinant in the selection of herbicide-resistant biotypes is the effective selection intensity that differentiates resistant individuals (more fit) from susceptible ones (less fit) in the face of selection (to the application of herbicide).

There are two ways in which resistance traits may arise within a weed population. A major resistant gene(s) may be present at low frequency so that selection acts to change a population, which is initially susceptible [Maxwell and Mortimer 1994]. Alternately, recurrent selection may occur continuously to achieve a progressive increase in average resistance from one generation to the other, with changes in gene frequency at many loci conferring resistance. Thus, herbicide resistance is developed by gene mutations or conferred by preexisting genes.

In general, gene mutations conferring resistance to a herbicide class are not induced by application of the herbicide, but rather, occur spontaneously. Spontaneous mutations at gene loci recur with characteristic frequency such that new mutations are continuously generated in natural populations of weeds. Mutations at some loci, particularly those encoding a specific herbicide site of action, may confer resistance. Typical spontaneous mutation rates in biological organisms are often cited as 1×10^{-5} or 1×10^{-6} gametes per locus per generation [Merrell 1981].

These values are for single, nuclear gene inheritance of evolution of herbicide resistance evolution. The rate of mutation to a single, dominant resistant allele of 1×10^{-6}, the probability of occurrence of at least one resistant plant in a 30 ha field with a weed density equal to, or exceeding, five plants m^{-1} is greater than 0.95 for a random mating species [Jasieniuk et al. 1996]. Thus, at least one resistant plant is almost certain to occur in a weed population of this size despite the low rate of mutation to resistance. Factors such as seed non-viability and seedling mortality may cause the plant's death prior to reproduction. However, should it survive and reproduce and if the corresponding herbicide or herbicide class is applied for several generations, the single resistant plant could give rise to a predominantly resistant population of weeds.

If the mutation rates are lower than 1×10^{-6}, i.e., 1×10^{-8} to 1×10^{-10}, the probability of occurrence of a resistant mutant is markedly reduced. It requires densities greater than 50 plants per m^2 for a resistant mutant to occur if the mutation rate is 1×10^{-8} [Jasieniuk et al. 1996]. The positive correlation between the size of a susceptible weed population and probability of occurrence of resistant mutant plants may partly explain why herbicide resistance has evolved in some species but not in others. The probability of herbicide-resistant plants occurring in a population is greater for weeds with high densities than for those which occur at low densities.

The frequency of mutation may be influenced by environment and dosage of herbicide. For example, atrazine applied at sub-lethal doses to certain susceptible genotypes of

Chenopodium album resulted in a progeny with triazine-resistant characteristics similar to highly resistant plants [Bettini et al. 1987]. This indicates that herbicide resistance could be induced by low doses of the chemical in certain genotypes. Most herbicides are extensively screened for efficacy in the field before release for commercial use. Herbicides are not accepted for market without at least 85–90 percent control of the target weeds. Therefore, it is very likely that resistance traits are present, if undetectable, in weed populations before large-scale selection with herbicides.

Mechanisms of Herbicide Resistance

The most common and important mechanisms of herbicide resistance are those which interrupt the transport of herbicides to biochemical sites of action, reduce the sensitivity of target sites, and detoxify the chemical or enhance repair can potentially confer resistance. These include the following:

1. Sequestration or Compartmentalization of the Herbicide in the Apoplast: Some plants restrict the movement of herbicides within the cells or tissues and prevent them from causing harmful effects. In this case, the herbicide may be inactivated either through binding (often to sugar moiety) or removed from metabolically active regions of the cell to inactive regions where it exerts no effect.
2. Altered Target Site: The herbicide has a specific site of action where it acts to disrupt a particular plant process or function. If the target site is altered, it no longer binds to the site and is unable to exert its phytotoxic effect. This is the most common herbicide resistance mechanism.
3. Differential Uptake and Translocation: In resistant biotypes, the herbicides are not taken up readily due to abnormal production of waxes, reduced leaf area, etc. Similarly, in resistant biotypes the apoplastic and symplastic transport of herbicide is reduced due to differential modifications.
4. Enhanced Metabolism: Weeds that have the ability to quickly degrade a herbicide may potentially inactivate it before it reaches its site of action within the plant, thus enhancing metabolism.
5. Over-expression of the Target Protein: If the target protein on which the herbicide acts is produced in large quantities by the plant, then the effect of herbicide becomes insignificant.
6. Enhanced Production of the Target Site: When production of the target site is enhanced, the herbicide will be unable to inactivate the enzyme. Thus, the enzyme spared by the herbicide will carry on the normal plant metabolic activities.
7. Modification of cell membrane function and structure.
8. Altered sensitivity of the key target enzyme caused by mutation(s).
9. Enhanced metabolic breakdown and conjugation of the herbicide.
10. Enhanced degradation of herbicide-generated toxic products.

These mechanisms, and consequently the expression of resistance, are controlled by genetic loci.

Inheritance of Herbicide Resistance

There are three modes of inheritance of herbicide resistance: Nuclear inheritance, Cytoplasmic inheritance, and Quantitative inheritance.

Nuclear Inheritance

Resistance to most classes of herbicides is caused by nuclear inheritance. These include auxinic herbicides, aryloxyphenoxypropionics, benzoics, bipyridiliums, dinitroanilines, sulfonylureas, substituted ureas, glycines, etc. In nuclear inheritance, the resistance-conferring alleles are transmitted through pollen and ovules. Adaptive evolution is achieved by the selection of phenotypes encoded by many genes (i.e., polygenes), each with a small additive effect. Generally, herbicide resistance is conferred by major genes present in weeds. In the majority of cases in which the number of genes has been determined, resistance is controlled by a single, major gene [Jasieniuk et al. 1995]. The predominance of major gene inheritance is attributed to the two following factors.

1. The most recently developed herbicides are highly target site-specific and interfere with single enzymes in major metabolic pathways. Mutation of the gene encoding for the enzyme may alter a plant's sensitivity to the herbicide and cause resistance.
2. Repeated application of these herbicides imposes strong selection, often causing 95–99 percent mortality, against susceptible phenotypes in weed populations.

Adaptation to herbicide is possible only if resistant genes are present in a population and have a significantly large phenotypic effect to allow the survival of a few individuals in a single generation [Mazur and Falco 1989]. With polygenic inheritance, recombination among individuals for many generations is required to bring together a sufficient number of favorable alleles to produce a highly resistant phenotype. Polygenic inheritance of resistance is thus more likely under conditions of weak selection as would occur with sublethal herbicide application [Jasieniuk et al. 1996].

Cytoplasmic Inheritance

Cytoplasmic inheritance of resistance occurs with triazine herbicides in several weed species. The gene conferring triazine resistance is located in the chloroplast genome [Hirschberg and McIntosh 1983]. Transmission of the chloroplast resistant gene mostly occurs by pollen, the paternal parent. For example, the mutation that confers maternally inherited triazine resistance involves a single base substitution in the *psbA* chloroplast gene which codes for a photosystem II (PS II) membrane protein to which triazine herbicides bind. The expected frequency with which a mutation occurs in the chloroplast genome and gives rise to gamete-transmissible triazine resistance is very low. The probability ranges from 1×10^{-9} to 1×10^{-12} mutations per gene locus.

Quantitative Inheritance

Quantitative patterns of inheritance of a phenotypic characteristic (trait) are controlled by polygenes. In this, the additive action of numerous genes, perhaps minor, results in a trait (e.g., height, seed production, etc.) showing continuous variability. The different minor genes that affect several processes will rapidly add up to a high level of resistance [Neve and Powels 2005a]. For instance, one gene may limit translocation of the herbicide, another may cause rapid metabolism, and yet another may affect the target site slightly [Gressel 2009]. Generally, differential resistance is quantitatively inherited.

Unlike monogenic traits, polygenic traits do not follow the Mendelian inheritance (separated traits). Genes, with each one causing a small increase in fitness under herbicide selection, may systematically promote increased fitness in genotypes as genetic recombination occurs over successive generations. Implicitly, the rate of evolution is likely to be slower than for single nuclear-encoded genes.

Awareness of the possibility of quantitative resistance has led to recommendations to apply labeled herbicide rates to weeds of the size recommended by the registration.

Resistance to chlorsulfuron by *Lolium perenne* is conferred by additive minor genes inherited in a quantitative manner [Mackenzie et al. 1995]. Resistance of *L. rigidum* to diclofop-methyl is controlled by at least three resistant genes and low herbicide dose can rapidly evolve polygenic broad-spectrum herbicide resistance by quantitative accumulation of additive genes of small effect [Busi et al. 2012].

Imposition of Selection Pressure by Herbicides

Selection pressure is an interaction between natural variation in a species and factors in its environment that cause a certain plant to have an advantage over the others. It pushes the evolution of a particular species toward a greater prevalence of this variation. Its effectiveness is measured in terms of differential survival and reproduction, and consequently in change in the frequency of alleles in a population.

In the case of weed resistance, herbicide application exerts a powerful selection pressure on huge populations of weed species in both herbicide-resistant transgenic crops and conventional crops, exposing individuals with a genetic ability to survive herbicide treatment. While the population as a whole suffers high mortality, the herbicide is effectively selecting individual plants that possess any of the genetically endowed traits (resistant genes) which enable them to survive the herbicide dosage used. These survivors produce seed and contribute to the gene pool of subsequent generations, enriching the population with resistance genes. Thus, herbicide resistance results from selection pressure working on genetic diversity. Selection by herbicides changes the population over time. If, in the first year, one plant in a million plants of a weed species treated with a herbicide exhibits resistance, in the second year it will multiply into more resistant plants, and the process repeats in the following years to end up in the evolution of even more resistant plants.

This evolution of herbicide resistance in weeds is dependent on the intensity of selection imposed by herbicides. Most herbicides are applied at rates that result in the mortality of 90–99 percent of the susceptible weeds. If genetic variation for resistance is present due to mutation or gene flow, even at very low frequencies, repeated herbicide application will normally result in a rapid increase in the frequency of resistant individuals until they dominate the population [Jasieniuk et al. 1996].

Selection pressure imposed by a herbicide is a primary factor that determines the rate of enrichment of herbicide resistance in a weed population. In general, selection pressure is a measure of the ability of a herbicide to differentiate between susceptible and resistant plants. The higher the intensity of selection imposed by a herbicide against susceptible species, the faster the expected rate of evolution and spread of resistance. Several herbicide characteristics and patterns of use result in higher mortality than others do and thus impose a more intense selection pressure for the development of resistance. These include a single target site and highly specific mechanism of action, long-term soil residual activity, and frequent applications [LeBaron and McFarland 1990].

Seed production is an essential component to assess selection pressure. Effective kill by a herbicide is measured as the percentage reduction in seed yield at the end of the growing season. Values obtained for both resistant and susceptible plants are then used to estimate the selection pressure. The selection pressure of a herbicide is calculated as the ratio of the fraction of resistant plants that survive its application to the corresponding fraction of susceptible plants [Gressel and Segel 1990].

Measurement of selection intensity exerted by herbicides is done theoretically and phenotypically. Population geneticists measure selection as the differential survival of alleles or change in gene frequency after the action of the selection agent. However, weed scientists measure selection frequency at the phenotype level. Selection coefficients may be variously defined at the genetic (gene) or zygotic (genotypic) level. The coefficient of selection, 'S', may be defined as the proportionate reduction in the contribution of a particular genotype (usually the most favored), whose contribution is usually taken to be unity [Maxwell and Mortimer 1994].

It is very important to take into account that the intensity of selection pressure depends on the type of treatment and/or herbicide, its formulation, frequency of application, and the biological characteristics of the weed and the crop. Herbicide selection pressure should be seen in the group of actions carried out in the field: tillage, crop rotation, use of other control methods and cropping. Thus, a herbicide with selection pressure, used sporadically and alternating with other non-chemical control methods, will have a low risk of causing problems of resistance. Some herbicide groups have a higher selection pressure than others. For example, selection pressure is high (within 4–8 yr of continuous application) with ACCase-(acetyl CoA carboxylase) and ALS-(acetolactate synthase) inhibiting herbicides, while it is medium (within 10–15 yr of application) with PS II-, PS I-, carotenoid biosynthesis-inhibiting, and auxin-inhibiting herbicides. For other herbicides the selection pressure is on the low, with resistance occurring only after continuous application for 15 years and more.

Alteration of Selection Pressure

When resistance is determined by major genes, a lowering of the selection pressure may delay the onset of resistance. This can be achieved [Mortimer 1993] by:

a. Reducing the rate of application of the herbicide selected.
b. Invoking mixtures of herbicides. In this, co-evolution of resistance to two different herbicide chemistries would be slow because the frequency of dual-resistant plant would be the compounded frequency of the two herbicides [Gressel and Segel 1990].
c. Adopting rotation of herbicides with non-chemical methods of control.
d. Using herbicides which have fundamentally different modes of action.

Reducing the selection pressure may delay the evolution of resistance for the following reasons:

a. Plants that are susceptible to selective herbicides may contribute progeny to the next generation and hence lower the frequency of resistant alleles in the total population.
b. Where a weed species has a persistent seed-bank, only a fraction of that weed population will be exposed to selection in each cropping season. Hence populations in successive seasons will include susceptible individuals recruited from the seed-bank, and the survival of these individuals may again result in a lowering of the frequency of resistant alleles.
c. Plants that escape mortality may be the recipients of immigrant pollen (from external sources), thereby enabling an influx of susceptible alleles and hence leading to lowering of resistance gene frequency. The effectiveness of gene flow will be

determined by the mode of inheritance of resistance and the frequency of resistant alleles at the time when gene flow occurs. The influence of such immigration will be noticeable if resistance is conferred as a recessive allele that is at low frequency in the population. When resistance is controlled by a dominant allele, the effect is likely to be small, and if maternally inherited, zero.

The reverse situation in which resistance alleles emigrate into the surrounding populations of susceptible genotypes is significant for the management of existing herbicide resistant populations. Where selection is more relaxed due to change in management, the frequency of resistance alleles in the susceptible population will reduce more slowly than the absence of pollen flow.

Effect of Polygenic Resistance on Selection Pressure

The response to selection based on polygenes depends on genetic recombination causing several or many genes (each contributing in a minor way) to 'coalesce' in a single genotype [Mortimer 1993]. Relatively rapid response to selection will occur if low selection pressures are applied since this will strongly select for genotypes showing elevated resistance as individual genes become combined within a genotype. Application of increasingly strong doses of herbicide will intensify the response to selection. If selection pressure is high initially, then genotypes with small enhancement of resistance will be lost from the population and the frequency of recombinations of polygenes or multiple gene amplifications will be greatly reduced.

Fitness

Fitness is the ability of the organism to survive and produce offspring in a given environment. It is the central idea in evolutionary theory. If differences between alleles of a given gene affect fitness, then the frequencies of the alleles will change over generations. In the theoretical plant population model constructed to predict herbicide resistance, two sets of biological processes serve as major factors. These are ecological fitness and gene flow. Knowledge about both factors is necessary to develop effective strategies for management of herbicide resistant weeds.

An individual plant is a unique genotype with variation at many loci affecting fitness. The fitness of a group of plants having a certain genotype is assessed in relation to the fitness of other genotypes lacking key traits of interest. Fitness is a measure of survival and ability of a given genotype (e.g., herbicide-resistant biotypes) to produce viable offspring in competition with the wild type (e.g., herbicide-susceptible biotypes) [Gressel 2002]. It describes the evolutionary advantage of a phenotype, based on its survival and reproductive success. Under conditions of natural selection, genotypes with greater fitness produce, on average, more offspring than less fit genotypes. It is measured over the whole life cycle of a plant, encompassing the effects of selection on mortality and seed production of survivors.

In a single interbreeding population of plants in a homogeneous environment, the genotypic response due to allelic changes at a single locus is considered to occur against a constant environmental and genetic background and the expression 'genotypic fitness' is used. With this approach, it is possible to measure and calculate 'genotypic fitness' in the field and laboratory for a given genotype (homozygote and heterozygote) [White and White 1981].

Ecological fitness of a given biotype indicates that it will leave a greater proportion of its genes in the future gene pool of the population. The most fit plant will leave the greatest number of offspring. Differences in ecological fitness between resistant-(R) and susceptible-(S) biotypes will influence the rate at which herbicide resistance appears, as well as development of resistant populations when not treated with the herbicide. Although herbicide-R biotypes should be less fit than S biotypes, the fitness of resistant populations may vary depending on the mechanism of resistance and the environmental conditions. For example, under agricultural field conditions, triazine-R biotypes than their normal-R counterparts have been shown to be less fit than the S biotypes. However, many studies have not been able to detect fitness penalty in biotypes resistant to ALS inhibitors (e.g., imazethapyr) under those conditions [Ashigh and Tardif 2009]. The existence of fitness penalty under field conditions could be exploited for management of those resistant biotypes affected by it.

For an annual species, the seed produced by a genotype per generation constitutes a fitness estimate for a given environment only at one point in the evolutionary time. While determining the fitness of a weed with a persistent seed-bank, the rate of loss of seed from the soil needs to be measured. Seed carryover from previous generations plus the seed produced in the current generation contribute to total seed production. In order to understand the rate of evolution of resistance or management of resistance, measurements of fitness need to be conducted only under field conditions with the crop, and with and without herbicide application. For plant species reproducing vegetatively, measuring fitness is intrinsically more difficult and may require measurement of biomass or plant parts over a time period appropriate to the species in question.

Fitness is expressed in relative terms whereby genotypes are compared amongst themselves relative to the most successful one but it is important to recognize the following:

a. Absolute fitness contributes to the rate of evolution in its own right. When all other factors are equal, evolutionary rate is proportional to per capita rates of increase of the weed population.
b. Fitness is a measure of genotypic performance in a particular environment. Early studies on resistant biotypes of weeds pointed to the fact that there may be a 'cost' to resistance reflected in traits such as growth and competitiveness. Thus susceptible genotypes have superior fitness to resistant ones in the absence of selection.

Gene Flow and Spread of Herbicide Resistance

Gene flow, also known as gene migration, is the transfer of genetic material or alleles from one plant to another and from one site to another. It results in a change in gene frequency in one population due to movement of gametes, individuals, or groups of individuals from one population to another [Slatkin 1987] and occurs both spatially and temporally [Mallory-Smith and Olguin 2011]. It is a natural process to which all genes are subject and this contributes to evolution of species. Gene flow is of two types: horizontal and vertical. Horizontal gene flow is the movement of genes between disparate, unrelated species as in the case of plants and microbes. On the other hand, vertical gene flow, which is of greater importance in evolution of species, is the exchange of genes between closely related species. It occurs in only one generation between varieties or types of plants within the same species, and sometimes even between species. Thus vertical gene flow is a natural process that occurs incessantly and permanently between biologically compatible organisms and to which all genes are subject.

Gene flow occurs via the movement of pollen, individual plants, seeds, or vegetative propagules [Mallory-Smith and Zapiola 2008; Slatkin 1987]. As mentioned earlier, gene flow, a major biological factor in predicting herbicide resistance, has a significant impact on the rate of evolution of herbicide resistance. If a single plant or set of plants survive a herbicide application, the allele for resistance is passed to other plants in the field. Gene flow is not unique to herbicide-resistant and transgenic crops. Rather, it occurs independently of the techniques used to produce the crops. Crops such as wheat and rapeseed are products of gene flow and natural hybridization [Kimber and Sears 1987; Woo 1935].

Rates of gene flow are generally believed to be higher than rates of mutation, and hence would result in a higher frequency of plants resistant to a particular herbicide prior to its initial application. Therefore, gene flow and the resultant increase in initial frequency of resistant genes would reduce the time required to reach a specific level of resistance within a field once a herbicide is applied [Jasieniuk et al. 1996].

Gene Flow via Pollen

Gene flow among plants can occur through pollen dispersal and the resultant fertilization. Breeding of herbicide resistant crops may be done by self-pollination, cross-pollination, and mixed. Cross-pollinated crops have a greater potential for gene flow via pollen than self-pollinated crops in which pollen flow is minimal. Some cross-pollinated plants exhibit self-incompatibility, which prevents self-pollination and promotes outcrossing. Crops with mixed mating patterns can produce seeds through either self-pollination or cross-pollination. The likelihood of gene flow via pollen also increases with the neighboring presence of highly compatible related species, synchronous flowering with compatible species, large pollen sources, and strong winds [Giddings 2000; Giddings et al. 1997a,b; Levin 1981].

The distance at which gene flow occurs via pollen is highly variable, and it is difficult to predict the farthest distance that viable pollen can move. In general, most gene flow via pollen occurs at relatively short distances because pollen is viable for only hours or days. Pollen is subject to desiccation, and its viability is influenced by environmental factors, such as temperature and humidity.

For example, outcrossing of resistant pollen to susceptible plants was 1.4 percent at a distance of 28.9 m in *Kochia scoparia*, resistant to sulfonylurea herbicide, and 1 percent at > 6.84 m distance in *Lolium multiflorum*, resistant to diclofop-methyl [Maxwell 1992; Stallings et al. 1993]. Mulugeta et al. [1992] recovered *Kochia scoparia* pollen as far as 50 m from its source, indicating that pollen dispersal could lead to long distance spread of herbicide resistance genes. Isolation of a herbicide-resistant crop is likely to lessen the gene flow via pollen. A physical buffer, such as increasing the distance between fields, reduces gene flow because most pollen remains close to the source. A temporal buffer, such as the use of staggered planting dates, reduces pollen-mediated gene flow between maize hybrids [Halsey et al. 2005] because flowering times across fields become asynchronous.

Despite efforts to mitigate the movement of pollen, natural dispersal of pollen via wind and insects cannot be prevented nor absolutely predicted, and it can occur over considerable distances [Beckie et al. 2003; Reichman et al. 2006; Watrud et al. 2004]. However, gene flow via pollen could be reduced greatly by placement of the herbicide-resistance gene in the DNA of chloroplasts, which are maternally inherited, or by using a male-sterile breeding system (vide Chapter 9).

Gene Flow via Vegetative Propagules

Vegetative propagules such as stolons, rhizomes, roots, crowns, and bulbs allow single plants to reproduce in isolation and can become a source of herbicide resistance genes [Vencill et al. 2012]. Short distance movement can occur between fields via natural means or on shared equipment as it is moved between fields. Long distance movement, however, is not likely except with human intervention or possibly via waterways, like the movement of seeds. Vegetative propagules left in the soil can result in an established plant in the following year and make eradication difficult. Reproduction via vegetative propagules must be considered a risk factor when developing management plans designed to prevent gene flow [Vencill et al. 2012].

Gene Flow via Seeds

Seed dispersal plays a greater role, hence a major mechanism, in the inter-population gene flow than pollen dispersal. Loss of seeds from herbicide-resistant crops may occur at any point from planting to post-harvest operations. This paves the way for seed-mediated gene flow. The mixing of herbicide resistant and conventional seeds is known as commingling or admixture [Mallory-Smith and Zapiola 2008]. Commingling may occur if seeds from volunteer plants (of previous crop) are harvested with the current crop or mixed during postharvest operations, such as seed cleaning, seed conditioning, transport, or storage. Gene flow via seeds can be reduced if strict post-harvest guidelines are followed. However, some seed loss at each step from planting through final use is inevitable [Mallory-Smith and Zapiola 2008].

Natural dispersal of seeds via wind, water, and animals contributes to gene flow and this cannot be prevented. Seeds, being more environmentally persistent than pollen, can move farther over time [Squire 2005].

Gene flow is greater when seeds are smaller, have extended viability and dormancy, and possess greater persistence. Dormancy affects soil seed-bank. Herbicide resistant crops that emerge from the persistent seed-bank will be a problem if a herbicide resistant crop with the same trait is planted in the rotation or if there is no good control option available. Herbicide resistant rice does not have a persistent seed-bank, but once the herbicide resistant trait escapes to weedy rice, resistance will persist in the weedy seed-bank. In Arkansas, USA, ALS-resistant weedy rice plants were detected in all sampled fields with imidazolinone-resistant non-transgenic 'Clearfield' rice (CL121, CL141, and CFX51 lines) cropping history [Singh et al. 2012].

Gene flow could occur from herbicide treated fields to adjacent unsprayed areas. Similar flow could also occur from resistant plants within the crop to susceptible plants at the field edge. Devlin and Ellstrand [1990] showed that a significant proportion of a population's seed crop can be fathered by plants from outside of the population. Such gene flow could lead to a rapid spread of a herbicide resistance gene among weed populations in a particular area.

For example, *Amaranthus palmeri* S. Wats (Palmer amaranth), which can produce a large amount of small seeds capable of floating in water, has the ability to spread rapidly throughout a production field in a short time span [Norsworthy et al. 2014]. This prolific seed producing annual broadleaf weed has rapid dispersal and is highly competitive with crops, making herbicide-resistant strains difficult to control. In a cotton field applied with only glyphosate used for weed management, Norsworthy et al. [2014] found in the first year a separate patch of *A. palmeri* emerging 375 ft (115 m) from the original location. In the second year, resistant plants expanded to reach field boundaries and infested 20 percent of the field area resulting in decreased yield and significant problems with cotton harvest. By the third growing season, glyphosate-resistant Palmer amaranth had completely colonized the fields, making the cotton crop impossible to harvest. This finding helps to explain the rapid takeover of many farms by glyphosate-resistant Palmer amaranth, particularly when glyphosate was the only means of weed control. This could happen with other weeds species and other herbicides.

Gene flow that occurs between adjacent populations by seed and pollen movement may result in a cline (change in fitness with distance) in tolerance. The steepness of the cline may be sharp and will depend on gene flow and the differential in fitness of the various genotypes along the gradient. The shape of the cline will also depend upon whether a steady state in genotype replacement has been reached. The effect of gene flow should be to flatten the cline but this will very much depend on ecological factors affecting seed movement and pollen dispersal. In the long term, gene flow may also exert considerable selection pressure for reproductive isolation.

Types of Herbicide Resistance

Herbicides target attack at one or more locations. These include enzyme proteins, non-enzyme proteins, cell division path, etc. One example is ALS enzyme required for the first step in the synthesis of branched chain amino acids (vide the later part of Chapter). Imidazolinones, pyrimidinyl oxybenzoates, sulfonylamino carbonyl triazolinones, sulfonylureas, triazolopyrimidines, etc. bind to this enzyme and prevent amino acid synthesis. When this happens, it leads to protein deficiency followed by death of the plant. Although the chemical structures of the above-mentioned herbicide families are different, their target site is the same. The plant that resists ALS herbicides has altered the enzyme in such a way that it does not bind with the herbicide. Now, the resistant weed biotype that has been evolved by selection pressure from one ALS-inhibitor will be resistant to all herbicides that act on this particular site.

There are different types of herbicide resistance. These include: single resistance, multiple-resistance, cross-resistance, target-site resistance, and non-target site resistance.

Single Resistance

When resistance is confined to only one herbicide or one with single site of action, it is called single herbicide resistance. Herbicides that interfere with a single site of action are generally more likely to select for resistant weeds because a change (mutation) in only one gene may be enough to affect the binding potential of a herbicide to the site of action. Therefore, it is more probable that a resistant weed population will develop if a difference of only one gene is required. This explains why development of resistance to single site of action herbicides is more likely than to multiple site of action herbicides.

Multiple Resistance

In multiple resistance, weed or crop biotype evolves resistance to two or more herbicides with different mechanisms of action and resistance. Commonly, after resistance to one herbicide chemistry has developed, the population is exposed to, and develops resistance to, a different herbicide. In extreme cases, a number of mechanisms endowing both target site and non-target site cross-resistance are accumulated within the same individual. For example, after a weed or crop biotype has developed resistance to herbicide A, then herbicide B is used. Consequently, resistance evolves to herbicide B as well. The plant is now resistant to both herbicides A and B, but through two separate selection processes. Multiple resistance occurs after sequential selection.

As mentioned earlier in this chapter, the populations of *Lolium rigidum*, a good example of multiple herbicide resistance, are superb accumulators of resistance mechanisms because resistance genes are quickly spread by outcrossing, producing a diverse progeny. It has developed resistance to 12 different herbicide families with as many sites of action over the past 30 years. These sites of action include inhibition of ACCase, ALS, microtubules, carotenoid biosynthesis, mitosis, lipids, long-chain fatty acid synthesis, photosystem II (two sites), EPSP synthase, and photosystem I [Heap 2014]. These different resistance mechanisms can be inherited from each parent. Its population expresses resistance to different herbicide chemistries at different target sites. In addition, genetic diversity in this annual grass weed populations allows them to respond successfully to rapid changes in selection pressure.

Cross-Resistance

Cross-resistance is the phenomenon of a plant population developing simultaneous resistance to more than one class of herbicides with similar mechanisms and sites of action. In this, herbicides of dissimilar chemistry bind to identical or overlapping domains of the same target site. Cross-resistance occurs when mutations within the target enzyme endow resistance to herbicides from various classes that inhibit the target site. For example, plants with different mutations conferring resistance to ALS-inhibiting herbicides have varying patterns of target site cross-resistance to three chemically dissimilar classes of ALS-inhibiting herbicides. Each herbicide has a subtly different binding domain within ALS. A single-point mutation in the ALS enzyme may provide resistance to both sulfonylurea and imidazolinone herbicide families, both being ALS-inhibitors [HRAC 2009].

Cross-resistance may be conferred either by a single gene or, as in the case of quantitative (polygenic) inheritance, by two or more genes. For example, after the extensive use of herbicide A in a field, selection of a weed biotype resistant to herbicide A would also be resistant to herbicide B, although herbicide B was never used in that field.

The reason for cross-resistance to occur is that there can be many different binding sites at a particular site of action (e.g., an enzyme) and these binding sites can be very herbicide specific. Therefore, different herbicides may bind to the same enzyme but at different sites on the enzyme. As a result, it is not possible to predict herbicide cross-resistance. However, the greatest potential for herbicide cross-resistance exists among herbicides of the same family and having the same site of action.

A biotype of the resistant *Lolium rigidum* selected with chlorsulfuron, metsulfuron, and eight other sulfonylurea herbicides was also selected for resistance to imazapyr and imazethapyr [Mallory-Smith et al. 1990a,b]. Similarly, a *Kochia scoparia* biotype resistant to five sulfonylurea herbicides is also resistant to one imidazolinone herbicide, imazapyr [Primiani et al. 1990]. While being resistant to ALS-inhibitors, these weed biotypes are not resistant to herbicides with alternate modes of action.

The phenomenon of cross-resistance is important from both practical and scientific viewpoints. Farmers can experience substantial economic loss and other problems if cross-resistance to a range of herbicides limits weed control options. Unraveling the biochemical and genetic basis of cross resistance, as well as implementing sustainable weed control programs, are important scientific challenges.

Weed biotypes selected for resistance to aryloxyphenoxypropionates which act on ACCase are also less sensitive to cyclohexanedione herbicides and vice versa [Peniuk et al. 1993; Tardif et al. 1993]. However, there are exceptions for cross-resistance in weeds. For example, the R biotype of *Xanthiumstrumarium* is sensitive to imazaquin and other imidazolinones, but not to chlorimuron, a sulfonylurea herbicide used in soya bean. As this biotype has a significantly different cross-resistance pattern, it probably contains a different ALS mutation than the one selected for resistance to sulfonylurea herbicides. This biotype is apparently controlled by herbicides with alternate mode(s) of action.

Saari et al. [1994] reported that all the weeds selected for resistance to one sulfonylurea herbicide are also cross-sensitive at the enzyme level to all other sulfonylurea herbicides. Triazolopyrimidines are also similarly affected by the mutation conferring resistance to the sulfonylureas. There is also a lower but consistent cross-insensitivity to imidazolinone herbicides, especially imazapyr, for weeds selected for sulfonylureas. A population of *Centaurea solstitialis* evolved resistance to picolinic acid, a synthetic auxinic herbicide. When this population was tested with another picolinic acid herbicide, clopyralid, the plant showed the same resistance [Prather et al. 2000].

Negative Cross-Resistance

When a plant becomes resistant to one herbicide, other physiological changes may occur that result in increased sensitivity to other herbicide families. The mutated, resistant plant that is more susceptible to the second herbicide displays the characteristic of negative cross-resistance. The second herbicide targets different functions of the plant. Negative cross-resistance can be a most useful preemptive, cost-effective tool for delaying the evolution of resistance as well as for resistance management, after resistant populations evolved. A plan of resistance management can be formulated to attack the weeds with different herbicides, controlling the resistant populations.

The triazine-resistant *Conyza canadensis* and *Echinochloa crus-galli* var. *crus-galli* biotype populations showed negative cross-resistance to 11 herbicides belonging to 18 herbicide families tested in Poland [Gadamski et al. 2000]. One *Kochia scoparia* accession, resistant to six alternative herbicides that attack different sites and growth processes of the plant, showed negative cross-resistance [Beckie et al. 2012]. This acetolactate synthase-resistant *K. scoparia* populations were 80, 60, and 50 percent more sensitive to pyrasulfotole (hydroxy phenyl-pyruvate-dioxygenase/HPPD inhibitor), mesotrione (HPPD inhibitor), and carfentrazone (protoporphyrinogen oxidase/PPO inhibitor) respectively [Beckie et al. 2012]. This may be due to a pleiotropic effect related to the Trp574 mutation.

Target-Site Resistance

Target-site resistance to a herbicide is achieved if changes in a gene encode a structural change in its gene product (enzyme), such that the herbicide no longer binds in an inhibitory manner. Such structural change in the enzyme of a weed involves either modification, by a genetic mutation, of the target site enzyme or protein, or decrease in herbicide concentration at the target site. This limits or resists the activity and effectiveness of the herbicide because it can no longer bind in an inhibitory manner. Target-site resistance is the primary (but not only) mechanism for ALS-inhibitors, ACCase-inhibitors, mitotic inhibitors, PPO-inhibitors, and some PS II-inhibiting herbicides [Powles and Preston 1995, 2006].

With a few exceptions, resistance to triazine herbicides, which inhibit photosynthesis at PS II is target-site based, entailing a single amino acid change in the D1 protein of PS II. The substitution of serine at position 264 with glycine dramatically reduces binding of triazine herbicides to the D1 protein and renders these plants resistant. This mutation endows triazine resistance without severe inhibition of Q_B binding to D1 protein, leaving enough photosynthetic capacity for survival. Only this mutation is observed in triazine-resistant weeds, indicating that other changes to the Q_B niche vastly reduce fitness.

Weed biotypes resistant to ALS-inhibiting herbicides are endowed with a resistant form of the target enzyme, ALS, which is present in many populations of *Lolium rigidum* and all ALS-resistant dicot weed populations examined thus far. The ALS-encoding gene contains two regions, Domain A and B, where mutations endowing resistance occur. Within Domain A, substitution of a conserved proline residue with a range of other amino acids confers resistance to some but not all ALS-inhibiting herbicides. Within Domain B, substitution of leucine for a conserved tryptophan endows strong resistance to a wider range of ALS-inhibitors. Hence, several mutations in the ALS-encoding gene confer herbicide resistance without compromising enzyme function.

The target-site resistance also exists in the case of herbicides inhibiting ACCase, a multi-functional enzyme (located in chloroplasts) that catalyzes the first step in fatty acid biosynthesis (vide the later part of this Chapter). The chloroplastic ACCase of dicot species is structurally different from that of grass species and is resistant to these herbicides.

Many cases of resistance to ACCase-inhibiting herbicides in *L. rigidum* and other resistant species are the result of a resistant target enzyme. One biotype (WLR 96) of *L. rigidum* developed resistance after a 10-year exposure to the aryloxyphenoxypropionate (AOPP) herbicide diclofop methyl, while the other biotype (SLR 3) developed tolerance after only three consecutive years of exposure to the cyclohexanedione (CHD) herbicide sethoxydim [Holtum and Powles 1991]. Both types exhibited target-site cross resistance to the AOPP and CHD herbicides. However, despite dissimilar periods of exposure, resistance to AOPPs was higher than resistance to CHDs.

Target-site cross-resistance, a variant of target site resistance, occurs when a change at the biochemical site of action of one herbicide also confers resistance to a herbicide from a different chemical class that inhibits the same site of action in the plant. Target site cross resistance does not necessarily result in resistance to all herbicide classes with a similar mode of action or indeed all herbicides within a given herbicide class. This type of resistance occurs in the case of ACCase herbicides. However, the patterns of cross-resistance resulting from mutations within ALS and ACCase are not predictable as they depend upon the particular mutation that has occurred.

Non-Target Site-Resistance

In addition to modifications of the target site, resistance can occur through restricted transport of herbicide to target sites. Plants with a herbicide-sensitive target enzyme can survive if the herbicide only reaches its target at sub-lethal concentrations. Non-target site-resistance (NTSR) can be achieved by rapid metabolism of a herbicide to non-toxic products. Enhanced metabolism is most often catalyzed by cytochrome P450-dependent microsomal oxidases acting on herbicides as substrates.

This large family of enzymes also catalyzes numerous reactions in plants in biosynthetic pathways for syntheses of lignins, gibberellins, carotenoids, steroids, and cutin. NTSR is particularly prevalent in *Lolium rigidum* in which resistance can be achieved by rapid metabolism of a herbicide to non-toxic products. Elevated activity of cytochrome P450-dependent microsomal oxidases confers resistance to PS II-inhibitors in some *L. rigidum* biotypes by accelerating de-alkylation of herbicide molecules at up to four times the rate in susceptible biotypes. The chemically unrelated, substituted-urea herbicide, chlorotoluron, is degraded by two enzymes, one producing a demethylated product and the other a ring-methyl-hydroxylated product. The demethylated product retains some activity against the chlorotoluron target site but the ring-methyl-hydroxylated product is an entirely inactive PS II inhibitor.

In another instance, resistant biotypes of *L. rigidum* metabolize certain ALS-inhibiting herbicides at twice the rate observed in susceptible biotypes through the action of cytochrome P450-dependent microsomal oxidase: only very low concentrations of the herbicide reach the active site. For example, chlorsulfuron is metabolized by hydroxylation of the phenyl ring, rendering the herbicide inactive [Atwell et al. 1999; Christopher et al. 1991].

In yet another instance, enhanced metabolism of ACCase-inhibitors occurs in many resistant populations of *L. rigidum*. In these, the ACCase-inhibiting herbicide diclofop is metabolized at about 1.5 times the rate observed in susceptible populations. In biotypes of *Lolium* and *Avena* spp. which are susceptible to diclofop, the herbicide can be directly conjugated to glucose. This reaction is believed to be reversible, hence providing a continuous pool of the active form of diclofop. Diclofop concurrently undergoes a slow aryl hydroxylation catalyzed by a microsomal oxidase, followed by sugar conjugation. Resistant biotypes of *L. rigidum* have an enhanced ability to detoxify diclofop acid by accelerating the rate of aryl hydroxylation.

Non-Target Site Cross-Resistance

Non-target site cross-resistance (NTSCR) is the cross resistance to dissimilar herbicide classes conferred by a mechanism(s) other than resistant enzyme target sites. It is often referred to as metabolic resistance. Certain weed biotypes of *L. rigidum* exhibit enhanced rates of herbicide metabolism, mediated by microsomal oxidases. In such cases, the degree of resistance at the whole plant level, while being sufficient to provide resistance at the recommended rates, is much less than that conferred by the target-site cross-resistance mechanism. Uptake of herbicides and/or its movement to the site of action are the other mechanisms that confer non-target site cross-resistance.

The *L. rigidum* population that is resistant following selection with a triazine herbicide is also resistant to the chemically dissimilar substituted ureas. Equally, resistance to substituted urea herbicides confers triazine resistance. Both populations metabolize both triazine and substituted ureas at enhanced rates. Similarly, plants resistant to an ACCase-inhibitor also have non-target site-resistance to ACCase and ALS-inhibiters. In this case, the enhanced metabolism detoxifies three distinct herbicides with two different target sites. Such complex metabolism is achieved by the action of several CYP450-dependent microsomal oxidases acting on herbicides as substrates rather than a single multifunctional enzyme.

Costs Associated with Resistance

Ever-increasing herbicide resistant weeds have become the bane of farmers worldwide. This problem needs to be tackled from the very first day that farmers see these weeds in their fields. Ignoring this will make them fall behind the numbers curve, with resistant weeds everywhere. Herbicide resistance, affecting the efficacy of major herbicides has economic, fitness, and social costs.

Economic Costs

The economics of managing herbicide resistance in weeds has focused on cost-effective responses by farmers to the development of resistance at the individual farm and field level. Economic analyses of optimal herbicide use have focused on optimizing farmer returns in the long run.

Economic costs include crop yield loss, reduced commodity price because of weed-seed contamination, reduced land value, cost of mechanical and cultural controls, and ultimately the additional expense of alternative herbicides or cropping systems or both for managing the resistant weeds. In the U.S., the occurrence of glyphosate-resistant *Conyza canadensis* resulted in a net increase in production cost of $28.42 ha^{-1} in soya bean [Mueller et al. 2005]. The normal range of herbicide-resistant costs varies from $20 to $100 ha^{-1}, depending on the level of infestation of resistant weeds. The additional cost of managing *Amaranthus palmeri* in Georgia and Arkansas (USA) cotton production systems was estimated at $48 ha^{-1} [Vencill et al. 2011]. In Arkansas, the additional cost involved in controlling propanil- and quinclorac-resistant *Echinochloa crus-galli* in rice was estimated at $64 ha^{-1} [Norsworthy et al. 2007].

Fitness Costs

In evolutionary genetics, a basic tenet is that adaptation to a new environment will often involve the 'cost of adaptation' [Strauss et al. 2002], which is called fitness cost. Some herbicide resistance genes impose a fitness cost upon the plant. There are important practical and theoretical implications of the presence or absence of a fitness cost associated with herbicide resistance. A fitness cost associated with resistance influences the time period over which resistance develops as well as the impact of resistance and the time period over which resistance persists if the particular herbicide usage ceases [Vila-Aiub et al. 2009]. For these reasons, establishing the fitness cost of a particular herbicide resistance mechanism is important.

The fitness costs can prevent the fixation of novel adaptive alleles [Tian et al. 2003] and contribute to the maintenance of genetic polymorphisms within populations [Antonovics and Thrall 1994]. However, some herbicide resistance alleles have no observable impact on plant fitness. The fitness consequences of herbicide resistance alleles in the presence and absence of a herbicide is important for predicting the evolutionary dynamics of herbicide resistance [Neve et al. 2003].

Natural mutations resulting in herbicide resistance in populations may be associated with fitness cost either due to pleiotropic effects of the resistance gene itself or due to linkage of the resistance gene with one or more other loci that impose the fitness cost [Mithila et al. 2011]. Vila-Aiub et al. [2009] reported that the observed fitness costs were evident from herbicide resistance-endowing amino acid substitutions in proteins involved in amino acid, fatty acid, auxin, and cellulose biosyntheses, as well as enzymes involved in herbicide metabolism. However, these costs are not universal and their expression depends on particular alleles and mutations.

When a herbicide resistance allele confers a fitness cost, there are at least three explanations for its origin. First, fitness costs may result when novel, resistance-conferring mutations in herbicide target enzymes (target-site resistance) also compromise or interfere at some level with normal plant function or metabolism. For example, a single amino acid substitution may not only cause a structural modification in the target enzyme that limits herbicide binding, but also compromise the efficiency of enzyme function and kinetics [Délye 2005; Powles and Preston 2006; Powles and Yu 2010].

Second, resource-based allocation theory predicts a trade-off between plant reproduction, growth and defense functions. Herbicide resistance is an evolved plant defense mechanism that could potentially divert resources away from growth and reproduction. For example, herbicide resistance endowed by enhanced metabolism may rely on the novel or increased production of CYP450 enzymes. According to the resource-based allocation model, when this novel enzyme production is constitutive, the additional energy and resource investments to synthesize these enzymes will divert resources away from growth and reproduction and may impose a resistance fitness cost in the absence of herbicide [Vila-Aiub et al. 2009].

Third, fitness costs may arise as a consequence of altered ecological interactions. If a resistance allele has pleiotropic effects such that the resistant phenotype becomes, for instance, less attractive to pollinators or more susceptible to diseases, fitness costs may occur independently or in addition to any energetic drain or alteration of normal metabolism [Salzmann et al. 2008]. When there is a fitness cost associated with recently evolved herbicide resistance alleles, co-adaptation and integration of the resistance allele into the genome can reduce the fitness cost over generations.

One example of fitness costs is the triazine-resistant weeds possessing the Ser-264-Gly mutation (vide this Chapter: PS II Inhibitors) that have reduced photosynthetic potential, growth rates, resource competitive ability, and sexual reproduction. The consequence of this mutation is the associated fitness cost as modulated by biotic and abiotic (viz., contrasting combinations of light and temperature) factors which may amplify, neutralize, or even reverse its negative effect on photosynthesis and plant growth. Besides, triazine-resistant plants are more susceptible to fungal infections and insect herbivory further contributing to the fitness cost of the resistance-endowing Ser-264-Gly mutation. Triazine-resistant weeds have been shown to possess higher nitrogen content in leaf tissues. It is speculated that the increase in leaf-level nitrogen is potentially a photosynthetic compensatory trait which translates into an ecologically based cost of triazine resistance resulting from a greater susceptibility to herbivores [Gassmann 2005].

In order to attribute fitness costs unequivocally, it is important to conduct fitness studies using herbicide resistant and sensitive plants with similar genetic backgrounds differing only for the alleles imparting herbicide resistance [Vila-Aiub et al. 2009], such as in near-isogenic lines (NILs) [Jasieniuk et al. 1996].

Factors Influencing Fitness Costs

There are five factors that influence fitness costs. These are: genetic variability, biochemical basis of herbicide resistance, life history of weed species, resource completion, and environmental gradient.

Genetic Variability. Herbicide-resistant and herbicide-susceptible individuals from different plant populations will probably exhibit genetic variability at a number of fitness-related loci. Therefore, in order to attribute costs to the herbicide resistance-endowing allele, relative fitness should be measured in resistant and susceptible individuals that share a similar genotype, except for the alleles endowing herbicide resistance.

Biochemical and Molecular Basis of Resistance. Where possible, it is desirable to characterize the biochemical and molecular bases of resistance before conducting fitness study. This knowledge is essential to ascribe identified pleiotropic effects to particular genes and mutations, and to comprehend their biochemical and physiological origins and causes [Vila-Aiub et al. 2009]. More than one resistance-endowing mechanism may be present at the population and individual level and different resistance mutations may be involved. For example, either or both enhanced detoxification and an insensitive target enzyme may endow resistance to ACCase- or ALS-inhibiting herbicides in *Alopecurus myosuroides* and *Lolium rigidum* populations. For both, target site ACCase and ALS herbicide resistance, there are several mutations, some associated with a fitness cost [Vila-Aiub et al. 2009].

Life History Traits. Plant fitness may be defined as the relative offspring contributing to future generations by one form compared with another. Evaluation of seed production, viability, growth, competition between resistant and susceptible pollen, and pollen discounting in species with both self-pollinated and outcrossing production reproductive systems need to be accounted to measure fitness. Ovule size and number may also be estimated to evaluate female reproductive fitness [Burd et al. 2009].

Resource Competition. Ecological fitness costs are differentially expressed depending on interactions with other organisms. Ecological costs may become evident as a consequence of biotic interactions, such as predation, disease and/or competition [Vila-Aiub et al. 2009]. If a herbicide resistance allele results in impaired ability to capture resources or a less efficient use of captured resources, ecological fitness costs should be more evident under intense resource competitive conditions. For this reason, it is easier to measure herbicide resistance fitness costs under competitive conditions.

Environmental Gradient. The expression and magnitude of the fitness costs associated with resistance alleles may be environmentally specific [Heidel et al. 2004; Martin and Lenormand 2006; Jessup and Bohannan 2008]. There is a general premise that fitness costs may be more evident when plants are growing under more extreme, stressful environmental conditions. Although this prediction is not always true, it may be more relevant when a mechanism endowing herbicide resistance depends on limited environmental resources to operate. For instance, if resistance depends on herbicide sequestration or detoxification, or cellular processes that require the synthesis of constitutively expressed nitrogen-rich proteins, plant growth may be compromised in nitrogen-poor environments [Vila-Aiub et al. 2009].

Social Costs

Herbicide resistance management has so far focused more on its cost-effectiveness at the individual farm level, much to the neglect of its effect at social level. The costs mostly included estimates of direct market, from economic standpoint, but not the non-market values, i.e., from social standpoint. Herbicide resistance is known to have resulted from multiple events. This includes gene mobility or flow through spread of pollen, seed, and plants. In cases where weed mobility is high, failure to recognize costs associated with externalities results in myopic behavior by individual agents, causing overuse of the herbicide resource [Marsh et al. 2006].

One example of social of costs is glyphosate which is closely associated with the use of conservation tillage techniques (no-till and minimum-till) which, in turn, reduce soil disturbance and therefore the probability of wind and water erosion. The cost-effectiveness is a factor that influences the increased adoption of conservation tillage. Both systems depend on glyphosate efficacy and have considerable economic value to farmers, and also to society through environmental benefits.

However, adoption of conservation tillage technology both in conventional and transgenic cropping systems have caused a rapid increase in the use of glyphosate, eventually leading to the rise of more weeds resistant to it. This rise was relatively more rapid from 2004, when the glyphosate was becoming widely used in conventional cropping and non-cropping systems as well as glyphosate-resistant transgenics after an initial lag period (Fig. 2.2) [Heap 2014].

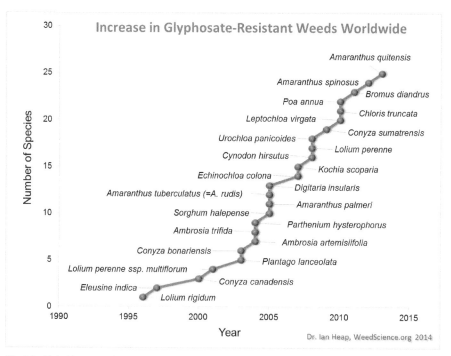

Fig. 2.2. Global increase of glyphosate-resistant weeds beginning with *Lolium rigidum* in Victoria, Australia in 1996 [Heap 2014].

Exacerbation of resistance in some situations by spread through resistance mobility has presented a case for social costs. This aspect needs to be considered when assessing optimal use of this herbicide resource at the farm level. Possible social costs include the loss of glyphosate efficacy, potential failure of herbicide-resistant crop systems, reduced use of conservation tillage techniques, and the possibility of more reliance on herbicides with greater environmental and health risks.

Management of Herbicide Resistance

There are two principles fundamental to herbicide-resistance management. These are: reducing the intensity of selection and preventing reproduction by the surviving, resistant individuals. When susceptible weeds are eliminated by a herbicide application before they reproduce, they facilitate a selective advantage for resistant individuals in the weed population. These resistant plants then transmit the resistance trait to their offspring, leading to their survival when exposed to the same herbicide. Repeated and sustained use of the same herbicide or one with same mechanism of action (MOA) over a period of time favors survival and reproduction of these resistant biotypes, leading to a weed population in which resistant plants predominate. Repeating the same control tactics at a given timing, whether a herbicide application or, for that matter, even a nonchemical control method, may also result in the evolution of avoidance mechanisms in a weed population by selecting for biotypes that have not emerged, or are outside the optimal growth stage, when control is implemented [Reddy and Norsworthy 2010]. Lack of diversification in management practice will inevitably lead to evolution of resistance. The factors that aid in weed population shift to predominantly resistant individuals are:

a. The intensity of selection, a function of herbicide use, frequency, and time of application;
b. Mutation rate and the initial frequency of resistant individuals in the population exposed to the herbicide;
c. The genetic basis of the resistance (mode of inheritance, dominance);
d. Life-history characteristics of the weed species (annual vs. perennial life cycle, self-fertilization vs. cross-pollination, fecundity, extent of seed dormancy, etc.);
e. The rate of reproduction and potential for recruitment of susceptible or resistant individuals from outside the population (from the soil seed-bank or by immigration).

Although introduction of glyphosate-resistant transgenic crops in 1996 helped solve some weed-management problems such as evolution of weeds resistant to ALS- and PPO-inhibitors, it created an altogether different, but a more potent problem. All herbicide technologies, if used repeatedly, are at risk of losing efficacy because of continuous evolution of herbicide resistance in weed populations. For example, *Eleusine indica* and *Amaranthus tuberculatus* (*A. rudis*), susceptible to glufosinate and HPPD-inhibitors respectively, have been reported to be resistant to them with two different modes of action.

Strategies

Effective herbicide-resistance management combines a variety of chemical and nonchemical management tactics to diversify selection pressure on weed populations and minimize spread of resistance genes. This can be achieved by sustainable herbicide-resistance management program on a long-term perspective. A single season effort will be insufficient. The basic objective of the program should be to prevent viable seed production and seed-bank replenishment. There are different strategies, also called best management practices, which may be adopted as suggested here.

Understanding the Biology of Weeds Present

Agro-ecosystems typically contain diverse weed species, each with different life histories and different responses to control measures. In order to implement an effective preventive weed management strategy, an understanding of the biology of the weeds present is a prerequisite. Biology includes seed/propagule germination, seedling emergence pattern, mode of seed dispersal, persistence and dynamics of seeds in soil seed-bank, lifecycle, and reproductive biology. Normally, many annual weed species have a seed-bank life of 3–5 years, with an annual rate of decline of 50–80 percent. Seed-bank may be exhausted by: a) disturbing the soil before and after crop planting to stimulate germination of some weed species, b) applying nonselective 'burn-down' herbicides like paraquat and glyphosate to kill the existing weed growth and prevent new additions to the seed-bank, and c) adopting chemical or fallow method before a crop is planted, so the germinating weed seedlings can be controlled [Rao 1981].

Repeated herbicide application on a highly prolific, annual weed species with a shorter life cycle exposes a large number of individuals to selection for rare resistance alleles [Gressel and Levy 2006]. In the case of perennial weed species, fewer individuals are exposed to such selection within the same period, potentially slowing the rate of selection [Norsworthy et al. 2012]. Self-pollinating and cross-pollinating weed species behave differently in the evolution and spread of herbicide resistance alleles. In self-pollinating weeds, dominant or recessive alleles are likely to increase equally in frequency under selection pressure. However, in cross-pollinated populations the frequency of dominant resistance alleles will increase more rapidly than recessive alleles under selection pressure. Weed species with a high degree of cross-pollination and recombination are also at greater risk of evolving multi-gene herbicide resistance [Neve 2008]. Hence, knowledge on factors affecting pollen viability and longevity, and dispersal distances, including vectors, is essential for managing gene flow.

Using Weed-Free Crop Seeds

The first step in preventing the spread of herbicide resistance into new areas is planting weed-free crop seeds. In eastern Australia about 73 percent of samples of seeds conserved by farmers were contaminated with weed seed, with an average of 62 weed seeds 10 kg^{-1} of wheat, barley, or lupin (*Lupinus* spp.) seed, substantially higher than the 28 seeds 10 kg^{-1} expected by farmers [Michael et al. 2010]. Among the weed seeds recovered, *Lolium rigidum* had resistance to four different herbicides while wild radish (*Raphanus raphanistrum*) and wild oat (*Avena fatua*) were each resistant to one herbicide. Herbicide-resistant *L. rigidum* seeds were also found as a contaminant of Australian wheat exported to Japan, demonstrating the potential for international dispersal of resistance genes via weed seed contamination of grain shipments. In these shipments, about 4,500 resistant *L. rigidum* seeds were detected in a 20-kg wheat sample in 2006 and 2007, about 35 and 15 percent of which were resistant to diclofop-methyl, 5 and 6 percent were resistant to sethoxydim, and 56 and 60 percent were resistant to chlorsulfuron, respectively [Shimono et al. 2010].

Another example was the invasive dicot weed *Parthenium hysterophorous* which entered India via wheat imports from the United States in the 1960s.

Routine Scouting for Resistant Weed Species. A single herbicide application is rarely effective against all weeds. As a result, a constant vigil for, and maintenance of inventory of resistant weed species within the plants survived is another component of a successful preventive herbicide-resistance management program. This information allows for recognizing the most efficacious herbicide for the weed spectrum present and following it with additional treatment options [Norsworthy et al. 2012]. Furthermore, weeds present at harvest are a good indicator of species that are not effectively controlled and they need to be monitored and taken into consideration in the following season. Routine scouting allows for early identification of a lack of or inadequate herbicide efficacy which may result in resistance evolution. If herbicide resistance is suspected, seed production from plants that survived herbicide application should be prevented from flowering and seed production by destroying them, and in some cases, even the crop.

Adoption of Modified Herbicide Practice

The modified herbicide approach includes rotation and combinations of herbicides with different modes of action (MOAs). The greatest risk factor for evolution of herbicide resistant weeds is using herbicides with the same MOA every year. Adoption of a different schedule involving herbicides with different MOAs in annual rotations, combinations, and sequential applications can delay the evolution of resistance by decreasing the frequency of resistant alleles and minimizing the selection pressure imposed on those weed populations by a particular herbicide.

Application of different MOAs either simultaneously, as tank mixtures, or sequentially every year greatly reduces the likelihood of survival and reproduction of the resistant weed population. In a mixture, weed biotypes resistant to one herbicide (the vulnerable one) would be destroyed by the mixing partner, or at least be rendered relatively unfit compared to the wild biotypes. Herbicide mixtures are of two types: broad spectrum mixture (BWM) and target weed mixture (TWM). In BWM, each member of the mixture affects different weed spectra, one on grasses and the other on broadleaf weeds. While BWMs provide good broad-spectrum control, they have no influence in delaying or inhibiting evolution of resistance in either of the weed groups. In TWM, both the vulnerable herbicide and the mixing partner aim at complete kill of the same weed species. In order for the TWM to be effective in preventing resistance, the mixing partner should have the following traits: a) control the same spectra of weeds, b) have the same persistence, c) have a different target site, d) be degraded in a different manner, and e) preferably exert negative cross resistance [Wrubel and Gressel 1994]. One good example of TWM is combination of ALS herbicide and PS II inhibitor or auxinic herbicide. Some TWMs exhibit synergistic activity, in which the combined activity of mixing herbicides is greater than the additive effect of each. However, contact and systemic herbicides with antagonistic activity are often not effective TWMs.

When both types of herbicides are a part of a resistance-management strategy, sequential usage will be superior to mixtures for reducing the risks of resistance. In a 'double-knock' tactic developed in Australia, application of glyphosate is followed seven days later by paraquat to eliminate *L. rigidum* plants that survived glyphosate application [Walsh and Powles 2007]. Similar success was also obtained with these herbicides in controlling the perennial *Imperata cylindrica* in tea crop [Rao 1981].

Application of Herbicides at Recommended Rates

When a herbicide is applied at lower rates, weed population which is under selection is likely to run the risk of accumulating minor genes. This confers only a slight increase in fitness, but it provides significant levels of herbicide resistance. Even within the population, variation for herbicide resistance exists. This variation may be attributed to different alleles at minor loci contributing to polygenic resistance or to different allele combinations at modified loci affecting expression of a major resistance gene [Norsworthy et al. 2012; Neve and Powles 2005a]. Repeated exposure of these populations to reduced doses may cause incremental enrichment of the gene pool with resistance alleles. Exposure of *Lolium rigidum* to sub-lethal doses of glyphosate caused a population shift to resistance in 3 to 4 cycles of selection [Busi and Powles 2009]. Lower rates of the ACCase-inhibiting diclofop allowed substantial *L. rigidum* survivors due to the potential in this cross-pollinated monocot to accumulate all minor herbicide traits in the population [Manalil et al. 2011]. At lower rates, the herbicide is subject to enhanced detoxification by the target weed species. This suggests that herbicides should be used at the recommended rates.

However, slow-degrading soil-applied herbicides, used over multiple growing seasons, may deliver sub-lethal doses to the later emerging weeds and allow for gradual accumulation of resistance alleles. Similarly, improper spraying, imprecise calibration, or inaccurate mixing of postemergence of herbicides may lead to inadequate coverage of weeds due to dense, over-size crop cover. This requires farmers to take adequate precautions to avoid delivering lower rates of herbicides.

Suppression of Weeds by using Competitive Crops and Cultural Methods

Better Cultivars and Cover Crops with Better Ground Coverage. Suppression of weed growth limits seed production of those weeds that escape control, thus gradually reducing future weed problems. Generally, plant height, tillering, leaf angle, and canopy formation are the criteria to be used for selecting competitive crop cultivars. Hybrid rice cultivars with greater tillering, faster growth, and faster row-space coverage characteristics are more competitive than regular non-hybrid lines. Additionally, choosing a herbicide-resistant crop cultivar that does not flower at the same time as, or is less cross-compatible with the weed relative will prevent transfer of the resistance trait to the weedy species [Shivrain et al. 2009]. Selection of a full-season cultivar to maximize the period with crop cover will suppress weed emergence for prolonged periods, whereas planting an early maturing cultivar may broaden the window of post-harvest seed production of escaped weeds that potentially contribute to seed-bank persistence and herbicide-resistance evolution [Reddy and Norsworthy 2010]. A cover crop, planted in the inter-row space of the main crop, enhances land productivity while suppressing weed growth for a substantial period during the cropping season and delaying the need for an early season herbicide application. Legumes serve as better cover crops because they also improve the structure, tilth, and fertility of the soil.

Cultural Methods. These include planting date, row spacing, seed rate, crop rotation, and nutrient and irrigation management. Manipulating planting dates without compromising yields may place the crop at a competitive advantage over specific weeds. For example, delaying crop planting by a week or two may shift peak weed emergence before crop establishment, allowing for effective weed control before crop emergence. One example is the implementation of a 2-week delay in planting of wheat in some regions of Australia [Walsh and Powles 2007]. Delayed planting allows for control of a large proportion of *Lolium rigidum* emerging after the initial spring rains, using non-selective herbicides before crop establishment. Double-cropping system (June-planted soya bean instead of May-planting following winter wheat harvest) in the southern USA reduced the density of glyphosate-resistant *Amaranthus palmeri* in soya bean because the one-month planting delay and the suppressive nature of the preceding wheat crop greatly reduced emergence of this weed [DeVore et al. 2011].

Reducing row space and/or increasing seed rate allows crop canopy to be dense and closer. Although higher seed rates increase input costs, it is a proven method for management of herbicide-resistance weeds. One example is suppression of herbicide resistant *L. rigidum* in Australia, where higher seeding rates have been readily adopted by wheat farmers to achieve more rapid crop canopy closure and increased crop competitiveness with this monocot weed, in turn reducing its growth and number of seed returned to the seed-bank [Walsh and Powles 2007].

Although crop rotation may not be as effective as tillage in exhausting the seed-bank, it has a greater effect in minimizing the density of seed-bank. Lack of crop rotation results in a weed community with less diversity and hence reduces herbicide options for controlling those weeds. Risks of resistance are generally greater in systems with limited crop rotation compared with those where crop rotation (including rotation of herbicide-resistance traits) is regularly practiced [Neve et al. 2011a,b]. Model simulations have shown that rotating glyphosate-resistant cotton with maize can reduce the risk of glyphosate-resistant *A. palmeri* almost twofold against growing glyphosate resistant cotton alone [Neve et al. 2011a,b].

Nutrient Management. Nutrient management is another useful herbicide resistance management tool. Quantity, timing, and placement of nutrients may be tailored to maximum crop growth and minimum weed growth. Improper placement of fertilizers favors greater weed growth rather than crop growth. Improper placement of fertilizers at a time when weeds can utilize them fast gives the crop a competitive disadvantage, leading to greater weed seed production and increased risk of evolution of herbicide resistance.

Irrigation. Irrigation before crop planting induces weed seed germination and reduces the germinable fraction of seed-bank that will otherwise emerge in the crop. Flooding the irrigated crop like rice before planting inhibits germination and emergence of many weed species, paving the way for a relatively weed-free situation for the crop. Planting pre-germinated rice seeds or seedlings and growing the crop in standing water is a technique followed in Asian countries including India and China. It has a smothering effect on annual weed species, but allows certain aquatic weeds to take their place and fill the void.

Many of the suggested cultural management methods above involve high production costs but these can only be adopted if the economic benefit outweighs the additional costs. However, the cost-benefit ratio depends on farm size, labor availability, crop duration, crop variety, crop production practice, management costs, weather during crop and weed growth, and crop produce price. Weeds respond differently to cultural management, and even evolve themselves to thrive in a changed situation. This requires a constant vigil over weeds and their survival mechanisms and adoption of the suggested cultural practices only when the situation warrants in combination with other seed-bank depletion methods.

Adoption of Mechanical and Biological Weed Management Methods

Tillage. Tillage prevents buildup of any particular species or group of weeds in the seed-bank of annual and perennial weed species by stimulating weed seed germination. However, weed shifts are affected less by tillage than by herbicides because the latter have a stronger constraint than tillage on the assembly of communities [Booth and Swanton 2002]. But combining with proper timing of herbicide application, tillage or cultivation can effectively control perennial grass weeds like *Cynodon dactylon* [Etheredge et al. 2009] and *Imperata cylindrical* [Rao 1981]. Inter-row cultivation is a valuable practice in reducing the risk of herbicide-resistance evolution in crops. For example, resistance of *Echinochloa crus-galli* var. *crus-galli* to glyphosate in cotton was 68 percent over a 30-yr period when a continuous glyphosate system with no inter-row cultivation was used; the risk, however, was reduced to 38 percent over the same period when integrating a single mid-season inter-row cultivation into the existing system [Norsworthy et al. 2011]. Deep tillage buries weed seed deep enough to prevent successful germination.

Mulch. Covering the inter-row space with organic mulch, living or nonliving, prevents or reduces availability of sunlight and suppress weed growth. A good mulch (crop straw, hay, stubble, etc.) is a good weed management option, aside from bestowing nutrient benefit on the soil and crop and arresting soil erosion. While organic mulch is suitable for seasonal field crops, black polythene mulch is effective for high-value, perennial crops against tough-to-control weeds such as sedges *Cyperus rotundus* and *Cyperus esculentus*. Polythene mulches have been successfully integrated with herbicides in many high value vegetable and perennial crops.

Prevention of Spread of Weed Seed and Vegetative Propagules

Containment of resistant alleles conferring herbicide resistance on weeds is one way of minimizing the spread of herbicide resistance. Movement of weed propagules from one farm or field to another leads to new herbicide-resistant weeds reaching non-infested areas. Common weed propagule carriers include organic manure, compost, organic mulches, straw, hay, harvest cleanings, etc. Furthermore, it is a common tendency among farmers to ignore or overlook a few resistant plants in a field until the population builds to a level that herbicide failure is recognizable. At this point, seed and pollen containing resistant alleles may have reached neighboring fields, exacerbating the problem in the whole farming area. Prevention of immigration of propagules and pollen from infested areas is an important weed management program.

Prevention of Seed-bank Buildup at Harvest

Weed seeds present at crop harvest often reach the soil seed-bank. This problem is more acute when mechanical harvesters are used. The chaff fraction of the harvested debris contains more weed seeds, suggesting the need to collect and destroy the chaff as well as standing stubble before letting the seed reach the seed-bank. This can be achieved by burning them at very high temperatures over long periods and/or herbivorous decay of seed. This is more apt for annual weed species. However, the extent of decay depends on the predatory herbivores used to feed on plant parts including seeds. Factors that influence seed decay by herbivores include tillage and depth of seed burial. Herbivorous seed decay is greater when the seeds are near the soil surface in a no-till system. Additionally, soil microbes can interact with the feeding herbivorous insects, leading to increased fungal infection of weed seed.

Prevention of Weed Influx into the Field

Cross-border influx of weed seeds and pollen, including the herbicide-resistant ones, is possible from unmanaged habitats including field margins, neighboring farms, roadsides, railroads, ditch-banks, etc. The density, diversity, and fecundity of wild and weed plant populations are generally greatest along field borders, with a gradual reduction with increasing distance into cropping area. Weed seed production in field borders can have a long-term effect on seed-bank within the crop production area, more so when cross-pollination occurs with herbicide-resistant populations near a field, allowing faster spread of resistance. Weed influx into the field can be prevented by regular cutting and mowing of the weeds, aside from maintaining a dense grass cover along the borders. But the more effective tactic is managing field margins by repeat applications of glyphosate or paraquat on weed growth and preventing seed production.

Implementation of Herbicide Resistance Management Programs

The herbicide resistance management programs that involve cultural, mechanical, and chemical options, may involve additional costs in short term, but long-term benefits far outweigh them. Although farmers are the primary focus, governments, research institutions, and industry bear greater responsibility in having the best and sustainable herbicide-resistance management strategies implemented effectively. Weed scientists need to monitor the development and spread of herbicide-resistant biotypes, and provide the farmers cost-effective and ecologically sound resistance-mitigating weed management programs. They should also develop different models to predict yield loss due to weed pressure and the rate of selection of herbicide-resistant biotypes. Weed management is local- and crop-specific, and it requires specific solutions that do not rely on one technique or one herbicide. Agriculture needs diverse weed management tools and strategies for every crop, including resistance management programs that use more than one herbicide or management technique in each cropping scenario.

Bagavathiannan et al. [2014] developed a computer simulation model that can analyze the simultaneous evolution of resistance in barnyard grass {*Echinochloa crus-galli* (L.) Beauv.} to herbicides that inhibit ALS and ACCase enzymes. This annual grass weed can reduce rice yield up to 30 percent. The model took into account three stages of growth (dormant seed-bank, emerged seedlings, and mature plants) extending over a 30-year period. It assumed the use of the non-transgenic imidazolinone-resistant 'Clearfield' rice cultivars (vide Chapter 6) in the Mississippi Delta region in the midsouthern U.S.

Under continuous three annual applications of IMI herbicides (imazethapyr and imazamox) in Clearfield rice, resistance was predicted within 4 years with 80 percent risk by year 30. However, weed management programs that consisted of both ALS- and ACCase-inhibiting herbicides (fenoxaprop and cyhalofop) greatly reduced the risk of ALS-inhibiting herbicide resistance, with only 12 percent risk by year 30. But the model found a higher risk for the weed developing resistance to ACCase-inhibitor by year 14. Resistance to both of these herbicide types when used together was predicted by year 16 [Bagavathiannan et al. 2014].

Similar simulation models can be beneficial in understanding the resistance evolution and developing strategies to prevent it. Minimizing the weed seed-bank should be the focus of successful weed management. The model developed by Bagavathiannan et al. [2014] showed that diversified management techniques combined with timely applications of herbicide(s) is the best way to achieve positive results.

As new herbicide discovery has slowed dramatically in recent years, with practically no new herbicide release over the past two decades, herbicide industry needs to encourage farmers to achieve sustainability with the most suitable of the herbicides available in combination with cultural and mechanical methods. Federal and state governments bear greater responsibility in supporting herbicide-resistance management projects. They also need to enact laws that prohibit or restrict inter-state, inter-county movement of seeds and propagules of herbicide-resistant weed biotypes. Additionally, governments, weed scientists, extension specialists, industry personnel, and media may coordinate public awareness programs through field days, press news releases, technical brochures, and websites.

Mechanisms of Herbicide Action and Weed Resistance

The global use of herbicides over the past seven decades has caused significant changes in farm production systems. It enabled farmers to raise crop yields significantly while lowering production costs. However, herbicides have not made weeds extinct. Rather, they have caused, along with other influencing factors, a continuous selection of plants to occur and this has enabled them both to survive and reproduce. Consequently, these resistant plants with survival properties are able to become dominant and distributed over increasingly large areas worldwide.

The relatively steady increase in the number of new cases of resistance since 1980 accounts for the increasing importance of herbicide resistance in weeds. During the period between 1970 and 1990, the most documented cases of resistance were concerned with the photosystem II-inhibiting triazines (HRAC Group C1). The introduction of new classes of herbicides such as acetyl-CoA carboxylase (ACCase inhibitors: Group A) and acetolactate synthase inhibitors (ALS inhibitors: Group B) with different sites of action in the early 1980s caused a significant shift (Fig. 2.3) [Heap 2014]. Additionally, rapid adoption of glyphosate-resistant transgenic crops in Australia, North America, and South America and the use of glyphosate as a preemergence herbicide in different cropping systems have resulted in increasing cases of weed species resistant to it [Menne and Köcher 2012]. The probability of resistance developing to glyphosate had been expressed as being likely but underestimated, though less frequently in comparison with other herbicides families of different sites of action [Heap and LeBaron 2001].

Currently, two systems of classification of herbicide-resistant weeds are in use. Both are based on a site of action (SoA), common name, and chemical family of herbicide. The one developed by and Retzinger and Mallory-Smith, first published in 1997 [Retzinger and Mallory-Smith 1997] and updated in 2003 [Mallory-Smith and Retzinger 2003], was adopted by Weed Science Society of America (WSSA). It uses a numbering system (1 to 27), while the one developed by Herbicide Resistance Action Committee (HRAC) for global use, is largely based on letters (A to Z), with numbers being added to indicate more specificity in SoA [Heap 2014]. For example, in the case of PS II inhibitors, subclasses C1, C2, and C3 indicate different binding behaviors at the binding site protein D1. Similarly, as bleaching can also be caused in different ways, bleachers are subdivided in F1, F2, and F3.

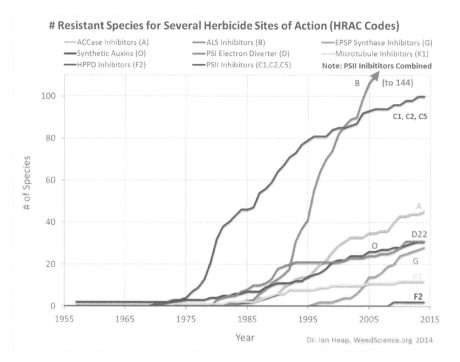

Fig. 2.3. Increase in weed species to herbicides of several sites of action beginning 1957 [Heap 2014].

Color image of this figure appears in the color plate section at the end of the book.

This part of the chapter deals with the biochemical aspects of herbicide resistance including mechanism of action and resistance of weed species to different herbicides and herbicide families. It followed the inhibitor groups based on HRAC classification (vide Appendix: Table 2).

Photosystem I Inhibitors

Photosystem I (PS I) inhibitors (Group D), include bipyridilium herbicides, paraquat and diquat. Paraquat has been one of the world's most widely used herbicides since 1962 in crop and non-crop lands. The first recorded evidence of resistance to paraquat was in 1980 when the dicot weed *Conyza canadensis* was found resistant in Japan. In all, 31 species (Fig. 2.3; Group D; 22 dicots and nine monocots) of weed resistance have been found to date in 16 countries [Heap 2014]. Other prominent paraquat-resistant weed species include dicots *Conyza bonariensis*, *Epilobium ciliatum*, and *Solanum nigrum*, etc. and monocots *Eleusine indica*, *Hordeum murinum* ssp. *glaucum*, *Lolium rigidum*, *Poa annua*, etc.

Mechanism of Action

PS I is a part of photosynthetic electron transport located in thylakoid membranes. Paraquat, also called methyl viologen, enters the chloroplast to reach its target site of action at PS I. Being a divalent cation, paraquat short-circuits the light-driven photosynthetic electron transport in PS I by accepting electrons that normally flow to ferredoxin (Fd). In normal case, when light harvesting complexes (LHC) bound to PS I transfer excitation energy to P700 (chlorophyll 'a' dimer), it undergoes charge separation and an exited electron is released. This e^- is received by 'Ao' (chlorophyll 'a' monomer). From 'Ao', the electron moves to Fe-S centers, Fx and Fa/Fb and finally to Fd which transfers electron to Fd-NADP$^+$ oxido-reductase (FNR) which, in turn, catalyzes the reduction of NADP to NADPH.

The cationic paraquat is applied as divalent cationic solution (PQT^{++}). The redox potential of PQ^{++} is -446 mv, that of Fa/Fb is -560 mv and that of Fd is higher than that of PQT^{++}. This enables PQ^{++} to act as a competitor for electron flow from Fa/Fb. So Fa/Fb donates the electron to PQT^{++} instead of Fd, as shown in the following equations. After receiving the electron, PQT^{++} becomes intensely blue colored monovalent cation (PQT$^{+\cdot}$). This PQT$^{+\cdot}$ is very reactive and will reduce oxygen to superoxide and in the process PQT^{++} is regenerated.

$$PQT^{++} + PSI\ (e^-) \longrightarrow PQT^{+\cdot}$$
$$PQT^{+\cdot} + O_2 \longrightarrow PQT^{++} + O_2^{-\cdot}$$
$$2H^+ + O_2^{-\cdot} + O_2^{-\cdot} \longrightarrow H_2O_2 + O_2$$
$$H_2O_2 + O_2^{-\cdot} \longrightarrow O_2 + OH + OH^-$$

In the reaction that ensues, H_2O_2 and hydroxyl radical are produced. This radical reduces oxygen to form superoxide, which is catalyzed by the enzyme superoxide dismutase yielding a highly toxic hydroxy radical. This extremely reactive toxic form of oxygen (superoxide) causes lipid peroxidation, rapid loss of membrane integrity (allowing cytoplasm to leak into intercellular spaces), photooxidation, leaf wilting, desiccation, and plant death. Sunlight, which enables photosynthetic electron transport, aids in the toxicity of paraquat. Bipyridiliums can be reduced/oxidized repeatedly.

Mechanism of Resistance

Although herbicide metabolism is a major selectivity and resistance mechanism in many plant species, it is unlikely to be responsible for resistance to paraquat [Holt et al. 1993]. In the resistant (R) weed biotypes, paraquat uptake and translocation are reduced or restricted, due to sequestration, compared to susceptible (S) biotypes. Preston et al. [1992] reported that there was no difference in gross translocation of paraquat in the R biotype of *Arctotheca calendula*, but movement to the target site was reduced, probably as a result of its reduced penetration into the leaf cells. These and other studies suggest that paraquat is detoxified or removed from the chloroplasts of the R biotype [Jansen et al. 1990; Shaaltiel and Gressel 1987]. Biotypes resistant to paraquat may be cross-resistant to other Group D herbicides.

Resistance of weeds is due to the presence of paraquat-resistant genes *mvr*A (methyl viologen-resistantA) and *mvr*C ((*mv-resistantC*) which encode enzyme protein and membrane protein respectively, while another gene *Pqr* reduces the permeability of cell membrane or enhancing the efflux of paraquat. The fourth gene that is involved in paraquat resistance is *emvE* which is involved in proton exchange (vide Chapter 5).

Photosystem II Inhibitors

There are three types of photosynthesis II (PS II) inhibitors: a) phenylcarbamates, pyridazinones, triazines, triazinones (amicarbazone), and uracils of Group C1; b) amides and ureas of C2; and c) benzothiadiazinone (bentazon, nitriles and phenylpyridazinone (pyridate) of C3 (vide Appendix, Table 1).

The dicot *Senecio vulgaris* infesting nurseries was found to be resistant to simazine and atrazine as early as in 1968 [Ryan 1970]. In all, 100 species (Fig. 2.3; Groups C1: 72; C2: 24; C3: 4) developed resistance (including those with multiple resistance to other herbicide families) to PS II herbicides, with with 60 and 40 of them being with dicots and monocots respectively [Heap 2014]. Of the 72 species that developed resistance to C1 group herbicides, the most prominent resistant biotypes arose from *Chenopodium album*, *Amaranthus retroflexus*, *Amaranthus hybridus*, *Kochia scoparia*, *Lolium rigidum*, *Poa annua*, *Senicio vulgaris*, *Solanum nigrum*, and *Echinochloa crus-galli* var. *crus-galli*. The ubiquitous monocot weed, *Lolium rigidum*, which first evolved resistance to the atrazine of group C1 in Israel in 1979 became cross-resistant to herbicides of several groups. The annual dicot *Chenopodium album* which first showed resistance to atrazine in Ontario, Canada, continued to do so at 38 other locations around the world until 2010 [Heap 2014]. Similarly, *Amaranthus retroflexus* also showed resistance to atrazine at 28 global locations between 1980 and 2011. Of all PS II-inhibitors, atrazine contributed to evolution of 66 resistant species, the most by any single herbicide [Heap 2014].

Mechanism of Action

PS II, which involves water-plastoquinone oxidoreductase, is the first protein complex in the light-dependent reactions of oxygenic photosynthesis. It is located in the thylakoid membrane of the chloroplasts. The enzyme captures photons of light to energize electrons that are then transferred through a variety of coenzymes and cofactors to reduce plastoquinone to plastoquinol. The energized electrons are replaced by oxidizing water to form hydrogen ions and molecular oxygen. By obtaining these electrons from water, PS II provides the electrons for all of photosynthesis to occur. The hydrogen ions (protons) generated by the oxidation of water help to create a proton gradient that is used by ATP synthase to generate ATP. The energized electrons transferred to plastoquinone are ultimately used to reduce $NADP^+$ to NADPH or are used in cyclic photophosphorylation.

The core of PSII consists of a pseudo-symmetric heterodimer of two homologous proteins D1 and D2 [Rutherford and Faller 2003]. Unlike the reaction centers of all other photosystems which have a special pair of closely spaced chlorophyll molecules, the pigment that undergoes the initial photo-induced charge separation in PS II is a chlorophyll monomer. As the positive charge is not shared across two molecules, the ionized pigment is highly oxidizing and can take part in the splitting of water [Rutherford and Faller 2003]. PS II is composed of around 20 subunits (depending on the organism) as well as other accessory, light-harvesting proteins. Each PS II contains at least 99 co-factors. These include 35 chlorophyll amolecules, 12 β-carotenes, two pheophytins, two plastoquinones (Q_A, Q_B; alternate acronyms: PQ_A, PQ_B), two hemes, one bicarbonate, 20 lipids, the Mn_4CaO_5 cluster (including chloride ion), and one non-heme Fe^{2+} and two putative Ca^{2+} ions per monomer [Guskov et al. 2009].

In the light reaction, one molecule of the pigment chlorophyll absorbs one photon and loses one electron. This electron is passed to a modified form of chlorophyll called pheophytin, which passes the electron to a quinone molecule, allowing the start of a flow of electrons down an electron transport chain that leads to the ultimate reduction of NADP to NADPH. In addition, this creates a proton gradient across the chloroplast membrane; its dissipation is used by ATP synthase for the concomitant synthesis of ATP. The chlorophyll molecule regains the lost electron from a water molecule through photolysis, which releases a dioxygen (O_2) molecule. The overall equation for the light-dependent reactions under the conditions of non-cyclic electron flow in green plants is shown below [Raven et al. 2005].

$$2\ H_2O + 2\ NADP^+ + 3\ ADP + 3\ P_i + light \rightarrow 2\ NADPH + 2\ H^+ + 3\ ATP + O_2$$

The photosynthetic action spectrum depends on the type of accessory pigments present. In green plants, the action spectrum resembles the absorption spectrum for chlorophylls and carotenoids with peaks for violet-blue and red light.

The light-dependent reactions have two forms: cyclic and non-cyclic. In the non-cyclic reaction, the photons are captured in the light-harvesting antenna complexes of PS II by chlorophyll and other accessory pigments. When a chlorophyll molecule at the core of the PS II reaction center obtains sufficient excitation energy from the adjacent antenna pigments, an electron is transferred to the primary electron-acceptor molecule, pheophytin, through a process called photoinduced charge separation. These electrons are shuttled through an electron transport chain that initially functions to generate a chemiosmotic potential across the membrane as shown in the so-called Z-scheme (Fig. 2.4) [Govindjee and Weit 2010]. An ATP synthase enzyme uses the chemiosmotic potential to make ATP during photophosphorylation, whereas NADPH is a product of the terminal redox reaction.

The Z scheme shows the pathway of electron transfer from water to $NADP^+$. Using this pathway, plants transform light energy into 'electrical' energy (electron flow) and hence into chemical energy as reduced NADPH and ATP.

Mechanism of Resistance

Resistance varies from 3- to 9-fold depending on the herbicide. The inheritance of resistance in these R biotypes is a polygenic trait under nuclear control. The R biotype of *Polygonum lapathifolium* was 35.5 times more resistant to atrazine than an S biotype [Padro et al. 1995]. Electron transport in the R biotype of this weed was 781 times less sensitive to atrazine than in S biotypes. Its resistance was due to reduced affinity for atrazine at the target site. There are three pathways in which PS II inhibiting herbicides show their resistance. These are: point mutation in *psbA* gene, glutathione conjugation, and oxidation.

Point Mutation in psbA Gene. As the D1 protein is a choloroplast gene product, triazine resistance is maternally inherited. A great majority of weed species exhibits greater than expected incidence of mutation rate of chloroplast *psbA* gene, resulting in higher resistance. The *psbA* gene encodes for D1 protein of the PS II. Due to mutation, the serine at 264th position is replaced by glycine (Ser-264-Gly) in the mutant D1 protein. Because of this, the herbicide molecule is deprived of one H bond as it cannot form H bond with glycine. So the affinity of herbicide molecule towards D1 niches is decreased considerably and now the normal Q_B molecule easily replaces them from the niches with the result that normal electron transport continues in the mutant even in the presence of herbicide [Hirschberg et al. 1984]. This leads to a reduced photosynthetic capacity as a result of an inefficiency of electron transfer within PS II complex. This is the resistant mechanism in most of the triazine-resistant weed species.

The triazine R biotypes of some weed species (e.g., *Chenopodium album*) rapidly develop resistance to other PS II-inhibiting herbicides since the presence of the mutator gene would increase the likelihood of other mutations in the Q_B-binding pocket [Solymosi

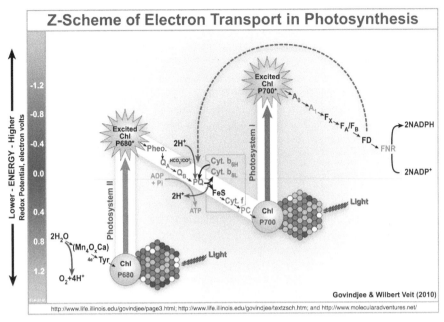

Fig. 2.4. Z-Scheme, an energy diagram for electron transfer in the 'light reactions' of plant photosynthesis. It shows the pathway of electron transfer from water to $NADP^+$. The process involves the oxygen-evolving complex, reduction potentials, and electron flow during photosynthesis, reduction of CO_2, and NADP phosphorylation. Light energy absorbed by special pair pigments, P680 of photosystem II (PS II) and P700 of photosystem I (PS I), drives electron flow uphill. Acronyms (from left to the right of the diagram): Mn: a manganese complex containing 4 Mn atoms, bound to PS II reaction center; Tyr: a particular tyrosine in PS II; O_2: oxygen; H^+: protons; P680: the reaction center chlorophyll (Chl) in PS I, the primary electron donor of PS II; P680*: excited (Chl) P680 that has the energy of the photon of light; Pheo: pheophytin molecule (the primary electron acceptor of PSII {it is like a chlorophyll *a* molecule where magnesium (in its center) has been replaced by two "H"s}; Q_A (also called PQ_A): a plastoquinone molecule tightly bound to PS II; Q_B (PQ_B): another plastoquinone molecule that is loosely bound to PS II; FeS: Rieske Iron Sulfur protein; Cyt. f: Cytochrome f; $Cytb_6$ (Cyt. b_6L and Cyt. b_6 H): Cytochrome b_6 (of Low and High Energy); PC: copper protein plastocyanin; ChlP700: the reaction center chlorophyll (Chl: actually a dimer, i.e., two molecules together) of PSI—it is the primary electron donor of PSI; Excited ChlP700*: P700 that has the energy of the photon of light; Ao: a special chlorophyll a molecule (primary electron acceptor of PSI); A_1: a phylloquinone (Vitamin K) molecule; F_X, F_A, and F_B: three separate Iron Sulfur Centers; FD: ferredoxin; and FNR: Ferredoxin NADP oxido Reductase (FNR). Three major protein complexes are involved in running the "Z" scheme: (1) PS II; (2) Cytochrome bf complex (containing $Cytb_6$; FeS; and Cytf); and (3) PS I. The diagram does not show where and how ATP is made [Govindjee and Veit 2010].

Color image of this figure appears in the color plate section at the end of the book.

and Lehoczki 1989]. The triazine-resistant *Poa annua* (annual bluegrass) populations with Ser-264-Gly mutation are cross-resistant to the triazolinone herbicide amicarbazone [Perry et al. 2012].

Glutathione Conjugation. The triazine resistance in the R biotypes of *Abutilon theophrasti* is not maternally inherited. It is controlled by a single, partially dominant nuclear gene that is not cytoplasmically inherited [Anderson and Gronwald 1987]. Unlike most triazine-R weed biotypes, the resistant *Abutilon* biotypes are only 10 times as resistant as susceptible biotypes. Resistance in this weed species is due to enhanced glutathione S-transferase enzyme activity, which confers enhanced capacity to detoxify atrazine via glutathione conjugation [Anderson and Gronwald 1991].

Oxidation of Herbicide. In simazine-resistant *Lolium rigidum*, the resistance is conferred by increased metabolism of herbicide. Here the herbicide is acted upon by cytochrome P450 monooxygenase enzyme and converted to herbicidally inactive de-ethyl simazine and di-de-ethyl simazine [Burnet et al. 1993]. In the case of *Alopecurus myosuroides*, the inheritance of resistance to urea herbicides chlorotoluron and isoproturon is polygenic and under nuclear control. The rapid degradation of these herbicides in the R biotypes is catalyzed by this enzyme.

Acetolactate Synthase Inhibitors

Herbicides that inhibit acetolactate synthase (ALS) are among the most widely used in the world. Unfortunately, these herbicides are also notorious for their ability to select resistant (R) weed populations. Currently, there are more weed species that are resistant to ALS-inhibiting herbicides than to any other herbicide group. Five chemically dissimilar classes of herbicides (Group B) viz. sulfonylureas, imidazolinones, triazolopyrimidines, pyrimidinyl-thiobenzoates, and sulfonyl-aminocarbonyltriazolinones share a common target site, ALS inhibition (vide Appendix, Table 1). These are commercially used for selective weed control in such diverse crops as wheat, maize, soya bean, and rice.

Resistance in weeds to ALS-inhibiting herbicides was first observed in the monocot *Alopecurus myosuroides* infesting wheat in the U.K. in 1984 within a few years after the commercial introduction of chlorsulfuron, a sulfonylurea herbicide. Shortly later, a few other weed species also evolved resistance to ALS inhibitors. These include monocot *Avena fatua* to diclofop methyl in South Africa in 1986 and dicots *Lactuca serriola* and *Kochia scoparia*, both resistant to chlorsulfuron in USA in 1987.

To date, 145 weed species (including those resistant to other herbicide families) have been found to be resistant to ALS-inhibiting herbicides (Fig. 2.3: Group B) [Heal 2014]. Of these, 46 were reported from USA alone, followed by Canada and Australia with 25 and 22 respectively. Majority of these cases were from dicots (88 species), with the remaining from monocots (57). The resistant species include *Amaranthus retroflexus*, *Amaranthus tuberculatus* (*A. rudis*), *Amaranthus retroflexus*, *Kochia scoparia*, *Avena fatua*, *Conyza canadensis*, *Cyperus difformis*, *Echinochloa crus-galli* var. *crus-galli*, *Lolium rigidum*, *Lolium perenne* ssp. *multiflorum*, *Sinapis arvensis*, *Stellaria media*, *Xanthium strumarium*, etc. Many of these weeds showed multiple cross-resistance. Resistance to ALS-herbicides evolved relatively quickly after only four to seven continuous applications in monoculture crop areas and non-crop areas. The long residual activity of many of these herbicides has also contributed to the rapid development of resistance by increasing the effective kill and hence the selection pressure.

Mechanism of Action

The enzyme ALS plays an essential role in branched-chain amino acid biosynthesis in chloroplasts leading to leucine, isoleucine, and valine. It catalyzes the formation of 2-acetolactate from two pyruvate molecules, and in the pathway to isoleucine the formation of 2-acetohydroxybutyrate from 2-ketobutyrate and pyruvate. Due to this double function, this enzyme is also referred to as acetohydroxyacid synthase (AHAS).

The four enzymes involved in the three biosynthetic pathways of branched-chain amino acids are: acetolactate synthase, acetolactate reductoisomerase, dihydroxyacid dehydratase, and branched-chain amino (or amino acid) transferase. Besides, isoleucine requires an additional enzyme, threonine deaminase whereas leucine requires three additional enzymes 2-isopropylmalate synthase, 3-isopropyl malate dehydratase, and 3-isopropylmalate dehydrogenase. Regulation of the pathway is caused by inhibition of, a) threonine dehydratase by isoleucine, b) 2-isopropylmalate synthase by leucine, and c) ALS by all three amino acids. This suggests that the biosynthetic pathway can be blocked through feedback inhibition from the end-products of the pathway.

Mechanism of Resistance

Target Site Sensitivity and Mutation at Amino Acid Residues. Resistance to ALS-inhibitors, usually inherited as a semi-dominant trait, is conferred by a single, dominant, or at least partially dominant, nuclear-encoded gene for less-susceptible ALS. The *ALS* gene codes for an enzyme of 670 amino acids with a molecular weight of about 73 kDa.

Resistance involves a reduction in the sensitivity of the target site to inhibition by herbicides. In the R biotypes of *K. scoparia* and other weed species, the resistant mechanism identified was an insensitive ALS [Saari et al. 1990]. The other possibilities such as enhanced metabolism (detoxification) and differential uptake and translocation had no role in the resistance mechanism exhibited by the R biotypes. In *Stellaria media*, *Lolium perenne*, and *Salsola iberica*, resistance against ALS inhibitors was due to a less sensitive ALS enzyme [Saari et al. 1992]. Eberlein et al. [1999] found that resistance in *Lactuca sativa* was conferred by a single point mutation that encodes a Proline-197-Histidine substitution in Domain-A of ALS protein.

Two ALS protein domains, Domain-A and Domain-B, appear to influence resistance. In Domain-A, consisting of 13 amino acids, substitution of serine, glutamine or alanine for proline residue at amino acid 173 confers resistance to ALS inhibitors [Wiersma et al. 1989]. Domain-B, with four amino acids, enhances the level of resistance conferred by mutation in Domain-A in tobacco [Lee et al. 1988]. Proline residue in Domain-A was pivotal for developing resistance to chlorsulfuron [Wiersma et al. 1989].

In response to the widespread use of these herbicides, several weed species have evolved target site-based resistance caused by multi-point mutations in the AHAS gene [Saari et al. 1994; Tranel and Wirght 2002]. Resistance-endowing mutations at seven highly conserved key amino acid residues viz., Ala-122, Pro-197, Ala-205 (located at the amino-terminal end), Asp-376, Trp-574, Ser-653, and Gly-654 (located near the carboxy-terminal end) in the AHAS gene have been documented. To date, 22 amino acid substitutions at these seven conserved amino acid residues in the AHAS gene have been identified to confer target-site resistance to ALS-(AHAS) inhibiting herbicides in biotypes of field-evolved resistant weed species (Table 2.1) [Yu et al. 2010; Heap 2014].

Amino acid substitutions encompass amino acid alterations in which an amino acid is replaced with a different naturally occurring amino acid residue. Such substitutions may be classified as 'conservative', in which an amino acid residue contained in the wild-type ALS protein is replaced with another naturally occurring amino acid of similar character as shown by: Gly<→Ala, Val<→Ile<→Leu, Asp<→Glu, Lys<→Arg, Asn<→Gln, or Phe<→Trp<→Tyr. In several resistant *Amaranthus powellii* populations, the Trp-574-Leu AHAS mutation was associated with thinner roots and stems besides a severe leaf area reduction, which led to a resistance cost of 67 percent (above-ground vegetative

Table 2.1. Amino acid substitutions in weed species showing target-site resistance to ALS-(AHAS) inhibiting herbicides [Heap 2014].

Amino Acid Substitution	No. of Sp.	Amino Acid Substitution	No. of Sp.
Trp 574 to Leu	27	Ala 205 to Val	4
Pro 197 to Ser	21	Pro 197 to Arg	3
Pro 197 to Leu	11	Ala 122 to Val	2
Pro 197 to Thr	11	Ala 122 to Tyr	1
Pro 197 to Ala	9	Arg 377 to His	1
Asp376 to Glu	7	Gly 654 to Asp	1
Pro 197 to Gln	7	Pro 197 to Asn	1
Ala 122 to Thr	6	Pro 197 to Ile	1
Pro 197 to His	6	Ser 653 to Ile	1
Ser 653 to Asn	5	Trp 574 to Gly	1
Ser 653 to Thr	5	Trp 574 to Met	1

biomass) as well as severe reduction in seed production [Tardif et al. 2006]. Similar AHAS resistance was also evident in imidazolinone-resistant rice crops with the Gly-654-Glu AHAS mutation, which showed 5 to 11 percent lower grain yield when compared with conventional rice cultivars [Sha et al. 2007].

The Pro-197-Ser and Trp-574-Leu mutations exhibited no significant effects on *L. rigidum* plant growth while the Pro-197-Arg mutation resulted in lower growth rates [Yu et al. 2010]. The five AHAS resistance mutations, Pro-197-Ala, Pro-197-Arg, Pro-197-Gln, Pro-197-Ser, and Trp-574-Leu, had no major impact on AHAS functionality and hence probably no plant resistance costs in *L. rigidum* [Yu et al. 2010]. These results, in part, explained why so many Pro-197 AHAS resistance mutations in AHAS have evolved and why the Pro-197-Ser and the Trp-574-Leu AHAS resistance mutations are frequently found in many weed species [Yu et al. 2010]. The resistant populations of *L. rigidum*, which showed non target-site-based resistance while metabolizing chlorsulfuron, were able to accumulate mutant alleles at positions 197 and 574, which reflected the outcrossing ability of the species [Kaundun et al. 2012].

The available evidence on amino acid substitutions indicated that the herbicide-binding site of the AHAS can tolerate these substitutions without causing any major consequence to normal catalytic functions. It was, therefore, speculated that the herbicide-binding site and the active site of AHAS are different, despite their being in close proximity [Menne and Köcher 2012]. There are two target site mutations in three ALS-resistant *Salsola tragus* biotypes: Trp-574-Leu mutation in all three biotypes and Pro-197-Gln in one biotype [Warwick et al. 2010].

Cross-Resistance. A biotype of *Monochoria vaginalis*, found in rice in Korea, showed high levels of cross-resistance to bensulfuron-methyl and pyrazosulfuron-ethyl (both sulfonylureas) as well as flumetsulam [Hwang et al. 2001], a triazolopyrimidine herbicide. In rice fields of Japan, a biotype of *Scirpus juncoides* was selected which exhibited a high degree of resistance to imazasulfuron [Tanaka 2003]. Genotype variations are generally associated with cross-resistance to ALS inhibitors within and between the resistant populations of *Anthemis cotula* as evidenced by different Pro197 substitutions [Intanon et al. 2011].

In a biotype of *Amaranthus retroflexus* from Israel, resistance was caused by a change of Pro197 to Leu. This biotype exhibited cross-resistance to sulfonylureas, imidazolinones, triazolopyrimidines, and also to pyrithiobac-sodium *in vivo* and on the ALS enzyme level [Sibony et al. 2001]. In mutations of *Amaranthus rudis*, Ser653 was found to be exchanged by Thr or Asn; such mutants were only resistant to imidazolinones [Patzold and Tranel 2001].

Certain herbicide-resistant biotypes of *K. scoparia* also exhibit negative cross resistance in which an herbicide-resistant biotype becomes more sensitive to, and more easily controlled by classes of herbicides other than the class to which it is resistant. Beckie et al. [2012] found one resistant *K. solaria* accession (MBK2) with the Trp574 mutation showing negative cross-resistance to protoporphyrinogen oxidase-(PPO: e.g., carfentrazone) and hydroxyphenyl pyruvate dioxygenase-(HPPD: e.g., pyrosulfotole, mesotrione) inhibiting herbicides.

Altered target site cross-resistance to ALS-inhibiting herbicides was also found in *Cyperus difformis* which has evolved rapidly in many rice areas worldwide. Merotto et al. [2009] reported that whole plant and ALS enzyme activity dose response assays indicated that the WA biotype of *C. difformis* was cross-resistant to all five classes of ALS-inhibiting herbicides. However, the IR biotype, resistant to bensulfuron-methyl (sulfonylurea), orthosulfamuron (sulfonylurea), imazethapyr (imidazolinone), and propoxycarbazone-sodium (sulfonylamino-carboxyl-triazolinone), was less resistant to bispyribac-sodium (pyrimidinyloxybenzoate) and halosulfuron-methyl (sulfonylurea) while being susceptible to penoxsulam (triazolo-pyrimidine sulfonamide).

Glucose Conjugation. The resistance pathway of *Lolium rigidum* to chlorsulfuron was inactivation of the herbicide by glucose conjugation [Cotterman and Saari 1992] while that of *Echinochloa crus-galli* to primisulfuron was by hydroxylation followed by glycosylation [Neighbors and Privalle 1990].

ACCase Inhibitors

At this time, three herbicide families, aryloxyphenoxypropionics (AOPP), cyclohexanediones (CHD), and phenylpyrazoline (PPZ: pinoxaden) belonging to Group A, have acetyl-CoA carboxylase (ACCase) as the target site of inhibition. Although AOPPs and CHDs are structurally dissimilar, they exhibit similarities in their activity at the whole plant level as well as physiological/biochemical level. They share some common structural element(s) responsible for their herbicidal activity. These post-emergence herbicides which control annual and perennial grass weeds in a wide variety of crops including rice, wheat, soya bean, barley, rapeseed (canola), onion, vegetables, pulses, lucerne (alfalfa), pastures, etc. have little activity against dicots or non-graminaceous monocots.

In 1982, just 4 years after the initial ACCase herbicide use in 1978, the first ACCase herbicide-resistant *Lolium rigidum* population was evident in wheat in South Australia. Since then, these ACCase herbicide-resistant populations have evolved across the Australian croplands, with 100,000 to 1,000,000 acres (40,500–405,000 ha) being currently infested with multiple resistant biotypes of this ubiquitous grass weed. The same year saw *Alopecurus myosuroides*, becoming resistant to Group A herbicides in wheat in the United Kingdom. In all, 46 weed species (including those with multiple resistance: unique cases), with all of them being annual grasses, have been reported to have shown resistance to ACCase-inhibiting herbicides (Fig. 2.3: Group A) [Heap 2014]. All of the grass weeds belonged to Poaceae. The more prominent of them include *Avena fatua, Avena sterili, Alopecurus myosuroides, Digitaria sanguinalis, Echinochloa colona, Echinochloa crus-*

galli var. *crus-galli*, *Lolium perenne* ssp. *multiflorum*, *Phalaris minor*, *Setaria viridis*, *Sorghum halepense*, etc. [Heap 2014].

Mechanism of Action

Plastids contain acetyl-CoA synthetase, acetyl-CoA carboxylase, and fatty acid synthetase, key components of fatty acid synthesis in plants. Inhibitors that affect any of the steps involving these enzymes can block glycerolipid and phospholipid synthesis. The most affected are young developing leaves and meristematic tissues, which depend on the efficient fatty acid supply.

Acetyl-CoA carboxylase is a multifunctional, biotinylated (biotin attached to a protein) protein located in stroma of plastids. It catalyzes the carboxylation of Acetyl-CoA (substrate for ACCase) to produce malonyl-CoA, the precursor of fatty acids. This is a two-step, reversible reaction.

The initial step of *de novo* fatty acid biosynthesis, consisting of the ATP-dependent carboxylation of the biotin group on the carboxyl carrier domain by the biotin-carboxylase activity, is followed by the transfer of the carboxyl group from biotin to acetyl-CoA by the carboxyl-transferase activity. ACCase catalyzes these partial reactions occurring at two different sites thus:

a. Reaction at carboxylation site
   Enzyme-biotin + HCO_3^- + ATP $\longrightarrow$ enzyme-biotin–CO_2^- + ADP + Pi
b. Reaction at carboxyltransferase site
   Enzyme-biotin-CO_2+acetyl CoA $\longrightarrow$ malonyl CoA+ enzyme-biotin

Enzyme with biotin prosthetic group serves as a mobile carboxyl carrier between the two sites [Gronwald 1991]. There are two isoforms of ACCase in plants. These are a) plastid ACCase (accounting for more than 80 percent of the total ACCase activity), essential in biosynthesis of primary fatty acids, and b) the cytosolic ACCase involved in biosynthesis of long-chain fatty acids and flavonoids (e.g., gibberellins, abscisic acid, carotenoids, and other isoprenoids) as well as in malonylation. The homomeric ACCase in the cytosol (cytoplasmic matrix or intercellular fluid) of nearly all plant species and the heteromeric ACCase in the chloroplasts of dicots are insensitive to AOPP, CHD, and PPZ herbicides. In contrast, the plastidic homomeric ACCase in nearly all grass species (particularly of Poaceae) is herbicide-sensitive, and this forms the basis for selective control of grass weeds by ACCase herbicides. This plastidic homomeric ACCase is encoded by a nuclear gene distinct from that coded for the cytosolic ACCase. All ACCase isoforms contain a) the biotin-carboxyl carrier protein (BCCP), b) biotin-carboxylase (BC), and c) carboxyl-transferase (CT) domains [Nikolau et al. 2003]. The CT domain of the plastidic homomeric ACCase is the primary target site for AOPP and CHD herbicides while the two regions of the CT domain of the plastidic ACCase are critical for sensitivity to these herbicides [Zhang et al. 2004]. ACCase inhibiting herbicides bind to the target enzyme in a near competitive manner with respect to the substrate acetyl-CoA.

Inhibition of ACCase enzyme by AOPP, CHD, and PPZ herbicides blocks the production of phospholipids used in building new membranes required for cell growth. Broadleaf weeds are naturally resistant to these herbicides because of an insensitive ACCase. Similarly, natural tolerance of some grass species appears to be due to a less sensitive enzyme [Stoltenberg 1989]. Mechanism of action of ACCase herbicides appears to involve destruction of the electrochemical potential of the cell membrane.

Both AOPPs and CHDs are foliage-active, systemic herbicides used for the control of annual and perennial grasses in broadleaf crops and in certain cereals. Selectivity in the case of dicot crops is based on low sensitivity of dicot ACCase, and in cereals it could be attributed to an enhanced herbicide detoxification. In susceptible grasses, AOPPs and CHDs are linear, noncompetitive inhibitors of grass ACCase for all 3 substrates (Mg, ATP, HCO_3^-, and acetyl CoA) [Burton et al. 1991]. As a result, the carboxylation of acetyl CoA is prevented and hence fatty acid synthesis is hampered.

Mechanism of Resistance

Alteration of Target Site Enzyme. The obligate cross-pollinated grass weed *L. rigidum* found in Australia has an ability to rapidly evolve resistance to ACCase and other herbicide groups. The biochemical basis of ACCase-herbicide resistance has been revealed in several weed populations to involve resistant ACCase [Matthews et al. 1990; Holtum et al. 1991; Tardif et al. 1993, 2006].

Target-site resistance (TSR). Resistance to ACCase herbicides can be due to enhanced metabolic degradation of the toxophore or insensitivity of the target enzyme. Metabolism is complex and it involves several genes that are gradually selected and combined over several generations. TSR of biotypes of ACCase inhibitors has been confirmed for several grass weed species including *Lolium multiflorum* [Stanger and Appleby 1989] and *Lolium rigidum* [Holtum and Powles 1991]. Selection with an AOPP or with a CHD herbicide resulted in TSR to both herbicide groups. However, the level of resistance in these biotypes was higher towards AOPP than towards CHD herbicides. The ACCase resistance factors were 30 to 85 for diclofop, > 10 to 216 for haloxyfop, and 1 to 8 for sethoxydim [Holtum and Powles 1991; Tardif et al. 1993; Powles and Preston 1995]. There was no cross-resistance of AOPP-resistant *Lolium multiflorum* (*Lolium perenne* ssp. *multiflorum*) biotype to the CHD herbicides, sethoxydim, or clethodim [Gronwald et al. 1992].

Biotypes with target-site-based resistance to ACCase inhibitors were also selected in wild oat species (*Avena fatua*, *Avena sterilis*), with the resistance patterns being variable. For example, the resistance factors for ACCase from the Canadian *A. fatua* biotype UM1 were 105 for sethoxydim, 10 for tralkoxydim, and 10 for diclofop and fenoxaprop, whereas for the *A. fatua* biotype UM33 the ratios were 10.5 for fenoxaprop, 1.2 for diclofop, 5 for sethoxydim, and 1.7 for tralkoxydim. It was proposed that this variable effect was due to different point mutations, each being associated with a characteristic resistance pattern [Devine 1997]. Another reason might be the frequency of homozygote and heterozygote resistant and susceptible plants within a tested population.

Genetic studies revealed that the TSR in the two *A. myosuroides* biotypes to fenoxafop, diclofop, fluazifop, and sethoxydim was monogenic and nuclear inherited, with the resistant allele showing complete dominance [Moss et al. 2003]. The AOPP and CHD herbicides do not bind to the target site in an identical manner; they have 'overlapping binding sites' at the ACCase enzyme. Further studies by Délye et al. [2004; 2005] supported this finding.

Target-site resistance results from a single amino acid change in the ACCase enzyme. The presence of different point-mutations at the target enzyme accounts for the variable resistance patterns. In the case of point-mutation, substitution of isoleucine (Ile) amino acid to leucine (Leu) at 1781 position in the chloroplastic ACCase CT (carboxyl-transferase) domain of the enzyme occurs [Zagnitko et al. 2001]. The mutant leucine-ACCase allele in the *Setaria viridis* species was characterized to be dominant, with no negative effect on ACCase function of the mutant [Délye et al. 2002]. Brown et al. [2002] reported that the leucine found in the plastidic homomeric ACCase of mutated resistant grass weeds was also found in the heteromeric plastidic enzyme of non-grass species, and also in the cytosolic homomeric enzymes that are 'naturally' resistant to these herbicides. Hence, the selective action of ACCase-inhibiting herbicides appears to reside at this enzyme site [Powles and Preston 1995].

In addition to position 1781 (involving Ile→Leu mutation), six other positions (1999, 2027, 2041, 2078, 2088, and 2096) have been reported [Powles and Yu 2010; Délye 2005] to involve other target-site point mutations in different grass weed species that gave rise to insensitive ACCase. These include Trp→Cys (position 1999), Trp→Cys (pos. 2027), Ile→Asn (pos. 2041), Ile→Val (pos. 2041), Asp→Gly (pos. 2078), Cys→Arg (pos. 2088), and Gly→Ala (pos. 2096) [Délye 2005; Délye et al. 2008]. The Ile→Leu mutation is most common and this causes resistance to mainly all ACCase herbicides. Resistance conferred by target-site mutations can be broad or specific and strong or weak, in part, reflecting the different binding modes of the three classes of ACCase herbicides [Powles and Yu 2010].

Point mutations causing resistance-endowing amino acid substitutions do occur at other positions too. Thus far, six distinct amino acid substitutions in the CT domain of the plastidic ACCase gene that individually endow resistance to certain ACCase herbicides have been characterized in different grass weed species [Délye et al. 2005]. These include: Ile-1781-Leu (in *Lolium rigidum, Setaria viridis, Alopecurus myosuroides, Avena fatua, Lolium perenne* ssp. *multiflorum, Avena sterilis*), Ile-2041-Asn (*Alopecurus myosuroides, Lolium rigidum, Avena sterilis*), Ile-2041-Val (*Lolium rigidum*), Trp-2027-Cys (*Alopecurus myosuroides, Avena sterilis*), Gly-2096-Ala (*Alopecurus myosuroides*), Asp-2078-Gly (*Alopecurus myosuroides, Avena sterilis*), and Trp-1999-Cys (*Avena sterilis*). The Ile-1781-Leu mutation is associated with resistance to AOPP and some CHD herbicides (excluding clethodim). The Trp-2027-Cys, Ile-2041-Asn, or Gly-2096-Ala mutations confer resistance only to AOPP herbicides. The Asp-2078-Gly mutation confers resistance to many AOPP and CHD herbicides including clethodim. The Trp-1999-Cys mutation confers resistance only to the AOPP herbicide fenoxaprop [Liu et al. 2007].

In the case of clethodim, used to control grass weeds infesting dicot crops, *Lolium rigidum* has evolved 14 diverse and complex resistant populations to ACCase [Yu et al. 2007]. Yu et al. [2007] identified a new mutation of Cys to Arg at position 2088 along with the known mutations of Ile-1781-Leu, Trp-2027-Cys, Ile-2041-Asp, and Asp-2078-Gly in plants surviving the clethodim field rate of 60 g ha⁻¹ in Australian wheat. They reported 12 combination patterns of mutant alleles in relation to clethodim existence. They also established that the mutation 2078-Gly or 2088-Arg endows sufficient level of resistance to this CHD herbicide. Additionally, combinations of two mutant 1781-Leu alleles or two different mutant alleles (1781-Leu/2027-Cys, 1781-Leu/2041-Asp) also confer clethodim resistance. Plants homozygous for the mutant 1781, 2078, or 2088 alleles were found to be clethodim-resistant and cross-resistant to a number of other ACCase inhibitor herbicides including clodinafop, diclofop, fluazifop, haloxyfop, butroxydim, sethoxydim, tralkoxydim, and pinoxaden. This showed that in addition to herbicide rates, the homo/heterozygous status of a specific mutation and combinations of different resistant alleles are important in contributing to the overall level of herbicide resistance in genetically diverse, cross-pollinated *Lolium* species [Yu et al. 2007].

Non-target site-based Resistance (NTSR). Many populations resistant to other herbicide families also have a non-target site-based resistance mechanism of enhanced rates of metabolism to ACCase herbicides [Tardif and Powles 1994; Preston et al. 1996; Preston and Powles 1998]. In the case of *A. myosuroides*, 18–40 percent of individual plants showed

NTSR to most of the herbicides approved for control of this annual grass [Délye et al. 2011]. Herbicide families include AOPPs (fenoxafop and clodinafop), PPZ (pinoxaden), sulfonylureas (iodosulfuron+mesosulfuron mixture), and sulfonamide (pyroxsulam). A segregation analysis showed that multi-resistant phenotypes are endowed by multiple NTSR genes, underlining the complexity of NTSR [Délye et al. 2011]. In the two R populations of *L. rigidum*, the multiple-resistance to glyphosate as well as ACCase and ALS herbicides is due to the presence of distinct NTSR mechanisms for each herbicide [Yu et al. 2009]. Resistance to ACCase and ALS herbicides is likely due to enhanced herbicide metabolism involving different cytochrome P450 enzymes [Yu et al. 2009].

The hexaploid wild oat (*Avena fatua*), the global weed which evolved herbicide resistance over the last three decades, has up to, a) three ulinked ACCase gene loci assorting independently, b) all three of these homoeologous ACCase genes were transcribed, with each able to carry its own mutation, and c) in a hexaploid background, each individual ACCase resistance mutation confers relatively low-level herbicide resistance, in contrast to high-level resistance conferred by the same mutations in unrelated diploid weed species of the Poaceae (grass) family [Yu et al. 2013]. Low resistance conferred by individual ACCase resistance mutations is likely due to a dilution effect by susceptible ACCase expressed by homoeologs in hexaploid wild oat and/or differential expression of homoeologous ACCase gene copies. Thus, polyploidy in hexaploid wild oat may slow resistance evolution. Evidence of coexisting non-target site-resistance mechanisms among wild-oat populations was also revealed [Yu et al. 2013].

Over-production of ACCase. A different mechanism of target-site resistance to ACCase exists in *Sorghum halepense*. In a quizalofop-resistant biotype of this dicot weed found in Virginia, USA, over-production of ACCase was the mechanism that conferred a moderate level of resistance to quizalofop, sethoxydim, and fluazifop. Owing to the overproduction of this enzyme, the resistant biotype was able, presumably, to sustain a level of malonyl-CoA production necessary for survival of herbicide treatment [Bradley et al. 2001].

Glucose Conjugation. In resistant biotypes in wheat, diclofop methyl (AOPP herbicide) is rapidly hydrolyzed to diclofop, a toxic form, which is then irreversibly detoxified by arylhydroxylation in the presence of cytochrome 450 monooxygenase to form ring OH group which, in turn, is rapidly conjugated to form herbicidally inactive O-glucoside [Romano et al, 1993]. Non-enzymatic conjugation of fenoxaprop-methyl (AOPP herbicide) with glutathione appears to be an important mechanism of tolerance of some resistant grass weeds to this herbicide [Tal et al. 1995]. Conjugation of glutathione *S*-transferase (GSH) and cysteine was the major mechanism of resistance of the R biotype of *Echinochloa phyllopogon* (*E. oryzicola*) to fenoxaprop-p-ethyl [Bakkali et al. 2007].

Synthetic Auxins

Four independent discoveries of hormone herbicides, chlorophenoxyacetic acids, in the early 1940s by three American scientist groups (led by William Templeman; Franklin Jones; Ezra Kraus and John Mitchell) and one British group (at Rothamsted) and their commercial production and application in 1945 not only initiated an agricultural revolution but also gave birth to modern weed science. Consequently, the use of 2,4-D in cereal crops transformed agricultural production throughout the world. This discovery "transformed agriculture and are considered to be amongst the greatest scientific discoveries" of the 20th century [Fryer 1980]. The commercial success of 2,4-D, a synthetic plant hormone auxin, paved the way for the discovery of many structurally similar herbicides, subsequently grouped as auxinic herbicides.

The auxinic group 'O' includes phenoxy-carboxy (phenoxyacetic) acids, benzoic acids, pyridine-carboxylic acids, and quinoline-carboxylic acids. These are applied postemergence to control dicot weeds in many graminaceous crops and certain dicot crops as well as turfgrass. 2,4-D is also used at preemergence in transplanted irrigated rice.

Ever since the biotypes of *Commelina diffusa* and *Daucus carota* were found resistant to 2,4-D in 1957 in Hawaii (USA) and Ontario (Canada) respectively, 29 more species showed resistance unique cases of resistant weed species (Fig. 2.3: Group O) [Heap 2014]. These include species with multiple resistance. Of the total 31 species, dicots accounted for 23 of them, in addition to eight monocots. The more prominent resistant species include dicots such as *Kochia scoparia*, *Raphanus raphanistrum Sphenoclea zelaynica*, and *Papaver rhoeas* besides the monocot *Echinochloa crus-galli* var. *crus-galli*.

Mechanism of Action

Auxinic herbicides are structurally similar to natural plant hormone auxin, discovered by Paàl in 1919 [Went 1926]. They mimic several of the same physiological and biochemical responses at low concentrations as those of the hormone indole-3-acetic acid (IAA), initially discovered in 1885 and rediscovered with structure identification 50 years later. However, higher doses of auxinic herbicides cause cell elongation, epinasty, hypertrophy, and excessive root formation, as well as ethylene and cyanide biosynthesis, followed by plant death. Indole compounds such as indole ethanol, indole acetaldehyde, and indoleacetonitrile, all derived from amino acid tryptophan, are the precursors of biosynthesis of the major auxin, IAA, in plants [Salisbury and Ross 1992].

After several decades of research to understand the auxin signal transduction pathway, the receptors for binding and the resultant biochemical and physiological responses in plants have been reported by Mithila et al. [2011]. However, the precise mode of action for the auxinic herbicides is not completely understood despite their extensive use in agriculture for about 70 years. The available evidence suggests that auxins affect plant and growth and development processes by regulating gene expression [Guilfoyle 2007]. At low concentrations, auxin-responsive genes are not expressed due to the presence of Aux/IAA repressor proteins (Aux/IAAs) that bind to the promoters of auxin-responsive genes [Chapman and Estelle 2009]. Following *de novo* IAA synthesis or release of IAA from stored conjugates [Ludwig-Müller 2011], auxin concentrations increase and promote gene expression by ubiquitin-mediated degradation of transcriptional Aux/IAA repressors, thereby activating gene expression by a novel "release from repression" mechanism. Thus at higher doses, auxinic herbicides accomplish this by binding directly to the bottom of the TIR1 (transport inhibitor response1) pocket and acting as a "molecular glue" that stabilizes the interaction between the auxin receptor protein, TIR1, and its homologs (ABPs: auxin-binding proteins) and its substrates (Aux/IAA repressors) in an auxin-dependent manner [Guilfoyle 2007; Tan et al. 2007]. This leads to rapid Aux/IAA degradation. ABPs are located in the lumen of the endoplastic reticulum for subsequent migration to the outer face of the plasma membrane.

Mithila [2011] summarized the biochemical pathway of the mechanism of action of auxinic herbicides in the following steps: 1) 2,4-D and IAA bind to the "promiscuous pocket" of TIR1 or its homologs, leading to auxin-responsive gene expression; 2) 2,4-D/IAA are actively transported into plant cells via a common carrier protein; 3) 2,4-D is not a substrate for conjugation to GH3 (Gretchen Hagen3) gene family of proteins, while IAA is; and 4) 2,4-D is not hydroxylated by cytochrome P450s or other metabolic enzymes in sensitive dicots. The fact that auxinic herbicides are substrates for proteins/enzymes in the first two steps (binding and transport) and not for enzyme in the last two steps (conjugation and detoxification) is likely to be responsible for the lethality of sensitive dicots [Kelly et al. 2004; Kelley and Riechers 2007]. This might also contribute as a potential resistance mechanism in auxinic herbicide-resistant dicot weeds if mutations in these proteins reduce herbicide binding (steps 1 and 2) or increase binding of auxinic herbicides (steps 3 and 4) [Mithila 2011].

Once the auxinic herbicide enters the plant and reaches a living cell, it is bound to a receptor (as elucidated above), triggering the cell to turn on several genes to cause a cascade of secondary responses. One of the induced genes helps loosen the cell-wall region, causing differential lengthening on one side of the stem compared to another. This leads to the characteristic elongation, epinasty, etc. mentioned earlier. The genes also turn on the enzyme ACCase that aids in the synthesis of ethylene, another plant growth regulator.

Ethylene, which also causes epinasty, can also induce production of abscisic acid (ABA) in sensitive plants. ABA is another hormone that causes closure of stomata to block access of CO_2 needed for photosynthesis. Another product of ethylene synthesis pathway is cyanide which injures sensitive grasses. Some plant species, however, are insensitive to ethylene. In insensitive species, other genes induced by auxinic herbicides may be responsible for plant injury. Regulation of gene expression may have a secondary effect on nucleic acid (mRNA) and protein syntheses by affecting transcription and translation. In the case of *Amaranthus tuberculatus*, synthetic auxins constitute the sixth mechanism-of-action group to which it evolved resistance [Bernards et al. 2012].

Mechanism of Resistance

Despite the fact that 2,4-D and many other auxinic group herbicides are being used extensively the world over for about seven decades, evolution of resistance to this group has been relatively slow. Reports of auxinic-resistance evolution were few and far between until the 1980s when only 10 weed species were found resistant, followed by 11 in the 1990s. In this new millennium, 10 more weed species have shown resistance thus far. The inability of resistant weed species to evolve and propagate in response to auxinic herbicides applied in crops may be due to the rare occurrence of resistant individuals (resistance alleles) in weed populations or that mutations conferring resistance are lethal [Jasieniuk et al. 1995]. Additionally, the relatively low selection pressure or short-lived residual activity (excluding pyridine-carboxylic acids) in the soil may contribute to the low occurrence of auxinic herbicide-resistant weeds [Mithila et al. 2011].

The relatively moderate evolution of weed resistance in field crops has been attributed earlier [Gressel and Segel 1982; Morrison and Devine 1994] to multiple sites of action of auxinic compounds. However, subsequent inheritance studies showed that resistance of *Sinapis arvensis* to 2,4-D and picloram was conferred by a single dominant gene [Jasieniuk et al. 1995; Jugulam et al. 2005].

Resistance of *Kochia scoparia* biotypes to dicamba was suggested to be due to a mutation in the auxin receptor(s) that may affect endogenous auxin binding and altered auxin-mediated responses, such as geotropism and root growth inhibition [Dyer et al. 2002; Goss and Dyer 2003]. But the involvement of genes/proteins in these and related responses to dicamba remains to be determined. However, resistance in a different biotype of this weed in Nebraksa, USA was reported to be determined by a single allele with a high degree of dominance [Preston et al. 2009]. In a *Galeopsis tetrahit* biotype, resistance to MCPA is controlled by two additive genes [Weinberg et al. 2006].

The mechanism of resistance in clopyralid- and picloram-resistant *Centaurea solstitialis* biotypes was apparently due to the recessive mutation conferring pyridine-specific resistance controlled by a single nuclear gene [Sabba et al. 2003]. This resistance could be due to an altered auxin receptor (AFB), similar to that identified in picloram-resistant *Arabidopsis* mutants [Walsh et al. 2006]. Resistance of *Galium spurium* biotypes to quinclorac might be due to the potential alteration in auxin signal transduction pathway [Van Eerd et al. 2005].

The rate at which the resistant trait can spread within a population is much higher if the trait is determined by a single gene or few major genes than with a polygenic trait. Since the chance of simultaneous occurrences of several mutations leading to resistance in a single plant is a product of its individual probabilities, the possibility of a polygenic trait occurring is much lower than for a qualitative trait [Jasieniuk et al. 1996]. Besides, inheritance of polygenic traits depends on outcrossing for accumulation of several genes with minor effects. This reduces the evolution of herbicide resistance.

Relative to sensitive biotypes of *Sinapis arvensis*, the resistant biotypes are 10-, 18-, and 104-fold resistant to MCPA, 2,4-D, and dicamba respectively [Heap and Morrison 1992]. Resistance to these herbicides is not due to an altered uptake, translocation, or metabolism [Hall et al. 1993, 1995; Peniuk et al. 1993]. Production of more ethylene in sensitive than resistant biotypes is attributed to differences at the primary site of action resulting in reduced expression of aminocyclopropane-1-carboxylic acid synthase (in resistant biotype), a key enzyme in the ethylene biosynthetic pathway [Grossman 2000, 2010]. Although lot of work was done to study the role of ABP1 (auxin binding protein 1) in auxinic herbicide resistance in *Sinapis arvensis*, no conclusive evidence is available to prove that resistance is due to an altered ABP1 homolog [Mithila et al. 2011].

Walsh et al. [2004] discovered *Daucus carota* populations in wheat fields of western Australia, where typical herbicide-use patterns had been practised for 17 seasons, showing multiple-resistance across many herbicides with four modes of action. One population exhibited multiple resistance to phytoene-desaturase (PDS)-inhibiting herbicide diflufenican (3-fold), the auxinic 2,4-D (2.2-fold), and the PS II-inhibiting metribuzin and atrazine. Another population showed resistance to the ALS-inhibiting herbicides, the PDS-inhibiting diflufenican (2.5-fold), and 2,4-D (amine: 2.4-fold). Long-term use of phenoxy herbicides may increase the frequency of phenoxy-resistant *D. carota* populations or may increase the magnitude of phenoxy resistance in individual biotypes [Walsh et al. 2009].

Resistance of *Arabidopsis thaliana* to 2,4-D and dicamba was due to a recessive mutation, caused by a single gene, *axrl*, with the mutant producing bushy and short plants having small thin roots and smaller leaves [Estelle and Somerville 1987]. The *axrl* gene (in *A. thaliana*) may encode for a different auxin receptor and resistance may be due to an alteration that has a greater effect on the affinity of this receptor for 2,4-D and IAA. Lincoln et al. [1990] reported that all the 20 *axrl* mutants of *A. thalinana*, with at least five different *axrl* alleles, had similar phenotypes, and the extent of auxin resistance of each mutant line could be directly correlated with severity of the morphological alterations.

Gleason et al. [2011] studied gene responses by comparing the auxin-sensitive and auxin-insensitive mutants of *A. thaliana* to dicamba and 2,4-D. The mutant *axr4-2*, which disrupted auxin transport into cells, was resistant to 2,4-D but susceptible to dicamba. Of the five auxin-signaling F-box receptor mutants (*tir1-1*, *afb1*, *afb2*, *afb3*, and *afb5*), only *tir1-1* and *afb5* were resistant to dicamba, with resistance being additive in the double *tir1-1/afb5* mutant. Mutant *tir1-1* was resistant to 2,4-D but not *afb5.* They also found that dicamba stimulated many stress-responsive and signaling genes, including those involved in biosynthesis or signaling of auxin, ethylene, and ABA with TIR1 and AFB5 required for the dicamba-responsiveness of some genes. Their work on dicamba-regulated gene expression and the selectivity of auxin receptors has provided molecular insight into dicamba-regulated signaling which could help in the development of novel herbicide resistance in crop plants.

Glycines

Glyphosate {*N*-(phosphonomethyl) glycine} belongs to herbicide group 'glycines' (Group G) which also includes glyphosate-trimesium. This broad-spectrum systemic herbicide was discovered by John Franz of Monsanto Company in 1970. This nonselective, translocated herbicide, available as an isopropylammonium salt since 1974, is used for effective control of several rhizamatous and deep-rooted perennial grasses, broadleaf weeds, and woody plants in plantation crops as well as in certain perennial crops and even widely spaced annual row crops when used as a directed spray. It is effective on actively growing plants. Glyphosate has moderate persistence in soil, with a half-life of 47 d. It is rapidly inactivated by soil particles, and the unbound herbicide can be degraded by bacteria.

There have been no reports of evolved glyphosate-resistant weeds during the initial 20-yr period of use. Its widespread application in conventional crops to facilitate no-till and minimum-till tillage practices and the introduction of transgenic crops changed the use pattern of this herbicide and weed management system. The first reported resistance to glyphosate came in 1996 from Victoria, Australia where *Lolium rigidum*, infesting wheat, rapeseed, and cereals, showed resistance. Since then, 30 more weed species (16 dicots and 14 monocots), including those with multiple resistant ones, have been reported from all over the world (Fig. 2.3: Group G) [Heap 2014]. The other major species included *Amaranthus palmeri*, *Amaranthus tuberculatus* (*A. rudis*), *Ambrosia artemisiifolia*, *Ambrosia trifida*, *Kochia scoparia*, *Conyza bonariensis*, *Conyza canadensis*, and *Lolium perenne* ssp. *multiflorum*.

Mechanism of Action

Glyphosate kills plants by inhibiting a reaction that occurs in the shikimic acid pathway, which is vital for their survival. The end products are several amino acids (tryptophan, tyrosine, and phenylalanine), flavonoids, ubiquinone, folic acid, *p*-aminobenzoic acids, lignins, anthocyanins, coumarins, vitamins E and K, etc. needed for protein synthesis or other biosynthetic pathways leading to plant growth. Glyphosate is a highly specific competitive inhibitor of the enzyme 5-enolpyruvyl-shikimate-3-phosphate synthase (EPSPS) involved in the shikimic acid biosynthesis. The EPSPS catalyzes the reaction between shikimate-3-phosphate (S3P) and 5-enolpyruvyl-shikimate-3-phosphate (EPSP).

The inhibition is centered at the ternary dead-end complex of EPSP synthase-shikimate-3-phosphate-glyphosate. A build-up of shikimate-3-phosphate due to inhibition of EPSPS by glyphosate has a deleterious effect on the formation of chorismic acid, a precursor required for the biosynthesis of aromatic compounds.

Mechanism of Resistance

In resistant plants, glyphosate accumulates in the leaves rather than being translocated throughout the plant. Target-site mutations, causing altered pattern of herbicide translocation, are by far the most common mechanism of resistance to glyphosate. These mutations alter the target site in such a way that it can still function but herbicide molecules can no longer bind to it. In many weed populations, resistance to glyphosate correlates with the reduced rates of translocation to active meristematic root and shoot tissues [Powles and Preston 2006; Preston and Wakelin 2008; Shaner 2009], but not in *Amaranthus palmeri* [Culpepper et al. 2006]. This suggests that reduced translocation is far less common.

As there is no single genetic alteration responsible in all of the resistant weeds, the genetic basis of many of the glyphosate-resistant weeds remains unknown. Some populations of *Eleusine indica* from Malaysia, *Lolium rigidum* from Australia, and *Lolium perenne* ssp. *multiflorum* from Chile exhibit target site-based resistance to glyphosate through mutation at Pro106 of the *EPSPS* gene. This causes production of enzyme molecules that glyphosate is less likely to bind to. Pro106 is not actually at the glyphosate binding site; but an amino acid substitution at this position is believed to reorient the binding site in such a way that it would cause a reduced affinity for glyphosate. Mutations change from Pro106 to Ser, Thr, or Ala, resulting in an *EPSPS* that confers modest degree of glyphosate resistance. This moderate resistance is sufficient for commercial failure of the herbicide to control these plants in the field. However, two other populations of *L. rigidum* infesting grape vinery in southern Australia were found more highly resistant to glyphosate as a consequence of expressing two different mechanisms concurrently [Bostamam et al. 2012].

Pline-Srnic [2006] suggested that at least one of the three different mechanisms appears to confer resistance to glyphosate: EPSPS overexpression, *EPSPS* gene amplification, or increased enzyme stability. Within the resistant population of *Amaranthus palmeri*, glyphosate resistance correlates with increases in a) genomic copy of EPSPS, b) expression of the EPSPS transcript, c) EPSPS protein level, and d) EPSPS activity [Gaines et al. 2010]. Resistance mechanism in this dicot weed involves increased glyphosate activity due to amplification (also known as gene duplication or chromosomal duplication) of *EPSPS* gene [Ganies et al. 2010]. Genomes of resistant plants contained 5-fold to more than 160-fold more copies of the EPSPS gene than did genomes of susceptible plants. Quantitative RT-PCR (vide Chapter 3) analysis of cDNA (complementary DNA) revealed that EPSPS expression was positively correlated with genomic EPSPS relative copy number [Powles 2010; Gaines et al. 2010]. The amplified genes were not clustered on the chromosomes but distributed among all of the chromosomes. These results suggest that the *EPSPS* genes were amplified through mobile genetic elements (jumping genes). The gene amplification, however, is not due to genome duplication. Dose response results from the resistant and an F_2 population suggest that between 30 and 50 EPSPS genomic copies are necessary to survive glyphosate rates between 0.5 and 1.0 kg ha^{-1} [Gaines et al. 2011].

The resistant *Convolvulus arvensis* had higher levels of 3-deoxy-D-arbino-heptulosonate 7-phosphate synthase, the first enzyme in the shikimate pathway, suggesting that increased carbon flow through the shikimate pathway can provide glyphosate resistance. Following glyphosate treatment, there was greater accumulation of shikimate (derived from shikimate-3-Pi) in resistant than in susceptible plants.

Another mechanism, a non-target site-resistance mechanism, is evident in glyphosate-resistant populations of *Conyza canadensis* and *Lolium rigidum* from the U.S. and Australia, respectively. In *Conyza canadensis*, the glyphosate-resistant biotype shuttled 85 percent of the herbicide into the vacuole as against only 15 percent in the sensitive weed [Ge et al.

2010]. The resistant biotype has a pump in the tonoplast membrane surrounding the vacuole and it shuttles glyphosate into this storage compartment where it can no longer interfere with the critical biological reactions taking place in the chloroplast. Once glyphosate gets to the vacuole it is trapped, because the resistant *C. canadensis* rapidly shuttles glyphosate into the vacuole, with less of it becoming available for translocation to rapidly growing parts of the plants [Ge et al. 2010].

The resistant populations of *Amaranthus palmeri* biotypes in Mississippi, USA showed multiple resistance to glyphosate and pyrithiobac (an ALS inhibitor) and that resistance to glyphosate was partly due to its reduced absorption and translocation [Nandula et al. 2012].

Ureas and Amides

The first weed to show resistance to substituted areas was the monocot *Alopecurus myosuroides* when it was found to be resistant to methabenzthiazuron in 1979 in Israel. Substituted ureas first came to light with the discovery of monuron in 1951. The Group C2 herbicides also includes amides.

Currently, 24 weed species (eight dicots and 16 monocots), including those with multiple resistance, have been reported to show resistance to chlorotoluron, diuron, linuron, monolinuron, isoproturon, propanil, pyroxsulam (sulfonamide herbicide), etc. Majority of these weeds are monocots. Aside from *A. myosuroides*, other major resistant weeds include *Echinochloa crus-galli* var. *crus-galli*, *E. colona*, *Amaranthus retorflexus*, *Apera spica-venti*, and *Conyza canadensis*. To date, *Phalaris minor* is the only weed species that showed resistance in India in 1991 to isoproturon being used in wheat.

Mechanism of Action

Urea herbicides inhibit photosynthesis by binding to the Q_B-binding niche on the D1 protein of the PS II complex in chloroplast thylakoid membranes, thus blocking electron transport from Q_A to Q_B. This stops CO_2 fixation and production of ATP and $NADPH^{++}$ (all needed for plant growth), but plant death occurs in most cases by other processes. Inability to reoxidize Q_A promotes the formation of triplet state chlorophyll molecule ($^3Chl^*$), which interacts with ground-state oxygen to form singlet oxygen ($^1O_2$). Both triplet chlorophyll and singlet oxygen can abstract hydrogen from unsaturated lipids, producing a lipid radical and initiating a chain reaction of lipid peroxidation. Lipids and proteins are attacked and oxidized, resulting in loss of chlorophyll and carotenoids as well as leaky membranes; consequently, cells and cell organelles dry and disintegrate rapidly.

Propanil, an acylanilide herbicide, is used extensively in rice to control dicot and monocot weeds, is rapidly degraded by the tolerant rice into non-phytotoxic compounds 3,4-dichloaniline and propionic acid due to the presence of high level of aryl acylamidase (aryl-acylamine amidohydrolase).

Mechanism of Resistance

Resistance of *Lolium rigidum* biotype to urea herbicides is not due to differences in the herbicide target site [Burnet et al. 1993]. In the case of chlorotoluron and isoporturon, it could be attributed to more rapid and enhanced metabolism of the herbicides via alkyl oxidation (4-methylphenyl hydroxylation), possibly involving cytochrome 450 enzymes, but not to alterations in absorption, translocation, or binding [de Prado et al. 1997]. In a resistant biotype of *Bromus tectorum* it was the cytochrome-450 monooxygenase inhibitor 1-aminobenzonitrile (ABT) that enhanced the ability of resistant biotype to degrade the molecule to -nontoxic ring-alkylhydroxylated intermediates at a much faster rate than its susceptible counterpart, followed by conjugation to form polar conjugates [Menendez et al. 2006].

In *Phalaris minor* resistant to isoproturon, the target site is in the photosynthetic apparatus at D1 protein of PS II. Nucleotide sequence of both susceptible and resistant *psbA* gene has been reported to have four point mutations. Singh et al. [2012] found that the resistance was due to alteration in secondary structure near the binding site, resulting in loss in cavity area, volume and change in binding position, loss of hydrogen bonds, hydrophobic interaction, and complete loss of hydrophobic sites. In an attempt to aid the resistant biotype to regain sensitivity, they reported that N-methyl triazole derivative substituted isoproturon as a potential substrate for the herbicide. Resistance of biotypes of *Echinochloa* (*colona* and *crus-galli* var. *crus-galli*) was due to increased and elevated levels of acylamidase activity [Leah et al. 1994; Hoagland et al. 2004].

Dinitroanilines and Others

Herbicide families, phosphoroamidates, pyridines, benzamides, and benzoic acid belong to Groups K1, K2, and K3 groups (vide Appendix, Table 1).The more prominent dinitroanilines, informally called 'yellow' herbicides, were first discovered in 1960. Twelve weed species (two dicots and 10 monocots) showed resistance to Group K1 herbicides (Fig. 2.3), with nine being to dinitroanilines. The first weed species which showed resistance was *Eleusine indica* in 1973 when it was infesting cotton continuously applied with trifluralin in North Carolina, USA. Besides this monocot weed (seven biotypes), other resistant prominent monocot weeds include *Lolium rigidum*, *Eleusine indica*, and *Setaria viridis*. The only dicot weeds to show resistance to K1 group herbicides include *Amaranthus palmeri* in USA and *Fumaria densiflora* in Australia.

Mechanism of Action

Dinitroaniline, benzamide, benzoic acid, phosphoramidate, and pyridine herbicides disrupt mitosis in the meristematic cells of seedling plants. Specifically, they inhibit synthesis of microtubulin necessary in the formation of cell walls and in chromosome movement to daughter cells during mitosis. The cell does not complete division and the affected cells remain as single cells with multiple nuclear chromosomes, i.e., multi-nucleated cells.

Herbicides belonging to this group bind to tubulin, the major microtubule protein. The herbicide-tubulin complex inhibits polymerization of microtubules at the assembly end of the protein-based microtubule but it has no effect on depolymerization of the tubule at the other end [Vaughn and Lehnen 1991], leading to a loss of microtubule structure and function. As a result, the spindle apparatus is absent, thus preventing the alignment and separation of chromosomes during mitosis. In addition, the cell plate cannot be formed. Microtubules also function in cell wall formation. Herbicide-induced microtubule loss may cause the observed swelling of root tips as cells in this region neither divide nor elongate. Microtubules are cylindrical polymers of tubulin constructed to form α-β-tubulin heterodimers. They are important for maintaining cell structure, providing platforms for intracellular transport, forming mitotic spindle, as well as other cellular processes. There are many proteins that bind to microtubules, including motor proteins such as kinesin and dynein, severing proteins like katanin, and other proteins important for regulating microtubule dynamics.

Mechanism of Resistance

Dinitroaniline resistance of *Eleusine indica* is inherited as a single, nuclear gene. This inheritance is displayed at multiple resistance levels, depending on the level of resistance showed by biotypes. Resistance is controlled by three alleles at a single locus: *Drp-S, Drp-i,* and *Drp-r* [Zeng and Baird 1994]. The highly resistant (R) and intermediately resistant (I) biotypes are 50-fold and 1000-fold more resistant than the susceptible (S) biotype [Vaughn et al. 1990]. The basis of resistance in the highly R-biotype was due to hyper-stabilization of microtubules, rendering them immune to dinitroaniline inhibition. This mechanism is not present in the I-biotype. The dicot *Daucus carota* naturally has this hyper-stabilized microtubule form and high levels of dinitroaniline resistance.

Thiocarbamates and Others

Herbicides belonging to Group N include thiocarbamates, phosphorodithioates, and benzofurans. Thiocarbamates are a well-established group of herbicides discovered since the late 1950s. The first weed species to show resistance in this group was *Lolium rigidum* found in 1982 in South Australia. This was followed by *Avena fatua* which first showed resistance to triallate in 1989 in Alberta, Canada, with this grass weed exhibiting multiple resistance to ACCase (A), ALA (B), and antimicrotubule mitotic disrupter (Z) [Heap 2014]. In all, nine species, all of them monocots, have been found to be resistant to this herbicide group.

Mechanism of Action

Thiocarbamate and phosphorodithioate herbicides affect surface lipids such as waxes, cutin, and suberin by inhibiting the biosynthesis of very long-chain fatty acids (VLCFA). VLCFAs are defined as fatty acids of chain length greater than 18 nucleotide phosphate (NADPH)-dependent elongation of selected acyl-CoA substrates, using malonyl-CoA as the condensing agent for greater than C18 fatty acids [Matthews and Powles 1992]. Prevention of acyl-CoA elongases, which catalyze the synthesis of VLCFA, or inhibition of their activity, is suggested to be affected by thiocarbamates. This inhibitory effect of acyl-CoA elongases could be attributed to the selective effect of thiocarbamate on *in vivo* synthesis of VLCFA or the metabolite of the thiocarbamate, such as its sulphoxide derivative catalyzed by either cytochrome 450 monooxygenase or a peroxygenase. It is commonly assumed that the sulphoxide derivatives of thiocarbamates alkylate (carbomylate) CoA and interfere with CoA metabolism. However, the thiocarbamate molecule itself (as opposed to its sulphoxide derivative) inhibits key enzymes involved in the synthesis of acetyl-CoA [Gronwald 1991].

Inhibition of VLCFAs would lead to reduction in epicuticular wax formation on the plant foliage which increases the leaf wettability, allowing the leaf to be more effectively treated by the foliage-applied herbicides.

Mechanism of Resistance

The mechanism of resistance in weeds to this group of thiocarbamates and related herbicides is unknown.

PPO Inhibitors

This Group E of herbicides inhibiting the enzyme protoporphyrinogen oxidase (PPO) include those belonging to diphenylether, N-phenylphthalimide, oxadiazole, phenylpyrazole, pyrimidinedione, thiadiazole, and triazolinone, and other families. These are applied in varied crops at preemergence and postemergence. All the six weed species found resistant to this group of herbicides since 2001 have been dicots, the more predominant being *Amaranthus retroflexus* (*A. rudis*). This dicot also showed reistance to three other herbicide groups, B (ALS), C1 (PS II), and G (glycine) [Heap 2014].

Mechanism of Action

The diphenyl ether (DPE) herbicides target the enzyme protoporphyrinogen oxidase (PPO or Protox) which oxidizes protoporphyrinogen (PPG IX), in the presence of light, to protoporphyrin IX (PP IX), a tetrapyrrole intermediate. This product is important because it is a precursor molecule for both chlorophyll (needed for photosynthesis) and heme (needed for electron transfer chains).

The PP IX released in the cell is excited to the triplet state with a high efficiency and interacts with molecular O_2 to produce singlet oxygen ($^1O_2$). Singlet oxygen is toxic to cells because it is much more destructive than molecular oxygen in the normal triplet state. Favorite targets of singlet oxygen include the double bonds of fatty acids and amino acids. Membranes, sites with high concentrations of fatty acids, are particularly vulnerable to peroxidation (molecular damage from free radicals). The plasma membrane of the plant cell is considered to be the vulnerable component most impacted by the photodynamic damage from herbicides that inhibit Protox.

In healthy tissues, the enzymatically-produced PP IX is normally channeled to the proper metal chelatase (a three-component enzyme to produce protoheme or Mg-Protox IX), thus preventing theaccumulation of photo-destructive PP IX. PPG IX destruction may be a mechanism for providing protection from the toxic effects of PP IX accumulation in healthy tissues.

Inhibition of Protox by DPE, N-phenylphthalimide, phenylpyrazole, oxadiazole, and triazolinone herbicides in organelles leads to the accumulation of PP IX which is thought to result from the translocation of the Protox substrate PPG IX outside the organelle followed with oxidation by non-enzymatic reactions or by herbicide-insensitive oxidases mainly in the plasma membrane. In the presence of light, the accumulated PP IX, which is now improperly compartmentalized, is unable to be utilized for further synthesis and thus reacts with oxygen to give rise to singlet oxygen. This singlet oxygen, in turn, causes lipid peroxidation, especially on membranes, damage of other cellular constituents, and eventually cell death and, thus, the herbicidal action of these chemicals.

Mechanism of Resistance

The evolved resistance of two biotypes of *Amaranthus retroflexus* (*A. rudis*) to diphenylether herbicides involves reduction of the photodynamic pigment protoporphyrin IX (Proto) accumulation; they no longer accumulate large quantities of this harmful pigment that is normally the cause of death in sensitive plants [Li et al. 2004].

There are at least two nuclear PPO genes present in plants *PPX1* and *PPX2* which encode plastid- and mitochondrial-targeted PPO isoforms respectively. These genes reduce accumulation of protoporphyrin IX. Lee et al. [2008] suggested that a) there are likely two genes *PPX1* and *PPX2L* in an *A. rudis* resistant biotype and b) the 3-bp (base pair) deletion allele ΔG210 of *PPX2L* correlated with whole-plant resistance to PPO inhibitors in each of four other resistant populations. Rousonelos [2010] found that of the multiple polymorphisms responsible for resistance, one polymorphism (R98L) could be responsible for resistance of *Ambrosia artemisiifolia* to PPO-inhibiting DPE herbicide lactofen. The resistance mechanisms of these two weed species are due to mutations resent in the *PPX2* gene. Resistance in *A. rudis* can be obtained by a deletion, but for *A. artemisiifolia* only a single base pair change is sufficient to confer resistance to PPO-inhibiting herbicides.

Triazoles and Isoxazolidiones

This F3 group includes herbicides of triazole, isoxazolidione, and diphenyl ether families. Ever since *Lolium rigidum* infesting orchards showed resistance in 1982 to this group herbicide in South Australia, four other species (one dicot and three monocots) followed suit. These include *Agrostis stolonifera*, *Echinochloa crus-galli* var. *crus-galli*, *Poa annua*, and *Polygonum aviculare*.

Mechanism of Action

Amitrole inhibits accumulation of chlorophyll and carotenoids in the light, although the specific site of action has not been determined. Precursors of carotenoid synthesis, including phytoene, phytofluene, carotene, and lycopene accumulate in amitrole-treated plants [Barry and Pallett 1990]. This suggests that phytoene desaturase, lycopene cyclase, imidazoleglycerol phosphate dehydratase, nitrate reductase, or catalase may be inhibited. However, other research [Heim and Larrinua 1989] indicates that the histidine, carotenoid, and chlorophyll biosynthetic pathways probably are not the primary sites of amitrole action. Instead, it may have a greater effect on cell division and elongation than on pigment biosynthesis.

Clomazone (isoxazolidione herbicide) is metabolized to the herbicidally active 5-keto form of clomazone. The 5-keto form inhibits 1-deoxy-D-xyulose 5-phosphate synthase (DOXP), a key component to plastid isoprenoid synthesis. Clomazone does not inhibit geranylgeranyl pyrophosphate biosynthesis [Croteau 1992; Weimer 1992].

Aclonifen, unlike other diphenylether herbicides, affects two biochemical sites concurrently. It inhibits chlorophyll and cytochrome synthesis, allowing cytoplasmic accumulation of singlet oxygen while adversely affecting biosynthesis of carotenoids which protect the membrane against the free radicals against singlet oxygen [Özgür 2011]. The inhibitory activity of aclonifen on carotenoid biosynthesis which is absent in diphenylethers, seems to depend on the specificity of the substitution pattern of the chemical structure, probably due to the presence of the $1\text{-}NH_2$.

Mechanism of Resistance

The exact mechanism of resistance of weeds to the herbicide Group F3 is not known.

Nitriles and Others

This group, C3, includes nitrile, benzothiadiazinone, and phenylpyridazine herbicide families. Of the four weed species that showed resistance to nitriles (bromoxynil) so far, the first one was the dicot *Senicio vulgaris* when it infested mint crop in 1995 in Oregon, USA. This was followed by *Amaranthus hybridus* in 2004 and *Amaranthus retroflexus* in 2005 besides the monocot *Sagittaria montevidensis* in 2009.

Mechanism of Action

Like phenylcarbamates, pyridazinones, triazines, triazinones, and uracils C1 group as well as amides and ureas of C2 group, the C3 group herbicides also inhibit photosynthesis by binding to the Q_B-binding niche on the D1 protein of the PS II complex in chloroplast thylakoid membranes, thus blocking electron transport from Q_A to Q_B. This stops CO_2 fixation and production of ATP and $NADPH^{++}$, all required for plant growth. But plant death may occur by other processes. Inability to reoxidize Q_A promotes formation of triplet chlorophyll (^3chl) which interacts with ground state oxygen to form singlet oxygen ($^1O_2$). Both ^3chl and $^1O_2$ extract hydrogen from unsaturated fatty acids, producing a lipid radical and initiating a chain reaction of lipid peroxidation. Lipids and proteins are attacked and oxidized, resulting in loss of chlorophyll and carotenoids, as well as leaky membranes; consequently cells and cell organelles dry and disintegrate rapidly.

Pyridate, a phenylpyridazine herbicide, is hydrolyzed to 3-phenyl-4-hydroxy-6-chloropyridazine, which then inhibits PS II electron transport.

Mechanism of Resistance

The mechanism of resistance in weeds to nitrile, benzothiadiazinone, and phenylpyridazine is yet to be determined.

Chloracetamides and Others

To date, four biotypes of as many weed species have been found resistant to K3 group herbicides including chloracetamides, acetamides, oxyacetamides, tetrazolinones, and others (anilofos). These weed species include *Lolium rigidum*, the first to show resistance to the chloroacetamide herbicide metolachlor in South Australia 1982, *Lolium perenne ssp. multiflorum* to flufenacet, *Echinochloa crus-galli* var. *crus-galli* to butachlor, and *Alopecurus myosuroides* to flufenacet.

Mechanism of Action

Herbicides of this group are generally applied at preemergence to inhibit seed germination by interfering with the metabolic activities related to it [Rao 2000]. The seedlings of annual grass species against which they are effective do not emerge following preemergence application.

Chloroacetamides have a strong inhibitory effect on the biosynthesis of protein and lipid biosynthesis in susceptible species. They interfere with fatty acid metabolism by alkylating key enzymes involved in fatty acid biosynthesis or by alkylating CoA and thereby interfering with CoA metabolism. CoA plays an important role in lipid metabolism and other metabolic processes that are inhibited by acetamides. Other evidence suggests that acetamide, chloroacetamide, oxyacetamide, and tetrazolinone herbicides are examples of those that are currently thought to inhibit very long chain fatty acid (VLCFA) synthesis localized in the microsomes [Böger et al. 2000], the primary mechanism of action. The direct consequence of this inhibition is a strong decrease of VLCFAs required as constituents of the plasma membrane and the substitution by shorter acyl chains. Apparently, physical properties and function of the plasma membrane are affected eventually leading to death of the plant [Matthes and Böger 2002].

Carotenoid Inhibitors

Carotenoid biosynthesis inhibitor group, F1, includes pyridazinone, pyridinecarboxyamide, and other herbicides. So far, three weed species have shown resistance to carotenoid biosynthesis. These include the dicots *Raphanus raphanistrum* and *Sisymbrium orientale* and the aquatic monocot *Hydrilla verticillata*. Of these, *R. raphanistrum* showed multiple resistance to B (ALS), G (glycine), and O (auxin) groups.

Mechanism of Action

Herbicides of this group inhibit carotenoid biosynthesis catalyzed by phytoene desaturase, resulting in accumulation of phytoene and other intermediates, which have short chromophores and cannot protect against photo-oxidation. As a result, the plant will be rapidly killed by light and oxygen.

Carotenoids play a role in photosynthesis by harvesting light and transferring the captured energy to chlorophyll molecules within the photosynthetic apparatus. They play three essential protective roles in the photosynthetic apparatus. They do this because they are highly effective quenchers, having the ability to absorb excitation energy and dissipate it harmlessly as heat. The first protective role is the ability to quench the energized triplet chlorophyll (3Chl*) molecules back to the ground state. The second role is to quench the destructive singlet oxygen ($^1O_2$) molecules back to the normal triplet state. The third role is in quenching the photosystem reaction centers when overexcited in very bright light. For this, zeaxanthin, a specific carotenoid, is produced from violaxanthin that is normally present in the chloroplast.

Inhibitors of carotenoid biosynthesis cause a general bleaching of the plant. This is because each time a chlorophyll molecule absorbs the energy from a photon there is a small, but finite chance that it will generate a triplet state. Without the presence of carotenoids to quench triplet chlorophyll, active oxygen species are generated and destroy the photosynthetic apparatus within the thylakoid membrane. Destruction of chlorophyll causes a bleaching of the leaf. Carotenoids are largely absent in fluridone-treated plants, allowing $^1O_2$ and 3Chl* to abstract a hydrogen from an unsaturated lipid (e.g., membrane fatty acid, chlorophyll) producing a lipid radical. The lipid radical interacts with O_2 yielding a peroxidized lipid and another lipid radical. Thus, a self-sustaining chain reaction of lipid peroxidation is initiated which functionally destroys chlorophyll and membrane lipids. Proteins also are destroyed by $^1O_2$. Destruction of integral membrane components leads to leaky membranes and rapid tissue desiccation.

Mechanism of Resistance

The potential target for herbicides in resistant weeds appears to be the genes encoding enzymes of carotenoid biosynthesis. Michel et al. [2004] reported that three independent fluridone-resistant *Hydrilla* biotypes arose from the selection of somatic mutations at the argenine 304 codon of PDS (phytoene desaturase) enzyme. The three-PDS variants had specific activities similar to the wild-type enzyme but were two to five times less sensitive to fluridone.

Glutamine Synthetase Inhibitors

The glutamine synthetase inhibiting phosphinic acid herbicides (glufosinate ammonium and bialophos) belong to H group. Two weed species have become resistant to this herbicide so far. These include *Eleusine indica* which showed resistance while infesting orchards in Malaysia and *Lolium perenne* ssp. *multiflorum* found resistant in 2010 in Oregon, USA. Of these, *L. perenne* ssp. *multiflorum* showed multiple-resistance to glyphosate GroupG which inhibits EPSP synthase.

Mechanism of Action

Glutamine synthetase (GS) catalyzes glutamate and ammonia to glutamine which is necessary for biosynthesis of several amino acids as shown below.

Glutamine synthetase
L-glutamate+ATP+NH$_3$ --------------------------> L-glutamine+ATP+ADP+Pi+H$_2$O

The glutamine formed by chloroplast GS is used as an amino donor for a variety of biosynthetic pathways, including amino acid and nucleotide biosynthesis. Inhibition of GS produces not only an accumulation of ammonia levels in cells but also a decrease in the rate of CO_2 fixation. Accumulation of ammonia in the plant destroys cells and directly inhibits PS I and PS II reactions. Ammonia reduces the pH gradient across the membrane which can uncouple photophosphorylation. Inhibition of GS leads to multiple deleterious effects, thus making it an exquisitely toxic site. Symptoms usually develop on the third or fourth day of treatment. Chlorosis and wilting are quickly followed by necrosis.

Mechanism of Resistance

The mechanism of resistance of weed species to glutamine synthetase inhibiting herbicides is yet to be determined.

HPPD Inhibitors

Group F2 includes HPPD inhibitors which encompass herbicide families of isoxazole, triketone, benzoylpyrazole, pyrazole, and callistemone. So far, *Amaranthus palmeri* and *Amaranthus tuberculatus* (*A. rudis*) are the only weed species to develop resistance to this group of herbicides. Both showed resistance to herbicides of three sites of action (Fig. 2.3) [Heap 2014].

Mechanism of Action

Herbicides of callistemone, isoxazole, benzoylpyrazole, pyrazole, and triketone families inhibit 4-hydroxyphenyl-pyruvate-dioxygenase (4-HPPD), an Fe(II)-containing non-heme oxygenase enzyme which converts 4-hydroxymethylpyruvate to homogentisate. This is a key step in plastoquinone biosynthesis and its inhibition gives rise to bleaching symptoms on new growth. Biosynthesis of plastoquinone, a cofactor of phytoene desaturase, is a key factor in the synthesis of carotenoid pigments which protect a plant's chlorophyll from decomposition by sunlight. Without carotenoid pigments, the sun damages chlorophyll pigments and the plant becomes 'bleached' and dies.

Mechanism of Resistance

The mechanism of resistance of *A. tuberculatus* to HPPD-inhibiting herbicides is yet to be determined.

Mitosis Inhibitors

The ubiquitous monocot *Lolium rigidum* is the only weed species to show resistance to mitosis-inhibiting carbanilate herbicides of Group K2. These include carbamates such as carbetamide, chlorpropham, and propham.

Herbicides of this group inhibit cell division and microtubule organization and polymerization, central to biological functions of cells. Microtubules (long polymers made of smaller units of monomers made of the protein tubulin) are the structures that pull the cell when it divides. They utilize the energy of GTP hydrolysis to fuel a unique polymerization mechanism called dynamic instability.

The mechanism of resistance of *L. rigidum* to these mitotic inhibitors is yet to be determined.

Cellulose Inhibitors

Group L herbicides inhibit biosynthesis of cellulose. This group is represented by quinolinecarboxylic acid (quinclorac, also of O group), nitrile, and benzamide families. The monocot weed *Echinochloa erecta* is the only species to evolve resistance to this quinclorac applied in rice in Italy in 2004 [Heap 2014]. It has also evolved resistance to PS II-inhibiting C2 group.

Quinclorac has an unusual action mechanism. It acts as a cellulose inhibitor in grasses while inhibiting dicots as auxinic herbicides dichlobenil and isoxaben. In *E. erecta*, it strongly inhibits cellulose and a hemicellulose fraction presumed to be glucuronoarabinoxylan which constitutes 5 percent of the primary walls of suspension-cultured sycamore cells. Inhibition of cellulose is catalyzed by cellulose synthase. Quinclorac inhibits glucose incorporation into cellulose and other cell wall polysaccharides.

The mechanism of resistance of *E. erecta* has not yet been determined.

Other Inhibitors

Herbicides belonging to this Z group have diverse structures. These include organoarsenicals (DSMA, MSMA); antimicrotubule mitotic disrupting arylaminopropionic acids (flamprop-methyl, flamprop-M-isopropyl); and cell elongation-inhibiting pyrazolium (difenzoquat). It also includes the unclassified herbicide endothal whose site of action is unknown. The weed species found resistant to these herbicide groups are *Xanthium strumarium* to organoarsenicals: *Avena fatua, A. sterilis* and *A. sterilis* ssp. *ludoviciana* to arylaminopropionic acids, and *A. fatua* to difenzoquat.

Mechanism of Resistance of Organoarsenicals

DSMA and MSMA affect chlorophyll synthesis, amino acid content, and respiration in sensitive plants. These arsenic herbicides are reduced by PS I to form sulfhydryl groups of enzymes involved in CO_2 fixation and its regulation, thereby inhibiting CO_2 fixation. Uptake, translocation, and metabolism are not involved in the mechanism of resistance of the R biotype of *Xanthium strumarium* to MSMA [Keese and Camper 1994; Nimbal et al. 1995a]. The R biotype is not cross-resistant to other herbicides. Inhibition of carbon assimilation is more rapid and of greater magnitude in the S biotype than in the R biotype of *X. strumarium* [Nimbal et al. 1995b]. MSMA has no direct effect on photosynthetic electron transport. The effects mentioned above might be due to inhibition of nucleic acids.

References

Anderson, R.N. and J.W. Gronwald. 1991. Atrazine resistance in velvetleaf (*Abutilon theophrasti*) biotype due to enhanced glutathione-*S*-transferase activity. Plant Physiol. 96: 104–109.

Anderson, R.N. and J.W. Gronwald. 1987. Noncytoplasmic inheritance of a triazine tolerance in velvetleaf (*Abutilon theophrasti*). Weed Sci. 35: 496–498.

Antonovics, J. and P.H. Thrall. 1994. Cost of resistance and the maintenance of genetic-polymorphism in host-pathogen systems. Proc. Royal Soc. of London Series B-Biological Sciences 257: 105–110.

Ashigh, J. and F.J. Tardif. 2009. An amino acid substitution at position 205 of acetohydroxyacid synthase reduces fitness under optimal light in resistant populations of *Solanum ptychanthum*. Weed Res. 49: 479–489.

Atwell, B.J., P.E. Kriedemann and C.G.N. Turnbull (eds.). 1999. Plants in Action: Adaptation in Nature, Performance in Cultivation. Macmillan Education Australia, 664 pp.

Bakkali, Y., J.P. Ruiz-Santaella, M.D. Osuna, J. Wagner, A.J. Fischer and R. De Prado. 2007. Late watergrass (*Echinochloa phyllopogon*): mechanisms involved in the resistance to fenoxaprop-p-ethyl. J. Agri. Food Chem. 55: 4052–4058.

Barry, P. and K.E. Pallett. 1990. Herbicidal inhibition of carotenogenesis detected by HPLC. Z. Naturforsch. 45c: 492.

Beckie, H.J., E.N. Johnson and A. Légère. 2012. Negative cross-resistance of acetolactate synthase inhibitor–resistant Kochia (*Kochia scoparia*) to protoporphyrinogen oxidase– and hydroxyphenylpyruvate dioxygenase–inhibiting herbicides. Weed Technol. 26: 570–574.

Beckie, H.J., S.I. Warwick, H. Nair and G. Seguin-Swartz. 2003. Gene flow in commercial fields of herbicide-resistant canola (*Brassica napus*). Ecological Applications 13: 1276–1294.

Bernards, M.L., R.J. Crespo, G.R. Kruger, R. Gaussoin and P.J. Tranel. 2012. A waterhemp (*Amaranthus tuberculatus*) population resistant to 2,4-D. Weed Sci. 60: 379–384.

Bettini, P., S. McNally, M. Sevingnac, H. Darmency, J. Gasquez and M. Dron. 1987. Atrazine resistance in *Chenopodium album*: low and high levels of resistance to the herbicide related to the same chloroplast *psbA* gene mutation. Plant Physiol. 84: 1442–1446.

Blackman, G.E. 1950. Selective toxicity and development of selective weed killers. J. Roy. Soc. Arts 98: 500–517.

Böger, P., B. Matthes and B.J. Schmalfu. 2000. Towards the primary target of chloroacetamides—new findings pave the way. Pest Manage. Sci. 56: 497–508.

Booth, B.D. and C.J. Swanton. 2002. Assembly theory applied to weed communities. Weed Sci. 50: 2–13.

Bostamam, Y., J.M. Malone, F.C. Dolman, P. Boutsalis and C. Preston. 2012. Rigid Ryegrass (*Lolium rigidum*) populations containing a target site mutation in EPSPS and reduced glyphosate translocation are more resistant to glyphosate. Weed Sci. 60: 474–479.

Bradley, K.W., J. Wu, K.K. Hatzios and E.S. Hagood, Jr. 2001. The mechanism of resistance to aryloxyphenoxy-propionate and cylohexanedione herbicides in a Johnsongrass biotype. Weed Sci. 49: 477–484.

Brown, A.C., S.R. Moss, Z.A. Wilson and L.M. Field. 2002. An isoleucine to leucine substitution in the ACCase of *Alopecurus myosuroides* (blackgrass) is associated with resistance to the herbicide sethoxydim. Pestic. Biochem. Physiol. 72: 160–168.

Burd, M., T.L. Ashman, D.R. Campbell, M.R. Dudash, M.O. Johnston, T.M. Knight et al. 2009. Ovule number per flower in a world of unpredictable pollination. Amer. J. Botany 96: 1159–1167.

Burnet, M.W.M., B.R. Loveys, J.A.M. Holtum and S.B. Powles. 1993. Increased detoxification is a mechanism of simazine resistance in *Lolium rigidum*. Pestic. Biochem. Physiol. 46: 207–218.

Burton, J.D., J.W. Gronwald, R.A. Keith, D.A. Somers, B.G. Gengenbach and D.L. Wyse. 1991. Kinetics of inhibition of acetyl co-enzyme A carboxylase by sethoxidim and haloxyfop. Pestic. Biochem. Physiol. 39: 100–109.

Busi, R., P. Neve and S.B. Powles. 2012. Evolved polygenic herbicide resistance in *Lolium rigidum* by low-dose herbicide selection within standing genetic variation. Evolutionary Applications. doi: 10.1111/j. 1752-4571.2012.00282.

Busi, R. and S.B. Powles. 2009. Exposure of *Lolium rigidum* to sub-lethal doses of glyphosate caused a population shift to resistance in 3 to 4 cycles of selection. Heredity 103: 318–325.

Chapman, E.J. and M. Estelle. 2009. Mechanism of auxin-regulated gene expression in plants. Ann. Rev. Genet. 43: 265–285.

Christopher, J.T., S.B. Powles, D.R. Liljegren and J.A.M. Holtum. 1991. Cross resistance to herbicides in annual ryegrass (*Lolium rigidum*). II. Chlorsulfuron resistance involves a wheat-like detoxification system. Plant Physiol. 95: 1036–1043.

Cotterman, J.C. and L.L. Saari. 1992. Rapid metabolic inactivation is the basis for cross-resistance to chlorsulfuron in diclofop-methyl-resistant rigid ryegrass (*Lolium rigidum*) biotype SR2/84. Pestic. Biochem. Physiol. 43: 182–192.

Croteau, R. 1992. Clomazone does not inhibit the conversion of isopentenyl pyrophosphate to geranyl, farnesyl, or geranylgeranyl pyrophosphate *in vitro*. Plant Physiol. 98: 1515–1517.

Culpepper, A.S., T.L. Grey, W.K. Vencill, J.M. Kichler, T.M. Webster, S.M. Brown, A.C. York, J.W. Davis and W.W. Hanna. 2006. Glyphosate-resistant Palmer amaranth (*Amaranthus palmeri*) confirmed in Georgia. Weed Sci. 54: 620–626.

Délye, C.J., J.A.C. Gardin, K. Boucansaud, B. Chauvel and C. Petit. 2011. Non-target-site-based resistance should be the center of attention for herbicide resistance research: *Alopecurus myosuroides* as an illustration. Weed Res. 51: 433–437.

Délye, C., A. Matejicek and S. Michel. 2008. Cross-resistance patterns to ACCase-inhibiting herbicides conferred by mutant ACCase isoforms in *Alopecurus myosuroides* Huds. (black-grass) re-examined at the recommended herbicide field rate. Pest Manage. Sci. 64: 1179–1186.

Délye, C. 2005. Weed resistance to acetyl coenzyme A carboxylase inhibitors: an update. Weed Sci. 53: 728–746.

Délye, C., X.Q. Zhang, S. Michel, A. Matejicek and S.B. Powles. 2005. Molecular bases for sensitivity to acetyl-coenzyme A carboxylase inhibitors in black-grass. Plant Physiol. 137: 794–806.

Délye, C., C. Straub, A. Matjicek and S. Miche. 2004. Multiple origins for black-grass (Huds) target-site-based resistance to herbicides inhibiting acetyl-CoA carboxylase. Pest Manage. Sci. 60: 35–41.

Délye, C., T.Y. Wang and H. Darmency. 2002. An isoleucine-leucine substitution in chloroplastic acetyl-CoA carboxylase from green foxtail (*Setaria viridis* L. Beauv.) is responsible for resistance to the cyclohexanedione herbicide sethoxydim. Planta 214: 421–427.

De Prado, R., J.L. De Prado and J. Menéndez. 1997. Resistance to substituted urea herbicide in *Lolium rigidum* biotypes. Pestic. Biochem. Physiol 57: 126–136.

Devine, M.D. 1997. Mechanisms of resistance to acetyl-coenzyme A carboxylase inhibitors: A Review. Pestic. Sci. 51: 259–264.

Devlin, B. and N.C. Ellstrand. 1990. The development and application of a refined method for estimating gene flow from angiosperm paternity analysis. Evolution 44: 248–259.

DeVore, J.D., J.K. Norsworthy, D.B. Johnson, C.E. Starkey, M.J. Wilson and G.M. Griffith. 2011. Palmer amaranth emergence as influenced by soybean production system and deep tillage. Proc. South. Weed Sci. Soc. 64: 239.

Dyer, W.E., G.A. Goss and P. Buck. 2002. Gravitropic responses of auxinic herbicide-resistant *Kochia scoparia*. Ann. Meet. Weed Sci. Soc. Amer. Abstr. 171. p. 49.

Eberlein, C.V., M.J. Guttieri, C.A. Mallory-Smith, D.C. Thill and R.J. Baerg. 1997. Altered acetolactate synthase activity in ALS-inhibitor resistant prickly lettuce (*Lactuca serriola*). Weed Sci. 45: 212–217.

Estelle, M.A. and C. Somerville. 1987. Auxin-dependent mutants of *Arabidopsis thaliana* with an altered morphology. Mol. J. Gent. 206: 200–206.

Etheredge, L.M., Jr., J.L. Griffin and M.E. Salassi. 2009. Efficacy and economics of summer fallow conventional and reduced-tillage programs for sugarcane. Weed Technol. 23: 274–279.

Franz, J.E., M.K. Mao and J.A. Sikorski. 1997. Glyphosate: A Unique and Global Herbicide. ACS Monograph No. 189. Amer. Chem. Soc., Washington, DC, 653 pp.

Fryer, J.D. 1980. Foreword. pp. 1–3. *In*: C. Kirby (ed.). The Hormone Weed Killers. A short history of their discovery and development. Croyden, Great Britain: Brit. Crop Prot. Coun.

Gadamski, G., D. Ciarka, J. Gressel and S. Gawronski. 2000. Negative cross-resistance in triazine-resistant biotypes of *Echinochloa crus-galli* and *Conyza canadensis*. Weed Sci. 48: 176–180.

Gaines, T.A., D.L. Shaner, S.M. Ward, J.E. Leach, C. Preston and P. Westra. 2011. Mechanism of resistance of evolved glyphosate-resistant Palmer Amaranth (*Amaranthus palmeri*). J. Agric. Food Chem. 59: 5886–5889.

Gaines, T.A., W. Zhang, D. Wang, B. Bukun, S.T. Chisholm, D.L. Shaner et al. 2010. Gene amplification confers glyphosate resistance in *Amaranthus palmeri*. Proc. Natl. Acad. Sci. USA 107: 1029–1034.

Gassmann, A.J. 2005. Resistance to herbicide and susceptibility to herbivores: environmental variation in the magnitude of an ecological trade-off. Oecologia 145: 575–585.

Ge, X., D.A. d'Avignon, J.J. Ackerman and R.D. Sammons. 2010. Rapid vacuolar sequestration: the horseweed glyphosate resistance mechanism. Pest Manage. Sci. (DOI) 10.1002/ps.1911.

Giddings, G. 2000. Modeling the spread of pollen from *Lolium perenne*. The implications for the release of wind-pollinated transgenics. Theor. Appl. Genet. 100: 971–974.

Giddings, G.D., N.R. Sackville Hamilton and M.D. Hayward. 1997a. The release of genetically modified grasses, Part 1: pollen dispersal to traps in *Lolium perenne*. Theor. Appl. Genet. 94: 1000–1006.

Giddings, G.D., N.R. Sackville Hamilton and M.D. Hayward. 1997b. The release of genetically modified grasses. Part 2: the influence of wind direction on pollen dispersal. Theor. Appl. Genet. 94: 1007–1014.

Gleason, C., R.C. Foley and K.B. Singh. 2011. Mutant analysis in *Arabidopsis* provides insight into the molecular mode of action of the auxinic herbicide dicamba. PLoS One 6(3): e17245.

Goss, G.A. and W.E. Dyer. 2003. Physiological characterization of auxinic herbicide-resistant biotypes of kochia (*Kochia scoparia*). Weed Sci. 51(6): 839–844.

Govindjee and W. Veit. 2010. Z-Scheme of Electron Transport in Photosynthesis. http://www.life.illinois.edu/ govindjee/Z-Scheme.html.

Gressel, J. 2009. Evolving understanding of the evolution of herbicide resistance. Pest Manag. Sci. 65: 1164–1173.

Gressel, J. and A.A. Levy. 2006. Agriculture: the selector of improbable mutations. Proc. Natl. Acad. Sci. USA 103: 12215–12216.

Gressel, J. 2002. Transgenic herbicide-resistant crops–advantages, drawbacks and failsafes. pp. 597–634. *In*: K.-M. Oksman-Caldentey and W.H. Barz (eds.). Plant Biotechnology and Transgenic Plants. Marcel Dekker, New York.

Gressel, J. and L.A. Segel. 1990. Modeling the effectiveness of herbicide rotation and mixtures as strategies to delay or preclude resistance. Weed Technol. 4: 186–198.

Gressel, J. and L.A. Segel. 1982. Interrelating factors controlling the rate of appearance of resistance: the outlook for the future. pp. 325–447. *In*: H.M. LeBaron and J. Gressel (eds.). Herbicide Resistance in Plants. John Wiley and Sons, New York.

Gronwald, J.W., C.V. Eberlein, K.J. Betts, R.J. Baerg, N.J. Ehlke and D.L. Wyse. 1992. Mechanism of diclofop resistance in an Italian ryegrass (*Lolium multiflorum* Lam.) biotype. Pesticide Biochem. Physiol. 44: 126–139.

Gronwald, J.W. 1991. Lipid biosynthesis inhibitors. Weed Sci. 39: 435–449.

Grossman, K. 2010. Auxin herbicides: current status of mechanism and mode of action. Pest Manage. Sci. 66: 113–120.

Grossman, K. 2000. The mode of action of auxin herbicides: a new ending to a long, drawn out story. Trends Plant Sci. 5: 506–508.

Guilfoyle, T. 2007. Sticking with auxin. Nature 446: 621–622.

Guskov, A.J. Kern, A. Gabdulkhakov, M. Broser, A. Zouni and W. Saenger. 2009. Cyanobacterial photosystem II at 2.9 Å resolution and the role of quinones, lipids, channels and chloride. Nat. Struct. Mol. Biol. 16(3): 334–42.

Hall, J.C., S.M.M. Alam and D.P. Murr. 1993. Ethylene biosynthesis following foliar application of picloram to biotypes of wild mustard (*Sinapis arvensis* L.) susceptible or resistant to auxinic herbicides. Pestic. Biochem. Physiol. 47: 36–43.

Halsey, M.E., K.M. Remund, C.A. Davis, M. Qualis, P.J. Eppard and S.A. Berberich. 2005. Isolation of maize from pollen-mediated gene flow by time and distance. Crop Sci. 45: 2172–2185.

Heap, I. 2014. International Survey of Herbicide Resistant Weeds. Online. Internet. 10 October. Available www.weedscience.com.

Heap, I. and H. LeBaron. 2001. Introduction and overview of resistance. pp. 1–22. *In*: S.B. Powles and D.L. Shaner (eds.). Herbicide Resistance and World Grains. CRC Press, Boca Raton, FL.

Heap, I.M. and I.N. Morrison. 1992. Resistance to auxin-type herbicides in wild mustard (*Sinapis arvensis* L.) populations in western Canada. Annul. Meet. Weed Sci. Soc. Amer. Abstr. 32: 164.

Heidel, A.J., J.D. Clarke, J. Antonovics and X. Dong. 2004. Fitness costs of mutations affecting the systemic acquired resistance pathway in *Arabidopsis thaliana*. Genetics 168: 2197–2206.

Heim, D.R. and I.M. Larrinua. 1989. Primary site of action of amitrole in *Arabidopsis thaliana* involves inhibition of root elongation but not histidine or pigment biosynthesis. Plant Physiol. 91: 1226.

Herbicide Resistance Action Committee. 2009. http://www.hracglobal.com/ Glossary/tabid /369/ Default. aspx.

Hilton, H.W. 1957. Herbicide tolerant strains of weeds. Hawaiian Sugar Plant.Assoc. Ann. Rep., p. 69.

Hirschberg, J., A. Bleeckes, D.J. Kyle, L. McIntosh and C.J. Arntzen. 1984. The molecular basis of triazine resistance in higher plant chloroplasts. Z. Naturforch. 39c: 412–419.

Hirschberg, J. and L. McIntosh. 1983. Molecular basis for herbicide resistance in *Amaranthus hybridus*. Science 22: 1346–1349.

Hoagland, R.E., J.K. Norsworthy, F. Carey and R.E. Talbert. 2004. Metabolically based resistance to the herbicide propanil in *Echinochloa* species. Weed Sci. 52(3): 475–486.

Holt, J.S., S.B. Powles and J.A.M. Holtum. 1993. Mechanisms and agronomic aspects of herbicide resistance. Ann. Rev. Plant Physiol. Plant Mol. Biol. 44: 203–209.

Holtum, J.A.M. and S.B. Powles. 1991. Annual Ryegrass: an abundance of resistance, a plethora of mechanisms. Brighton Crop Protec. Conf. – Weeds (Farnham, U.K.: British crop Protection Council), Weeds 1071–1078.

Holtum, J.A.M., J.M. Matthews, D.R. Liljegren and S.B. Powles. 1991. Cross-resistance to herbicides in annual ryegrass (*Lolium rigidum*). III. On the mechanism of resistance to diclofop-methyl. Plant Physiol. 97: 1026–1034.

Hwang, I.T., K.H. Lee, S.H. Park, B.H. Lee, K.S. Hong, S.S. Han and K.Y. Cho. 2001. Resistance to acetolactate synthase inhibitors in a biotype of *Monochoria vaginalis* discovered in Korea. Pestic. Biochem. Physiol. 71: 69–76.

Intanon, S., A. Perez-Jones, A.G. Hulting and C.A. Mallory-Smith. 2011. Multiple Pro197 ALS Substitutions Endow Resistance to ALS Inhibitors within and among Mayweed Chamomile Populations. Weed Sci. 59: 431–437.

Jansen, M.A.K., C. Malan, Y. Shaalitiel and J. Gressel. 1990. Mode of evolved photooxidant resistance to herbicides and xenobiotics. Z. Naturforsch. 45c: 463–469.

Jasieniuk, M., A.L. Brule-Babel and I.N. Morrison. 1996. The evolution and genetics of herbicide resistance in weeds. Weed Sci. 44: 176–193.

Jasieniuk, M., I.N. Morrison and A.L. Brule-Babel. 1995. Inheritance of dicamba resistance in wild mustard (*Brassica kaber*). Weed Sci. 43: 192–195.

Jessup, C.M. and B.J.M. Bohannan. 2008. The shape of an ecological trade-off varies with environment. *Ecology Letters* 11: 1–13.

Jugulam, M., M.D. McLean and J.C. Hall. 2005. Inheritance of picloram and 2,4-D resistance in wild mustard (*Brassica kaber*). Weed Sci. 53: 417–423.

Kaundun, S.S., R.P. Dale and G.C. Bailly. 2012. Molecular basis of resistance to herbicides inhibiting acetolactate synthase in two rigid ryegrass (*Lolium rigidum*) populations from Australia. Weed Sci. 60: 172–178.

Keese, R.J. and N.D. Camper. 1994. Uptake and translocation of [^{14}C] MSMA in cotton and MSMA-resistant and -susceptible cocklebur. Pestic. Biochem. Physiol. 49: 138–143.

Kelley, K.B. and D.E. Riechers. 2007. Recent developments in auxin biology and new opportunities for auxinic herbicide research. Pestic. Biochem. Physiol. 89: 1–11.

Kelly, K.B., K.N. Lambert, A.G. Hager and D.E. Riechers. 2004. Quantitative expression analysis of *GH3*, a gene induced by plant growth regulator herbicides in soybean. J. Agric. Food Chem. 52: 474–478.

Kimber, G. and E.R. Sears. 1987. Evolution of the gene *Triticum* and the origin of cultivation of wheat. pp. 164–165. *In*: E.G. Heyne (ed.). Wheat and Wheat Improvement. Agronomy Monograph, No. 13. Madison, WI: ASA, CGGA, SSSA.

Leah, J.M., J.C. Caseley, C.R. Riches and B. Valverde. 1994. Association between elevated activity of aryl acylamidase and propanil resistance in jungle-rice, *Echinochloa colona*. Pesticide Sci. 42: 281–289.

LeBaron, H.M. and M. McFarland. 1990. Herbicide resistance in weeds and crops: an overview and prognosis. pp. 336–352. *In*: M.B. Green, H.M. LeBaron and W.K. Moberg (eds.). Managing Resistance to Agrochemicals: From Fundamental Research to Practical Strategies. Amer. Chem. Soc., Washington, D.C.

Lee, K.Y., J. Townsend, J. Tepperman, M. Black, C.F. Chui, B. Mazur et al. 1988. The molecular basis of sulfonylurea herbicide resistance in tobacco. EMBO J. 7: 1241–1248.

Lee, R.M., A.G. Hager and P.J. Tranel. 2008. Prevalence of novel mechanism to PPO-inhibiting herbicides in waterhemp (*Amaranthus tuberculatus*). Weed Sci. 56: 371–375.

Levin, D.A. 1981. Dispersal versus gene flow in plants. Ann. Mo. Bot. Gard. 68: 233–253.

Li, J., R.J. Smeda, K.A. Nelson and F.E. Dayan. 2004. Physiological basis for resistance to diphenyl ether herbicides in common waterhemp (*Amaranthus rudis*). Weed Sci. 52: 333–338.

Lincoln, C., J.H. Britton and M. Estelle. 1990. Growth and development of the *axr*1 mutants of *Arabidopsis*. Plant Cell 2: 1071–1076.

Liu, W.J., D.K. Harrison, D. Chalupska, P. Gornicki, C.C. O'Donnell, S.W. Adkins et al. 2007. Single-site mutations in the carboxyltransferase domain of plastid acetyl-CoA carboxylase confer resistance to grass-specific herbicides. Proc. Natl. Acad. Sci. USA 104: 3627–3632.

Ludwig-Müller, J. 2011. Auxin conjugates: their role for plant development and in the evolution of land plants. J. Expt. Bot. 62: 1757–1773.

Mackenzie, R., A.M. Mortimer, P.D. Putwain, I.B. Bryan and T.R. Hawkes. 1995. The inheritance of chlorsulfuron resistance in perennial ryegrass: strategic implications for management of resistance. Brighton Crop Protection Conf: weeds. Proc. Internatl. Conf., Brighton, UK, 20–23 Nov. 1995. 2: 769–774.

Mallory-Smith, C.A. and E.S. Olguin. 2011. J. Agric. Food Chem. 59(11): 5813–5818.

Mallory-Smith, C.A. and M. Zapiola. 2008. Gene flow from glyphosate-resistant crops. Pest Manag. Sci. 64: 428–440.

Mallory-Smith, C.A. and E.J. Retzinger, Jr. 2003. Revised classification of herbicides by site of action for weed resistance management strategies. Weed Technol. 17: 605–619.

Mallory-Smith, C.A., D.C. Thill and M.J. Dial. 1990a. Identification of sulfonylurea herbicide-resistant prickly lettuce (*Lactuca serriola*). Weed Technol. 4: 163–168.

Mallory-Smith, C.A., D.C. Thill, M.J. Dial and R.S. Zemetra. 1990b. Inheritance of sulfonylurea herbicide resistance in *Lactuca* spp. Weed Technol. 4: 787–790.

Manalil, S., R. Busi, M. Renton and S.B. Powles. 2011. Rapid evolution of herbicide resistance by low herbicide dosages. Weed Sci. 59: 210–217.

Marsh, S.P., R.S. Llewellyn and S.B. Powles. 2006. Social Costs of Herbicide Resistance: The Case of Resistance to Glyphosate. Proc. 50th Ann. Conf. of the Australian Agri. and Resource Econ. Soc., Manly Pacific Sydney, Australia.

Martin, G. and T. Lenormand. 2006. The fitness effect of mutations across environments: a survey in light of fitness landscape models. Evolution 60: 2413–2427.

Matthes, B. and P. Böger. 2002. Chloroacetamides affect the plasma membrane. Zeitschrift fur Naturforschung Teil C Biochemie Biophysik Biologie Virologie 57: 843–852.

Matthews, J.M. and S.B. Powles. 1992. Aspects of population dynamics of selection for herbicide resistance in *Lolium rigidum* (Gaud.). Proc. Intnl. Weed Cont. Cong., Melbourne. Weed Sci. Soc. Victoria, Australia, pp. 318–320.

Matthews, J.M., J.A.M. Holtum, D.R. Liljegren, B. Furness and S.B. Powles. 1990. Cross-resistance to herbicide in annual ryegrass (*Lolium rigidum*). I. Properties of the herbicide target enzyme acetyl-coenzymeA carboxylase and acetolactate synthase. Plant Physiol. 94: 1180–1186.

Maxwell, B.D. and A.M. Mortimer. 1994. Selection for herbicide resistance. pp. 1–25. *In*: S.B. Powles and J.A.M. Holtum (eds.). Herbicide Resistance in Plants. Lewis Publishers, Ann. Arbor., Michigan, USA.

Maxwell, B.D. 1992. Predicting gene flow from herbicide resistant weeds in annual agriculture systems. Bull. Ecol. Soc. Am. Abstr. 73: 264.

Maxwell, B.D., M.L. Roush and S.R. Radosevich. 1990. Predicting the evolution and dynamics of herbicide resistance in weed populations. Weed Technol. 4: 2–13.

Mazur, B.J. and S.C. Falco. 1989. The development of herbicide resistant crops. Ann. Rev. Plant Physiol. Plant Mol. Biol. 40: 441–470.

McCall, B. 1954. Are our weeds becoming more resistant to herbicides? Hawaiian Sugar Technol. Rep. p. 146.

Menendez, J., F. Bastida and R. de Prado. 2006. Resistance to chlortoluron in a downy brome (*Bromus tectorum*) biotype. Weed Sci. 54: 237–245.

Menne, H. and H. Köcher. 2012. HRAC classification of herbicides and resistance development. pp. 5–28. *In*: W. Kr̈amer, U. Schirmer, P. Jeschke and M. Witschel. (eds.). Modern Crop Protection Compounds, (2/ed.). Wiley-VCH Verlag GmbH & Co. KGaA.

Merotto, A. Jr., M. Jasieniuk, M.D. Osuna, F. Vidotto, A. Ferrero and A.J. Fischer. 2009. Cross-resistance to herbicides of five ALS-inhibiting groups and sequencing of the ALS gene in *Cyperus difformis* L. J. Agric. Food Chem. 57: 1389–1398.

Merrell, D.J. 1981. Ecological Genetics. Univ. of Minn. Press, Minneapolis, Minnesota, USA, Ph.D. Thesis. 512 pp.

Michael, P.J., M.J. Owen and S.B. Powles. 2010. Herbicide-resistant weed seeds contaminate grain sown in the Western Australian Grainbelt. Weed Sci. 58: 466–472.

Michel, A., R.S. Arias, B.E. Schefler, S.O. Duke, M. Netherland and F.E. Dayan. 2004. Somatic mutation-mediated evolution of herbicide resistance in the nonindigenous invasive plant hydrilla (*Hydrilla verticillata*). Molecular Ecol. 13: 3229–3237.

Mithila, J., C. Hall, W.G. Johnson, K.B. Kelley and D.E. Riechers. 2011. Evolution of resistance to auxinic herbicides: historical perspectives, mechanisms of resistance, and implications for broadleaf weed management in agronomic crops. Weed Sci. 59: 445–457.

Morrison, I.N. and M.D. Devine. 1994. Herbicide resistance in the Canadian prairie provinces: five years after the fact. Phytoprotection 75(Suppl.): 5–16.

Mortimer, A.M. 1993. A review of graminicde resistance. Monograph No. 1. Herbicide Research Action Committee–Graminicide Working Group.

Moss, S.R., K.M. Cocker, A.C. Brown, L. Hall and L.M. Field. 2003. Characterization of target-site resistance to ACCase-inhibiting herbicides in the weed *Alopecurus myosuroides* (blackgrass). Pest Manage. Sci. 59: 190–201.

Mueller, T.C., P.D. Mitchell, B.G. Young and A.S. Culpepper. 2005. Proactive versus reactive management of glyphosate-resistant or tolerant weeds. Weed Technol. 19: 924–933.

Mulugeta, D., P.K. Fay and W.E. Dyer. 1992. The role of pollen in the spread of sulfonylurea resistant *Kochia scoparia* L. (Schrad.). Weed Sci. Soc. Amer. Abstr. 32: 16.

Nandula, V.K., K.N. Reddy, C.H. Koger, D.H. Poston, A.M. Rimando, S.O. Duke et al. 2012. Multiple resistance to glyphosate and pyrithiobac in Palmer Amaranth (*Amaranthus palmeri*) from Mississippi and response to flumiclorac. Weed Sci. 60: 179–188.

Neighbors, S. and L.S. Privalle. 1990. Metabolism of primisulfuron by barnyardgrass. Pest. Biochem. Physiol. 37: 145–153.

Neve, P., J.K. Norsworthy, K.L. Smith and I.A. Zelaya. 2011a. Modelling evolution and management of glyphosate resistance in *Amaranthus palmeri*. Weed Res. 51: 99–112.

Neve, P., J.K. Norsworthy, K.L. Smith and I.A. Zelaya. 2011b. Modelling glyphosate resistance management strategies for Palmer amaranth in cotton. Weed Technol. 25: 335–343.

Neve, P. 2008. Simulation modeling to understand the evolution and management of glyphosate resistance in weeds. Pest Manag. Sci. 64: 392–401.

Neve, P. and S.B. Powles. 2005a. Recurrent selection with reduced herbicide rates results in the rapid evolution of herbicide resistance in *Lolium rigidum*. Theor. Appl. Genet. 110: 1154–1166.

Neve, P. and S.B. Powles. 2005b. High survival frequencies at low herbicide use rates in populations of *Lolium rigidum* result in rapid evolution of herbicide resistance. Heredity 95: 485–492.

Neve, P., A.J. Diggle, F.P. Smith and S.B. Powles. 2003. Simulating evolution of glyphosate resistance in *Lolium rigidum* 1: population biology of a rare resistance trait. Weed Res. 43: 404–417.

Nikolau, B.J., J.B. Ohlrogge and E.S. Wurtele. 2003. Plant biotin-containing carboxylases. Arch. Biochem. Biophys. 414: 211–222.

Nimbal, C.I., J.J. Hettholt, D.R. Shaw and S.O. Duke. 1995a. Photosynthetic performance of MSMA-resistant and -susceptible Mississippi biotypes of common cocklebur. Pestic. Biochem. Physiol. 53: 129–137.

Nimbal, C.I., G.D. Wills, S.O. Duke and D.R. Shaw. 1995b. Uptake, translocation, and metabolism of ^{14}C-MSMA in organic arsenical-resistant and -susceptible Mississippi biotypes of common cocklebur (*Xanthium strumarium*). Weed Sci. 43: 549–554.

Norsworthy, J.K., T. Griffith, T. Griffin, M. Bhagavathiannan and E.E. Gbur. 2014. In-field movement of glyphosate-resistant palmer amaranth (*Amaranthus palmeri*) and its impact on cotton lint yield: evidence supporting a zero-threshold strategy. Weed Sci. 62(2): 237–249.

Norsworthy, J.K., S.M. Ward, D.R. Shaw, R.S. Llewellyn, R.L. Nichols, T.M. Webster et al. 2012. Reducing the risks of herbicide resistance: Best management practices and recommendations. Weed Sci. 60 (sp. 1 issue): 31–62.

Norsworthy, J.K., M.V. Bagavathiannan, P. Neve, K. Smith and I. Zelaya. 2011. Integrating nonchemical practices into simulation modeling for herbicide resistance. A proactive strategy. Proc. Ann. Weed Sci. Soc. Amer. Abstract 226.

Norsworthy, J.K., N.R. Burgos, R.C. Scott and K.L. Smith. 2007. Consultant perspectives on weed management needs Arkansas rice. Weed Technol. 21: 832–839.

Özgür, K. 2011. Aclonifen: The identikit of a widely used herbicide. African J. Agri. Res. 6: 2411–2419.

Padro, R.D., E. Romera and J. Menendaz. 1995. Atrazine detoxification in *Panicum dichotomiflorum* and target site in *Polygonum lapathifolium*. Pestic. Biochem. Physiol. 52: 1–11.

Patzoldt, W.L. and P.J. Tranel. 2001. ALS mutations conferring herbicide resistance in waterhemp. Proc. N. Cent. Weed Sci. Soc. 56: 67.

Peniuk, M.G., M.L. Romano and J.C. Hall. 1993. Physiological investigations into the resistance of wild mustard (*Sinapis arvensis* L.) biotype to auxinic herbicides. Weed Res. 33: 431–440.

Perry, D.H., J.S. McElroy, F. Dane, E. van Santen and R.H. Walker. 2012. Triazine-resistant annual bluegrass (*Poa annua*) populations with Ser$_{264}$ mutation are resistant to amicarbazone. Weed Sci. 60: 355–359.

Pline-Srnic, W. 2006. Physiological mechanisms of glyphosate resistance. Weed Technol. 20: 290–300.

Powles, S.B. 2010. Gene amplification delivers glyphosate-resistant weed evolution. Proc. Natl. Acad. Sci. USA 107: 955–956.

Powles, S.B. and Q. Yu. 2010. Evolution in action: plants resistant to herbicides. Ann. Rev. Plant Biol. 61: 317–347.

Powles, S.B. and C. Preston. 2006. Evolved glyphosate resistance in plants: biochemical and genetic basis of resistance. Weed Technol. 20(2): 282–289.

Powles, S.B. and C. Preston. 1995. Herbicide cross-resistance and multiple resistance in plants. Monograph No. 2. The Herbicide Resistance Action Committee. Univ. of Adelaide.

Prather, L.A., C. Ferguson and R.K. Jansen. 2000. Polemoniaceae phylogeny and classification: implications of sequence data from the chloroplast gene ndhF. Am. J. Bot. 87: 1300–1308.

Preston, C.D., D.S. Belles, P.H. Westra, S.J. Nissen and S.M. Ward. 2009. Inheritance of resistance to the auxinic herbicide dicamba in kochia (*Kochia scoparia*). Weed Sci. 57: 43–47.

Preston, C. and A.M. Wakelin. 2008. Resistance to glyphosate from altered herbicide translocation patterns. Pest Manage. Sci. 64: 372–376.

Preston, C. and S.B. Powles. 1998. Amitrole inhibits diclofop metabolism and synergises diclofop-methyl in a diclofop-methyl-resistant biotype of *Lolium rigidum*. Pestic. Biochem. Physiol. 62: 179–189.

Preston, C., F.J. Tardif, J.T. Christopher and S.B. Powles. 1996. Multiple resistance to dissimilar herbicide chemistries in a biotype of *Lolium rigidum* due to enhanced activity of several herbicide degrading enzymes. Pestic. Biochem. Physiol. 54: 123–134.

Preston, C., J.A.M. Holtum and S.B. Powles. 1992. On the mechanism of resistance to paraquat in *Hordeum glaucum* and *H. laporium*: Delayed inhibition of photosynthetic CO_2 evolution after paraquat application. Plant Physiol. 100: 630–636.

Primiani, M.M., J.C. Holtum and L.L. Saari. 1990. Resistance of kochia (*Kochia scoparia*) to sulfonylurea and imidazolinone herbicides. Weed Technol. 4: 169–172.

Rao, V.S. 2000. Principles of Weed Science. Science Publishers, Enfield, NH, USA, 555 pp.

Rao, V.S. 1981. Some aspects of efficient use of herbicides in tea. 29th Conf. of Tocklai Tea Res. Station. Jorhat, India. 17–19 December 1981.

Raven, P.H., R.F. Evert and S.E. Eichhorn. 2005. Biology of Plants (7th ed.). W.H. Freeman and Company Publishers. New York.

Reddy, K.N. and J.K. Norsworthy. 2010. Glyphosate-resistant crop production systems: impact on weed species shifts. pp. 165–184. *In*: V.K. Nandula (ed.). Glyphosate Resistance in Crops and Weeds: History, Development, and Management. J. Wiley, Singapore.

Reichman, J.R., L.S. Watrud, E.H. Lee, C.A. Burdick, M.A. Bollman, M.J. Storm et al. 2006. Establishment of transgenic herbicide-resistant creeping bentgrass (*Agrostis stolonifera* L.) in nonagronomic habitats. Mol. Ecol. 15: 4243–4255.

Retzinger, E.J., Jr. and C.A. Mallory-Smith. 1997. Classification of herbicides by site of action for weed resistance management strategies. Weed Technol. 11: 384–393.

Romano, M.L., G.R. Stephenson, A. Tal and J.C. Hall. 1993. The effect of monooxygenase and glutathione S-transferase inhibitors on the metabolism of diclofop methyl and fenoxaprop-ethyl in barley and wheat. Pestic. Biochem. Physiol. 46: 181–189.

Rousonelos, S.L. 2010. Mechanism of resistance in common ragweed to PPO-inhibiting herbicides. M.S. Thesis. University of Illinois at Urbana, IL, 108 pp.

Rutherford, A.W. and P. Faller. 2003. Photosystem II: evolutionary perspectives. Philosophical Transactions of the Royal Society of London. Series B: Biological Sciences 358(1429): 245–253.

Ryan, G.F. 1970. Resistance of common groundsel to simazine and atrazine. Weed Sci. 18: 614–616.

Saari, L.L., J.C. Cotterman and D.C. Thill. 1994. Resistance to acetolactate synthase inhibiting herbicides. pp. 141–170. *In: S.B.* Powles and J.A.M. Holtum (eds.). Herbicide Resistance in Plants, Biology and Biochemistry. CRC Press Press, Boca Raton, FL, USA.

Saari, L.L., J.C. Cotterman, W.F. Smith and M.M. Primiani. 1992. Sulfonylurea herbicide resistance in common chickweed, perennial ryegrass, and Russian thistle. Pestic. Biochem. Physiol. 42: 110–118.

Saari, L.L., J.C. Cotterman and M.M. Primiani. 1990. Mechanism of sulfonylurea herbicide resistance in the broadleaf weed, *Kochia scoparia*. Plant Physiol. 93: 55–61.

Sabba, R.P., I.M. Ray, N. Lownds and T.M. Sterling. 2003. Inheritance of resistance to clopyralid and picloram in yellow starthistle (*Centaurea solstitialis*) is controlled by a single nuclear recessive gene. J. Hered. 94: 523–527.

Salisbury, F.B. and C.W. Ross. 1992. Plant Physiology (4th Edition). Wadsworth Publishing Company, Belmont, CA, USA, 682 pp.

Salzmann, D., R.J. Handley and H. Mueller-Scharer. 2008. Functional significance of triazine-herbicide resistance in defence of *Senecio vulgaris* against a rust fungus. Basic Appl. Ecol. 9: 577–587.

Sha, X.Y., S.D. Linscombe and D.E. Groth. 2007. Field evaluation of imidazolinone-tolerant clearfield rice (*Oryza sativa* L.) at nine Louisiana locations. Crop Sci. 47: 1177–1185.

Shaaltiel, Y. and J. Gressel. 1987. Kinetic analysis of resistance to paraquat in *Conyza*. Evidence that paraquat transiently inhibits leaf chloroplast reactions in resistant plants. Plant Physiol. 85: 869–871.

Shaner, D.L. 2009. Role of translocation as a mechanism of resistance to glyphosate. Weed Sci. 57: 118–123.

Shimono, Y., Y. Takiguchi and A. Konuma. 2010. Contamination of internationally traded wheat by herbicide-resistant *Lolium rigidum*. Weed Biol. Manag. 10: 219–228.

Shivrain, V.K., N.R. Burgos, M.A. Mauromoustakos, D.R. Gealy, K.L. Smith, H.L. Black and M. Jia. 2009. Factors affecting the outcrossing rate between Clearfield™ rice and red rice (*Oryza sativa*). Weed Sci. 57: 394–403.

Sibony, M., A. Michel, H.U. Haas, B. Rubin and K. Hurle. 2001. Sulfometurn-resistant *Amaranthus retroflexus*: Cross-resistance and molecular basis for resistance to acetolactate synthase–inhibiting herbicides. Weed Res. 41: 509–522.

Singh, D.V., K. Adepa and K. Misra. 2012. Mechanism of isoproturon resistance in *Phalaris minor*: *in silico* design, synthesis and testing of some novel herbicides for regaining sensitivity. J. Mol. Modeling 18: 1431–1445.

Singh, V.N., N.R. Burgos, T.M. Tseng, H.L. Black, L. Estorninos, Jr., R.A. Salas et al. 2012. Differentiation of weedy traits in ALS-resistant red rice. The 65th Ann. Meeting of the Southern Weed Sci. Soc. Conf.

Slatkin, M. 1987. Gene flow at the geographic structure of natural populations. Science 236: 787–792.

Solymosi, P. and E. Lehoczki. 1989. Characterization of a triple (atrazine-pyrazon-pyridate) resistant biotype of common lambsquarters (*Chenopodium album* L.). J. Plant Physiol. 134: 685–690.

Squire, G.R. 2005. Contribution to gene flow by seed and pollen. pp. 73–77. *In*: A. Messean (ed.). Proc. of 2nd International Conference on Coexistence of Genetically Modified and non-GM based Agricultural Supply Chains. Montpellier, France: Agropolis.

Stallings, G.P., D.C. Thill and C.A. Mallory-Smith. 1993. Pollen-mediated gene flow of sulfonylurea-resistant kochia {(*Kochia scoparia* L.) Schrad.}. Weed Sci. Soc. Amer. Abstr. 33: 60.

Stanger, C.E. and A.P. Appleby. 1989. Italian ryegrass (*Lolium multiflorum*) accessions tolerant to diclofop. Weed Sci. 37: 350–352.

Stoltenberg, D.E., J.W. Gronwald, D.L. Wyse and J.D. Burton. 1989. Effect of sethoxydim and haloxyfop on acetyl-CoA Carboxylase activity in *Festuca* species. Weed Sci. 37: 512–516.

Strauss, S.Y., J.A. Rudgers, J.A. Lau and R.E. Irwin. 2002. Direct and ecological costs of resistance to herbivory. Trends in Ecology & Evolution. 17: 278–285.

Switzer, C.M. 1957. The existence of 2,4-D resistant strains of wild carrot. Proc. of the North Eastern Weed Control Conf. 11: 315–318.

Tal, J.A., J.C. Hall and G.R. Stephenson. 1995. Non-enzymatic conjugation of fenoxaprop-ethyl with glutathione and cysteine in several grass species. Weed Res. 35: 133–139.

Tan, X., L.I.A. Calderon-Villalobos, M. Sharon, C. Zheng, C.V. Robinson, M. Estelle and N. Zheng. 2007. Mechanism of auxin perception by the TIR1 ubiquitin ligase. Nature 446: 640–645.

Tanaka, Y. 2003. Properties of acetolactate synthase from sulphonylurea-resistant *Scirpus juncoides* Roxb. Var. *ohwianus* T. Koyama. Pesti. Biochem. Physiol. 77: 147–153.

Tardif, F.J., I. Rajcan and M. Costea. 2006. A mutation in the herbicide target site acetohydroxyacid synthase produce morphological and structural alterations and reduces fitness in *Amaranths powellii*. New Phytol. 169: 251–264.

Tardif, F.J. and S.B. Powles. 1994. Herbicide multiple-resistance in a *Lolium rigidum* biotype is endowed by multiple mechanisms: isolation of a subset with resistant acetyl-CoA carboxylase. Physiol. Plant. 91: 488–494.

Tardif, F.J., J.A.M. Holtum and S.B. Powles. 1993. Occurrence of a herbicide resistant acetyl-coenzymeA carboxylase mutant in annual ryegrass (*Lolium rigidum*) selected by sethoxydim. Planta 190: 176–181.

Tian, D., M.B. Traw, J.Q. Chen, M. Kreitman and J. Bergelson. 2003. Fitness costs of R-gene-mediated resistance in *Arabidopsis thaliana*. Nature 423: 74–77.

Tranel, P.J. and T.R. Wright. 2002. Resistance of weeds to ALS-inhibiting herbicides: what have we learned? Weed Sci. 50: 700–712.

Van Eerd, L.L., G.R. Stephenson, J. Kwiatkowski, K. Grossmann and J.C. Hall. 2005. Physiological and biochemical characterization of quinclorac and resistance in a false cleavers (*Galium spurium*) biotype. J. Agric. Food Chem. 53: 1144–1151.

Vaughn, K.C. and L.P. Lehnen, Jr. 1991. Mitotic disrupter herbicides. Weed Sci. 39: 450–457.

Vaughn, K.C., M.A. Vaughn and B.J. Gossett. 1990. A biotype of goosegrass (*Eleusine indica*) with an intermediate level of dinitroaniline resistance. Weed Technol. 4: 157–162.

Vencill, W.K., R.L. Nichols, T.M. Webster, J.K. Soteres, C. Mallory-Smith, N.R. Burgos et al. 2012. Herbicide resistance: toward an understanding of resistance development and the impact of herbicide-resistant crops. Weed Sci. (Special Issue No. 1): 60: 2–30.

Vencill, W.K., T. Grey and S. Culpepper. 2011. Resistance of weeds to herbicides. pp. 585–594. *In*: A. Kortekamp (ed.). Herbicides and Environment. InTech, Available from: http://www. intechopen. com/ articles /show/title/resistance-of weeds-to-herbicides.

Vila-Aiub, M.M., P. Neve and S.B. Powles. 2009a. Fitness costs associated with evolved herbicide resistance alleles in plants. New Phytologist. 184: 751–767.

Walsh, M.J., N. Maguire and S.B. Powles. 2009b. Combined effects of wheat competition and 2,4-D amine on phenoxy herbicide resistant *Raphanus raphanistrum* populations. Weed Res. 49: 316–325.

Walsh, M.J. and S.B. Powles. 2007. Management strategies for herbicide-resistant weed populations in Australian dryland crop production systems. Weed Technol. 21: 332–338.

Walsh, M.J., S.B. Powles, B.R. Beard, B.T. Parkin and S.A. Porter. 2004. Multiple-herbicide resistance across four modes of action in wild radish (*Raphanus raphanistrum*). Weed Sci. 52: 8–13.

Walsh, T.A., R. Neal, A.O. Merlo, M. Honma, G.R. Hicks, K. Wolff et al. 2006. Mutations in an auxin receptor homolog AFB5 and in SGT1b confer resistance to synthetic picolinate auxins and not to 2,4-dichlorophenoxyacetic acid or indole-3-acetic acid in *Arabidopsis*. Plant Physiol. 142: 542–552.

Warwick, S.I., C.A. Sauder and H.J. Beckie. 2010. Acetolactate synthase (ALS) target-site mutations in ALS inhibitor-resistant Russian thistle (*Salsola tragus*). Weed Sci. 58: 244–251.

Watrud, L.S., E.H. Lee, A. Fairbrother, C. Burdick, J.R. Reichman, M. Boltman et al. 2004. Evidence for landscape-level, pollen-mediated gene flow from genetically modified creeping bentgrass with CP4 EPSPS as a marker. Proc. Natl. Acad. Sci. USA 101: 14533–14538.

Weed Science Society of America. 1998. "Herbicide resistance" and "herbicide tolerance" defined. Weed Technol. 12: 789.

Weimer, M.R., N.E. Balke and D.D. Buhler. 1992. Herbicide clomazone does not inhibit *in vitro* geranylgeranyl synthesis from mevalonate. Plant Physiol. 98: 427.

Weinberg, T., G.R. Stephenson, M.D. McLean and J.C. Hall. 2006. MCPA (4-chloro-2-ethylphenoxyacetate) resistance in hemp-nettle (*Galeopsis tetrahit* L.). J. Agric. Food Chem. 54: 9126–9134.

Went, F.W. 1926. Growth accelerating substances in coleoptile of *Avena sativa*. Proc. K. Akad. Wet. Amsterdam 30: 10–19.

White, R.J. and R.M. White. 1981. Some numerical methods for the study of genetic changes. pp. 295–341. *In*: J.A. Bishop and L.M. Cook (eds.). Genetic Consequences of Man-made Change. Academic Press, USA.

Wiersma, P.A., M.G. Schmiemann, J.A. Condie, W.L. Crosby and M.M. Moloney. 1989. Isolation, expression, and phylogenetic inheritance of an acetolactate synthase gene from *Brassica napus*. Mol. Gen. Genet. 219: 413–420.

Woo, J.C. 1935. Genome analysis in *Brassica* with special reference to the experimental formation of *Brassica napus* and molecular mode of fertilization. Japan. J. Bot. 7: 389–452.

Wrubel, R.P. and J. Gressel. 1994. Are herbicide mixtures useful for delaying the rapid evolution of resistance? A case study. Weed Technol. 8: 635–648.

Yu, Q., M.S. Ahmad-Hamdani, H. Han, M.J. Christoffers and S.B. Powles. 2013. Herbicide resistance-endowing ACCase gene mutations in hexaploid wild oat (*Avena fatua*): insights into resistance evolution in a hexaploid species. Heredity 110(3): 220–231.

Yu, Q., H. Han, M. Martin, Vila-Aiub and S.B. Powles. 2010. AHAS herbicide resistance endowing mutations: effect on AHAS functionality and plant growth. J. Exp. Botany. 61: 3925–3934.

Yu, Q., I. Abdallah, H. Han, M. Owen and S.B. Powles. 2009. Distinct non-target site mechanisms endow resistance to glyphosate, ACCase and ALS-inhibiting herbicides in multiple herbicide-resistant *Lolium rigidum*. Planta 230: 713–723.

Yu, Q., A. Collavo, M.-Q. Zheng, M. Owen, M. Sattin and S.B. Powles. 2007. Diversity of acetyl-coenzymeA carboxylase mutations in resistant *Lolium* populations: Evaluation using clethodim. Plant Physiol. 145: 547–558.

Zagnitko, O., J. Jelenska, G. Tevzadze, R. Haselkorn and P. Gornicki. 2001. An isoleucine/leucine residue in the carboxyltransferase domain of acetyl-CoA carboxylase is critical for interaction with aryloxyphenoxypropionate and cyclohexanedione inhibitors. Proc. Natl. Acad. Sci. USA 98: 6617–6622.

Zeng, L. and Wm. Vance Baird. 1999. Dinitroaniline herbicides in an "Intermediate" Resistant Biotype of *Eleucine indica* (Poaceae). Amer. J. Botany. 86: 940–947.

Zhang, H., B. Tweel and L. Tong. 2004. Molecular basis for the inhibition of the carboxyltransferase domain of acetyl-coenzymeA carboxylase by haloxyfop and diclofop. Proc. Natl. Acad. Sci. USA 101: 5910–5915.

CHAPTER 3

Gene, Genome, and Crop Improvement

Modern genetics, the branch of biology that deals with heredity, began with Gregor Mendel (1822–84), the Augustian monk from Brünn, Austria. His pioneering work of several years studying genetic traits in garden pea (*Pisum sativum*) plants was first presented under the German title "*Versushe über Pflanzen-Hybriden*" (Experiments on Plant Hybridization) to the Natural History Society of Brünn in two lectures on 8 February and 8 March, 1865 [Mendel 1866]. This work, based on his testing of 29,000 pea plants between 1856 and 1863 (published in the Proceedings of the Natural History Society of Brünn in 1866) ultimately led to the science of genetics in later years. When he cross-bred pea plants over many generations, many of them showed different traits (tall and short plants; round and wrinkly peas; purple and white flowers; terminal-positioned and axil-positioned flowers; yellow and green seeds) which are heritable. However, his work was largely ignored by biologists of the time, and even Mendel himself seemed apprehensive of the utility of his findings.

Mendel made three important conclusions from his experimental results: a) the inheritance of each trait is determined by 'units' or 'factors' that are passed on to descendants unchanged (these units were later called '**genes**'), b) an individual inherits one such unit from each parent for each trait, and c) a trait may not show up in an individual but can be passed on to the next generation. The trait, later called 'Mendelian trait', is controlled by a single locus. The two principles that arose from his work included the "law of segregation" and the "law of independent assortment". These two principles of inheritance, along with the understanding of unit inheritance and dominance, were the beginnings of modern genetics.

According to the theory of Mendelian inheritance, variations in phenotype (observable physical and behavioral characteristics of a plant) are due to variations in genotype, or the plant's particular set of genes, each of which specifies a particular trait. Different forms of a gene, which may give rise to different phenotypes, are known as alleles. Many plants, including pea, have two alleles for each trait, one inherited from each parent. Alleles may be dominant or recessive. Dominant alleles give rise to their corresponding phenotypes when paired with another copy of the same allele. For example, if the allele specifying tall stems in pea plants is dominant over the allele specifying short stems, then pea plants that inherit one tall allele from one parent and one short allele from the other parent will also have tall stems. Mendel's work demonstrated that alleles assort independently in the production of gametes, or germ cells, ensuring variation in the next generation.

In 1889, the Dutch botanist Hugo DeVries published his book *Intracellular Pangenesis* (pan: 'whole'; genesis: 'birth') which was based on a modified version of Charles Darwin's 1868 "Theory of Pangenesis", a hypothetical mechanism for heredity. He postulated that different characters have different hereditary carriers. He specifically postulated that inheritance of specific traits in organisms comes in particles. He called these particles '*pangenes*' [Stamhuis et al. 1999]. In 1900, De Vries, known for his 'mutation theory', became aware of Mendel's work and he altered some of his own terminology (dominance and recessiveness, and independent assortment explaining the 3:1 ratio of phenotypes in the second generation) to match Mendel's. However, he neglected to mention and acknowledge Mendel's work, but after criticism by the German geneticist Carl Correns he conceded Mendel's priority. Correns restated Mendel's results and his laws of segregation and independent assortment. Thus, Mendel's 'Laws of Heredity' came to be rediscovered universally after his death [Corcos and Monaghan 1987].

In 1909, the word '*pangenes*' was shortened to '**genes**' ('gen' in Danish and German) by the Danish botanist Wilhelm Johannsen who coined the terms 'phenotype' and 'genotype' in his paper "*Om Om arvelighed i samfund og i rene linier*" [Johannsen 1903] and book "*Arvelighedslærens Elementer*" [Johannsen 1905]. The related word "genetics" (from the Greek 'gennō', '*γεννώ*'; to give birth) was first used in 1905 by William Bateson, the English geneticist and outspoken Mendelian, to describe the study of heredity and biological inheritance. Thus was born the discipline of genetics, the study of inheritance and the science of variation.

In 1910, Thomas Morgan, the American biologist and Nobel laureate (1933) showed that genes reside on specific chromosomes. He and his students began the first chromosomal map of the fruitfly *Drosophila melanogaster*. They eventually elucidated many basic principles of heredity, including sex-linked inheritance, epistasis, multiple alleles, and gene mapping. These "drosophilists" extended Mendel's work by describing X-linked inheritance and by showing that genes located on the same chromosome do not show independent assortment. In 1913, Alfred Sturtevant, biologist at Caltech, California, constructed the first genetic map of a chromosome. This classical method of chromosome mapping that is being used today, shows genes are arranged on chromosomes in a linear fashion, like beads on a necklace. He also showed that the gene for any specific trait was in a fixed location (locus) [Lewis 1961].

A series of subsequent discoveries led to the realization that chromosomes within cells are the carriers of genetic material, and they are made of deoxyribonucleic acid (DNA), a polymeric molecule found in all cells on which the 'discrete units' of Mendelian inheritance are encoded. DNA was discovered in 1869 by Freiderich Miescher of the University of Basel, Tübingen, Switzerland. In 1933, the Belgian biochemist Jean Brachet was able to show that DNA was found in chromosomes and that ribonucleic acid (RNA) is present in the cytoplasm of all cells [Pirie 1990]. In 1941, George Wells Beadle and Edward Lawrie Tatum, both American geneticists and Nobel laureates in 1958, showed that genes control steps in metabolism, and mutations in genes caused errors in specific steps in metabolic pathways. This showed that specific genes code for specific proteins, leading to the "one gene, one enzyme" hypothesis [Beadle and Tatum 1941].

Geneticists Oswald Avery (Canada-born American), Colin MacLeod (Canada-born American), and Maclyn McCarty (American) showed in 1944 that DNA holds the gene's information. In March 1953, James Watson (American molecular biologist) and Francis Crick (English molecular biologist), both 1962 Nobel laureates, demonstrated the molecular structure of DNA, which they resolved as double-helix [Watson 1968]. These discoveries established the central dogma of molecular biology, which states that proteins are translated from RNA which is transcribed from DNA and that this can be reversed. This dogma has since been shown to have exceptions, such as reverse transcription in retroviruses.

Barbara McClintock, Cornell cytogeneticist and the 1983 Nobel laureate, produced the first genetic map for maize, linking regions of the chromosome with physical traits, and demonstrated the role of the telomere and centromere, regions of the chromosome that are important in the conservation of genetic information. She also discovered "transposition" and used it to show how genes (called "jumping genes") are responsible for turning physical characteristics on and off. She also developed theories to explain repression or expression of genetic information from one generation of maize plants to the next [Comfort 2001]. Alongside genetical experimental work, mathematicians developed the statistical framework of population genetics, brining genetic explanations into the study of evolution.

In order to understand the processes related to transgenic engineering, brief descriptions of gene, genome, and their related functions and processes is given in this part of the Chapter.

Gene

As mentioned earlier, a gene is a molecular unit of heredity of a living organism. Pearson [2006] defined gene as "a locatable region of genomic sequence, corresponding to a unit of inheritance, which is associated with regulatory regions, transcribed regions, and or other functional sequence regions". Plants encode genes in long strands of DNA. Genes hold information to build and maintain the cells of an organism and pass genetic traits to offspring, although some organelles like mitochondria are self-replicating and not coded for the organisms's DNA. Plants may have many genes corresponding to various biological traits. Although genes comprise over a million base pairs (bp), they are usually much smaller, averaging around 3,000 bp (30 kb).

When proteins are synthesized, the genetic code—the set of rules by which information is encoded—is first copied or transcribed into RNA, an intermediate product. In other cases, the RNA molecules are the actual functional products. The genetic code defines each amino acid in a protein, or polypeptide, in terms of a series of specific sequences of three nucleotides, called **codons**, in the DNA. Therefore, the genetic code is called a triplet code. The codons serve as the *words* in the genetic *language*. The genetic code specifies the correspondence during protein translation between codons and amino acids. The vast majority of living organisms encode their genes in long strands of DNA. RNA molecules, which contain the base of uracil in place of thymine, are less stable than DNA and are typically single-stranded. The four different nucleotides in DNA can form 64 different triplet codons. As there are only 24 amino acids found in proteins, some amino acids are encoded by more than one codon. Three of the triplet codons that do not encode any amino acid are known as **stop codons**, which identify the end of the message (like the period '.' at the end of a sentence) encoded in genes. The genetic code is nearly universal, which means specific codons code for the same amino acids in nearly all organisms.

All genes have regulatory regions in addition to regions that explicitly code for a protein or RNA product. A regulatory region shared by almost all genes is known as the **promoter**. It provides a position that is recognized by the transcription machinery when a gene is about to be transcribed and expressed. A gene can have more than one promoter, resulting in RNAs that differ in how far they extend in the 5′ end [Mortazavi et al. 2008]. Although promoter regions have a consequence sequence that is the most common sequence at this position, some genes have 'strong' promoters that bind the transcription machinery well, and others have 'weak' promoters that bind poorly. These weak promoters usually permit a lower rate of transcription than the strong promoters. Other regulatory regions include **enhancers**, which can compensate for a weak promoter. Most regulatory regions are before or towards the 5' end of the transcription initiation site. Many eukaryotic (plants) genes are transcribed only one at a time, but may include long stretches of non-encoding DNA called **introns** which are transcribed but never translated into protein. They are spliced out before translation.

Deoxyribonucleic Acid

Deoxyribonucleic acid is the hereditary or genetic material present in all cells. It carries information for the structure and function of living things. This linear structure consists of a pair of tightly-held molecules encoded with genetic instruction. DNA, together with RNA and proteins, is one of the three major macromolecules essential for all known forms of life. It consists of two long strands entwined like vines, in the shape of a 'double helix' (Fig. 3.1A). The repeats of nucleotide contain both the segment of the backbone of the molecule, which holds the chain together, and a **nucleobase** which interacts with the other DNA strand in the helix. When a nucleobase is linked to a sugar it is called a **nucleoside** and if it is linked to a sugar and one or more phosphate groups it is known as a **nucleotide**. In a polynucleotide, multiple nucleotides are linked.

The backbone of the DNA strand is made from alternating phosphate and sugar residues [Ghosh and Bansal 2003]. The sugars, composed of 2-deoxyriboses (the monosaccharide pentose sugars), are joined together by phosphate groups that form phosphodiester bonds between the third and fifth carbon atoms of adjacent sugar rings (Fig. 3.1B). In a double helix, the direction of the nucleotides in one strand is opposite to their direction in the other strand, in an antiparallel fashion. The asymmetric ends of DNA strands are called the 5' (five prime) and 3' (three prime) ends, with the 5' end having a terminal phosphate group and the 3' end a terminal hydroxyl group (Fig. 3.1A). The major difference between DNA and RNA is the sugar, with the 2-deoxyribose in DNA being replaced by the alternative pentose sugar ribose in RNA [Berg et al. 2002]. DNA double helix is stabilized by hydrogen bonds between nucleotides and base-stacking interactions among the aromatic nucleotides [Yakovchuk et al. 2006]. The four nucleobases (or simply called bases) found in DNA are adenine (A), cytosine (C), guanine (G), and thymine (T) and these are attached to the sugar/phosphate to form the complete nucleotide as in the case of adenosine monophosphate (AMP).

Nucleobases are of two types: **purines** and **pyrimidines** (Fig. 3.1A). In purines, A and G are being fused to form five- and six-membered heterocyclic compounds. Purines, consisting of a pyrimidine ring fused with imidazole ring, are the most widely-occurring nitrogen-containing heterocycle in nature. In the pyrimidine nucleobase, the six-membered rings of C and T are held together. Uracil (U), a fifth pyrimidine nucleobase is not usually found in DNA, but it takes the place of thymine in RNA. It differs from thymine by lacking a methyl group on its ring. Uracil, though not usually found in DNA, occurs as a breakdown product of cytosine. Pyrimidines, similar to benzenes and pyridines, contain two nitrogen atoms at 1 and 3 of the six-member ring. In bacteriophages like *Bacillus subtilis* and *Bacillus thuringiensis* (*Bt*, the one used to develop insect-resistant transgenic crops), thymine is replaced by uracil.

In the DNA, double helix occurs due to complementary base pairing in which each type of nucleobase on one strand normally interacts with one type of nucleobase on the other strand. In this, purines form hydrogen bonds to pyrimidines, with A bonding only to T, and C bonding only to G. This arrangement of two nucleotides binding together across

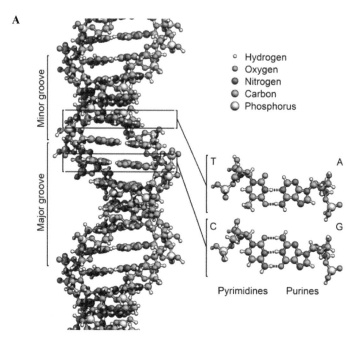

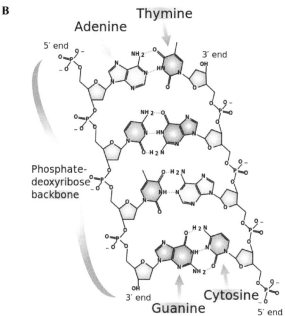

Fig. 3.1. A. The structure of the DNA double helix showing the four bases (adenine, cytosine, guanine, and thymine), and the location of the major and minor groove. The atoms of the structure are color-coded by element and the detailed structure of two base pairs is shown in the bottom right [Wikipedia Commons 2012]. B. The chemical structure of the DNA with colored label identifying the four bases and the phosphate and deoxyribose components of the backbone [Zephyris 2011].

Color image of this figure appears in the color plate section at the end of the book.

the double helix is called a **base pair** (bp). As hydrogen bonds are non-covalent, the two DNA strands in a double helix can be pulled apart, like a zipper, by either mechanical force or high temperature [Clausen-Schaumann et al. 2000]. Consequently, all the information in the double-stranded sequence of a DNA helix is duplicated on each strand. This is vital in DNA replication. This reversible and specific interaction between complementary base pairs is critical for all the functions of DNA [Bruce et al. 2002].

Although the two polymer strands of DNA are bound together by noncovalent bonds in a helical fashion, this double stranded structure (**dsDNA**) is maintained largely by the intrastrand base stacking interactions. The G-C stacks are stronger than A-T stacks. These strands can come apart in a process known as melting caused by high temperature, low salt, and high pH. The stability of dsDNA depends not only on the GC-content but also on sequence and length. It is both the percentage of GC base pairs and the overall length of the DNA double helix that determine the strength of the association between the two strands of DNA. Long DNA helices with a high GC-content have stronger-interacting strands, while short helices with high AT content have weaker-interacting strands [Chalikian et al. 1999].

DNA can be in a **supercoil** state when it is twisted like a rope. When it is in a **relaxed** state, a strand usually circles the axis of the double helix once every 10.4 base pairs. If it is twisted, the strands become more tightly or more loosely wound. If the DNA is twisted in the direction of the helix, this is positive supercoiling, with the bases being held more tightly together. If they are twisted in the opposite direction, it is negative supercoiling, and the bases come apart more easily. In nature, most DNA has slight negative supercoiling that is introduced by enzymes and topoisomerases [Champoux 2001]. These enzymes are also needed to relieve the twisting stresses introduced into DNA strands during transcription and DNA replication [Wang 2002]. Replication allows cell division, and thus continuity of growth and repair. A series of enzyme-mediated steps allows the double helix to unzip, separating the two strands.

Within cells, DNA is organized into long structures called chromosomes. During cell division, these chromosomes are duplicated in the process of DNA replication, providing each cell its own complete set of chromosomes. Eukaryotes, including plants, store most of their DNA inside the cell nucleus and some in organelles, such as mitochondria or chloroplasts [Russell 2001]. Within chromosomes, chromatin proteins such as **histones** compact and organize DNA, and these guide the interactions between DNA and other proteins, and help control which parts of the DNA are transcribed.

Plant DNA

In the plant kingdom, DNA is contained within the membrane-bound cell structures of the nucleus, mitochondria, and chloroplasts. It has several properties that are unique among chemical molecules. This carrier of genetic code undergoes changes in chemical structure, from both environmental and internal causes, called 'mutations', which contribute to evolution, diversity, and disease. DNA is often compared to a set of blueprints or a recipe, or a code, since it contains the instructions needed to construct other components of cells, such as proteins and RNA molecules. The genetic information on DNA segments is carried by genes. Other DNA sequences have structural purposes, or are involved in regulating the use of genetic information. Plant geneticists use this sequencing of DNA to their advantage as they splice and delete certain genes and regions of DNA molecules to produce different or desired genotype (as in the case of producing transgenic plants) and thus, also approaching a different phenotype.

The genetic information encoded in DNA allows for all the development and maintenance of the plant cell. The language of this code lies in a linear reading of adjacent nucleotides on each strand. Every three nucleotides specify or fit a particular amino acid, the individual units of proteins. The second molecule, RNA, copies the molecular structure of DNA and brings the information outside the nucleus into the surrounding cytoplasm of the cell, where the amino acids are assembled, in specified order, to produce a protein. Post-production modifications of these proteins, such as the addition of sugars, fats, or metals, allow a vast array of functional and structural diversity. Plant-DNA codes for a variety of products, which sustain not only the plants but also the entire ecological niche as well as mankind.

Mitochondrial and Chloroplast DNA

An independently functioning set of DNA exists in mitochondria and chloroplast, the two organelles outside the cell's nucleus. In mitochondria, the powerhouse of cells, carbohydrates, fats, and proteins are broken down to their raw elements with the release of stored chemical bond energy in the form of heat (calories). In chloroplasts, photosynthesis transforms carbon dioxide, water, and solar energy to produce sugars and, later, fats and proteins, with the release of oxygen. This critical process undertaken by plants sustains most life on earth. Both mitochondrial and chloroplast DNAs (mtDNA; cpDNA) replicate separately from nuclear DNA during cell division. It is postulated that these organelles once, billions of years ago, may have been independently living organisms that were incorporated into their cells to form the eukaryotic cells which have internal, membrane-bound organelles and a distinct nucleus that physically separates the genetic material of the cell from all of the other parts of the cell. All plants, fungi, and animals are composed of eukaryotic cells.

Plant Proteins and other Products

A large number of proteins that are unique to plants are encoded on plant DNA. These include phytochemicals, flavonoids, phytosterols, carotenoids, indoles, coumarins, oganosulfurs, terpenes, saponins, lignins, and isothiocyanates. Each group contains specific proteins that are both antioxidants and anticarcinogens protecting human and animal cells from cancer-causing agents. Large segments of plant DNA are devoted to coding for specialized plant hormones. These substances are produced by one group of cells, circulate to another site, and affect the DNA of the target cells. In plants, these hormones control cell division, growth, and differentiation. The five well-described classes of plant hormones include auxins, gibberellins, cytokinins, ethylene, and abscisic acid. Auxins allow phototropism in plants aside from making cells elongate. The gaseous ethylene ripens fruits and causes them to drop from the plants. Abscisic acid contributes to the ageing and abscission of leaves.

Transgenic Manipulation

As plants are easy to manipulate, plant DNA is second only to bacterial DNA as a primary experimental object for bioengineers. The direct modification of DNA by adding or removing a particular segment of genes that code for specific traits is the focus of bioengineering and biotechnology. In the recent past, crops have been modified genetically (transgenically) to develop new varieties that resist more importantly herbicides aside from insects, bacteria, and viruses to help reduce pesticide use. Transgenic crops are also designed to enhance a variety of characteristics like faster-growth, higher yield, good taste, seedless fruits, slower fruit ripening, etc. Future transgenic crops may contain human vaccines, human hormones, and other pharmaceutical products.

DNA Barcode

A DNA barcode, proposed in 2003 by Paul Herbert of University of Guelph, Canada [Herbert et al. 2003, 2004], refers to a sequence-based identification systems that may be constructed of one locus or a combination of two or more loci used together as a complementary unit [Kress et al. 2005; Kress and Erickson 2007].

DNA barcoding is a technique, which provides quick identification of species without involving the morphological cues [Selvaraj et al. 2013]. DNA Barcode aids in taxonomic identification of a plant species based on a standard short genomic region. In practice, a DNA sequence—a short fragment—from such a standardized gene region can be generated from a small tissue sample taken from an unidentified species. This sequence is then compared to a library of reference sequences from known species. A match of the sequence from an unknown species to one of the reference sequence can provide a rapid and reproducible identification [Kress and Erickson 2007].

The main aim in barcoding—similar to UPC (Universal Product Code) barcoding on products—is to establish a shared community resource of DNA sequences that can be used for organizational identification and taxonomic clarification [Hollingworth et al. 2011]. Besides, it also aids in the process of providing insights into the taxonomy of over 400,000 species and identifying unknown specimens to known species. Barcoding was successfully pioneered in animals using a 648-base region the mitochondrial gene *cytochrome-C oxidase-1 (CO1)* [Herbert et al. 2003]. In plants, establishing a standardized DNA barcoding system is more challenging [Hollingworth et al. 2011] largely because of a relatively low rate of nucleotide substitution in plant mitochondrial genomes than in animals, and paucity of comparative data encompassing all candidate markers and a broad taxonomic example.

Finding no agreement among plant scientists on common barcode required for barcoding, the Plant Working Group of the Consortium for the Barcode of Life (COBOL), an international initiative of more than 200 organizations from over 50 countries, proposed in 2009 two genetic sequences, or loci, taken from chloroplast genes called *rbcl* (ribulose-bisphosphate carboxylase) and *matK*(maturaseK), as the standard barcode regions for land plants [COBOL 2009]. The *matK* gene, a highly conserved gene in plant systematics, is involved in Group II intron splicing. This gene contains approximately 1500 bp located within the intron of the *trnK* [Selvaraj et al. 2013]. The *rbcL* gene in higher plant presents as a single copy per chloroplast genome, but many copies of the genome are present in each plastid, and the actual *rbcL* copy number per chloroplast can be high. It contains only exons and polypeptide with ~475 amino acids.

This two-locus (*rbcl+matK*) standard barcode gives a better discrimination than single marker barcodes. Using DNA markers *rbcl* and *matK*, de Vere et al. [2012] assembled 97.7 percent coverage for *rbcl* and 90.2 percent for *matK*, and a dual-locus barcode for 89.7 percent of the 1,143 archaeophyte flowering plants and conifers of Wales, U.K. The recoverability of DNA barcode for *rbcl* was much higher at 77.3 percent than for *matK* at 56.6 percent. Both *rbcl* and *matK* performed better using DNA extracted from freshly collected material compared to herbarium specimens [de Vere et al. 2012].

Genome Expression and Role of RNA

Gene expression refers to the process by which the gene produces the protein it encoded. Changes in gene expression underlie the responses to environmental cues and stresses, the response against pathogens, the regulation of metabolic pathways, and the regulation of photosynthesis. A major controlling step in gene expression is transcription, the regulation of which determines the tissue-specific and developmental stage-specific activity of many genes. Regulation by transcription factors is an integral part of a highly complex network.

Transcription

There are two distinct steps in the conversion of DNA sequences into protein sequences: transcription and translation. The production of RNA copies of the DNA is called transcription, the first step in gene expression. It is also described as a process by which the nucleotide sequence of a gene (in the DNA of a chromosome) is used to make complementary copy of RNA. This protein-making process is initiated by a complex of enzymes including RNA polymerase, which bind to the DNA strand at a specific location known as the promoter or enhancer region. Transcription results in the formation of a single strand, or polymer, of ribonucleotides, the order of which is based on the order of the chain of deoxyribonucleotides that constitute the **template** DNA strand.

In the double-stranded molecule, the genes are arranged along DNA on each strand. When there is a gene on one strand, called the coding strand, the other strand (non-coding strand) opposite the gene contains a nucleotide sequence that is complementary. During transcription process, it is the coding strand that is transcribed into a complementary strand of RNA. The transcribed RNA is complementary to the template $3' \rightarrow 5'$ strand, which is complementary to the coding $5' \rightarrow 3'$ DNA strand. Therefore, the resulting $5' \rightarrow 3'$ strand is identical to the coding DNA strand with the exception that thymines (T) are replaced with uracils (U) in the RNA. A coding DNA strand reading 'ATG' is indirectly transcribed through the non-coding strand as 'AUG' in RNA.

Messenger RNA. Transcription of plant genes leaves a primary transcript of RNA, known as precursor mRNA (pre-mRNA), an immature single strand of messenger ribonucleic acid (mRNA) synthesized from a DNA template. Once pre-mRNA has been completely processed, it is termed **mature messenger RNA** or simply **mRNA**. Pre-mRNA exists in plants only briefly before it is fully processed into mRNA. Pre-mRNAs include two different types of segments: **exons** and **introns**. Exons are segments retained in the final mRNA, while introns are removed in a process called splicing which is performed by the spliceosome (except for self-splicing introns). In the RNA-protein catalytical complex, spliceosome catalyzes two transesterification reactions which remove an intron and release it in the form of lariat structure (the $5'$ G of the intron is joined in a $2'$, $5'$-phosphodiester bond to an adenosine near the $3'$ end of the intron), and then splice the neighboring exons.

Once the DNA is transcribed into mRNA, the mRNA strand outside the nucleus of the cell to an organelle is called the **ribosome**, which is composed of structural RNA. The mRNA then attaches itself to the ribosome and begins producing proteins.

Transfer RNA. Transfer RNA (tRNA) is an adaptor molecule composed of RNA. This 73–93 nucleotides-long tRNA serves as a physical link between the nucleotide sequence of nucleic acids (DNA and RNA) and the amino acid sequence of proteins. It does this by carrying an amino acid to the protein synthetic machinery of the cell ribosome as directed by a codon (three-nucleotide sequence) in mRNA. This makes tRNA a necessary component of protein translation in the biosynthesis of proteins. The specific nucleotide sequence of an mRNA specifies which amino acids are incorporated into protein product of the gene from which the mRNA is transcribed. The role of tRNA is to specify which sequence from the genetic code corresponds to which amino acid.

During protein biosynthesis, the anticodon forms three base pairs with a codon in mRNA. The mRNA encodes a protein as a series of contiguous codons, each of which is recognized by a particular tRNA. On the other end of the tRNA is a covalent attachment to the amino acid that corresponds to the anticodon sequence. The covalent attachment to the tRNA 3′ end is catalyzed by enzymes called **aminoacyl-tRNA synthetases**. During protein synthesis, aminoacyl-tRNAs (tRNAs with attached amino acids) are delivered to the ribosome by proteins, called elongation factors (eEF-1α in plants), which aid in decoding the mRNA codon sequence. If the tRNA's anticodon matches the mRNA, another tRNA already bound to the ribosome transfers the growing polypeptide chain from its 3′ end to the amino acid attached to the 3′ end of the newly-delivered tRNA. Elongation factors facilitate translational elongation beginning with the formation of the first peptide bond to the formation of the last one, at a rate of two amino acids per second in plants.

Ribosomes. Ribosomes, mentioned earlier, are macromolecular assemblies composed of various proteins and a second type of RNA, 'ribosomal RNA' (rRNA). They link amino acids together in the order specified by mRNA molecules. Ribosomes consist of two major subunits. Of the two, the small ribosome subunit reads and binds to the mRNA pattern, while the large subunit joins the tRNA and amino acids to form a polypeptide chain. The ribosomes and associated molecules are also known as the *translational apparatus*. Each ribosomal subunit is composed of one or more rRNA molecules and a variety of proteins. When ribosome finishes reading an mRNA molecule, these two subunits split apart.

Linking of amino acids during protein synthesis is catalyzed by ribosome enzymes, called 'ribozymes' (RNA enzymes), RNA molecules capable of catalyzing a variety of RNA processing reactions, including RNA splicing, viral replication, and tRNA biosynthesis. Ribozyme, also termed 'catalytic RNA', functions within ribosomes as part of large subunit ribosomal RNA.

Ribosomes bind to an mRNA chain and use it as a template for determining the correct sequence of amino acids in a particular protein. Amino acids are selected, collected, and carried to the ribosome by tRNA molecules which enter one part of the ribosome and bind to the mRNA-chain. The attached amino acids are then linked together by another part of the ribosome. Once the protein is produced, it can then fold to produce a specific three-dimensional structure before it can carry out its cellular function. The ribosomes in the mitochondria of eukaryotic cells, including those of plants, functionally resemble many features of those in bacteria, reflecting the likely evolutionary origin of mitochondria [Benne 1987].

Translation

Translation is a process by which mature mRNA molecule is used as a template for synthesizing a new protein. This third stage of protein biosynthesis is carried out by ribosomes. In plants and other eukaryotes, translation occurs across the membrane of the rough endoplasmic reticulum (ER). The translated ribosome-mRNA complex binds to the outer side of ER. The nascent protein polypeptide is then released inside for later vesicular transport and secretion outside the cell. Translation proceeds in four phases: initiation, elongation, translocation, and termination, describing the growth of the amino acid chain, or polypeptide, which is the product of translation.

Sorting of Polypeptides. A plant cell contains thousands of proteins. For the cell to function properly, each of them must be localized to the correct cellular membrane or cellular compartment. The process of protein sorting, also called protein targeting, is critical to the organization and functioning of plant cells. Protein sorting relies upon the presence of special signal sequences at one end of the protein molecule. These signal sequences direct proteins to various sites. For example, some proteins synthesized on ribosomes in the cytoplasm are targeted to organelles, such as the mitochondria or chloroplast. Other proteins, such as those found in the plant cell wall, are targeted to the cytoplasmic membrane for transport out of the cell to the cell wall. The signal sequences are frequently removed once the protein has reached its intended destination.

Reverse Transcription

Certain plants have the ability to transcribe RNA into DNA. The enzyme **reverse transcriptase** (RT) catalyzes the formation of double-stranded DNA from a single-stranded RNA genome. It is called reverse transcriptase because it reverses the usual direction of information flow, from DNA to RNA. RT has several enzymatic activities, including the RNA-dependent DNA polymerase, DNA-dependent DNA polymerase, Rnase H (a ribonuclease that degrades RNA in RNA-DNA structure), and the ability to unwind DNA-DNA and RNA-RNA duplexes. Each of these activities is required during the process of reverse transcription to convert the single-stranded RNA genome into a double DNA copy which, in turn, becomes integrated into the host chromosome of the infected cell catalyzed by a second pol(polymerase II: POL II) gene-encoded enzyme, called 'integrase'. RT lengthens the end of linear chromosomes.

Plant cells contain **telomerase**, a specialized reverse transcription enzyme. Telomerase is ribonucleoprotein that maintains telomere ends by the addition of the telomere repeat TTTAGGG, as in *Arabidopsis thaliana* (5' to 3' toward the end). The enzyme consists of a protein component with reverse transcriptase activity, encoded by this gene, and an RNA component that serves as a template for the telomere repeat. Telomerase expression plays a role in cellular senescence, as it is normally repressed in postnatal somatic cells, resulting in progressive shortening of telomeres. Telomerase also participates in chromosomal repair, since *de novo* synthesis of telomere repeats may occur at double-stranded breaks.

Telomerase carries an RNA template from which it synthesizes DNA repeating sequence, or 'junk' DNA. This repeated sequence of DNA is important because, every time a linear chromosome is duplicated, it is shortened in length. With "junk" DNA at the ends of chromosomes, this shortening eliminates some of the non-essential, repeated sequence rather than the protein-encoding DNA sequence farther away from the chromosome end.

Transcription Factors

Transcription of DNA to make mRNA is highly controlled by the cell. In order for a plant to function, genes must be turned 'on' and 'off' in coordinated groups in response to a variety of stresses, both 'biotic' (herbicides, insects, viral or bacterial infection, etc.) and 'abiotic' (light, temperature, drought, salinity, flooding, metals, low nutrients, etc.). The job of coordinating the functions of groups of genes falls to proteins called **transcription factors** (TFs). They perform these functions alone or with other proteins in a complex, by promoting (as an activator), or blocking (as a repressor) the recruitment of RNA polymerase (the enzyme that performs the transcription of genetic information from DNA to RNA) to specific genes. As master regulators of cellular processes, TFs are expected to be excellent candidates for modifying complex traits in crops plants [Century et al. 2008].

While regulating gene expression, each gene is preceded by a promoter region that includes a binding site for the RNA polymerase (which will copy DNA to RNA) and a variety of other features, including the 'TATA box' (a short segment of repeating thymidine and adenine residues) and one or more enhancer sites, which serve as a binding location for TFs [Goldstein et al. 2009]. In order for an mRNA to be created, an initiation complex must assemble in the promoter region. This complex consists of over 40 proteins, including the RNA polymerase, TATA binding protein, and one or more TFs.

TFs bind to DNA at the enhancer site and/or other proteins in the initiation complex. Through these protein-protein interactions, TFs are able to control whether RNA polymerase moves forward along the DNA to produce a message. In effect, TFs serve as the 'traffic cops' regulating mRNA production.

TFs often contain features that help the cell respond to the internal or external environment. These features are binding sites that interact with chemicals within the cell ('ligands') that modulate the activity of TFs. For example, TFs bind to hormones, chemicals like glucose, or to other proteins in order to 'sense' and respond to the environment [Goldstein et al. 2009]. In order to allow coordinated gene function, a particular TF may bind to multiple genes, and each gene may be controlled by multiple TFs, which often regulate other TFs. These TFs form complex networks that may control from one to many thousands of genes in response to conditions inside and outside of the cell [Goldstein et al. 2009].

The genes for TFs can be inserted into plants in the same manner as other genes such as the glyphosate-resistant *CP4 EPSPS* gene and insect-resistant *Bt* gene (vide Chapters 5 and 6). As with other genes, TF genes may carry non-specific promoters which allow them to be expressed at all times in all plant tissues, or they may carry tissue selective or other promoters to allow expression only in particular tissues, at particular time, or under particular conditions [Goldstein et al. 2009].

Although transcriptional regulators are being proposed as the wave of the future for agricultural biotechnology, there is strong evidence that TFs have already played a major role in the origin of agriculture through the domestication of various crop plants [Century et al. 2008]. Besides, some of the major yield gains achieved by conventional breeding have been subsequently attributed to alterations in TF activity. This includes global increase in wheat yields beginning in the 1960s with the availability of semi-dwarf varieties that resist lodging while responding to nitrogen fertilizer [Century et al. 2008].

Gene Promoters

There are numerous factors which influence gene expression, but promoters constitute the most important component of the regulatory control process at the transcriptional level. They serve as a sort of "On" switch to initiate transcription for the genes which follow the promoter DNA sequence. Promoters are DNA sequences just upstream from a gene that acts as a binding site for transcription factors and RNA polymerase II enzyme, which is responsible for the generation of RNA. The promoter region is usually assumed to be the key *cis*-acting regulatory region that controls the transcription of the adjacent coding region(s) into mRNA, which is then directly translated into proteins. RNA polymerase II binds to the promoter sequence and begins to work its way down the segment, constructing RNA to match DNA nucleotides over which the enzyme passes.

Besides DNA sequences, promoters contain regulatory elements, called response elements, that provide a secure initial binding site for RNA polymerase and proteins. These transcription factors have specific activator or repressor sequences of corresponding nucleotides that attach to specific promoters and regulate gene expression. They help recruit the transcriptional machinery to gene promoters. This typically involves conformational changes in chromatin structure to facilitate access to the DNA. Plant genomes contain large numbers of genes capable of encoding transcription factors, which can be grouped into families based upon the structure of their DNA-binding domains.

Understanding the function of their multiple components and factors associated with their performance has opened up the possibility of modulation of genes in plants in which foreign promoters together with genes of interest are inserted. This makes promoters playing a key role in controlling transgenes in transgenic plants (vide Chapter 4). Plants have potentially as many promoters as are genes. There are different types of promoters. These include: constitutive promoters, tissue-specific promoters, inducible promoters, and chemically-regulated promoters.

Constitutive Promoters. Constitutive gene promoters are essential components of crop transgenic engineering. They are the unregulated promoter segments of DNA that allow continuous transcription of their associated genes. Constitutive promoters are required to ensure that a specific gene transferred into a plant will be functional in all plant tissues and to export reporter and selectable marker genes needed for establishing a reliable transportation system in a particular plant species [Kadir et al. 2009]. They can be used to express any gene in any tissue of a plant species, in addition to the one from which it is isolated.

Constitutive promoters perform several vital functions during transgenic engineering. These include, a) the high-level expression of reporter genes such as β-glucuronidase (*GUS*), green fluorescent protein (*GFP*), and chloramphenicol acetyltransferase (*CAT*) to monitor transformation protocols, b) expression of marker genes for transgenic cell selection, c) expression of herbicide tolerance genes, d) repression of endogenous and/or pathogenic gene expression through antisense and co-suppression technologies, e) overproduction of biomolecules, and f) overexpression of disease and stress tolerance genes prior to the employment of the targeted gene expression strategies.

The first constitutive promoters used for expression of transgenes in plants have been isolated from plant pathogens. The more widely used of them includes the cauliflower mosaic virus (CaMV 35S), a pararetrovirus, used for high-level constitutive gene expression in the transformation of dicots. CaMV contains a circular double-stranded DNA (dsDNA) molecule of about 8 kb, interrupted by site-specific discontinuities resulting from its replication by reverse transcription. This DNA is transcribed into a full length, terminally redundant, 35S RNA and a sub-genomic 19S RNA. The CaMV 19S is a weak promoter compared to the 35S promoter. The 35S RNA, a very strong constitutive promoter used in plant transformation, is responsible for the transcription of the whole CaMV genome. It causes high levels of gene expression in dicot plants. However, it is less effective in monocots, especially in cereals. The differences in behavior are probably due to differences in quality and/or quantity of regulatory factors.

The 35S RNA is particularly complex (Fig. 3.2), containing a highly structured 600 nucleotide-long leader sequence with six to eight short open reading frames (ORFs) [Fütterer et al. 1988; Pooggin et al. 1998; Hemmings-Mieszczak et al. 1999]. It is made up of distinct domains, with each involved in organ-specific expression. Within these promoter domains, two *cis*-acting elements exist. The *cis* element controls gene expression in roots besides functioning in synergy with upstream domains to potentiate promoter activities in other tissues.

The other plant pathogen-related promoters include cassava vein mosaic virus (CsVMV), figwort mosaic virus (FMV), cestrum yellow leaf curling virus (CmYLCV), peanut chlorotic streak virus (PCISV), and tobacco etch virus (TVS). A discussion on these gene promoters is presented in Chapter 5.

Opine promoters also drive the expression of opines (mannopine, octopine, and nopaline), the hormone-like compounds generated by the soil bacterium *Agrobacterium* through the use of the plant's expression machinery. Opines are utilized by the bacterium as a source of carbon, nitrogen, and energy. Promoters from the nopaline synthase (*nos*), octopine synthase (*ocs*) and mannopine synthase (*mas*) genes have been isolated and inserted into transformation vectors upstream of foreign genes to control the expression of those genes. They are frequently used for transformation of dicot plants.

In monocots, the highly active constitutive promoters include *Ubi* (ubiquitin), *Adh-1* (alcohol dehydrogenase I), and *Zmubil* (*Zea mays* ubiquitin 1) from maize; *actin 1*,

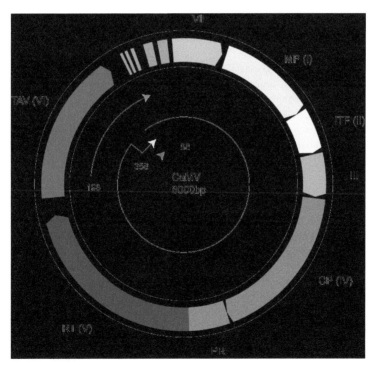

Fig. 3.2. Genomic CaMV with the 35S RNA, containing a highly structured 600 nucleotide long leader sequence with six to eight short open reading frames (ORFs) [Fütterer et al. 1988; Pooggin et al. 1998; Wikipedia Commons 2005; free use under The GNU Free Documentation License].

Color image of this figure appears in the color plate section at the end of the book.

OsTubA1 (*Oryza sativa* α-tubulin), *OsCc1* (*Oryza sativa* cytochrome c gene 1), *RUBQ1* (rice polyubiquitin gene 1), *rubi3* (rice polyubiquitin gene 3), and *OsAct2* (*Oryza sativa* actin2 gene), and *GOS2* from rice; *ubi7* (ubiquitin gene 7) from potato; *Gumbi* (*Glycine max* ubiquitin) from soya bean; *ubi4* (ubiquitin gene 4) and *ubi9* (ubiquitin gene 9) from sugarcane; and *tCUP* (**t**obacco **c**onstitutive **p**romoter) from tobacco.

Promoters such as *APX* (ascorbate peroxidase), *SCP1* (super core promoter 1), *PGD1* (phosphogluconate dehydrogenase 1), *R1G1B*, and *E1F5* are found in transgenic rice and they are linked to the *gfp* (green fluorescent protein) reporter gene and transformed into rice [Park et al. 2010]. The *R1G1B*, active in the whole grain including the embryo, endosperm, and aleurone layer, represents a constitutive promoter with activity in whole seeds. The promoters that work well in dicots, which lack introns, do not generally work well in monocots. For this reason, regulatory sequences from monocots need to be cloned into vectors for control of transgene expression.

Tissue-Specific Promoters. Sometimes, ubiquitous overexpression and mis-direction of gene products may cause undesirable pleiotropic effects on the plant. This could lead to generation of phenotypes not directly correlated with the recombinant protein itself [Himmelbach et al. 2007]. In order to avoid unwanted pleiotropic effects, transgene expression can be controlled by the use of cell- and tissue-specific promoters. They operate at certain developmental stages of a plant. They might specifically influence expression of genes in the roots, fruits, or seeds, or during the vegetative, flowering, or seed-setting stage.

The transgenes driven by these tissue-specific promoters will only be expressed in tissues where the transgene product is desired, leaving the rest of the tissues in the plant unmodified by transgene expression [Patent Lens]. Since tissue-specific promoters, also called **spatiotemporal promoters**, may be induced by endogenous or exogenous factors, so they may, sometimes, be classified as inducible promoters as well. Tissue-specific expression is the result of several interacting levels of gene regulation. As such, it is then preferable to use promoters from homologous or closely related plant species to achieve efficient and reliable expression of transgenes in particular tissues.

This group of promoters includes root promoters, fruit promoters, and seed promoters. Root promoters enhance or suppress the expression of a linked gene in root cells. Fruit promoters control the expression of genes in mature ovary tissue of a fruit and in the receptacle tissue of accessory fruits such as strawberry, apple, and pear. They also influence fruit development and ripening. Seed promoters provide favorable characteristics for expression control of seed products, including storage proteins. When a gene of interest needs to be expressed in more than one tissue type, for example the root, anthers and egg sac, etc. then multiple tissue-specific promoters may have to be included in the gene construct.

Several tissue-specific promoter genes used in transformation of rice include a) cereal promoters such as maize *Adh1* (alcohol dehydrogenase 1), wheat *His3* (histidine 3), and rice *rbcS* (ribulose-1,5-bisphosphate carboxylase small subunit); b) dicot promoters, tomato *rbcS* and potato *pinII* (proteinase inhibitor II); c) bacterial promoter *Agrobacterium rhizogenesrolC*; and d) viral promoters (rice tungro virus). Besides, the *Adh-1* promoter used in cereals such as rice, oat, and barley is also used in the dicot tobacco, to drive the expression of genes of interest. But it provides very low levels of expression in this dicot crop. The *Adh-1* promoter in conjunction with the intron has proved to be much superior for transformation of some cereals such as rice [Patent Lens]. The *His3* gene, found in the *Saccharomyces cerevisiae* yeast, encodes a protein called imidazoleglycerol-phosphate dehydratase which catalyzes the sixth step in histidine biosynthesis.

Inducible Promoters. These promoters are induced by the presence or absence of biotic and abiotic factors. They are used to control transgene expression externally. Many herbicide-resistant and disease-resistant transgenic plants have resistance transgenes placed under the control of a promoter that is induced by herbicides and pathogens respectively. Inducible promoters are a very powerful tool in genetic engineering because the expression of genes operably linked to them can be turned on or off at certain stages of development of an organism or in a particular tissue [Patent Lens]. There are two types of inducible promoters: a) chemically-regulated promoters, whose transcriptional activity is regulated by the presence or absence of chemicals, and b) physically-induced promoters, whose transcriptional activity is regulated by environmental factors such as light, temperature, water or salt stress, anaerobiosis, and wounding. These promoters contain regulatory elements that respond to environmental stimuli. The light-responsive promoter includes ribulose-1,5-bisphosphate carboxylase-oxygenase (*rbcS*) gene, the chlorophyll a/b binding protein. The advantage of herbicide-induced promoters is that they can be controlled by the application of herbicides in the field.

Chemically-Regulated Promoters. The activity of this group of promoters is regulated by non-plant chemical compounds that either turn off or turn on gene transcription. An ideal chemically controlled system should allow expression of a target gene to be precisely regulated temporally, spatially, and quantitatively.

In order for the chemicals to influence promoter activity and modulate gene expression, they should not be naturally present in the plant where expression of the transgene is sought, should not be toxic, should affect only the expression of the gene of interest, should be easy to apply or remove, and should induce a clearly detectable expression pattern of either high or very low gene expression. The chemically-regulated promoter may be derived from an organism distant in evolution to the plants where its action is required. These include bacteria (*E. coli*), algae, insects (Drosophila), virus, yeast, humans, etc. The promoters may be a) alcohol-regulated (*alcA*: alcohol dehydrogenase I gene promoter; *AlcR*: transactivator protein); b) tetracycline-regulated (*TetR*: a tetracycline repressor protein; *tetO*: a tetracycline operator sequence; *tRA*: a tetracycline transactivator fusion protein); c) steroid-regulated (*GT*: rat glucocorticoid receptor; *ER*: human estrogen receptor; EcR: ecdysone receptors from moths); d) metal-regulated (metallothionein, a protein that bind and sequester metal ions, from yeast, mouse, and human); or pathogenesis-related proteins (induced by salicylic acid, ethylene, benzothiadiazole, etc.) [Patent Lens].

Terminator and Terminator Gene

Terminator, also called transcription terminator, is a short base sequence at the 3' end of a gene which causes the RNA polymerase to terminate transcription. This terminator region signals RNA polymerase to release the newly made RNA molecule, which then departs from the gene. If the promoter signals initiation of transcription, terminator signals its end.

A terminator gene is that part of the cell designed to make a plant that would be unable to produce new young plants. It is a specific sequence of nucleotides inserted into the DNA of seeds, and they signal the end of transcription or translation and completion of synthesis of a nucleic acid or protein molecule. Once activated by a synthetic chemical catalyst, the sequence renders the seed and crop it produces sterile. The end result is that farmers using these seeds are forced to buy fresh seeds for the following season rather than seeds from the current crop. This means a travesty of justice to farmers who are accustomed to use seeds saved by them for the next crop, a practice that has existed for countless millennia (vide Chapter 9).

Some of the terminator genes used in engineering for herbicide and insect resistance (vide Chapter 6) include CaMV 35S 3' poly-adenylation signal; 3'-UTR region of the nopaline synthase (*nos*) gene from *Agrobacterium tumefaciens* (used as the 3'-UTR for transgene constructs); 3'-UTR of wheat HSP 17.3 (3' untranslated region of wheat heat shock protein 17.3); PINII (*S. tuberosum* proteinase inhibitor II); tahsp173' (*Triticum aestivum* heat shock protein 17.3); and ORF25 polyA (3' poly-adenylation signal from open reading frame 25 from *A. tumefaciens*), etc.

Antisense RNA

It is a single-stranded RNA that is complementary to a messenger RNA (mRNA) strand transcribed within a cell. Its sequence of nucleotides is called '**sense**' because it results in a gene product (protein). Normally, its unpaired nucleotides are 'read' by transfer RNA anticodons as the ribosome proceeds to translate the message. However, RNA can form duplexes just as DNA does. All that is needed is a second strand of RNA whose sequence of bases is complementary (mirror image) to the first strand (mRNA: 5' C A U G 3'; antisense RNA; 3' G U A C 5'). The second strand is called the **antisense RNA strand** because its sequence of nucleotides is the complement of message sense. When mRNA forms a duplex with a complementary antisense RNA sequence, translation is blocked. This may occur because either the ribosome cannot gain access to the nucleotides in the mRNA or duplex RNA is quickly degraded by ribonuclease in the cell. With recombinant DNA methods (vide Chapter 4), synthetic genes (DNA) encoding antisense RNA molecules can be introduced into the organism.

Plant cells containing genes that are naturally translated into antisense RNA molecules are capable of blocking the translation of other genes in the cell. These seem to represent another method of regulating gene expression. However, antisense RNA may be introduced into a cell to inhibit translation of a complementary mRNA by base pairing to it and physically obstructing the translation machinery [Weiss et al. 1999]. Antisense RNA has been used to artificially modulate gene expression in plants and animals. The effects of antisense RNA have often been confused with the effects of RNA interference, a related process. There is a stark difference. As discussed earlier, antisense RNA blocks translation of mRNA into protein, while in RNA interference a double strand RNA silences post-transactional function of the gene. However, attempts to genetically engineer plants by using antisense RNA pathway does activate the RNA interference pathway, although these processes result in differing magnitudes of the same downstream effect, gene silencing.

The most important example of antisense technology is in crop improvement regarding the development of Flavr Savr tomato by Calgene of Davis, California in 1992 only to be discontinued after five years. This involves the inhibition of polygalacturonase (*PG*) gene, which codes for the polygalacturonase enzyme. This enzyme caused degradation of cell wall, resulting in fruit ripening. Inhibition of *PG* gene expression greatly reduces fruit ripening, leading to improvement in the shelf life of tomatoes. Besides, antisense technology has also been applied for potato crop to reduce discoloration of tubers after bruising and to produce ring spot-resistant papaya.

RNA Interference

RNA interference (RNAi) is a dsRNA-induced gene-inhibiting/blocking phenomenon. In this, short sequences of RNA that match a part of the target gene's sequence is inserted, thus blocking gene function. In this process, two types of small RNA molecules—microRNA (miRNA) and small interfering RNA (siRNA)—bind to other specific mRNA molecules and prevent protein production. The RNAi pathway is found in many eukaryotes including plants and it is initiated by the enzyme Dicer (an endoribonuclease in the RNase III family) which cleaves long dsRNA molecules into short double stranded fragments of 20–25 nucleotides-long siRNAs, with a two-base overhang on the 3′ end. Each siRNA is unwound into two single-stranded RNAs (ssRNAs) which are called the passenger strand and the guide strand. The passenger strand is degraded, while the guide strand is incorporated into the RNA-induced silencing complex (RISC). SiRNA is more stable than ssRNA since it is double-stranded. It is not subject to degradation by RNAse enzymes.

The RNAi-related phenomena include suppression of transposon activity, resistance to virus infection, post-transcriptional regulation of gene expression, and epigenetic regulation of chromatin structure [Hannon 2002]. RNAi technology is used as a natural defense mechanism against molecular parasites such as jumping genes and viral genetic elements that affect genome stability. RNAi is very much related to functional genomics, which has been used since 2003 to find out the function of genes, thereby improving traits that various crops display.

RNAi also has immense practical application in the transgenic improvement of crop plants. Using RNAi, plant scientists have developed novel crop varieties. These include [HiRNA Works 2012; Crop View 2012]:

a) reduced nicotine content in cured tobacco leaves (target gene: *CYP82E4*);
b) allergen-reduced groundnut (*Arah 2*: *Arachis hypogea* 2);
c) decaffeinated coffee (*CaMXMT1*: *Coffea arabica* methylated xanthosine methyltransferase 1);
d) lysine-fortified maize (*ZLKR/SDH*: *Zea* lysine-ketoglutarate reductase/saccharopine dehydrogenase);
e) increased carotenoid oxidant lycopene (*Lyc*);
f) early ripening in tomato (*LeETR4*: *Lycopersicon esculentum* ethylene receptor 4);
g) 'tearless' onion (lachrymatory factor synthase gene);
h) non-narcotic alkaloid containing opium poppy (COR);
i) increased oleic acid content in rapeseed (FAD2: Fatty Acid Desaturation 2);
j) groundnut (FAD2) and cotton (FAD2);
k) slow ripening tomato (ACC dioxygenase gene);
l) arsenic hyperaccumulator in Arabidopsis for phytoremediation (ACR2);
m) high-flavonoid and β-carotene tomato (DET1); etc.

RNAi methodology involves the following steps:

a) Inserting a long dsRNA, such as an introduced transgene, a rogue genetic element, or a viral intruder. This triggers the RNAi pathway of cells, resulting in recruitment of the enzyme Dicer.
b) Cleaving of dsRNA by Dicer into short, 20–25 bp-long siRNA fragments.
c) Distinguishing of two siRNA strands as either sense or antisense by an RNA-induced silencing complex. This is followed by degradation of sense strands (with exactly the same sequence as the target gene).

d) Incorporation of antisense strands to the RNA-induced silencing complex; these strands are used as a guide to target mRNA in a sequence-specific manner.

e) Cleaving of mRNA which codes for amino acids by the RNA-induced silencing complex (RISC). The activated RISC can repeatedly participate in mRNA degradation, inhibiting protein synthesis.

MicroRNAs

MicroRNAs (miRNAs) are an extensive, highly conserved class of small (~22 nucleotides) genomically encoding noncoding RNAs that regulate gene expression by post-transcriptional or translational repression [Carrington and Ambros 2003] during plant development. Mature miRNAs are structurally similar to siRNAs produced from exogenous dsRNA, but before reaching maturity, miRNAs must first undergo extensive post-transcriptional modification. They behave more like siRNAs. MicroRNAs play an important regulatory role in such processes as leaf development, auxin signaling, phase transition, flowering, stress response, and genome maintenance. The expression of miR398, linked to stress tolerance, is transcriptionally down-regulated by oxidative stresses. In *A. thaliana*, miR398 targets two closely related Cu/n superoxide dismutase coding genes, cytosolic CSD1, and chloroplastic CSD2. Besides, a reduced level of miR398 leads to improved tolerance of transgenic lines compared with the wild-type plants under oxidative stress conditions [Sunkar et al. 2006]. Additionally, miR395 and miR399 are involved in sulphate and inorganic phosphate starvation responses, respectively [Jones-Rhoades and Bartel 2004; Fujii et al. 2005]. Of the various miRNAs found in *Arabidopsis*, 10 are related to high-salinity, four to drought, and 10 to cold regulation. Expression profiling by RT-PCR analysis showed great cross-talk among the high-salinity, drought, and cold stress signaling pathways [Liu et al. 2008].

In rice, miR169 is induced by drought [Zhao et al. 2007]. Several miRNA families target genes encoding nucleotide binding site leucine-rich repeat (NBS-LRR) plant innate immune receptors in legumes [Zhai et al. 2011]. Shivaprasad et al. [2012] found miR482 and miR2118 responsible for a regulatory cascade affecting disease resistance in tomato. All variants of this miRNA super family target the coding sequence for the P-loof motif in the mRNA sequences for disease resistance proteins with NBS and LRR motifs. In *Nicotiana benthamiana*, miR482 targets mRNAs for NBR-LRR disease resistance proteins with coiled-coil domains at their N terminus. The targeting causes mRNA decay and production of secondary siRNAs in a manner that depends on RNA-dependent RNA polymerase 6 [Shivaprasad et al. 2012]. These miRNAs are abundant in Rutaceae, Solanaceae, and Fabaceae families. The NBS-LRR proteins contribute to a novel layer of defense against pathogen attack on tomato.

Measuring Gene Expression

Plants make drastic changes to their transcriptome (set of all RNA molecules, including mRNA, rRNA, tRNA, and other non-coding RNA) in their adaptive responses of thousands of genes simultaneously to abiotic and biotic stresses. The coordination of defense gene transcription is often coupled with significant adjustments in the levels of expression of primary metabolic and structural genes to relocate resources, repair damage and/or induce senescence [Rehrig et al. 2011].

Gene expression can be measured by quantifying mRNA, DNA, and specific proteins. Traditional spectrophotometric method cannot distinguish DNA, rRNA, and tRNA from mRNA. Various mRNA and DNA quantification methods include Northern blotting, Southern blotting, polymerase chain reaction (PCR), Rt-PCR, RT-qPCR, RNAse protection, and microarray hybridization. On the other hand, detection and quantification of specific proteins involved in gene expression is done by Western blotting and enzyme immunoabsorbent assay (ELISA). These methods are also used to identify, measure, and quantify transgene expression in transgenic plants following backcrossing of transformed plants with elite cultivars, discussed later in this Chapter.

Northern Blotting

Northern Blotting (NB) method [Trayhurn 1996] reveals information about identity, size, abundance, and sequence information of RNA (or isolated mRNA), allowing a deeper understanding of gene expression levels. Messenger RNA, much smaller than genomic DNA, can be isolated from cells in its intact form, free from significant amounts of DNA.

N-blot: The Northern blotting procedure begins with extraction of total RNA from a homogenized sample or from cells. RNA samples are then denatured and separated by agarose gel electrophoresis without the enzymatic digestion steps required for the analysis of high molecular weight DNA. RNA is then electrophoretically separated by size in the presence of a denaturing agent, such as formamide or glyoxal/DMSO (dimethyl sulfoxide). After electrophoresis, RNA is transferred to a nitrocellulose or nylon-based membrane through a capillary or vacuum blotting system. This is followed by hybridization and blot washing. In this manner, specific RNA sequences corresponding to those in cloned DNA probes can easily be identified. Northern blotting is less sensitive than the other techniques. However, it is a less complex and straight forward quantification method compared with other methods.

The nylon membrane, with a positive charge, is very effective for use in Northern blotting because the negatively charged nucleic acids have a high affinity for them. The transfer buffer, formamide, lowers the annealing temperature of the probe-RNA interaction, thus preventing RNA degradation by high temperatures [Yang et al. 1993]. Once the RNA has been transferred to the membrane, it is immobilized through covalent linkage to the membrane by UV light or heat. After a probe has been labeled, it is hybridized to the RNA on the membrane. Experimental conditions that can affect the efficiency and specificity of hybridization include ionic strength, viscosity, duplex length, mismatched base pairs, and base composition [Streit et al. 2009]. The membrane is washed to ensure that the probe is bound specifically and to avoid background signals from arising. The hybrid signals are then detected by X-ray film and can be quantified by densitometry. To create controls for comparison in a Northern blot, samples not displaying the gene product of interest can be used after determination by microarrays or RT-PCR (reverse transcription polymerase chain reaction) [Streit et al. 2009].

Southern Blotting

In Southern blot analysis, named after its inventor, the British biologist Edwin Southern [Southern 1975], DNA strands of high molecular weight are cut into smaller fragments by using restriction endonucleases followed by electrophoresing of the fragments on an agarose gel to separate them by size. Fragments larger than 15 kb are broken into smaller pieces by treating them with an acid (dilute HCl) for de-purination. Thereafter, a sheet of nitrocellulose (or nylon) membrane is placed on top of (or below, depending on the direction of the transfer) the gel. Pressure is applied evenly to the gel either by using suction, or by placing a stack of paper towels and a weight on top of the membrane and gel. This absorbent material draws salt solution from the dish into the wick and through the gel by capillary action, which transfers the DNA fragments into the filter. The filter now contains the DNA fragments in the same pattern as the gel, but is more easily manipulated.

The membrane is now exposed to a hybridization probe—a single DNA fragment with a specific sequence whose presence in the target DNA is to be determined. The probe DNA is labeled so that it can be detected, usually by incorporating radioactivity or tagging the molecule with a fluorescent or chromogenic dye. In some cases, the hybridization probe may be made from RNA, rather than DNA. To ensure the specificity of the binding of the probe to the sample DNA, most common hybridization methods use salmon or herring sperm DNA for blocking of the membrane surface and target DNA, deionized formamide, and detergents such as SDS to reduce non-specific binding of the probe.

After hybridization, the unbound excess probe is washed from the membrane, typically using SSC (saline-sodium citrate) buffer, a piece of X-ray film is placed over the hybridized filter and left for several hours to several days. The pattern of hybridization is visualized on X-ray film by autoradiography in the case of a radioactive or fluorescent probe, or by development of color on the membrane if a chromogenic detection method is used.

Southern transfer may be used for homology-based cloning on the basis of amino acid sequence of the protein product of the target gene. Oligonucleotides (short, single-stranded DNA or RNA molecules) are chemically synthesized, radiolabeled, and used to screen a DNA library, or other collections of cloned DNA fragments. Sequences that hybridize with the hybridization probe are further analyzed, for example, to obtain the full length sequence of the targeted gene. Second, Southern blotting can also be used to identify methylated sites in particular genes. Particularly useful are the isoschizomeric restriction nucleases MspI and HpaII, both of which recognize and cleave within the same etranucleotide sequence (5'-CCGG-3') but display differential sensitivity to DNA methylation. HpaII is inactive when any of the two cytosines is methylated, but it digests the hemimethylated 5'-CCGG-3' at a lower rate compared with the unmethylated sequences. On the other hand, MspI digests 5'-C^{m}CGG-3' but not 5'-^{m}CCGG-3' [New England Biolabs 2005–06], which means that it cleaves only DNA unmethylated at that site. Therefore, any methylated sites within a sequence analyzed with a particular probe will be cleaved by the former, but not the latter, enzyme [Matthews 1990]. These features of the restriction enzymes are used to assess the global DNA methylation status of DNA preparations.

Western Blotting

The Western blot (sometimes called the protein immunoblot) is a widely used analytical technique used to separate and identify specific proteins in a sample of tissue homogenate or extract. It uses gel electrophoresis to separate native proteins by 3-D structure or denatured proteins by the length, molecular weight, and thus by type, of the polypeptide. The proteins are then transferred to a membrane (typically nitrocellulose or PVDF), where they are stained with antibodies specific to the target protein [Towbin et al. 1979; Renart et al. 1979]. The gel electrophoresis step is included to resolve the issue of the cross-reactivity of antibodies.

Samples can be taken from whole tissue or from cell culture. Solid tissues are broken down using a blender (for larger sample volumes), using a homogenizer (smaller volumes), or by sonication. Detergents, salts, and buffers are used to lyse cells and to solubilize proteins. This is followed by addition of protease and phosphate (alkaline phosphatase) inhibitors to prevent the digestion of the sample by its own enzymes. The sample is filtered and centrifuged to separate different cell compartments and organelles.

The proteins of the sample are separated using gel electrophoresis. Separation of proteins may be by isoelectric point (pI), molecular weight, electric charge, or a combination of these factors. Western blot uses two types of agarose gel: stacking and separating gel. The higher, stacking gel is slightly acidic (pH 6.8) and has a lower acrylamide concentration making a porous gel, which separates protein poorly but allows them to form thin, sharply defined bands. The lower gel, called the separating, or resolving gel, is basic (pH 8.8), and has a higher polyacrylamide content, making the gel's pores narrower. Proteins are thus separated by their size more so in this gel, as the smaller proteins travel more easily, and hence rapidly, than larger proteins [Mahmood and Young 2012].

The proteins when loaded on the gel have a negative charge, as they have been denatured by heating, and will travel toward the positive electrode when a voltage is applied. Gels are usually made by pouring them between two glass or plastic plates. The samples and a marker are loaded into the wells, and the empty wells are loaded with sample buffer. The gel is then connected to the power supply and allowed to run [Mahmood and Young 2012]. After separating the protein mixture, it is transferred to a membrane for electroblotting. It uses an electric current to pull proteins from the gel into the PVDF (polyvinylidene fluoride) or nitrocellulose membrane. The proteins move from within the gel onto the membrane while maintaining the organization they had within the gel.

As the membrane has been chosen for its ability to bind protein, steps must be taken to prevent the interactions between the membrane and the antibody used for detecting the target protein. Blocking of the non-specific binding is achieved by placing the membrane in a dilute solution of 3–5 percent of Bovine serum albumin (BSA) or non-fat dry milk in Tris-Buffered Saline (TBS) or I-Block, with 0.1 percent of detergent such as Tween 20 or Triton X-100. The membrane is then used using the label antibody, usually with an enzyme horseradish peroxidase (HRP), which is detected by the signal it produces corresponding to the position of the target protein. This signal is captured on a film which is usually developed in a dark room [Mahmood and Young 2012]. As Western blot method does not reveal protein only at one band in a membrane, size approximations are taken by comparing the stained bands to that of the marker or ladder loaded during electrophoresis.

The process is repeated for a structural protein, such as actin or tubulin. The amount of target protein is normalized to the structural protein to control between groups. This practice ensures correction for the amount of total protein on the membrane in case of errors or incomplete transfers. The data produced with a western blot is typically considered to be semi-quantitative. This is because it provides a relative comparison of protein levels, but not an absolute measure of quantity.

Polymerase Chain Reaction

The polymerase chain reaction (PCR) amplifies *in vitro* a single or a few copies of a segment of DNA across several orders of magnitude, generating thousands to millions of copies of a particular DNA sequence. Developed in 1983 by Karry Mullis [Bartlett and Stirling 2003], PCR has become an indispensable technique used in plant biological research.

The method relies on thermal cycling, consisting of cycles of repeated heating and cooling of the reaction for DNA melting and enzymatic replication of the DNA. Primers (short DNA fragments) containing sequences complementary to the target region along with a DNA polymerase (after which the method is named) are key components to enable selective and repeated amplification. As PCR progresses, the DNA generated is itself used as a template for replication, setting in motion a chain reaction in which the DNA template is exponentially amplified. PCR can be extensively modified to perform a wide array of genetic manipulations PCR requires: a) DNA template that contains the DNA target region to be amplified; b) two primers (short pieces of a single stranded DNA of 18–28 nucleotides-long), complementary to the 3′ ends of each of the sense and antisense strands of DNA target; c) taq polymerase (made from the bacterium *Thermus aquaticus*) or another DNA polymerase (used for polymerization or building new strands of DNA identical to the DNA template) with a temperature optimum of 70 C; d) four deoxynucleotide triphosphates (dNTPs: dATP, dTTP, dCTP, dGTP), the building blocks from which the DNA polymerase synthesizes a new strand; e) reaction buffer solution to provide suitable environment for optimum activity and stability of DNA polymerase; f) divalent cations (Mg^{2+} and Mn^{2+}) to utilize for PCR-mediated DNA mutagenesis; and g) monovalent cationic potassium ions.

PCR technique proceeds in four steps. In the first, initialization step, the two strands of the DNA double helix are physically separated at a high temperature in a process called DNA melting. In this, the reaction mixture is heated to 94–96°C for 1–9 min before adding DNA polymerase. This is followed by the denaturing step in which the target genetic material is heated at 94–96°C for 20–30 sec. Denaturing melts DNA template, unwinds its helix, and disrupts the hydrogen bonds between complementary bases, yielding single-stranded molecules.

In the next step, the temperature is lowered and the two DNA strands become templates for DNA polymerase to selectively amplify the target DNA. For this, the reaction temperature is lowered to 50–65 C for 20–40 sec allowing annealing of the primers to their complementary bases on the single-stranded DNA template. Stable DNA-DNA hydrogen bonds are only formed when the primer sequence very closely matches the template sequence. The polymerase (*Taq* polymerase) binds to the primer-template hybrid and begins DNA synthesis.

In the final step, elongation/extension, DNA polymerase synthesizes (at 72°C) a new DNA strand complementary to the DNA template strand by adding dNTPs that are complementary to the template in 5′ to 3′ direction and condensing the 5'-phosphate group of the dNTPs with the 3'-hydroxyl group at the end of the nascent (extending) strand.

The end result is that two new helixes are in place of the first, each composed of one of the original strands plus the newly assembled complementary strand. The PCR is usually carried out in a reaction volume of 10–200 μl in small reaction tubes (0.2–0.5 ml) in a thermal cycler. The thermal cycler heats and cools the reaction tubes to achieve the temperatures required at each step of the reaction. The versatility of PCR has led to a large number of variants. The ones more applicable to determine transgenic expression include reverse transcription PCR and real time quantitative PCR discussed below.

RT-PCR

Like Northern blot analysis, reverse-transcription polymerase chain reaction is commonly used to analyze transgene expression. This variant of polymerase chain reaction (PCR) detects RNA expression levels. RT-PCR is often confused with real-time quantitative polymerase chain reaction (qPCR). RT-PCR is used to qualitatively detect gene expression through creation of complementary DNA (cDNA) transcripts from RNA, while qPCR is used to quantitatively measure the amplification of DNA using fluorescent probes.

RT-PCR is used to clone expressed genes by reverse transcribing the RNA of interest into its DNA complement through the use of reverse transcriptase. Subsequently, the newly synthesized cDNA is amplified using traditional PCR (Fig. 3.3). Ever since its development, RT-PCR has displaced Northern blot as the method of choice for RNA detection and quantification [Bustin 2000].

The quantification of mRNA using RT-PCR can be achieved as either a one-step or a two-step reaction. In the one-step approach, the entire reverse transcriptase reaction from cDNA synthesis to PCR amplification occurs in a single tube. On the other hand, the two-step method involves creating cDNA first by means of a separate reverse transcription reaction and then adding the cDNA to the PCR separately.

The first reaction in both approaches uses reverse transcriptase and a primer to anneal and extend a desired mRNA sequence. If the mRNA is present, the reverse transcriptase and primer will anneal to the mRNA sequence and transcribe a complementary strand of DNA. This strand is then replicated with primers and *Taq* polymerase, and the standard PCR protocol is followed. This protocol copies the single stranded DNA (ssDNA) millions of times in a small amount of time to produce a significant amount of DNA. The PCR products (the DNA strands) are then separated with agarose gel electrophoresis. If a band shows up for the desired molecular weight, then the mRNA was in fact present in the sample, and the associated transgene was being expressed.

The advantages to one-step real-time RT-PCR is that it is quicker to set up, less expensive to use, and involves less handling of samples, thereby reducing pipetting errors, contamination, and other sources of error [Wacker and Godard 2005]. With the one-step method, gene-specific primers are used and both the RT and PCR occur in one reaction tube; therefore, other genes of interest cannot be amplified for later analysis. The RNA from the original sample must be initially aliquoted for archival storage and future testing. On the other hand, the main advantage of two-step RT-PCR is that it typically uses random

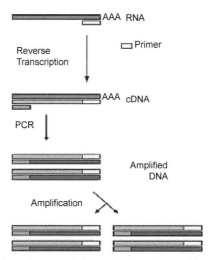

Fig. 3.3. Reverse transcription polymerase chain reaction [JPark623; Wikipedia Commons 2012].

Color image of this figure appears in the color plate section at the end of the book.

hexamer or oligo dT primers in an RT reaction in a separate tube [Wacker and Godard 2005]. This allows for the ability to convert all the messages in an RNA sample into cDNA, which would allow for archiving of samples and future testing of other genes.

The advantages of RT-PCR, whether approached in one-step or two-steps, include its simplicity in comparison to techniques such as RNase protection and the ability to detect rare transcripts or, similarly, mRNA from small tissue samples. The problem associated with RT-PCR, because of its sensitivity, is that the detection of a transcript from transgenic plants is complicated by cDNA copies in genomic DNA contamination of RNA. Therefore, special care must be taken during RNA isolation to ensure that the sample RNA is DNA-free. Despite this limitation, RT-PCR is currently the accepted method for quantitation of extremely rare transcripts from minute samples.

RT-qPCR

A quantitative polymerase chain reaction (qPCR), also called real-time polymerase chain reaction [Radonić et al. 2004] and real-time quantitative PCR [Livak and Schmittgen 2001], is based on the polymerase chain reaction used to amplify and simultaneously quantify a targeted DNA molecule. For one or more specific sequences in a DNA sample, qPCR enables both detection and quantification. The quantity can be either an absolute number of copies or a relative amount when normalized to DNA input or additional normalizing genes.

The procedure follows the general principle of polymerase chain reaction. Its key feature is that the amplified DNA is detected as the reaction progresses in "real time". This is a new approach compared to standard PCR, where the product of the reaction is detected at its end. The two common methods used for the detection of products in qPCR are: a) non-specific fluorescent dyes that intercalate with any double-stranded DNA, and b) sequence-specific DNA probes consisting of oligonucleotides that are labeled with a fluorescent reporter which permits detection only after hybridization of the probe with its complementary sequence to quantify mRNA and non-coding RNA in cells or tissues.

RT-qPCR is carried out in a thermal cycler with the capacity to illuminate each sample with a beam of light of a specified wavelength and detect the fluorescence emitted by the excited fluorophore. The thermal cycler is also able to rapidly heat and chill samples, thereby taking advantage of the physicochemical properties of the nucleic acids and DNA polymerase.

In this technique, reverse transcription (RT) is followed by real-time qPCR. Reverse transcription first generates a DNA template from the mRNA. This single-stranded template is called cDNA which is amplified in the quantitative step, during which the fluorescence emitted by labeled hybridization probes or intercalating fluorescent dyes changes as the DNA amplification process progresses. With a carefully constructed standard curve, qPCR can produce an absolute measurement of the number of copies of original mRNA, in units of copies per nanoliter of homogenized tissue or copies per cell. The technique of qPCR is very sensitive, but can be expensive. Fluorescently labeled oligonucleotide probes are more expensive than non-specific fluorescent dyes that intercalate.

RT-qPCR, a variant of PCR and the widely used technique, is performed for hundreds of genes simultaneously in the case of low-density arrays. It differs from RT-PCR which is used to qualitatively detect gene expression through creation of complementary DNA (cDNA) transcripts from RNA.

ELISA

The enzyme-linked immunoabsorbent assay (ELISA) is a test developed as a diagnostic tool in medicine to identify the presence of a substance, usually an antigen, in a liquid sample or wet sample by using antibodies and color change. Later, it was found that enzyme-labeling could yield quantitative assays with sensitivity comparable to radioimmunoassays. Currently, ELISA test is one of the tests used to detect the presence of the protein encoded by the transgene in a sample of plant tissue.

ELISA is ideal for qualitative as well as quantitative detection of many types of proteins. In this test, the antigen-antibody reaction takes place on solid plates (microliter plates). Of the various forms of ELISA test, the qualitative ELISA Immunostrip assay and the double-antibody sandwich form (DAS-ELISA) are most commonly used. The qualitative assay provides a simple positive or negative result for a sample in a few hours with detection limits. This test can be quickly done in the field using a kit and a small sample of tissue and will give a simple positive or negative response.

Antigen and antibody react and produce a stable complex, which can be visualized by addition of a second antibody linked to an enzyme. Addition of a substrate for that enzyme results in a color formation, which can be measured photometrically or recognized by naked eye. The advantage of ELISA is in addition to qualitative diagnosis, that it can also quantify the targeted protein, provided a standard curve is employed. Thus, the availability of antibodies with the desired affinity and specificity is the most important factor for setting-up immunoassay systems.

There are two formats in ELISA: the microwell plate (or strip) format and coated tube format [Pan 2002]. The antibody-coated microwells with removable strips of 8–12 wells are quantitative and highly sensitive. They are ideal for quantitative high-volume laboratory analysis, provided the protein is not denatured. The average run time for a plate assay is about 90 min, and an optical plate reader determines concentration levels in the samples. Generally, ELISA test kits provide the quantitative results in hours with detection limits less than 0.1 percent, although some companies operate with slightly higher quantification levels as, e.g., 0.3 percent. The second format is suitable for field-testing, with typical run times ranging from 15–30 min. Tubes can be read either visually or by an optical tube reader and results are qualitative. As there is no quantitative internal standard within the assay, no extra information can be obtained concerning the presence of transgenes at the ingredient level in food [Rogan 1999].

Nuclease Protection Assay

Also called 'transcript mapping', nuclease protection assay (NPA) includes both ribonucleaseprotection (RPA) assays and S1 nuclease assays. It is an extremely sensitive method for the detection, quantification, and mapping of specific RNAs in a complex mixture of total cellular RNA. The basis of NPAs is a solution hybridization of a single-stranded, discrete sized antisense probe(s) to an RNA sample. The small volume solution hybridization is far more efficient than more common membrane-based hybridization, and can accommodate up to 100 µg of total or poly(A) RNA. After hybridization, any remaining un-hybridized probe and sample RNA are removed by digestion with a mixture of nucleases.

NPA uses a single-stranded radioactive DNA or RNA probe. The nucleotide sequence of the probe contains at least some nucleotides that are complementary to the mRNAs being analyzed. The probe is annealed to the target mRNA by base-pairing, and the regions of the probe that are complementary to the target mRNA now become double-stranded, while the noncomplementary regions of the probe remain single-stranded. The annealed mixture is then subjected to digestion with an enzyme specific for single-stranded DNA (S1 nuclease), when using a DNA probe, or RNA (a mixture of RNase A and RNase T1: RNase A/T1 mix), when using an RNA probe. The double-stranded annealed areas resist digestion, while all the single-stranded noncomplementary parts of the probe are digested away. In essence, areas in the probe that anneal to the mRNA are "protected" from digestion by the nucleases. The surviving, undigested parts of the probe can then be analyzed by electrophoresis through an agarose or polyacrylamide gel followed by visualization by audioradiography. The amount of radiolabeled probe resistant to digestion is proportional to the amount of target mRNA in the sample.

NPA is used to map introns and 5' and 3' ends of transcribed gene regions. It is also used for simultaneous detection of several RNA species. Quantitative results can be obtained regarding the amount of the target RNA present in the original cellular extract. If the target is mRNA, it indicates the level of transcription of the gene in the cell. This technique is also used to detect the presence of double-stranded RNA whose presence could mean RNA interference. However, its utility is limited to the size of initial probes due to the destruction of the non-hybridized RNA during the nuclease digestion step. Besides, NPA lacks size of information on transcript and flexibility of probe.

Hybridization Microarray

Hybridization microarray is a widely used throughput analytical method to measure the expression of genes in a single experiment. The principle behind microarrays is hybridization between two single stranded DNA strands. In this, cDNA (complementary DNA) sequences are paired with each other by forming hydrogen bonds between complementary nucleotide base pairs. A greater number of complementary base pairs in a nucleotide sequence suggest tighter non-covalent bonding between two strands. After washing off of non-specific bonding sequences, only strongly paired (double) strands will remain hybridized. Fluorescently labeled target sequences that bind to a probe sequence generate a signal that depends on the hybridization conditions (such as temperature), and washing after hybridization. Strength of the signal depends upon the amount of target sample binding to the probes present in the spot.

The protocol involves five steps. The first one is isolation of RNA and preparing it for hybridization to microarray. This is followed by conversion of RNA into 'cDNA' and 'cRNA'. This allows modifying mRNA so that its signal may later be amplified before scanning. In the third step, the mRNA material is hybridized, with the unbound material washed away. The next step involves amplification of the signals followed by microarray scanning. The scan will report how much mRNA is bound to particular regions of the microarray. Once the data is collected, it can be analyzed by bioinformatics tools.

A microarray is a pattern of ssDNA probes (oligonucleotide sequences) which are immobilized on a surface, called a chip (or a microscope slide). A single array or "chip" may contain probes to determine transcript levels for every known gene in the genome of one or more organisms. The probe sequences are designed and placed on an array in a regular pattern of spots. This chip or slide is usually made of glass or nylon membrane. It is manufactured using technologies developed for silicon computer chips. Each microarray chip is arranged as a checkerboard of 10^5 or 10^6 spots or features, each spot containing millions of copies of a unique DNA probe (often 25 nucleotides long).

Like Southern and Northern blots, microarrays use hybridization to detect a specific DNA or RNA in a sample. Unlike in Southern blot, which uses a single probe to search a complex DNA mixture, a DNA microarray uses a million different probes, fixed on a solid surface, to probe such a mixture. The exact sequence of the probes at each feature/location is known. Wherever some of the sample DNA hybridizes to the probe in a particular spot, the hybridization can be detected because the target DNA is labeled (the unbound target is washed away). Therefore one can determine which of the million different probe sequences are present in the target.

Microarrays are broadly classified according to, a) the length of the probes, b) manufacturing method, and c) number of samples that can be simultaneously profiled on one array. Based on length of the probes, arrays can be classified into 'complementary DNA (cDNA) arrays', which use long probes of hundreds or thousands of basepairs, and 'oligonucleotide arrays', which use short probes (50 bp or less). Currently, there are several microarray platforms.

Genetic Recombination

Genetic recombination, an important process in producing transgenic plants by genetic engineering, is the breaking and rejoining of DNA strands to form new molecules of DNA encoding a novel set of genetic information (Fig. 3.4; Eccles 2006). It refers to the exchange between two DNA molecules, resulting in new combinations of genes on the chromosome. Recombination can occur in meiosis between similar molecules of DNA, as in the homologous recombination of chromosomal crossover, or dissimilar molecules, as in non-homologous end joining. It allows chromosomes to exchange genetic information and produces new combination of genes, which increases the efficiency of natural selection and can be important in the rapid evolution of new proteins [Pál et al. 2006]. Genetic recombination can also involve in DNA repair, particularly in the cell's response to double-strand breaks [O'Driscoll and Jeggo 2006].

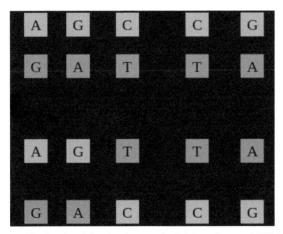

Fig. 3.4. Genetic recombination involving breakage and rejoining of parental DNA molecules (M, F), leading to new combinations of genes (C1, C2) on the chromosome that share DNA from both parents [Eccles 2006; Wikipedia Commons].

Color image of this figure appears in the color plate section at the end of the book.

The crossover process (exchange of genetic material) leads the offspring to have different combinations of genes from those of their parents, and can occasionally produce new chimericalleles. The recombination reaction is catalyzed by many different enzymes, called recombinases. The first step in recombination is a double-stranded break caused by either an endonuclease or damage to the DNA [Neale and Keeney 2006]. A series of steps catalyzed, in part, by the recombinase leads to joining the two helices by at least one 'Holliday junction', in which a segment of a single strand in each helix is annealed to the complementary strand in the other helix. The Holliday junction is a tetrahedral junction structure (between four strands of DNA) that can be moved along the pair of chromosomes, swapping one strand for another. The recombination reaction is then halted by cleavage of the junction and re-ligation of the released DNA [Dickman et al. 2006].

The most common form of chromosomal crossover is homologous recombination (HR). In this, the two chromosomes involved share similar sequences. Non-homologous recombination can be damaging to cells, as it can produce chromosomal translations and genetic abnormalities. Of the two, HR is important for generation, genetic diversity, maintaining the genome integrity as well as repairing the double strand breaks which are generated by the action of ionizing radiation, exposure to genotoxic chemicals, repairs in replication, and during cell development. In plants, HR occurs in meiosis and during somatic development. The meiotic HR is confined to coding regions. Its frequency, however, is not equal for all the genes. Different loci show different recombination frequency [Xu et al. 1995]. Homologous recombination can lead to either crossover or gene conversion event depending on the mechanism of the resolution of the Holliday junction. Generally, HR is more due to gene conversions than crossovers.

In plants, there are several genes related to HR process. The *Xr2* mutant of *Arabidopsis thaliana* plants are hypersensitive to DNA damaging agents like X-ray radiation (xrs), MMC (mitomycin C), and MMS (methyl methane sulfonate) with decreasing somatic HR and increasing meiotic HR [Masson et al. 1997; Masson and Paszkowski 1997]. In rice plants, *OsDMC1* (*Oryza sativa* disrupted meiotic complementary DNA1) gene is essential for meiosis. Two homologues of *OsDMC1* genes include *RiLIM15A* and *RiLIM15B* [Shimazu et al. 2001]. In several other plants including maize and tomato, orthologues of *Rad51* genes are present and these are required during meiotic recombination.

Genetical Change

Plants have the genetic flexibility to respond to biotic and abiotic stresses. Polyploid genomes display dynamic and pervasive changes in DNA sequence and gene expression probably as a response to certain stress and 'genome shock' (release of genome-wide constraints on gene expression and sequence organization) [McClintock 1984] to intergenomic organizations [Comai 2000]. Genome shock and stress activate some of the quiescent transposons in the allopolyploids. Song et al. [1995] detected non-additive inheritance of genomic fragments in the synthetic allotetraploids. The changes include the absence of parental genomic fragments and the presence of novel fragments that were absent from both parents. Many of these changes in *Brassica* allotetraploids are likely to be caused by reciprocal translocations and non-reciprocal exchanges (or transposition) between homologous chromosomes [Chen and Ni 2006].

Wheat allotetraploids displayed 10–15 percent genomic changes, mainly genome-specific sequence deletions, immediately after hybrid (F_1) formation and the homologous genomes showed little changes in the third generation (S_3) in selfing progeny [Feldman et al. 1997; Shaked et al. 2001] Allopolyploids undergo 'revolutionary phase' or rapid genetic and epigenetic changes immediately after allopolyploid formation followed by 'evolutionary phase' (changes in DNA sequence and arrangement) in later generations [Levy and Feldman 2002]. Interestingly, cotton allopolyploids display negligible amount of changes in genomic sequences. Wheat, cotton, and *Brassica* represent a diverse array of molecular evolutionary phenomena in polyploids.

Plants have genetic flexibility to respond to climate change. Climate in which a plant grows determines the suite of genes that gives it the best chance of surviving and reproducing throughout its natural range. In an effort to unlock the molecular basis for a plant's adaptability to climate change, Fournier-Level et al. [2011] found that a set of genes determines the fitness of *Arabidopsis thaliana*, and it varies according to the prevalent climatic conditions in the plant's region. Its adaptation to environment change is due to the presence of loci with different molecular functions. They identified a SNP (single nucleotide polymorphism) allele in a water-stress tolerance, called SAG21 (senescence-associated gene 21). The fitness-associated loci exhibited both geographic and climatic signatures of local adaptation. Relative to genomic controls, high-fitness alleles were generally distributed closer to the site where they increased fitness, occupying specific and distinct climate spaces. Independent loci with different molecular functions contributed most strongly to fitness variation in each site.

Gene Transfer

In gene transfer, the essential feature of transgenic engineering, a segment of DNA containing a gene sequence isolated from organism or plant and is introduced into a different organism or plant. This non-native DNA segment may retain the ability to produce RNA or protein in the transgenic plant, or it may alter the normal function of the transgenic plant's genetic code. It can be a cDNA (complementary DNA) segment, a copy of mRNA, or the gene itself residing in its original region of genomic DNA. The difference between the two lies in the fact that the cDNA has been processed to remove introns and also, usually, do not include the regulatory signals that are embedded around and in the gene. The advent of annotated cloned regions of the genome alongside the genome sequence, in particular as large clones in BACs (bacterial artificial chromosomes) or fosmids, and recombineering, which is the method that permits the engineering of these large clones, has changed the practice of transgenesis from its origins with cDNA-based constructs towards the more reliable genome-based constructs. This subject is discussed in greater detail in Chapter 4.

Genome and Genomics

A genome is an organism's complete complement of DNA, including all of its genes and the non-coding sequences of the DNA/RNA. The word **genome** was first used by German botanist Hans Winkler in 1920 when he combined the two words 'gene' and 'chromosome'. A genome contains all of the genetic information needed to build and maintain that organism. This information is carried by a single set of chromosomes in a haploid nucleus. In plant cell, nucleus, mitochondria and chloroplasts each have their own DNA. Thus, plant cell can have up to three different genomes. Of the three, nuclear genome is the largest and most complex. In a genome, each type of chromosome is represented only once.

Genome

Until very recently, the molecular analysis of plants often focused at the single gene level. Recent advances in genomic research have enabled detailed analysis of organisms in terms of genome organization, expression, and interaction. Plant nuclear genomes exhibit extensive structural variation in size, chromosome number, number and arrangement of genes, number of genome copies per nucleus, and the extent of repetitive sequences and polyploidy/duplication events. This variation is the outcome of a set of highly active processes, including gene duplication and deletion, chromosomal duplication, gene loss, amplification of retrotransposons separating genes, and genome rearrangement, the latter often following hybridization and/or polyploidy [Kellogg and Bennetzen 2004]. Nuclear genomes are largely colinear among closely related species, but more rearrangements are observed with increasing phylogenetic distance. However, the correlation between amount of rearrangement and time since divergence is not perfect. By changing the patterns of gene expression and triggering genome rearrangements, novel combinations of genomes may be a driving force in evolution [Kellogg and Bennetzen 2004].

Plant genomes contain various repetitive sequences and retrovirus-like retrotransposons and retroelements. Retrotransposons, also called **transposons**, which contain long terminal repeats, are mobile genetic elements that transpose through reverse transcription of an RNA intermediate. They are ubiquitous in plants and play a major role in plant gene and genome evolution. Transposons are particularly abundant in plants, where they are often constitute a principal component of nuclear DNA. In many cases, retrotransposons comprise over 50 percent of nuclear DNA content. In maize, 49–78 percent of the genome is made up of transposons [SanMiguel and Bennetzen 1998]. In wheat, about 90 percent of the genome consists of repeated sequences and 68 per cent of transposable elements (Tes) [Li et al. 2004].

Retroelements are elements transcribed into RNA, reverse-transcribed into DNA, and then inserted into a new site in the genome. There are two types of retroelements: long-interspersed and short-interspersed. They make a major contribution to host genome organization, function, and evolution because of their abundance, broad dispersion, and hypervariability. Retroelement insertions contribute to large differences between colinear genome segments in different plant species and to the 50 per cent or more difference in total genome size among species with relatively large genomes, such as maize. They also contribute to a smaller percentage of genome size in plants with smaller genomes such as *Arabidopsis* [Arabidopsis Genome Initiative 2000). If other repetitive sequences are accounted for, the maize genome is comprised of over 70 percent repetitive sequences and of 5 percent protein encoding regions [Meyers et al. 2001].

Many plant genomes also contain supernumerary, also called accessory, chromosomes, often known as **B chromosomes** [Kellogg and Bennetzen 2004]. They are a major source of intraspecific variation in nuclear DNA amounts in numerous plant species. B chromosomes favor large genomes and create polymorphisms for DNA variation in natural populations. By studying B chromosomes', useful knowledge can be gained about the organization, function, and evolution of genomes. These highly condensed chromosomes are usually small and largely or completely devoid of functional genes. In maize, pollen grains that carry B chromosomes exhibit an advantage in fertilization thereby selfish chromosomes will persist in subsequent generations [Carlson 1986]. Within a single species, individuals may have anywhere from zero to several of these accessory chromosomes, thereby altering nuclear chromosome number and genome size but having little other effect upon the biology of the organism.

Despite these dramatic differences in size and number, all seed plants appear to have fairly similar general organizations of their chromosomes [Kellogg and Bennetzen 2004]. Most angiosperm nuclear chromosomes have centromeric (kinetochore) regions (genetic loci that direct the behavior of chromosomes) that are necessary for efficient chromosome segregation. These regions show extensive chromatin condensation and are flanked by large regions of additional heterochromatin (the condensed chromosomal segments which appear in the interphase nucleus) that is enriched for tandem repeats and transposable elements. In large genomes like barley and wheat, these heterochromatic pericentromeric regions can make up to more than 50 percent of the physical length of the chromosome [Kellogg and Bennetzen 2004].

Plant nuclear genomes exist in an organelle that has significant three-dimensional structure. In hybrids of grass species, two parental chromosomes occupy different regions of the nucleus and these patterns are consistent for any given pair of parents [Bennett 1987; Heslop-Harrison and Bennett 1990]. Research in animals indicated that the patterns of three-dimensional structure presumably evolved as important components of regulated gene expression, nuclear packaging, and/or chromosomal mechanics [Belmont 2003]. These patterns of three-dimensional structure presumably evolved as important components of regulated gene expression, nuclear packaging, and/or chromosomal mechanics. In plants, the conservation of one predicted genome structure component (matrix attachment regions, or MARS) at specific locations in orthologous genes has been observed [Avramova et al. 1998; Tikhonov et al. 2000].

Genome Size

Genome size is the total amount of DNA contained within one copy of a single genome. It is measured in terms of mass in one trillionth (10^{-12}) of a gram, i.e., pictogram (pg) and the total number of nucleotide base pairs (bp), expressed in millions of base pairs (Mb). One picogram equals 978 Mb [Dolezel et al. 2003]. In diploid species, genome size is used, interchangeably, as C-value.

Two primary factors contribute to genome size: polyploidy and repetitive DNA sequences. Polyploidy is accumulation of additional sets of chromosomes through autopolyploidy (autoploidy, having more than two sets of chromosomes derived from the same species) or allopolyploidy (alloploidy, having two chromosomes derived from different species). Increased chromosome number and DNA content are immediate consequences of polyploidy. Most of the land plants have undergone polyploidy events at various times in their evolution [Soltis et al. 2004]. Polyploidy is a recurrent process that molds and shapes plant genomes during evolution.

Repetitive DNA sequences and, in particular, transposable elements (TEs) compose large fractions of most plant genomes, and they are impediments to efficient genome sequencing. TE is a DNA sequence that can change its position within the genome, sometimes creating mutations. TEs contribute to genome obesity in plants. Rapid amplification of TE families results in increased genome size. For example, *Oryza australiensis* is approximately twice the size of its nearest relative as a result of amplification of three TE families [Piegu et al. 2006]. Genome obesity in maize is due to TE amplification [SanMiguel et al. 1996]. The complicating factor of TE amplification on genome sequencing is not primarily the increase in the amount of DNA to sequence, but rather the effect of many copies of the same sequence throughout the genome that make mapping and assembly difficult.

In plants, the size of genome ranges from the smallest 63 Mb in *Genlisea margaretae* (carnivorous plant that lacks chlorophyll) to the largest 124,852 Mb in *Fritillaria assyriaca* (the bulbous perennial of the family Liliaceae) [Greilhuber et al. 2006], a 2,000-fold difference. This diversity of genome size has generated considerable interest in the nature of sequence variation among genomes. Similarly, chromosomes are also highly variable, but they do not generally relate to overall genomic size. For instance, the 430-Mb haploid rice genome (the first genome sequenced) is distributed across 12 chromosomes, while the 5,100-Mb haploid barley genome is present on only seven chromosomes [Morrell 2012]. The 2,300-Mb maize genome is spread across 10 haploid chromosomes as against 17,100-Mb-wheat (bread) hexaploid genome (derived from three related ancestral genomes, each having seven chromosomes, giving 42 in diploid cells) 42 chromosomes. The average plant genome is 6,000 Mb per haploid genome for angiosperms [Gregory et al. 2007], approximately twice the human genome size.

Over the last 15 yr, genome sequence analysis in plants has shifted from studies of single genes in isolation to detailed studies of larger chromosomal regions, including whole genomes. These studies have shown that plant genes are relatively compact and often clustered, even in large genomes. Plant introns are usually small, averaging less than 200 bp, so that the average transcribed portion of a gene is less than 2.5 Kb [Arabidopsis Genome Initiative 2000]. Upstream and downstream regulatory elements are usually small as well, amounting to no more than a few hundred additional bases in most genes. There are exceptions in that a regulatory element can be more than 90 Kb upstream of a locus [Stam et al. 2002]. However, regulation at a distance (common in many animal genes) appears to be rare in plants, so that the average gene plus its regulatory components will normally occupy only about 1–5 Kb of genomic space [Kellogg and Bennetzen 2004].

Genomics

The word 'genomics', coined by American geneticist Thomas Roderick in 1986, refers to a new scientific discipline of sequencing and analyzing of genomes. It essentially deals with the study and mapping of genes on the chromosomes and functioning and sequencing of genes as well as metabolic pathways in an organism. It also facilitates simultaneous analysis patterns of differential expression of all or thousands of genes in the genome representing different cells and tissues, and/or different treatments and conditions.

Genomics is the study of the organization, evolution, and function of the genes and non-coding regions of the genome. It helps in identifying all the genes in a plant as well as genetic properties and networks that contribute to the development of a superior plant. It also aids in understanding the origins and domestication of crop plants. The domesticated plants provide a model for studying their adaptation mechanisms. The demographic history of domestication of crop varieties, each with a set of special architecture traits, will help in geneticists' ability to identify casual genetic variants for crop improvement. Domestication is an evolutionary phenomenon. Most of the genealogical history at any locus will be shared between a domesticated crop and its wild progenitor. Comparison of alleles within and between domesticated and wild taxa will reveal divergence times that greatly predate the origin of cultivated form. This enables the geneticists to pinpoint the time to most recent common ancestor of the species rather than the time of divergence of the domesticate species. A detailed understanding of domestication history requires a large number of loci in conjunction with modeling of population demography.

Genomics has now become a major player in the production of enough food for a growing world population. It has several practical applications in crop improvement. These include, determining a) genome size, b) gene number in the genome, c) gene mapping, d) gene sequencing, e) tracing evolution of crop plants, f) gene cloning, g) identification of gene markers, h) marker-assisted selection, i) transgene breeding, j) construction of linkage maps, and k) quantitative trait locus (QTL) mapping.

The basic tools needed for genomic analysis include whole genome sequences, large scale mapping of genes, physical maps, bacterial artificial chromosome (BAC) libraries, collections of sequences of expressed genes (development of EST databases), methods to assess gene expression (at transcriptomic, proteomic and metabolomics level), as well as the databases for storage and computational programs needed to analyze and compare these huge datasets.

A number of sub disciplines of genomics can be combined to provide a powerful approach to studying adaptive genetic variation. They focus on specific attributes of genome analysis. These are structural, functional, comparative, statistical, and associative genomics.

Structural Genomics

Structural genomics is a term that refers to high-throughput three-dimensional structure determination and analysis of biological macromolecules, including every protein encoded by a given genome. It takes advantage of completed genome sequences in several ways in order to determine protein structure. Structural genomics determines the size of the genome and the number of genes present in the entire genome by sequencing individual genes, gene segments, or entire genomes. This study involves, a) high-resolution genetic and physical mapping, b) sequencing of DNA using computer algorithms, c) identifying the complete set of proteins, and d) determining the structure of the concerned protein. Besides, it can also make use of modeling-based approach that relies on homology between the unknown protein and a solved protein structure.

Determination of structure of the target protein is done by utilizing the completed genome sequences in several ways. One approach in structure determination is *de novo* method. In this, completed genome sequences allow every open reading frame (ORF; a part of the reading frame that contains no stop codons), that is likely to contain the sequence from the mRNA and protein, to be cloned and expressed as protein. These proteins are then purified and crystalized, followed by structure determination by either X-ray crystallography or nuclear magnetic resonance (NMR). The whole genome approach allows for the design of every primer required in order to amplify all of the ORFs, clone them, and then express them. It facilitates structural determination of every protein that is encoded by the genome. *De novo* technique obtains a structure more directly from a sequence without the need for a template.

Another technique, sequence-based modeling, compares the gene sequence of an unknown protein with sequences of proteins with known structures. Depending on the degree of similarity between the sequences, the structure of the known protein can be used as a model for solving the structure of the unknown protein. Highly accurate modeling is considered to require at least 50 percent amino acid sequence identity between the unknown protein and the solved structure. Sequence identity of 30–50 percent gives a model of intermediate-accuracy while that below 30 percent gives low-accuracy models. One disadvantage of this method is that structure is more conserved than sequence and thus sequence-based modeling may not be the most accurate way to predict protein structures.

When complete sequence of an entire genome is not available, the location of genes can be determined either by direct physical mapping or genetic mapping of the entire genome using numerous genetic markers. One of the most prominent applications of structural genomics for the study of adaptive genetic variation is QTL analysis via genome mapping. However, this approach aims to explain genomic structure and gene interaction at the genomic rather than functional level, unlike functional genomics.

Functional Genomics

Functional genomics, the study of genes, function, and regulation, assigns functions to each and every gene identified through structural genomics. It refers to the development and application of global (genome-wide and system-wide) experimental approaches to assess gene function by making use of the information and regaents provided by structural genomics [Xu 2010]. It focuses on the dynamic aspects such as gene transcription, translation, and protein-protein interactions while dealing with the study of function of DNA at gene level, RNA transcripts, and protein products. The goal of functional genomics is to understand the relationship between a plant's genome and its phenotype. It involves high-throughput methods, combined with the results of statistical and computational analysis (bioinformatics), rather than a more traditional 'gene-by-gene' approach.

Functional genomics includes not only function-related aspects of genome itself such as mutation and polymorphism (such as a single-nucleotide polymorphism: SNP) but also measurement of molecular activities. The latter comprise transcriptomes (gene expression), proteomes (protein expression), and metabolomes. Functional genomics uses mostly multiplex techniques to measure the abundance of many or all gene products such as mRNAs or proteins within a biological sample. Together these measurement modalities endeavor to quantitate the various biological processes and improve our understanding of gene and protein functions and interactions.

Transcriptome refers to the complete set of RNA molecules (including mRNA, rRNA, tRNA, and other non-coding RNAs produced in a cell) transcribed from a genome. Proteome, a blend of words 'protein' and 'genome', is a complete set of proteins encoded by a genome. Proteomics is important as proteins are active agents in cells and they execute the biological functions encoded by genes. The metabolome represents the collection of all metabolites in a biological cell, cellular compartment, plant tissue, organ, or organism, which are the end products of cellular processes [Jordan et al. 2009].

Comparative Genomics

Comparative genomics is the study of the relationship of genome structure and function across different biological species or strains. This young field of genomics uses information from different species and assists in understanding gene organization and expression and evolutionary differences. It takes advantage of the high level of gene conservatism in structure and function and applies this principle in an interspecific manner in the search for functional genes and their genomic organization. The long-term goals of comparative genomics are to establish relationships between map, sequence and functional genomic information across all plant species and to facilitate taxonomic and phylogenetic studies in higher plants [Xu 2010]. Comparative maps lay the groundwork for asking questions about whether specific 'linkage blocks' or gene arrangements are statistically associated with increased fitness or have a relationship between polyploidy and plant adaptation [Xu 2010].

Comparative genomic sequencing presents an opportunity to study the evolution of plant genome structure and the dynamics of molecular evolutionary process. The two species with completed genome sequences *Arabidopsis thaliana* and *Oryza sativa* last shared a common ancestor 150–200 million yr ago.

Associative Genomics

Associative genomics searches for mutations in populations by linkage disequilibrium analysis as also by direct assessment of association between alleles and phenotypes. This approach is used to find adaptive mutations such as herbicide resistance, disease resistance, drought tolerance, cold hardiness, etc. DNA variants or mutations (inherited differences in DNA sequence) can either directly contribute to phenotype variation, influencing phenotype characteristics (e.g., risk of herbicide-resistance, disease-resistance, and response to environment), or can be tightly linked to the genes causing this variation. In the latter case, the alleles serve as markers of the selective genes and can be in linkage disequilibrium with alleles of this gene due to the limited population size, recent origin, low combination rate and/or strong selection acting on alleles of the linked selective gene. Once candidate alleles responsible for adaptive traits are detected via QTL (quantitative trait loci) and comparative mapping, it will be possible to perform association studies to estimate effects of alleles or haplotypes (a particular combination of alleles or sequence variations that are closely linked on the same chromosome) on phenotypes. It should be practical to define common haplotypes using a dense set of polymorphic markers, and to evaluate each haplotype for association with disease or any particular adaptive trait.

Statistical Genomics

Statistical genomics is an integrated sub-discipline and it serves all other areas of genomics. It provides statistical tools for genome and QTL mapping in structural genomics, bioinformatics tools for gene search, comparison and annotation in functional genomics, and statistical population genetic methods in associative genomics. Statistical genomics is also very important in developing computerized comprehensive interactive biological databases. New computer tools are required to compose genetic data at all levels of biological organization (from gene to population, species, and ecosystems) and for multiple purposes, including gene conservation.

Genome Sequencing

Genome sequencing, also called DNA sequencing, which determines the precise order of nucleotide bases (A, G, C, and T) in a DNA strand and illustrates how the genes are encoded within the genome, has become the fundamental resource for plant biology. Although the pace of genome sequencing in plants lags behind that in human, bacteria, and mammalian systems, application of genomics and the associated data, expertise and hypotheses are rampant among such sub-disciplines of plant science as weed science, entomology, biochemistry, forestry, genetics, horticulture, plant pathology, and systematics.

Recent advances in genomics have accelerated the sequencing of new genomes, far outpacing the generation of gene and plant resources needed to annotate them. Plant genome sequencing offers two primary benefits. It identifies genes and other functional elements and provides useful data for annotation of plant genomes and provides an important tool to pursue gene isolation in target species.

Whole-genome sequencing (WGS), also known as full-genome sequencing, complete-genome sequencing, or entire-genome sequencing, is a laboratory process that determines the complete DNA sequence of a plant's genome at a single time. This entails sequencing a plant's DNA contained in mitochondria and chloroplasts. A biological sample containing the full copy of the DNA, even if small amount of DNA or ancient DNA, can provide the genetic material necessary for full genome sequencing. The term 'full genome' is sometimes loosely used to mean 'greater than 95 percent'.

Complete chloroplast genome sequences provide a valuable source of molecular markers for studies in molecular ecology and evolution of plants. Unlike full genome sequencing, DNA profiling, also known as 'DNA fingerprinting' or 'genotyping' (discussed later in the Chapter), only determines the likelihood of genetic material coming from a particular individual or group. It does not contain additional information on genetic relationships, origin, or susceptibility to specific disease. The plastid genome contains many essential genes, especially those required for photosynthesis. For example, plant leaf cells often contain 400 to 1,600 copies of the plastid genome. In angiosperms, most chloroplast genomes are circular DNA molecules ranging from 120 to 160 Kb. Mitochondrial genomes, especially those in seed plants, are exceptionally varied in size and structure, and their sequence contents accumulate many repetitive sequences [Alverson et al. 2010, 2011].

The advent of DNA sequencing technology has accelerated determination of the sequence of individual genes, gene identification, larger genetic regions (i.e., clusters of genes and operons), entire genomes, and the number of genes tested by 10-fold. This information helps geneticists and plant breeders to understand the function of specific genes and the genome and to improve plant breeding and development systems. There are several DNA sequencing methods currently available. With burgeoning DNA sequencing technology, chiefly used in medical research, biopharma, and plant biology, the one billion dollar (US)-market in 2011 is expected to double by 2016 [DeciBio 2011].

Currently, three strategies are being used for plant-genome sequencing methods. These include, a) physical-map, b) whole-genome shotgun-based, and c) gene-based methods. However, all these strategies either require isolation of chloroplast or mitochondrial DNA from nuclear DNA [Atherton et al. 2010]. Isolating mitochondria and their DNA is often challenging. This makes it imperative to develop better methods for sequencing and assembling these genomes that do not include experimental sample enrichment [Zhang et al. 2011]. The sequencing methods described here are not to be comprehensive, but to aid readers become familiar with them.

There are two groups of sequencing: **first-generation** and **next-generation**. The first-generation sequencing methods include Sanger Sequencing and Maxam-Gilbert Sequencing. These are described here briefly.

Sanger Sequencing

Sanger sequencing, developed by the British biochemist Frederick Sanger (Nobel laureate in 1958 and 1980) and his group in 1977 [Sanger et al. 1977] enabled the first large-scale gene discovery effort via sequencing, i.e., expressed sequence tags (ESTs) [Adams et al. 1993]. The single-end sequencing of cDNA clones permitted discovery of genes from discrete tissues of interest. This approach, while initially controversial, was embraced by plant biologists, and included sequencing of ESTs from the model species *Arabidopsis thaliana* [Newman et al. 1994].

This paradigm-changing technology requires a single-stranded DNA template, a DNA primer, a DNA polymerase, deoxynucleotidetriphosphates (dNTPs), and modified nucleotides (dideoxy NTPs: ddNTPs) that terminate DNA strand elongation. These chain-terminating ddNTPs lack a 3'-OH group required for a phosphodiester bond formation between two nucleotides, resulting in DNA polymerase to cease extension of DNA when a dNTP is incorporated. The ddNTPs may be labeled radioactively or fluorescently for detection in automated sequencing machines (Fig. 3.5).

The Sanger chain-termination method developed soon became the method of choice due to its relative ease and reliability [Sanger and Coulson 1975]. It uses fewer toxic chemicals and lower amounts of radioactivity than the Maxam and Gilbert method. Because of its comparative ease, the Sanger method was soon automated to be used by the first generation of DNA sequencers. The first plant genome to be sequenced was that of *Arabidopsis thaliana* was completed by using Sanger's method in 2000 [Arabidopsis Genome Initiative 2000]. This was followed by whole genome sequencing of *Orzya sativa* in 2002 [Yu et al. 2002] and 2005 [International Rice Genome Sequencing Project 2005]. Then came the draft sequence of the 2,300 Mb maize genome in 2008 [Pennisi 2008]. All were sequenced using 'traditional' sequencing approaches in which sequencing libraries are constructed from individual segments of the genome (such as BAC clones) and are sequenced via gel electrophoresis and dideoxy terminator chemistry outlined in Sanger

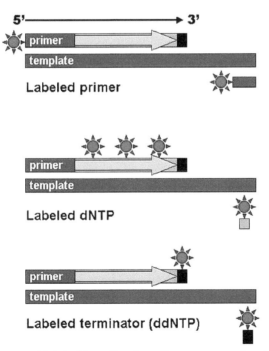

Fig. 3.5. DNA fragments are labeled with a radioactive or fluorescent tag on the primer (1), in the new DNA strand with a labeled dNTP, or with a labeled ddNTP [Lakdawalla 2007; Wikipedia Commons 2007].

Color image of this figure appears in the color plate section at the end of the book.

sequencing. Although Sanger sequencing was popular until about 10 yr ago as it can read up to 1,000 bp with 99.9 percent accuracy in 20 min to 3 hr time, it is, however, the most expensive of those in use and impractical for larger sequencing projects.

Maxam-Gilbert Sequencing

Maxam-Gilbert sequencing, also known as chemical sequencing, uses a method similar to the Sanger method in using polyacrylamide gels to resolve bands that terminated at each base throughout the target sequence, but very different in the way that products ending in a specific base were generated. This method starts with a double-stranded DNA restriction fragment radiolabeled at one end with ^{32}P. The fragment was then cleaved by base-specific chemical reactions. One reaction cleaves at both purines (the 'A + G' reaction), one preferentially at adenines (A > G), one at pyrimidines (C + T) and one at cytosines (C) only. The concentration of the modifying chemicals is controlled to introduce on average one modification per DNA molecule. Thus, a series of labeled fragments is generated, from the radiolabeled end to the first "cut" site in each molecule. The fragments in the four reactions are electrophoresed side by side in denaturing acrylamide gels for size separation.

To visualize the fragments, the gel is exposed to X-ray film for autoradiography, yielding a series of dark bands (each corresponding to a radiolabeled DNA fragment) from which the sequence may be inferred. The chemical method produces bands for every sequence position, including those within homopolymer runs. This advantage led to early widespread adoption of the chemical method following its publication in February 1977 [Maxam and Gilbert 1977].

The demand for low-cost sequencing has driven the development of high-throughput sequencing, also called **next-generation sequencing**. These technologies can produce thousands or millions of sequences in a single round, increase read accuracy, and lower the cost of DNA sequencing significantly. Next generation sequencing enables the *de novo* sequencing of a plant genome without any reference genome sequence in a fast and efficient way. Beginning 1995, various technologies have been developed to provide a cost-effective alternative to the traditional Sanger and Maxam-Gilbert methods. This part of the chapter covers a brief description of each of some of these methods. These include 454 Pyrosequencing, Illumina Sequencing, Sequencing by Ligation, Single Molecule Real Time Sequencing, Ion Semiconductor Sequencing, Heliscope Single Molecule Sequencing, Nanopore-based Sequencing by Synthesis, Chromosome Walking & Shotgun Sequencing methods.

454 Pyrosequencing

This first commercially successful next generation method, developed and made available in 2005 by the Branford, Connecticut-based Life Sciences Co. [Margulies et al. 2005] utilizes pyrosequencing technology. It relies on the detection of pyrophosphate released during nucleotide incorporation. In this, a single DNA fragment is clonally amplified by polymerase chain reaction (PCR) within a water-in-oil microreactor. Sequencing occurs on a flow cell with picoliter wells, in which addition of a nucleotide to the growing strand by DNA polymerase results in the release of a pyrophosphate. The pyrophosphate is then used in a coupled reaction with ATP sulfurylase, luciferase, and luciferin that emits light, which is captured by a CCD (charge-coupled device for digital imaging) camera. Apyrase (a calcium-activated plasma membrane-bound enzyme that catalyzes the hydrolysis of

ATP to yield AMP and inorganic phosphate) is used to degrade unused nucleotides and ATP before the next deoxyribonucleotide triphosphate is added. This platform has evolved with respect to throughput and read length. Reads may be single-end or fragments may be circularized, ligated, and selected to generate mate-pair sequences to provide scaffolding information [Hamilton and Buell 2012].

This technology provides intermediate read length (400–600 bp of DNA; with one million reads 24 hr run at 99.9 percent accuracy) and lower price per base compared to Sanger sequencing. 454 pyrosequencing has been utilized in transcript sequencing [Bajgain et al. 2011] and identification of novel transcripts in *Arabidopsis thaliana* genome [Weber et al. 2007].

Illumina Sequencing

In Illumina "sequencing by synthesis" method, developed by the Hayward, California-based Solexa, which was acquired in 2007 by Illumina [Bentley et al. 2008], DNA molecules and primers are first attached on a slide and amplified with polymerase so that local clonal colonies (DNA colonies), are formed. In order to determine the sequence, four types of reversible terminator-bound deoxynucleotide triphosphate (dNTPs: dATP, dCTP, dGTP, dTTP) are added, each fluorescently labeled with a different color and attached with a blocking group. The four bases then compete for binding sites on the template DNA to be sequenced, and the non-incorporated molecules are washed away. After each synthesis, a laser is applied resulting in the removal of the 3′ terminal blocking group and the probe. A detectable fluorescent color specific to one of the four bases is then visible, allowing for sequence identification and the beginning of the next cycle. The process is repeated until the full DNA molecule is sequenced [Meyer and Kircher 2010]. This platform has been widely used in plant genomics, for expression profiling, *de novo* sequencing, and re-sequencing [Hamilton and Buell 2012]. This very low-cost "sequencing by synthesis" method has a read length of 50–250 bp, with up to three million reads in one to 10 d run, depending upon sequencer and specified read length at 98 percent accuracy [van Vliet 2010].

Illumina's method powerfully combines the flexibility of single reads, short- and long-insert paired-end reads, enabling the broadest range of genomic applications, including whole-genome sequencing, targeted resequencing, *de novo* sequencing, amplicon sequencing, SNP discovery, identification of copy number variations, and chromosomal rearrangement.

Sequencing by Ligation

Sequencing by oligonucleotide ligation and detection (SOLiD) is another high throughput next-generation DNA sequencing technology developed by the Foster City, California-based Applied Biosystems of Life Biotechnologies. In this method, oligonucleotides are annealed and ligated by the enzyme DNA ligase to identify the nucleotide present at a given position in a DNA sequence. The ligation cycle is repeated until the desired read length is achieved. All possible oligonucleotides of a fixed length are labeled according to the sequenced position.

Unlike some currently popular DNA sequencing methods, SOLiD does not use a DNA polymerase to create a second strand. Before sequencing, the DNA is amplified by emulsion PCR. In this platform, two-color encoding of every base in color space provides a high-quality sequence, which is a major advantage of this platform. The resulting beads, each containing single copies of the same DNA molecule, are deposited on a glass slide [Valouev et al. 2008]. SOLiD method has a maximum read length of 85 bp, with 30 Gbp per run of 7 d [Liu et al. 2012] at 99.99 percent accuracy. It is applicable for whole-genome sequencing, targeted resequencing, transcriptome research (including gene expression profiling, small RNA analysis, and whole transcriptome analysis), and epigenome (like CHIP-Seq and methylation) [Liu et al. 2012].

Single Molecule Real Time (SMRT) Sequencing

SMART sequencing, developed by the Menlo Park, California company Pacific Biosciences, is based on the sequencing by synthesis approach. DNA sequencing is done on a chip that contains many zero-mode waveguides (ZMWs). A single DNA polymerase is affixed at the bottom of a ZMW with a single molecule of single stranded DNA as a template. The ZMW is a nanostructure array that creates an illuminated observation volume that is small enough to observe only a single nucleotide of DNA (dNTP) being incorporated by DNA polymerase. Each of the four DNA bases is attached to one of four different fluorescent dyes. When a nucleotide is incorporated by the DNA polymerase, the fluorescent tag is cleaved off and diffuses out of the observation area of the ZMW where its fluorescence is no longer observable. A detector detects the fluorescent signal from the phospho-linked nucleotide incorporated by the DNA polymerase at a single molecule level. Sequence data were aligned with the known reference sequence to assay biophysical parameters of polymerization for each template position [Eid et al. 2009]. This method has a mean read length of 2,900 bp, with 35,000–75,000 reads per run [Rasko et al. 2011] of 30 min to 2 hr at 99.3 percent accuracy.

Ion Semiconductor Sequencing

Developed in 2010 by the Gilford, Connecticut-based Ion Torrent Systems, later owned by Life Technologies, ion semiconductor sequencing technology uses standard sequencing chemistry, but with a novel, semiconductor based detection system, known as personal genome machine (PGM) sequencer. When a nucleotide is incorporated into the DNA molecules by polymerase enzyme, a proton is released. PGM detects these hydrogen ions. A microwell containing a template DNA strand to be sequenced is flooded with a single type of nucleotide. If the introduced nucleotide is complementary to the leading template nucleotide, it is incorporated into the growing complementary strand. This causes the release of a hydrogen ion that triggers a hypersensitive ion sensor, which indicates that a reaction has occurred. If homopolymer repeats are present in the template sequence, multiple nucleotides will be incorporated in a single cycle. This leads to a corresponding number of released hydrogens and proportionally higher electronic signals [Rusk 2011]. This low-cost ion semiconductor sequencing has 200 bp read length, with up to five million reads per run of 2 hr at 98 percent accuracy.

Heliscope Single Molecule Sequencing

This is one of the first techniques for sequencing from a single DNA molecule introduced by Braslavsky et al. [2003] and licensed by the Cambridge, Massachusetts-based Helibios Biosciences in 2007. It is based on "true single molecule sequencing" (tSMS) technology, in which no clonal amplification is required. Instead, it uses highly sensitive fluorescence detection system to directly interrogate single DNA molecules via sequencing by synthesis.

The process begins with library preparation through DNA shearing and addition of poly-A tailors to generation of DNA fragments [Ozsolak et al. 2010]. This is followed by hybridization of DNA fragments to the poli-T oligonucleotides which are attached to the flow cell and simultaneously sequenced in parallel reactions. The sequencing cycle consists of DNA extension with one, out of four fluorescently labeled nucleotides, followed by nucleotide detection with the Heliscope Sequencer. The subsequent chemical cleavage of fluorophores allows the next cycle of DNA elongation to begin with another fluorescently labeled nucleotide, which enables the determination of the DNA sequence [Harris et al. 2008]. The Heliscope sequencer is capable of sequencing up to 28 Gb in a single sequencing run of 8 d. It can generate short reads with a maximal length of 55 bases. In 2011, Helicos developed a new generation of 'one-base-at-a-time' nucleotides which allow more accurate homopolymer and direct RNA sequencing [Ozsolak and Milos 2011a,b].

Nanopore-based Sequencing by Synthesis

Contrary to all DNA sequencing technologies mentioned earlier, sequencing a DNA molecule with the nanopore sequencing is free of nucleotide labeling and detection. This method, under development since 1995 and which came to fruition in 2012, was developed from studies on translocation of DNA through various artificial nanopores (of one nanometer in diameter). These pores make holes in a special membrane, allowing charged particles in the solution around it to pass through the other side. Special electrodes on the other side of the membrane can sense this ion flow. When a molecule like a DNA nucleotide passes through the pore, the number of ions flowing through it changes.

During this novel single molecule nanopore-based sequencing by synthesis (Nano-SBS) method [Kumar et al. 2012], electrical signal of nucleotides is converted by passing it through a nanopore, an alpha-hemolysin (αHL) pore covalently attached with cylodextrin molecule, the binding site for nucleotides. During the sequencing process, the ionic current that passes through the nanopore is blocked by the nucleotide, previously cleaved by exonuclease from a DNA strand that interacts with cyclodextrin. The time period of current block is characteristic for each base and it enables the DNA sequence to be determined. The change in ionic current depends on the shape, size, and length of the DNA sequence. Recently, Oxford Nanopore Technologies has developed a nanopore technology that comes in a cartridge with all of the chemicals needed to process a DNA sample. The cartridges come in a larger desktop version of 2,000-nanopore capacity and a smaller USB stick with 512-nanopore capacity. The company also proposed that creating two recognition sites within a αHL pore may confer advantages in base recognition [Stoddart et al. 2010].

Chromosome Walking, Shotgun Sequencing

Since Sanger's chain termination method can only be used to sequence fairly short strands, longer sequences need to be subdivided into smaller fragments, and subsequently re-assembled to give the overall sequence. This can be accomplished by chromosome walking and shotgun methods. Chromosome walking progresses piece by piece through the entire strand. In shotgun sequencing [Staden 1979; Anderson 1981], DNA is broken up randomly into numerous small segments, which are sequenced using the chain termination method to obtain *reads*. Multiple overlapping reads for the target DNA are obtained by performing several rounds of fragmentation and sequencing. Computer programs are then employed by using the overlapping ends of different ends to assemble them into a continuous sequence [Staden 1979]. Shotgun sequencing is one of the precursor technologies that enable full genome sequencing.

Currently, many more DNA sequencing methods are under development. They make use of DNA polymerase by labeling it and reading the sequence as a DNA strand transits through nanopores [The Harvard Nano Group 2009]. Besides, they also make use of microscopy-based techniques, such as atomic force microscopy or transmission electron microscopy to identify the positions of individual nucleotides within long fragments (> 5,000 bp) by nucleotide labeling with heavier elements (e.g., halogens) for visual detection and recording [Xu et al. 2009]. These third generation sequencing technologies aim to increase throughput while reducing cost and the time to get results.

DNA sequences are important sources of data for phylogenetic analysis. It is a routine technique in molecular biology research. Recent advances in DNA sequencing technologies have dramatically reduced cost and time needed to sequence a plant's entire genome. Although DNA sequence information is available for 125,426 plant species, as of April 2012 [NCBI Taxonomy 2012], most of it is only for one or two genes. This information is used to construct genetic maps of mitochondrial or chloroplast DNA rather than the actual genomic DNA of plants.

Crop Improvement

Crop improvement refers to the genetic alteration of plants to satisfy human needs. In prehistory, human forebears in various parts of the world brought into cultivation a few hundred species from the hundreds of thousands available. In the process, they transformed elements of these species into crops through genetic alterations that involved conscious and unconscious selection. Through a long history of trial and error, a relatively few plant species have become the mainstay of agriculture and thus the world's food and fiber supply. This process of domestication involved the identification of certain useful wild species combined with a process of selection that brought about changes in appearance, quality, and productivity. The exact details of the process that altered the major crops are not fully understood, but it is clear that the genetic changes were enormous in many cases. In fact, some crop plants have been so changed that for many of them their origins are obscure, with no extant close wild relatives.

The mainstay of crop improvement in the modern era is breeding for better and newer varieties. This is made possible by conventional, mutation breeding, and molecular plant breeding techniques.

Plant Breeding

Plant breeding involves changing the genetic make-up of plants in order to produce the characters desired. This is accomplished through different techniques ranging from selecting plants with desirable characteristics such as faster growth, larger seeds or sweeter fruits, etc. for propagation, to more complex molecular techniques. This dramatically changed domesticated plant species compared to their wild relatives. Remarkably, many of the modern crops have been developed in the past by people who apparently lacked the scientific basis of plant breeding. Farmers used the selected plants as seed source for subsequent generations, resulting in accumulation of the desired characteristics over time.

The process of plant breeding essentially involves changing the genes of a plant to develop a new and better variety. In this process, desirable traits are identified and selected and combined them into one individual plant. As all traits are controlled by genes located on chromosomes, plant breeding can be considered as the manipulation of the combination of chromosomes. This is done in four steps: pure-line selection (or selection), crossing, hybridization, and polyploidy.

Selection

The selection of naturally occurring variants is the basis of crop improvement. Other selection techniques for asexual (vegetative) propagation, such as by using natural offshoots, rooting stem cuttings, or various grafting techniques, made it possible to 'fix' genetic variants. In the 18th and 19th centuries, an attempt was made to predict the performance of plants that could be expected from one seed generation to the next. The concept that ancestry was important in crop improvement led to refinement in the selection process. A new type of selection, 'pedigree' selection, increases the efficiency of the process. Progeny testing to evaluate the genetic utility by progeny performance increases efficiency of this process.

Selection, the basic procedure in plant breeding, involves three distinct steps. First, a large number of selections are made from the genetically variable original population. Second, progeny rows are grown from the individual selections for observational purposes. After eliminating undesirable ones, the selected plants are grown for several years to observe their performance under different environmental conditions for making further eliminations. Finally, the selected inbred lines are compared to existing commercial varieties in their performance for higher yields and other characters of agronomic importance.

Crossing

Plant breeding uses deliberate interbreeding, known as crossing, of closely or distantly related individual plants to produce new crop varieties or lines with desirable properties. Plants are crossbred to introduce traits/genes from one variety or line into one with a different genetic background. For example, a disease-resistant crop variety is crossed with a high-yielding but susceptible variety to introduce disease resistance into the latter but without losing its high-yielding characteristics. Progeny from the cross would then be crossed back with the high-yielding parent to ensure that the progeny were most like the high-yielding parent. This second process is called backcrossing. The resultant high-yielding and disease-resistant plants may either be used for further development or to produce inbred varieties/lines for breeding. Inbred lines are genotypes developed to be used as parents in the production of hybrid cultivars.

Before undertaking a breeding program, it is important to test for the combining ability of inbred lines. Combining ability is of two types: general combining ability and specific combining ability. General combining ability is the additive type of gene action while specific combining ability is due to non-additive (dominant or epistatic) gene action. The crossbred plants are generally superior to their parents due to a phenomenon called heterosis.

Heterosis. Heterosis refers to the increased vigor of crosses between species, or between distantly related variants within a species, compared with parents. It is an effect achieved by crossing highly inbred lines of crop plants. Heterosis, also called hybrid vigor, is more frequent and intense in cross-pollinated plants like maize than in self-pollinated plants like rice and wheat. This is mainly due to an effect of natural selection, which in cross-pollinated plants increases the genetic load. Heterosis, also called hybrid vigor, occurs when two homozygous individuals are cross-pollinated. This causes all loci to become heterozygous, and increased heterozygosity causes increased plant vigor. The most notable and successful heterosis-exhibiting hybrid was produced in maize. The first hybrid maize using a four-way cross made in 1917 by Donald Jones, geneticist at the Connecticut Agricultural Experiment Station, USA. In 1919, he invented a double cross pollination method, which allows for commercial production of hybrid maize. Two years later, the first commercial double-cross hybrid maize (Burr-Leaming) was released and recommended by him.

Farmers need to buy new hybrid seed every year because the heterosis effect is lost in the first generation after hybridization of the inbred parental lines. Crosses between inbreds from different heterotic groups yield F_1 hybrids with significantly more heterosis than F_1 hybrids within the same heterotic group. Heterotic groups are created by plant breeders to classify inbred lines, and improve them progressively by reciprocal recurrent selection.

Hybridization

Hybridization is the most frequently employed breeding technique to produce a hybrid which may have a distinct physical and visual trait (phenotype) than its parents. The first step is to generate homozygous inbred lines. This is normally done by using self-pollinating plants where pollen from male flowers pollinates female flowers from the same plants. Once a pure line is generated, it is combined with another inbred line through cross-pollination. The progeny is then selected for combination of desired traits. If a trait from a wild relative of a crop species is resistant against a disease, it is transferred into the genome of the crop.

There are two types of hybrid plants: interspecific and intergeneric. Sexual incompatibility limits the possibilities of introducing desired traits into crop plants through hybridization. The hybrid, being the offspring resulting from the mating of two distinctly homozygous plants, is a heterozygous plant. It inherits two alleles, dominant and recessive, one from each parent. The F_1 generation is phenotypically homogenous, producing offspring that are all similar to each other. There are different types of hybrids.

Single Cross Hybrids. These are generated by crossing homozygous inbred (genetically pure) parental lines, A and B. So, the hybrid population, AB, is genetically heterozygous and self-incompatible, but there is no genetic variation between siblings. The simple act of different strains/inbreds results in stronger and vigorous plants, and higher yields.

Double Cross Hybrids. If hybrid vigor is very strong, plant breeders may then produce double cross hybrids by crossing two different single F_1 hybrid plants, AB and CD. The seeds of the resultant double cross hybrid, ABCD, have two generations of hybrid vigor bred into them. Double-cross hybridization permits four successful unrelated inbred parents with desirable characteristics to be brought together into one hybrid. Double crosses have a low frequency of yielding recombinants in the F_2 that possesses a significant number of desirable parental genes. The double-cross hybrid is more broad-based than the single-cross hybrid, but is more time-consuming to make. Double-cross method was used almost exclusively from 1926 until the 1960s when scientists developed better inbred lines and found they could get better hybrids with a single cross, despite its being time-consuming.

Three-way Cross Hybrids. They arise out of crossing between one parent, an F_1 hybrid (of two inbred parents), and an inbred line. In this, crosses are made between a single-cross hybrid (AxB), the seed parent, and an inbred line (C), the pollen parent to give the pedigree

{(AxB)xC}. Three-way inbred seed is produced on single-cross plants so that yield and quality may be equal, or nearly so, to double-cross seed. These hybrids are more variable than single cross hybrids and less variable than double cross hybrids.

Triple Cross Hybrids. The triple cross hybrid is produced by crossing two three-way crosses. With inbred parents S_1S_1, S_2S_2, S_3S_3, S_4S_4, S_5S_5, and $S_6S_6S_6S_6$, the triple cross hybrid is produced thus [Roy 2000]:

$$S_1S_1 \text{ x } S_2S_2 \qquad\qquad S_4S_4 \text{ x } S_5S_5$$
$$\downarrow \qquad\qquad\qquad\qquad \downarrow$$

F₁: $\qquad\qquad S_1S_2 \text{ x } S_3S_3 \qquad\qquad S_4S_5 \text{ x } S_6S_6$
$\downarrow$
$\qquad\qquad\qquad\qquad\qquad \downarrow$

Three-way cross: $\qquad\qquad S_1S_3 \text{ x } S_2S_3 \text{ x } S_4S_6 \text{ x } S_5S_6$
$\downarrow$

2) Use of Triple cross: S_1S_4, S_1S_6, S_3S_4, S_3S_6, S_2S_4, S_2S_6, S_1S_5, S_1S_6, S_3S_5, S_2S_6, S_2S_5, S_2S_6.

Top Cross Hybrids. In this cross, one of the parents is an open-pollinated cultivar, the pollen parent, and the other a single-cross hybrid or an inbred line. Top cross hybrid is made by using an inbred line that is self-incompatible. When a cross is made using a variety and a single cross hybrid, a double top cross hybrid is produced. Most of the cabbage hybrids produced in USA includes top cross hybrids. A three-way top cross hybrid can be produced by crossing F₁ of the two highly self-incompatible inbreds with an open-pollinated line.

Classical plant breeding facilitates homologous recombination between chromosomes to generate genetic diversity. It makes use of a number of *in vitro* techniques such as protoplast fusion, embryo rescue or mutagenesis to generate diversity and produce hybrid plants that do not exist in nature. The traits that have been incorporated into crop plants over the past 100 years included increased quality and yield, increased tolerance to salinity, drought, herbicides, and insect pests, and resistance to viruses, bacteria, and fungi.

Polyploidy

Polyploidy is pervasive in plants, and many plant lineages show evidence of ancient paleopolyploidy in their genomes [Otto 2007]. Polyploid plants can arise spontaneously in nature by several mechanisms including meiotic (commonly due to metaphase 1) or mitotic failures and fusion of unreduced (2n) gametes. As mentioned earlier, there are two types of polyploid plants: autopolyploid and allopolyploid. Autopolyplods (e.g., potato) have more than two sets of chromosomes, both derived from the same species. Allopolyploid plants (e.g., wheat, cotton, rapeseed, etc.) consist of two or more sets of chromosomes derived from two or more different diploid (two sets of chromosomes) species. Most polyploids display heterosis relative to their parental species. Triploid (3x) crops include banana, apple, citrus, watermelon, ginger, etc. while tetraploids (4x) encompass such major crops as maize, durum wheat (*Triticum durum*), cotton, potato, groundnut (peanut) tobacco, cabbage, etc. Bread wheat (*Triticum aestivum*), oat, and kiwifruit are hexaploids (6x), whereas sugarcane and strawberry are octaploids (8x).

Mutation Breeding

Mutation, whether spontaneous (natural) or induced (artificial, with the aid of agents), is a heritable change in the genetic material. Mutations involve large sections of DNA becoming duplicated, usually through genetic recombination. In a plant species, genetic variability can be introduced through spontaneous or artificially-induced mutations during DNA repair. Spontaneous mutation, by way of molecular decay, occurs naturally due to a low level of natural mutagens present in the environment. One such mutation occurs in every million to one billion divisions. Induced mutation, as in the case of mutation breeding, involves exposure of seeds or plant parts to chemicals and radiation in order to generate mutants with desirable traits to be bred with other cultivars.

Depending on the type of structural change, mutations may be grouped into three: **genomic mutation**, **structural mutation**, and **gene mutation** [Acquaah 2007]. Genomic mutation includes changes in chromosome number (gain or loss of complete sets of chromosomes or parts of a set). Changes in chromosome structure, including duplications of segments, translocation of segments, etc. constitute structural mutation. In gene mutation, there will be changes (deletion or substitution) in the nucleotide constitution of DNA. Many mutations occur in the nuclear DNA or chromosomes, or in extranuclear (cytoplasmic) systems. In terms of gene action, a mutation may be recessive or dominant. In a recessive mutation, a dominant allele changes into recessive allele (A to a) and in dominant mutation, the recessive allele mutates into a dominant allele (a to A).

Mutagenesis is described as the exposure or treatment of biological material to a mutagen, i.e., a physical or chemical agent that raises the frequency of mutation above the spontaneous rate. Plants created using mutagenesis, aimed at disruption or alteration of genes, are sometimes called mutagenic plants or seeds. Over 2,500 mutagenic crop varieties have been released worldwide between 1930 and 2007, with 27 percent of them in China, 12 percent in India, and 6 percent in USA. The crops included rice, wheat, barley, soya bean, potato, onion, etc.

The mutagenesis procedures differ between self-fertilizing species, cross-fertilizing species, and vegetatively-propagated plants. The parental plants of the latter two types of species are often heterozygous at many loci. Consequently, mutations in dominant alleles at these loci are recognizable as mutants in the mutation-treated plants. In the case of vegetatively-propagated plants, plant parts instead of seeds are treated. As mutations are induced in single cells, mutations in these multi-cellular structures will appear as mutant sectors (chimeras) deriving from the cell that contains mutation. Non-chimeric mutants can be obtained by propagating the mutant sectors vegetatively. This can be done by using *in vitro* culture techniques. Mutations in self-fertilized species are visible in plants derived from mutagenized seed, but recessive mutations can only be obtained in their progeny. Cross-pollinators like maize can be selfed and treated by mutagens as self-pollinators, although generating selfed progenies is often more laborious. When selfing is not possible, other inbreeding procedures such as sib-mating may help reveal the presence of homozygous recessive mutants in the progeny of the mutagen-treated plants.

Mutations can be induced by chemical and physical means. They may act directly on DNA, causing direct damage resulting in replication error. Some mutagens, called promutagens form mutagenic metabolites through cellular processes.

Chemical Mutagens

The potency of chemical mutants varies in their ability to enter the cell, reactivity with DNA, and the type of chemical changes they introduce into the DNA. Base analog mutagens have structures similar to the nucleotides, which may be incorporated into the DNA replication system. Many analogs have an increased tendency for 'mispairing' during replication, which leads to mutations. DNA replication is required for the base analog-induced mutations to be incorporated into the DNA. The base analog mutagen 5-bromouracil (5BU) has two tautomeric forms. Its keto form (BU_k) form is a thymine (T)-mimic, which pairs with adenine (A) while its enol form (BU_e), a cytosine (C)-mimic, pairs with guanine (G). In the first replication, the keto form is incorporated into a new DNA strand. During the second replication, if the keto form undergoes a tautomeric shift to the enol form, it will cause A-T to G-C mutation. The base pair substitution may change after a number of replication cycles depending on whether the tautomeric form is within the DNA molecule or is an incoming base when it is enolized or ionized.

Some chemical mutagens such as alkylators, deaminators, and hydroxylamines, work by causing chemical modifications of purine and pyrimidine bases that alter their hydrogen-bonding properties. The alkylating mutagens include ethyl methanesulfonate (EMS), methyl methanesulfonate (MMS), diethylsulfate (DES), dimethylsulfate (DMS), N-nitrosoguanidine {N-methyl-*N'*-nitro-*N*-nitrosoguanidine (MNNG)}, etc. These agents donate alkyl groups (ethyl or methyl) to the nucleotide bases, resulting in mispairing of the alkylated base to induce TR5 (tryptophan synthetase mutation at locus 5) mutations. For example, EMS converts guanine to 7-ethylguanine which pairs with thymine. The mispairing will lead to mutation. Some alkylating agents may also cross-link DNA, resulting in chromosome breaks. The effect of these chemicals may, or may not, involve replicating genes.

The deaminating agent like nitrous acid deaminates cytosine, producing uracil which then pairs (by H-bonding) with adenine instead of guanine. It converts cytosine to uracil, adenine to hypoxanthine, and guanine to xanthine. The hydrogen-bonding potential of the modified base is altered, resulting in mispairing. The hydroxylamine (NH_2OH) mutagen modifies base structures by adding a hydroxyl group to cytosine, producing hydroxylaminocytosine. This hydroxylated cytosine has an increased tendency to undergo tautomeric shifts, which allow pairings with adenine, resulting in GC→AT transition.

A different group of mutagens, ethidium bromide, proflavin, acridine orange, etc. function as intercalating agents. These three-ringed molecules are about the same size as a nucleotide base pair. During DNA replication, these compounds can insert or intercalate between adjacent base pairs. This insertion causes a 'stretching' of the DNA duplex far enough that the DNA polymerase is 'fooled' into inserting an extra nucleotide opposite an intercalated molecule on the growing polypeptide chain.

Promutagens are chemical agents that are not mutagenic, but can metabolically be transformed into mutagens by plant systems. Many aromatic amines (sodium azide, nitrosamines, aflatoxins, maleic hydrazide, pyrrolizidine alkaloids, nitrofurans, etc.) are promutagens and plants can activate these agents into stable mutagens. The biochemical and molecular mechanisms involved in bioactivation of promutagens into mutagens have been partially revealed in plant systems to understand their mutation induction mechanisms. The mutagenic metabolite of sodium azide (NaN_3), L-azidoalanine, causes base substitutions, resulting in GC→AT transition, thereby modifying the function of proteins.

Physical Mutagens

Physical mutagens include electromagnetic (EM) radiation, ionizing radiation (gamma rays, X-rays), ultraviolet (UV) radiation, and particle radiation. Of these, radiation is the first agent known. Its effects on genes were first reported in the 1920s. Discovery of X-rays by the German physicist and the first Nobel laureate (1901) in physics Wilhelm Roentgen in the 1890s, radioactivity by the French physicist and Nobel laureate (1903) Antoine Henri

Becquerel in 1896, and radioactive elements by the Poland-born French physicists and Nobel laureates (1903) Marie and Pierre Curie in 1898 led to the birth of atomic physics and the understanding of electromagnetic radiation.

Visible light and other forms of radiation constitute electromagnetic radiation. It consists of electric and magnetic waves. Its wavelength varies widely and it is inversely proportional to the energy they contain. Ionizing radiation produces high-energy gamma rays and X-rays. The resultant reactive ions (charged ions or molecules)—streams of atomic and subatomic particles—react with biological molecules. There are two types of radioactive elements: alpha(α)- and beta(β)-particles. Alpha particles include helium nuclei, two protons, and two neutrons, while beta particles constitute electrons. UV radiation, which reacts with DNA and other biological molecules, is also important as a mutagen. Particle radiation is the radiation energy by means of fast-moving subatomic particles. It is referred as a particle beam if all the particles move in the same direction, similar to a light beam.

Ionizing radiation, which promotes formation of extremely reactive hydroxyl radicals (molecular fragments with unpaired electrons) and ions, break phosphodiester linkages within the DNA (both single- and double-strand) molecule and alter purine and pyrimidine bases. Double-strand breaks are difficult to repair accurately, resulting in the deletion of genetic information. The nonionizing UV radiation promotes the formation of pyrimidine dimers between adjacent pyrimidine bases in a DNA strand. These dimers cause a bend in DNA double helix (causing the loss of DNA polymerase ability to read the DNA template), block transcription and DNA replication. They are lethal if unrepaired. They can also stimulate mutation and chromosome rearrangement. Longer-wave UV may cause oxidative damage to DNA.

Molecular Breeding

Remarkable developments in molecular biology and biotechnology over the past 25 years have led plant breeders and geneticists to develop more efficient selection systems to complement with and enhance the effectiveness and efficacy of the traditional phenotypic-pedigree-based selection and breeding systems. Molecular breeding is currently the standard practice in many crops. The areas of molecular breeding includes **marker-assisted breeding** and **transgenic breeding**.

Marker-Assisted Breeding

Marker-assisted breeding involves the use of molecular markers, sometimes called DNA or genetic markers, to track the makeup of plants during variety development process. It provides a dramatic improvement in the reliability and efficiency with which breeders can select plants with desirable combination of genes. A molecular marker is a genetic tag that identifies a particular location within a plant's DNA sequences. Markers are used to transfer a single gene into a new cultivar or to test plants for the inheritance of many genes at once. They aid selection for target alleles that are not easily assayed in individual plants, minimize linkage drag around the target gene, and reduce the number of generations required to recover a very high percentage of the recurrent parent genetic background. Improvements in marker detection systems and in the techniques used to identify markers linked to useful traits, has enabled great advances to be made in recent years.

Although both DNA- and protein-markers have been widely used in plant breeding, the DNA-based markers by far predominate. Greater numbers of DNA-markers can be identified to cover all regions of an organism's DNA, and they are not based on the developmental stage of the plant as many protein-based markers are. DNA-based markers can be derived from seeds or seedlings in rapid screening tests performed by automated robotic systems. Plants lacking the desired traits can be eliminated before moving on to more expensive or lengthy greenhouse or field trials. Marker-based breeding is a non-invasive biotechnological method, and it is considered to be an alternative to transgenic breeding.

Unlike transgenic breeding, marker-assisted breeding does not involve the transformation of isolated, foreign genetic material into genomes. It works like conventional breeding, but it is endowed with speed and accuracy which could dramatically fast-track the entire process of plant breeding. Marker-assisted plant breeding involves several steps: genotyping, phenotyping, QTL mapping, marker-assisted selection, marker-assisted backcrossing, marker-assisted recurrent selection, and genome selection.

Marker-assisted selection (MAS) is the first step in marker-assisted breeding. In this, a trait of interest is selected, not based on the trait itself, but on a marker linked to it. The assumption is that linked allele associates with the gene and/or quantitative trait locus (QTL) of interest. MAS can be useful for traits that are difficult to measure, exhibit low heritability, and/or are expressed late in development. The molecular markers deployed for a successful MAS should a) co-segregate or map as close to the target gene as possible in order to have low recombination frequency between the target gene and the marker, b) display polymorphism between genotypes that have or do not have the target gene, and c) be cost-effective, simple, and high-throughput to ensure genotyping power needed for the rapid screening of large populations [Xu 2003; Mohler and Singrün 2004].

Genotyping

Plant genotyping, also known as DNA fingerprinting of plants, is used for identifying genetic diversity within a breeding population. Plant genotype analysis is used for the identification of plants in commerce, plant breeding, and research. Differences in the genetic makeup and genotype of a plant are determined by examining its DNA sequence using bioassays and comparing it with another plant's sequence or a reference sequence. It reveals the alleles it has inherited from its parent. Plant genotyping uses DNA sequences to define biological populations by using molecular markers.

A molecular marker is a fragment of DNA that is associated with a certain location within the genome. It may arise due to mutation or alteration in the genome loci. The DNA fragment is either a short sequence, with a single base-pair change (single nucleotide polymorphism), or a long one, like minisatellites. Variations in DNA sequence, called polymorphisms, can be associated with different forms (alleles) of nearby genes involved with particular traits. The polymorphism is the clue geneticists and plant breeders need to find the genes of interest and use them for crop improvement.

Molecular markers perform a number of tasks, including the genetic fingerprinting of plant varieties, determining similarities among inbred varieties, mapping of plant genomes, and establishing phylogeny among plant species. New techniques for the extraction, purification, and amplification of plant DNA are being developed on a regular basis, enabling researchers to reduce preparation time and obtain readily reproducible results. Plants can now be compared at the molecular level in several ways, via examination of restriction fragments, identification of isoenzymes (protein/gel electrophoresis), or products of the polymerase chain reaction.

DNA marker systems, introduced in the 1980s, have many advantages over the traditional morphological and protein markers that are used in genetic and ecological analyses of plant populations. These advantages include, that a) an unlimited number of DNA markers can be generated, b) DNA marker profiles are not affected by the environment, and c) DNA markers, unlike isozyme markers, are not constrained by tissue or developmental stage specificity.

Molecular markers need to have the following properties: a) high level of polymorphism; b) co-dominant inheritance; c) unambiguous designation of alleles; d) frequent occurrence in the genome; e) even distribution throughout the genome; f) selectively neutral behavior (no pleiotropic effect); g) easy access; h) easy and fast assay (amenable to automation); i) high reproducibility; j) easy exchange of data between laboratories; and k) development at reasonable cost.

The basic methodology of genotyping involves extraction of DNA from plant cells followed by quantification and quality assessment of extract. The methodologies are of two types: non-PCR (polymerase chain reaction)-based (RFLP) and PCR-based (RAPD, ISSR, SSR, SNP). Some of these belong to the first generation (RFLP) while the others second (RAPD, SSR) and third or next (ISSR, AFLP, SNP) generations (Table 3.1).

Restriction Fragment Length Polymorphism. Restriction fragment length polymorphism (RFLP) is a Southern blot-based marker [Xu 2010]. It is one of the first genetic fingerprinting marker techniques that detect variations in homologous DNA sequences. The principle behind the technology rests on the possibility of comparing band profiles generated after having the plant DNA molecules digested by restriction enzymes (endonucleases). RFLP has now become obsolete due to the rise of inexpensive DNA sequencing methods.

The technique involves fragmenting a sample of DNA by a restriction enzyme (types I, II, III, or IV), which can recognize and cut DNA wherever a specific short sequence occurs, in a process known as a restriction digest. The resultant double-stranded DNA fragments, ranging from 500 to 2,000 bp, are separated lengthwise by agarose gel electrophoresis and transferred to a filter membrane for subsequent fragment detection by probe hybridization using Southern blotting procedure. Hybridization (binding of two nucleic acid chains by base pairing) of the membrane to a labeled DNA probe (e.g., cyclooxygenase 1: cox1; cyclooxygenase 2: cox2; adenosine triphosphate 1: atp1; adenosine triphosphate 6: atp6; restriction endonuclease or restriction enzyme from *Providencia stuartii*: PstI, etc.) then determines the length of the fragments. DNA probe picks up sequences that are complementary and homologous to the thousands or millions of undetected fragments that migrate through gel. An RFLP occurs when the length of a detected fragment varies between individuals. Each fragment length is considered an allele which is used in genetic analysis. Most RFLP markers are co-dominant and highly locus-specific.

A probe is a piece of DNA or RNA used to detect specific nucleic acid sequences by hybridization. It is radioactively labeled so that the hybridized nucleic acid can be identified by autoradiography. The size of a probe ranges from a few nucleotides to hundreds of kilo-basepairs. Long probes are usually made by cloning. Originally, they may be double-stranded, but the working probes must be single-stranded. Short probes (oligonucleotide probes) can be made by chemical synthesis. They are single-stranded. Once a particular DNA fragment is identified, it can be isolated and amplified to determine its sequence. Sources of DNA probes include: a) genomic libraries, b) complementary DNA (cDNA) libraries, and c) cytoplasmic DNA libraries belonging to mitochondrial and chloroplast DNA. The genome library contains DNA fragments representing the entire genome of a

Table **3.1.** Comparison of the widely-used molecular systems of plant genome analyses. [Source: Korzun 2003; Semagn et al. 2006].

Character	RFLP*	RAPD*	AFLP*	SSR*	SNP*
Abundance	Medium	Very High	Very High	High	Very high
Types of Polymorphism	Single base change, insertion, deletion, inversion	Single base change, insertion, deletion, inversion	Single base change, inversion, deletion, inversion	Repeat Length Single base	Single base change
No. of polymorph loci analyzed	1.0–3.0	1.5–5.0	20–100	1.0–3.0	1.0
PCR-based	No	Yes	Yes	Yes	Yes
DNA required (µg)	10	0.02	05–1.0	0.05	0.05
DNA quality	High	Medium	High	Medium	Medium
DNA sequence information	Not required	Not required	Not required	Required	Required
Level of polymorphism inheritance	Medium	High	High	High	High
Reproducibility	High	Low	Medium	High	High
Technical complexity	High	Low	Medium	Low	Medium
Developmental cost	High	Low	Moderate	High in start	High
Cost/analysis	High	Low	Moderate	Low	Low
Species transferability	Medium	High	High	Medium	Low
Automation	Low	Medium	Medium	High	High

*RFLP: Restriction Fragment Length Polymorphism; RAPD: Random Amplified Polymorphic DNA; AFLP: Amplified Fragment Length Polymorphism; SSR: Simple Sequence Repeats; SNP: Single Nucleotide Polymorphism.

plant. The cDNA library stores complementary DNA molecules synthesized by mRNA molecules in a cell.

The RFLP analysis technique requires large amount of sample DNA. Besides, the combined process of probe labeling, DNA fragmentation, electrophoresis, blotting, hybridization, washing, and autoradiography take a long time, often several weeks, to complete. Although this method is used to identify the origins of a particular plant species, it is not much favored for DNA fingerprinting.

Several methods alternative to RFLP have been developed between 1985 and 1989. These include Variable Number Tandem Repeats (VNTR), Allele Specific-Polymerase Chain Reaction (AS-PCR), Oligonucleotide Polymorphism (OP), Single-Stranded Conformational Polymorphism (SSCP) and Sequence Tagged Sites (STS). However, these conventional hybridization-based assay techniques to detect DNA level variations gave way to the PCR-based second-generation methods in the early 1990s.

Random Amplified Polymorphic DNA. The second generation DNA marker, randomly amplified polymorphic DNA (RAPD), is a type of PCR reaction. It detects DNA segments that are amplified at random. The geneticist performing RAPD creates several arbitrary, short primers (8–12 nucleotides), and then proceeds with the PCR using a large template of genomic DNA, hoping that fragments will amplify. By resolving the resulting patterns, a semi-unique profile can be gleaned from a RAPD reaction. The RAPD process does not require knowledge of the DNA sequence for the targeted gene because the exact binding location of the primers is uncertain. This makes RAPD popular for comparing the DNA of plant species that have not received attention of geneticists.

In the RAPD technique, DNA fragments are amplified by PCR using 10-base pair synthetic oligonucleotide primers (also called decamers) of random sequence. Because of the short primers used, the re-annealing temperature in the PCR must be low (35–40°C) for the primer to bind. However, due to low temperature, the binding is not very specific, which means that primers will also bind to sequences which are not completely complementary. These oligonucleotides serve as both forward and reverse primer and usually are able to amplify fragments from 3–10 genomic sites simultaneously. The amplified fragments (usually within 100–3,000 bp band range), separated by gel-electrophoresis and polymorphism, are detected as the presence or absence of bands of particular size. These polymorphisms are considered to be primarily due to variation in the primer annealing sites. RAPD markers, however, are limited in their usefulness for mapping, in that they are dominant alleles. Hence, it necessary to prepare many closely linked markers to insure reliable comparisons among plant populations.

RAPD method helps in screening the differences in DNA sequences of two species of plants and diversity of germplasm. Although this technique is widely used in many analyses and cost-effective than RFLPs, it lacks specificity due to low annealing temperatures and easier reaction conditions. Other drawbacks include its low reproducibility of results and little use for comparative mapping as also the possibility mismatching between the primer and the template.

Amplified Fragment Length Polymorphism. Amplified fragment length polymorphism (AFLP, also called AFLP-PCR) tool is used in gene research, population genetics, molecular evolution, genetic engineering, and plant breeding. This PCR-based AFLP uses two different restriction enzymes (the four-base restriction enzyme **MseI**{A *E. coli* strain that carries the cloned MseI gene from *Micrococcus* species (R. Morgan)} and the six-base restriction enzyme **EcoRI**('*E. coli*' 'R'Y13' strain, 'first' Identified) to digest genomic DNA to produce, by ligation, well-defined restriction fragments with sticky ends [Chial 2008]. A subset of the restriction fragments is then selected for amplification. This selection is achieved by using primers complementary to the adaptor sequence, the restriction site sequence, and a few nucleotides inside the restriction site fragments. The amplified fragments are separated and visualized on denaturing polyacrylamide gels, either through autoradiography or fluorescence methodologies, or via automated capillary sequencing instruments.

In this process, synthetic double-stranded adapters, also called linkers, of 18–20 bp with matching sticky ends are ligated on all the restriction fragments. The ligated fragments are subsequently amplified in PCR with 18–20 nucleotide-long primers that recognize linkers in each end of the fragments. A primer used for amplification in AFLP is normally labeled with P^{33} or some non-radioactive labeling system. In order to avoid that all restriction fragments from the genome amplification in the same PCR, which will produce a smear because of too many DNA fragments, a two-step amplification procedure is used. In 'pre-amplification' step, part of the total number of restriction fragments are amplified, with primers containing one extra 'selective' nucleotide on their 3' end. This selective nucleotide allows amplification of restriction fragments with a matching nucleotide next to the linker. With one selective nucleotide on both primers, only 1/16th of all fragmentation fragments are used in the mixture during pre-amplification. During 'selective' amplification, additional one or two selective nucleotides on each primer will further reduce the number of fragments amplified. A good AFLP amplification will show 90 to 100 different fragments in one analysis.

Compared with RAPD and RFLP, AFLP has higher reproducibility, resolution, and sensitivity at the whole genome level. Furthermore, it has capability to amplify 50 to 100 fragments at a time, without the need to have no prior sequence formation for amplification [Meudt and Clarke 2007]. As a result, AFLP has become extremely beneficial in the study of taxa in plants and to determine a large number of polymorphisms.

Simple Sequence Repeats. Simple sequence repeats (SSRs), also known as microsatellite, are second generation molecular markers. Thse repeat-sequence based markers are simple tandem repeats (STR) of mono-, di-, tri-, tetra-, penta-, and hexa-nucleotides. They are highly polymorphic and informative markers. The high level of polymorphism is due to mutation affecting the number of repeat units. The value of SSRs is due to their genetic co-dominance, abundance, dispersal throughout the genome, multiallelic variation, and high reproducibility. These properties provide a number of advantages over other molecular markers, namely, that multiple SSR alleles may be detected at a single locus using a simple PCR-based screen, very small quantities of DNA for screening, and automated allele detection and sizing [Schlötterer 2000]. The hypervariability of SSRs among related organisms makes them excellent markers for a wide range of applications, including genetic mapping, molecular gene tagging, genotype identification, genetic diversity analysis, phenotype mapping, and marker-assisted selection. SSRs demonstrate a high degree of transferability between species, as PCR primers designed to an SSR within one species frequently amplify a corresponding locus in related species, enabling comparative genetic and genomic analysis.

Specific microsatellites can be isolated using hybridized probes followed by their sequencing. Like any DNA fragment, they can be detected by specific dyes or by radiolabelling using agarose of polyacrylamide gel electrophoresis. The advantage of using SSRs as molecular markers is the extent of polymorphism shown, which enables the detection of differences at multiple loci between strains. Coupled with chemical and morphological data, one can identify the plant species or strain of interest.

This method requires only small amounts of template DNA compared to the RFLP method. This is due to the large amounts of SSRs present in any genome. The level of genetic variation detected by SSRP analysis is almost two times higher than that detected by RFLPs [Morgante et al. 1994]. Furthermore, assays involving SSRs are more robust than RAPDs, making them up to seven times more efficient. Drawbacks to using SSRs include the need to develop separate SSR primer sets for each species, homoplasy (a character shared by a set of species but not present in their common ancestor), and high mutation rate. The latest research suggests that SSRs will be involved in new methods of detection of alterations of specific sequences in the DNA.

ISSR. Inter-simple sequence repeat (ISSR) technique, a third or next generation molecular marker, is based on variation found in the genome region between microsatellite loci. It involves amplification of DNA segments present at an amplifiable distance in between two identical microsatellite repeat regions oriented in opposite direction. This PCR-based inexpensive genotyping technique uses microsatellites as primers in a single primer PCR reaction targeting multiple genome loci to amplify ISSRs of different sizes. ISSR-PCR uses a single fluorescently labeled primer to target the regions between identical microsatellites.

An ISSR-PCR primer comprises three parts: a fluorescent tag, eight dinucleotide repeat units (or six trinucleotide repeat units), and one or more anchor nucleotides (tetra-nucleotide or penta-nucleotide) designed to target the end of microsatellite region and to prevent primer dimerization. ISSRs have high reproducibility possibly due to the use of longer primers (16–25 nucleotides) than RAPD primers (10 nucleotides), and this permits the subsequent use of high annealing temperature leading to higher stringency. The annealing temperature depends on the G-C content of the primer used and it ranges from 45 C to 65°C. The amplified products are usually 200–3,000 bp long and amenable to detection by both agarose and polyacrylamide gel electrophoresis (PAGE) [Segman et al. 2006]. The level of polymorphism detected varies with the detection method used. PAGE in combination with labelled nucleotide in PCR reaction is the most sensitive, followed by PAGE with silver staining and then agarose-ethidium bromide system of detection. Markedly higher number of bands was resolved per primer when polyacrylamide was used when compared to agarose [Moreno et al. 1998].

ISSR-PCR is a simple, fast, cost-effective, highly discriminant, and highly reliable technique widely applied in plant genetic analyses. The primers are not proprietary (as in SSR-PCR) and can be synthesized by anyone. Variations in primer length, motif, and anchor are possible. The primers of 16–25 bp long result in higher stringency. The amplified products (ISSR markers) are usually 200–2000 bp long and they are amenable to detection by both agarose and polyacrylamide gel electrophoresis [Reddy et al. 2002].

It has a wide range of uses, including characterization of genetic relatedness among plant populations including gene tagging, detection of clonal variation, cultivar identification, phylogenetic analysis, detection of genomic instability, and assessment of hybridization in many plant species, including rice, wheat, *Vigna*, sweet potato, etc. [Reddy et al. 2002].

Single Nucleotide Polymorphism. Single nucleotide polymorphism (SNP; pronounced as *snip*) marker is a DNA sequence variation occurring when a single base pair (nucleotide)—A, T, C, or G—in the genome (or other shared sequence) differs between members of a species. In effect, an SNP represents a single nucleotide difference between two species at a defined location. SNPs, the third generation molecular markers, are responsible for differences in genetic traits of plants. Their occurrence, distribution, and frequency vary with and plant species. There are three different forms of SNPs: transitions (C/T or G/A), transversions (C/G, A/T, C/A, or T/G), or small insertions/deletions (indels) [Edwards et al. 2007]. SNPs represent the most frequent type of genetic polymorphism and may therefore provide high density markers.

The high-throughput, high density SNP genotyping has become an essential tool for QTL mapping, association genetics, map-based gene isolation, gene discovery, germplasm characterization, molecular breeding, plant breeding, and population genomics studies in several plants and crops. The abundance of SNPs in plant genomes together with the rapidly falling costs and increased accessibility of genotyping technologies, have prompted an increasing interest to develop panels of SNP markers to expand resolution and throughput of genetic analysis in less-domesticated plant species with uncharacterized genomes such as those of legume crops [Muchero et al. 2009], forest [Eckert et al. 2009] and fruit trees [Myles et al. 2010].

SNPs comprise the most abundant molecular markers in the genome and they are numerous in plant genomes. For example, there is one SNP in every 170 bp in *Oryza indica* rice as against 350 bp in *O. japonica* rice [Yu et al. 2002]. In rice, SNPs for *japonica* and *indica* varied from as little as 3.0 SNP/kb in the coding regions to 27.6 SNP/kb in the transposable elements [Yu et al. 2005]. In a genome wide survey of 877 unigenes, SNPs were estimated to be present in every 200 bp in barley [Rostoks et al. 2005]. The cross-pollinating maize contains a higher frequency of SNPs at an average of one per 31 bp in noncoding regions and one per 124 bp in coding regions [Ching et al. 2002]. In soya bean, 280 SNPs are present per 76.3 kb of genomic DNA [Zhu et al. 2003]. This suggests that the frequency of the SNPs in crop plants may range one per 30 bp to one per 500 bp. Complete-genome sequences for an increasing number of plant species are being enlisted each year with rapid technical advancements in DNA sequencing strategies. As a result, molecular marker systems are being utilized more frequently in plant genomics and plant genetics.

SNP genotyping technologies have two components: determination of the type of base present at a given SNP locus (allele discrimination) and reporting the presence of the allele(s) signal detection. There are three allele discrimination methods: hybridization/annealing (with or without a subsequent enzymatic discrimination step); primer extension; and enzyme cleavage. In each case, the technology platform may be homogeneous (in solution) or heterogeneous (involving both a liquid and a solid phase, such as a microarray). Some of the assays require prior amplification of the genomic target, whereas others are sensitive enough to work directly on genomic DNA or cDNA.

Dynamic Allele-specific Hybridization (DASH). This technique takes advantage of the differences in the melting temperature in DNA that results from the instability of the mismatched base pairs. In the first step, a genomic segment is amplified and attached to a bead through a PCR reaction with a biotinylated 5′ primer. In the second step, the amplified product is attached to a streptavidin (a 60 kDa protein purified from the bacterium *Streptomyces avidinii*) column and washed with NaOH to remove the unbiotinylated strand. In the third step, an allele-specific oligonucelotide (a short piece of synthetic DNA complementary to the sequence of a variable target it acts as a probe) is added along with hybridization buffer containing a fluorescent double-strand-specific intercalating eye. The sample is then rapidly heated to above-denaturing temperature and gradually cooled to allow hybridization of the probe to target DNA.

Later, the hybridization solution is removed and detection buffer added. The sample is heated slowly from room temperature to above denaturing (melting) temperature (T_m) while monitoring fluorescence. A drop in fluorescence indicates the T_m point of the target/probe. These steps are repeated for subsequent allele-specific probe(s). In order to determine the T_m point, fluorescence values are plotted as a function of temperature. By plotting the negative derivative (slope of the fluorescence versus T_m), denaturation points are clearly seen as peaks. Peak T_m values can be used for final allele determination [Howell et al. 1999]. A SNP will result at a lower than expected T_m. One benefit of DASH process is that it is capable of measuring all types of mutations, not just SNPs. Other benefits include its ability to work with label-free probes and its simple design and performing conditions.

Oligonucleotide Ligase Assay. A variant of allele-specific hybridization method for genotyping SNPs is oligonucleotide ligase assay (OLA). In this method, the target sequences immediately surrounding the variable position are simultaneously hybridized (annealed) with two adjacent oligonucleotide probes so that their junction occurs at the site of variation. If the sequence of the target DNA is perfectly complementary with that of the oligonucleotides, they can be covalently joined by DNA ligase. If the target DNA is mismatched at the junction, the ligase will fail to join the oligonucleotides.

The process involves two steps. In the first one, an allele-specific oligonucleotide (ASO) probe (detector) is hybridized to the target DNA so that its 3' base is situated directly over the SNP nucleotide. In the next one, the oligonucleotide probe (3-fluorescein-labeled) hybridizes the template upstream (downstream in the complementary strand) of the SNP polymorphic site providing a 5' end for the ligation reaction. If the ASO probe matches the target DNA, it will fully hybridize to the target DNA and ligation can occur. Ligation does not generally occur in the presence of a mismatched 3' base. Ligated or unligated products can be detected by gel electrophoresis, matrix-assisted laser desorption/ionization time of flight (MALDI-TOF) mass spectrometer, or by capillary electrophoresis for large-scale applications [Rapley and Harbron 2004]. With appropriate sequences and tags on the oligonucleotides, high-throughput sequence data can be generated from the ligated products and genotypes determined [Curry et al. 2012]. The OLA is a rapid, easy, inexpensive, and single-tube method with high throughput capability.

SNP Microarray. In this SNP-based genotyping method (a type of DNA microarray), hundreds of thousands of probes are arrayed on a small chip, allowing for many SNPs to be interrogated simultaneously [Rapley and Harbron 2004]. Microarrays are used to identify alleles for known SNPs in the genomic material. The core principle behind this high throughput method is hybridization between two complementary strands which allow nucleic acid to form bonds between nucleotides. The fluorescently-labeled genetic material that binds to a probe fixed onto microarray generates a signal. Total strength of the signal depends on how much the genetic material is bound to the probe. This, in turn, depends on the quantity of the genetic material within the sample. Quantified strength of the signal is called 'expression level'.

The SNP microarray process involves hybridization and fluorescence microassay. The three mandatory components are: immobilized nucleic acid sequences of target, one or more labeled allele-ASO probes, and a detection system that records and interprets the hybridization signal. Since SNP alleles only differ in one nucleotide and because it is difficult to achieve optimal hybridization conditions for all probes on the array, the target DNA has the potential to hybridize to mismatched probes. This is addressed somewhat by using several redundant probes to interrogate each SNP. Probes are designed to have the

SNP site in several different locations as well as containing mismatches to the SNP allele. By comparing the differential amount of hybridization of the target DNA to each of these redundant probes, it is possible to determine specific homozygous and heterozygous alleles [Rapley and Harbron 2004]. Although oligonucleotide microarrays have a comparatively lower specificity and sensitivity, the scale of SNPs that can be interrogated is a major benefit. The expression levels of a large number of genes can be determined simultaneously with a high degree of sensitivity.

Transgenic Breeding

Currently, a number of transgenic crop varieties have been developed using the recombinant DNA technology by inserting the desired genes from non-plant sources (vide Chapters 4 and 6). Transgenic technology is considered as a supplementary tool to plant breeding and hence coupling the two will bring economically viable, consumer-oriented products to the marketplace.

In conventional breeding, the breeder selects for desirable recombination among a large segregating population, while in transgenic breeding he looks for a defined trait phenotype and then introgresses the transgenic event into a broad range of desirable genetic backgrounds [Visarada et al. 2009]. Once the transformed plant tissues leave the laboratory after tissue culture, they need to be extensively evaluated to determine if the inserted transgene has been stably incorporated and the inserted desired trait is expressing up to the threshold levels as in the case of tolerance to biotic and abiotic stresses. Evaluation is also done for stable inheritance of the trait and elimination of undesirable disruption of the genetic background. This is generally carried out first in the greenhouse or screenhouse followed by extensive field testing. If these transformed plants (T_0) which are hemizygous, passed this stage successfully, they will be subjected to plant breeding techniques to test their agronomic performance. In many cases, transgenic lines cannot be directly used for cultivation since they are lower yielding. Hence, it is important to have the line conferred with elite backgrounds, free of somaclonal variation, by subjecting it to backcross breeding.

Backcross Breeding

Backcross breeding, used for decades to incorporate specific traits into elite lines, enables transfer of transgenes from transformed lines into parental lines. Backcrossing involves crossing the transgenic line with an elite line of choice. The progeny of F_1 will have one copy of the transgene and 50 percent elite genes. In the next, the selected transgenic offspring are crossed back to the elite line. The progeny will have 75 percent elite genes. This process of crossing back to the elite line is repeated until the desired percentage of elite genes in the offspring has reached usually 98–99 percent elite genes and the transgene. The resulting crop plant, which exhibits the characteristics of an elite crop line and the desired trait from the transformed line, is called a transgenic event or line.

Successful backcross breeding requires two to seven backcrosses. This, however, depends on, a) the unwanted level of chromatin linked to the transgene, b) the importance of the recurrent parent phenotype, c) the amount of selection imposed during recurrent backcrosses, and d) and the level of eliteness of the donor parent [Fehr 1987]. Once a transgene is transferred to an elite line, it needs to be stably integrated and expressed in the final transgenic product before commercialization. This aspect is discussed in greater detail in Chapter 4.

Forward Breeding

A different strategy in transgenic breeding is forward breeding. It takes advantage of improved cultivars and genetic knowledge that may have been developed during the process of backcross breeding. In the case of forward breeding, the 'best' cultivar of interest, not the original cultivar of interest, is used as the recurrent parent. This allows for the inclusion of recent breeding advances into the backcross breeding program. Marker assisted selection is particularly useful when selecting for multiple quantitative traits.

As per the methodology of forward breeding, described by Fredrick Bliss in the U.S, Patent Application US2009/0064358 A1 of 5 March 2009, the donor parent which possesses the trait(s) of interest is selected and crossed to a first hybrid parent. The selected first generation progeny of the cross which consists of the trait(s) of interest is crossed with a second hybrid parent to produce a hybrid back-cross. However, forward breeding is useful only if the transgene donor also contains favorable allele(s). Therefore, germplasm genotype is often a consideration in forward breeding. In case elite, transformable genotypes are not available, backcross breeding would be the preferred strategy.

This method has been used for the development of glyphosate-resistant 'Roundup Ready' soya bean lines (vide Chapter 6). This method with the glyphosate-resistance trait on a large commercial breeding scale is relatively straightforward and inexpensive [Crosbie et al. 2006].

Yield Lag and Yield Drag

One point that needs to be highlighted is the possibility of 'yield lag' and 'yield drag' transgenic varieties derived via transgenic breeding. **Yield lag** is the term used for yield suppressions resulting from base genetic differences between transgenic and non-transgenic hybrids used in the comparison. Thus, transgenic hybrids may yield less than their non-transgenic counterparts simply because of the time required to get them to market. In the case of transgenic line, there is no selection for increased yield during the 3–5 years of backcross breeding while non-transgenic lines have had selection for improved yield potential every year. Therefore, the lines coming out of a backcrossing program have gone through a 'lag period' in which the plant breeder has not imposed selection for yield. These lines would be expected to experience a yield lag.

Once a transgene has been backcrossed into an elite inbred background, it is no longer necessary for it to undergo backcrossing again. It can be used in the same manner as non-transgenic lines in breeding programs being mated to other lines and undergoing selection for improved qualities including yield. Thus, over a time period, yield lag will no longer be an issue in a particular transgenic event. Testing for yield lag is done by comparing the yield potential of the backcrossed transgenic line to that of the newest elite inbreds.

Yield drag, the term used to describe yield suppression resulting from the inserted transgene itself and/or the gene insertion process, is a loss in productivity in transgenic as compared to non-transgenic crop lines. This phenomenon is common but not universal. Yield drag exerts a negative effect on yield potential associated with crop plants that have a specific gene or trait. It can be caused by the transgene inserting into a gene important for plant growth and yield disrupting its expression, or it can be caused by a drain in the limited pool of amino acids by having to produce large quantities of new proteins. This limits the amount of amino acids available for the production of other proteins important to plant growth and yield. Testing for yield drag is done by comparing the yield potential of the backcrossed transgenic line to that of the original elite line (isoline).

Genetically engineered crops are not necessarily more prone to yield drag or yield lag than non-genetically engineered crops. However, genetically engineered crops are unique in that an additional gene is being placed into the chromosome.

Transgene Expression

Transgene expression is of vital importance when transgenic plants are produced. Levels of transgene expression are influenced by many factors which include the site of integration of transgene within the plant genome, gene silencing, and the promoter attached. Gene expression can be measured by quantifying mRNA. This can be done by Northern blot analysis, RT-PCR, and RNase protection. It can also be done by quantifying DNA via Southern blotting analysis, PCR, RT-qPCR, and microarray hybridization, while analysis of specific transgenic proteins involved in gene expression is performed by Western blotting and ELISA methods. These techniques have been discussed earlier in this chapter (vide Measuring Gene Expression).

References

Acquaah, G. 2007. Principles of Plant Genetics and Plant Breeding. Blackwell Publishing, Malden, MA, USA, 570 pp.

Alverson, A.J., S. Zhuo, D.W. Rice, D.B. Sloan and J.D. Palmer. 2011. The mitochondrial genome of the legume *Vigna radiata* and the analysis of recombination across short mitochondrial repeats. Plos One 6(1): e16404. doi:10.1371/journal.pone.0016404.

Alverson, A.J., X.X. Wei, D.W. Rice, D.B. Stern, K. Barry and J.D. Palmer. 2010. Insights into the evolution of mitochondrial genome size from complete sequences of *Citrullus lanatus* and *Cucurbita* pepo (Cucurbitaceae). Mol. Biol. Evol. 27: 1436–1448.

Anderson, S. 1981. Shotgun DNA sequencing using cloned DNase I-generated fragments. Nucleic Acids Res. 9(13): 3015–3027.

Antisense RNA. 2001. www.dls.ym.edu.tw/ol_biology2/ultranet/AntisenseRNA.html. 30 July 2001.

Arabidopsis Genome Initiative. 2000. Analysis of the genome sequence of the flowering plant *Arabidopsis thaliana*. Nature 408: 796–815.

Atherton, R.A., B.J. McComish, L.D. Shepherd, L.A. Berry, N.W. Albert and P.J. Lockhart. 2010. Whole genome sequencing of enriched chloroplast DNA using the Illumina GAII platform. Plant Methods 6: 22.

Avramova, Z.A., M. Tikhonov, J. Chen and L. Bennetzen. 1998. Matrix-attachment regions and structural colinearity in the genomes of two grass species. Nucleic Acids Res. 26: 761–767.

Bajgain, P., B.A. Richardson, J.C. Price, R.C. Cronn and J.A. Udall. 2011. Transcriptome characterization and polymorphism detection between subspecies of big sagebrush (*Artemisia tridentata*). BMC Genomics 12: 370.

Bartlett, J.M.S. and D. Stirlin. 2003. A Short History of the Polymerase Chain Reaction. PCR Protocols 226: 3–6.

Beadle, G.W. and E.L. Tatum. 1941. Genetic control of biochemical reactions in neurospora. Proc. Natl. Acad. Sci. USA 27(11): 499–506.

Belmont, A.S. 2003. Dynamics of chromatin, proteins, and bodies within the cell nucleus. Current Opinion in Cell Biol. 15: 1–7.

Bennett, M.D. 1987. Ordered disposition of parental genomes and individual chromosomes in reconstructed plant nuclei, and their implications. Somatic Cell and Mol. Genet. 13: 463–466.

Benne, R.S. 1987. Evolution of the mitochondrial protein synthetic machinery. Biosystems 21: 51–68.

Bentley, D.R., S. Balasubramanian, H.P. Swerdlow, G.P. Smith, J. Milton, C.G. Brown et al. 2008. Accurate whole human genome sequencing using reversible terminator chemistry. Nature 456: 53–59.

Berg, J., J. Tymoczko and L. Stryer. 2002. Biochemistry. W.H. Freeman and Company, New York.

Braslavsky, I., B. Hebert, E. Kartalov and S.R. Quake. 2003. Sequence information can be obtained from single DNA molecules. Proc. Natl. Acad. Sci. USA 100: 3960–3964. doi: 10.1073/pnas.0230489100.

Bruce, A., A. Johnson, J. Lewis, M. Raff, K. Roberts and P. Walters. 2002. Molecular Biology of the Cell (4th ed.). Garland Science, New York and London.

Bustin, S.A. 2000. Absolute quantification of mRNA using real-time reverse transcription polymerase chain reaction assays. J. Mol. Endocrinol. 25(2): 169–193.

Carlson, W.R. 1986. The B chromosome of maize. Critical Reviews in Plant Sciences 3: 201–226.

Carrington, J.C. and V. Ambros. 2003. Role of microRNAs in plant and animal development. Science 301: 336–338.

CBOL Plant Working Group. 2009. A DNA barcode for land plants. Proc. Natl. Acad. Sci. USA 106(31): 12794–12797.

Century, K., T.L. Reuber and O.J. Ratcliffe. 2008. Regulating the regulators: the future prospects for transcription factor-based agricultural biotechnology products. Plant Physiol. 147(1): 20–29.

Chalikian, T., J. Völker, G. Plum and K. Breslauer. 1999. A more unified picture for the thermodynamics of nucleic acid duplex melting: a characterization by calorimetric and volumetric techniques. Proc. Natl. Acad. Sci. USA 96(14): 7853–7858.

Champoux, J. 2001. DNA topoisomerases: structure, function, and mechanism. Ann. Rev. Biochem. 70: 369–413.

Chen, J.Z. and Z. Ni. 2006. Mechanisms of genomic rearrangements and gene expression changes in plant polyploids. Bioassays 28(3): 240–252.

Chial, H. 2008. DNA fingerprinting using amplified fragment length polymorphisms (AFLP): no genome sequence required. Nature Education 1(1): 176.

Ching, A., K.S. Caldwell, M. Jung, M. Dolan, O.S.H. Smith, S. Tingey, M. Morgante and A.J. Rafalski. 2002. SNP frequency, haplotype structure and linkage disequilibrium in elite maize inbred lines. BMC Genet. 3: 19.

Clausen-Schaumann, H., M. Rief, C. Tolksdorf and H. Gaub. 2000. Mechanical stability of single DNA molecules. Biophys. J. 78(4): 1997–2007.

Comai, L. 2000. Genetic and epigenetic interactions in allopolyploid plants. Plant Mol. Biol. 43: 387–399.

Comfort, N.C. 2001. The Tangled Field: Barbara McClintock's Search for the Patterns of Genetic Control. Harvard University Press, Cambridge, MA, USA.

Corcos, A.F. and F.V. Monaghan. 1987. Correns, an independent discoverer of Mendelism? I. An historical/critical note. J. Hered. 78(5): 330.

Crop View. 2012. iRNA Technology and the Benefits. World Product Development.

Crosbie, T., S. Eathington, M. Edwards, R. Reiter, S. Stark, R. Mohanty, M. Oyervides et al. 2006. Plant breeding: past, present, and future. pp. 3–50. *In*: L. Kandall and L. Michael (eds.). Plant Breeding: The Arnel R. Hallauer Internatnl. Symp. Blackwell Publishing, Iowa, USA.

Curry, J.D., A.K. Lindholm-Perry, M. Shin, M., J.T. Liu, N. Bulsara, P. Dier et al. 2012. Mass genotyping by sequencing technology. Plant & Animal Genome XX Conference, January 14–18, 2012, San Diego, CA, USA.

DeciBio. 2011. The Next Generation Sequencing Market, Segmentation, Growth, and Trends by Provider. 124 pp. SKU: DCB3790143.

de Vere, N., T.C.G. Rich, C.R. Ford, S.A. Trinder, C. Long et al. 2012. DNA barcoding the native flowering plants and conifers of Wales. PloS One 7(6): e37945. Doi:10.1371/journal.pone0037945.

Dickman, M., S. Ingleston, S. Sedelnikova, J. Rafferty, R. Lloyd, J. Grasby and D. Hornby. 2002. The RuvABC resolvasome. Eur. J. Biochem. 269(22): 5492–501.

Dolezel, J., J. Bartoš, H. Voglmayr and J. Greilhuber. 2003. Nuclear DNA content and genome size of trout and human. Cytometry A 51(2): 127–128.

Eccles, D. 2006. Chromosomal Recombination svg. Wikimedia Commons. 4 July 2006.

Eckert, E., B. Pande, E. Ersoz, M. Wright, V. Rashbrook, C. Nicolet and D. Neale. 2009. High-throughput genotyping and mapping of single nucleotide polymorphisms in loblolly pine (*Pinus taeda* L.). Tree Gene. & Genomi. 5(1): 225–234.

Edwards, D., J.W. Forster, D. Chagné and J. Batley. 2007. What are SNPs? pp. 41–52. *In*: N.C. Oraguzie, E.H.A. Rikkerink, S.E. Gardiner and H.N. De Silva. (eds.). Association Mapping in Plants. Springer, NY.

Eid, J., A. Fehr, J. Gray, K. Luong, J. Lyle et al. 2009. Real-time DNA sequencing from single polymerase molecules. Science 323(5910): 133–138.

Fehr, W.R. 1987. Principles of Cultivar Development: Theory and Technique, Vol. 1. Macmillan Publishing Company, New York.

Feldman, M., B. Liu, G. Segal, S. Abbo, A.A. Levy et al. Rapid elimination of low-copy DNA sequences in polyploid wheat: a possible mechanism for differentiation of homoeologous chromosomes. Genetics 147: 1381–1387.

Fournier-Level, A., A. Korte, M.D. Cooper, M. Nordborg, J. Schmitt and A.M. Wilczek. 2011. A map of local adaptation in *Arabidopsis thaliana*. Science 334(6052): 86.

Fujii, H., T.J. Chiou, S.I. Lin, K. Aung and J.K. Zhu. 2005. A miRNA involved in phosphate-starvation response in *Arabidopsis*. Curr. Biol. 15: 2038–2043.

Fütterer, J., K. Gordon, J.-M. Bonneville, H. Sanfaçon, B. Pisan, J. Penswick and T. Hohn. 1988. The leading sequence of caulimovirus large RNA can be folded into a large stem-loop structure. Nucleic Acids Res. 16: 8377–8390.

Ghosh, A. and M. Bansal. 2003. A glossary of DNA structures from A to Z. Acta Crystallogr. D 59(4): 620–626.

Goldstein, D.A., D. Songstad, E. Sachs and J. Petrick. 2009. Transcription Factors in Plants. Monsanto. http:// www.monsanto.com.

Gregory, T.R., J.A. Nicol, H. Tamm, B. Kullman, K. Kullman, I.J. Leitch et al. 2007. Eukaryotic genome size databases. Nucleic Acids Res. 35: D332–D338.

Greilhuber, J., T. Borsch, K. Muller, A. Worberg, S. Porembski and W. Barthlot. 2006. Smallest angiosperm genomes found in Lentibulariaceae with chromosomes of bacterial size. Plant Biol. 8(6): 770–777.

Hamilton, J.P. and R.C. Buell. 2012. Advances in plant genome sequencing. The Plant J. 70: 177–190.

Hannon, G.J. 2002. RNA interference. Nature 418: 244–251.

Hebert, P.D.N., A. Cywinska, S.L. Ball and J.R. deWaard. 2003. Biological identifications through DNA barcodes. Proc. of the Royal Soc. London B. 270: 313–321. doi:10.1098/rspb.2002.2218.

Harris, T.D., P.R. Buzby, H. Babcock, E. Beer, I. Braslavsky et al. 2008. Single-molecule DNA sequencing of a viral genome. Science 320: 106–109.

Hemmings-Mieszczak, M., G. Steger and T. Hohn. 1997. Alternative structures of the cauliflower mosaic virus 35S RNA leader: implications for viral expression and replication. J. Mol. Biol. 267(5): 1075–1088.

Herbert, P.D.N., M.Y. Stoeckle, T.S. Zemlak and C.M. Francis. 2004. Identification of birds through DNA barcodes. Plos Biol. 2(10): e312.

Heslop-Harrison, J.S. and M.D. Bennett. 1990. Nuclear architecture in plants. Trends in Genetics 6: 401–405.

Himmelbach, A., U. Zierold, G. Hensel, J. Riechen, D. Douchkov, P. Schweizer and J. Kumlehn. 2007. A set of modular binary vectors for transformation of cereals. Plant Physiol. 145: 1192–1200.

HiRNA Works. 2012. iRNA Technology and The Benefits. Cropview 31 October. cropview.wordpress. com/2012/ 10/31/irna-technology.

Hollingworth, P.M., S.W. Graham and D.P. Little. 2011. Choosing and using a plant DNA barcode. PloS One 6(5): e19254.

Howell, W., M. Jobs, U. Gyllensten and A. Brookes. 1999. Dynamic allele-specific hybridization. A new method for scoring single nucleotide polymorphisms. Nat. Biotechnol. 17(1): 87–88.

International Rice Genome Sequencing Project. 2005. The map-based sequence of the rice genome. Nature 436: 793–800.

International Service for Acquisition of Agro-Biotech Applications: Pocket K No. 34: RNAi for Crop Improvement. 2012. HRNA Works. www.isaaa.org › Resources › Publications › Pocket K.

Johannsen, W.L. 1903. Om arvelighed i samfund og i rene linier. Oversigt over det Kongelige Danske Videnskabernes Selskabs Forhandlinger German ed. Erblichkeit in Populationen und in reinen Linien .Gustav Fischer, Jena 3: 247–270.

Johannsen, W.L. 1905. Arvelighedslærens elementer (The Elements of Heredity). Copenhagen.

Jones-Rhoades, M.W. and D.P. Bartel. 2004. Computational identification of plant microRNAs and their targets, including a stress-induced miRNA. Mol. Cell 14: 787–799.

Jordan, K.W., J. Nordenstam, G.Y. Lauwers, D.A. Rothenberger, K. Alavi, M. Garwood and L.L. Cheng. 2009. Metabolomic characterization of human rectal adenocarcinoma with intact tissue magnetic resonance spectroscopy. Diseases of the Colon & Rectum. 52(3): 520–525.

Kadir, A.P.G., S.M. Subhi and L.L. Ent Ti. 2009. A constitutive promoter for expressing foreign genes in plants ubiquitin extension protein. MPOB Information Series. ISSN 1511–7871.

Kellogg, E.A. and J.L. Bennetzen. 2004. The evolution of nuclear genome structure in seed plants. Am. J. Botany 91(10): 1709–1725.

Korzun, V. 2003. Molecular markers and their applications in cereals breeding. A paper presented during the FAO international workshop on "Marker assisted selection: A fast track to increase genetic gain in plant and animal breeding?" 17–18 October. Turin, Italy.

Kress, W.J. and D.L. Erickson. 2007. A two-locus global DNA barcode for land plants: the coding rbcl gene complements for the non-coding trnH-psbA spacer region. PLoS One 2(6): e50B: 1–10.

Kress, W.J., K.J. Wurdack, E.A. Zimmer, L.A. Weigt and D.H. Janzen. 2005. Use of DNA barcodes to identify flowering plants. Proc. Natl. Acad. Sci. USA 102(23): 8369–8374.

Kumar, S., C. Tao, M. Chien, B. Hellner, A. Balijepalli et al. 2012. PEG-labeled nucleotides and nanopore detection for single molecule DNA sequencing by synthesis. Scientific Reports 2: 684.

Lakdawalla, A. 2007. Sanger Sequencing. Wikipedia Commons .

Levy, A.A. and M. Feldman. 2002. The impact of polyploidy on grass genome evolution. Plant Physiol. 130: 1587–1593.

Lewis, E.B. 1976. Alfred Henry Sturtevant: 21 November 1891–5 April 1970. Dictionary of Scientific Biography Chas. Scribner's Sons, New York 13: 133–38.

Li, W., P. Zhang, J.P. Fellers, B. Friebe and B.S. Gill. 2004. Sequence composition, organization, and evolution of the core Triticeae genome. Plant J. 40(4): 500–511.

Liu, L., Y. Li, S. Li, N. Hu, Y. He, R. Pong et al. 2012. Comparison of next-generation sequencing systems. J. Biomed. Biotechnol. 2012: 251364. doi: 10.1155/2012/251364.

Liu, H.-H., X. Tian, Y.-J. Li, C.-A. Wu and C.-C. Zheng. 2008. Microarray-based analysis of stress-regulated micro RNAs in *Arabidopsis thaliana*. RNA. 14(5): 836–843.

Livak, K.J. and T.D. Schmittgen. 2001. Analysis of relative gene expression data using real-time quantitative PCR and the 2(-Delta Delta C(T)) Method. Methods 25(4): 402–408.

Mackay, Ian. 2007. Real-time PCR in Microbiology: From Diagnosis to Characterization. Norfolk, England: Caister Academic Press, 440 pp.

Margulies, M., M. *Egholm, W.E. Altman, S. Attiya, J.S. Bader, L.A. Bemben et al. 2005. Genome sequencing in microfabricated high-density picolitre reactors. Nature 437: 376–380.*

Masson, J.E., P.J. King and J. Paszkowski. 1997. Mutants of *Arabidopsis thaliana* hypersensitive to DNA-damaging treatments. Genetics 146: 401–407.

Masson, J.E. and J. Paszkowski. 1997. *Arabidopsis thaliana* mutants altered in homologous recombination. Proc. Natnl. Acad. Sci. USA 94(21): 11731–11735.

Mathews, C.K., K.E. Van Holde and K.F. van Holde. 1990. Biochemistry. The Benjamin/Cummings Publishing Company, Subs of Addison Wesley Longman, Inc. Chicago, IL, USA, 1168 pp.

Maxam-Gilbert Sequencing. 2013. Wikipedia Commons/Wikipedia.com 16 July.

Maxam, A.M. and W. Gilbert. 1977. A new method for sequencing DNA. Proc. Natl. Acad. Sci. USA 74(2): 560–564.

McClintock, B. 1984. The significance of responses of the genome to challenge. Science 226: 792–801.

Mendel, G. 1866, Versuche über Pflanzen-Hybriden. Verh. Naturforsch. Ver. Brünn 4: 3–47: (English: *In:* 1901, J. R. Hortic. Soc. 26: 1–32.

Meudt, H.M. and A.C. Clarke. 2007. Almost forgotten or latest practice? AFLP applications, analyses and advances. Trends. Plant Sci. 12(3): 106–117.

Meyer, M. and M. Kircher. 2010. Illumina Sequencing Library Preparation for Highly Multiplexed Target Capture and Sequencing. Cold Springs Harbor Protocols. doi:10.1101/pdb.prot5448.

Meyers, B., S.V. Tingey and M. Morgante. 2001. Abundance, distribution, and transcriptional activity of repetitive elements in the maize genome. Genome Res. 11: 1660–1676.

Mohler, V. and C. Singrun. 2004. General considerations: marker-assisted selection. pp. 305–317. *In:* H. Lorz and and G. Wenzl (eds.). Biotechnology in Agricultural and Forestry, Vol. 55. Molecular Marker Systems in Plant Breeding and Crop Improvement. Springer-Verlag, Berlin.

Moreno, S., J.P. Martin and J.M. Ortiz. 1998. Inter-simple sequence repeats PCR for characterization of closely related grapevine germplasm. Euphytica 101: 117–125.

Morgante, M., A. Rafalski, P. Biddle, S. Tingey and A.M. Olivieri. 1994. Genetic mapping and variability of seven soybean simple sequence repeat loci. Genome 37: 763–769.

Mortazavi, A., B.A. Williams, K. McCue, L. Schaeffer and B. Wold. 2008. Mapping and quantifying mammalian transcriptomes by RNA-Seq. Nat. Methods 5(7): 621–628.

Muchero, W., N. Diop, P. Bhat, R. Fenton, S. Pottorf, S. Hearne et al. 2009. A consensus genetic map of cowpea (*Vigna unguiculata* L.) Walp. and synteny based on EST-derived SNPs. Proc. Natnl. Acad. Sci. USA 106(43): 18159–18164.

Myles, S., J.M. Chia, B. Hurwitz, C. Simon, G.Y. Zhong, E. Buckler and D. Ware. 2010. Rapid genomic characterization of the genus Vitis. Plos One 5(1): e8219.

NCBI. 2012. http: //www. ncbi.nlm.nih.gov/Taxonomy/txstat.cgi. Retrieved 11-4-2012.

NCBI. 2012. http://www.ncbi.nlm.nih.gov/genomes/PLANTS/PlantList.html. Retrieved 11-4-2012.

Neale, M.J. and S. Keeney. 2006. Clarifying the mechanics of DNA strand exchange in meiotic recombination. Nature 442(7099): 153–158.

New England Biolabs Catalog and Technical Reference. 2005–06. New England Biolabs, Ipswich, MA. 266–267.

Newman, T., F.J. de Bruijn, P. Green, K. Keegstra, H. Kende, L. McIntosh et al. 1994. Genes galore: a summary of methods for accessing results from large-scale partial sequencing of anonymous Arabidopsis cDNA clones. Plant Physiol. 106: 1241–1255.

O'Driscoll, M. and P. Jeggo. 2006. The role of double-strand break repair—insights from human genetics. Nat. Rev. Genet. 7(1): 45–54.

Otto, S.P. 2007. The evolutionary consequences of polyploidy. Cell. 131(3): 452–62.

Ozsolak, F. and P.M. Milos. 2011. Transcriptome profiling using single-molecule direct RNA sequencing. Methods Mol. Biol. 733: 51–61.

Ozsolak, F., P. Kapranov, S. Foissac, S.W. Kim, E. Fishilevich, A.P. Monaghan et al. 2010. Comprehensive polyadenylation site maps in yeast and human reveal pervasive alternative polyadenylation. Cell 143: 1018–1029.

Ozsolak, F. and P.M. Milos. 2011a. RNA sequencing: advances, challenges and opportunities. Nat. Rev. Genet. 12: 87–98.

Pál, C., B. Papp and M. Lercher. 2006. An integrated view of protein evolution. Nat. Rev. Genet. 7(5): 337–348.

Pan, T.-M. 2002. Current status and detection of genetically modified organism. J. Food and Drug Analysis 10(4): 229–241.

Park, S.-H., N. Yi, Y.S. Kim, M.-H. Jeong, S.-W. Bang, Y.D. Choi and J.-K. Kim. 2010. Analysis of five novel putative constitutive gene promoters in transgenic rice plants. J. Exp. Botany 61(9): 2459–2467.

Pearson, H. 2006. Genetics: what is a gene? Nature 441(7092): 398–401.

Pennisi, E. 2008. Plant sciences. Corn genomics pops wide open. Science 319(5868): 1333.

Piegu, B., R. Guyot, N. Picault, A. Roulin, A. Saniyal, H. Kim, K. Collura et al. 2006. Doubling genome size without polyploidization: dynamics of retrotransposition-driven genomic expansions in *Oryza australiensis*, a wild relative of rice. Genome Res. 16: 1262–1269.

Pirie, N.W. 1990. Jean Brachet. 19 March 1909–10 August 1988. Biographical Memoirs of Fellows of the Royal Soc. 36: 84.

Pooggin, M.M., T. Hohn and J. Fütterer. 1998. Forced evolution reveals the importance of short open reading frame A and secondary structure in the cauliflower mosaic virus 35S RNA leader. J. Virol. 72(5): 4157–4169.

Radonić, A., S. Thulke, I.M. Mackay, O. Landt, W. Siegert and A. Nitsche. 2004. Guideline to reference gene selection for quantitative real-time PCR. Biochem. Biophys. Res. Communi. 313(4): 856–862.

Rapley, R. and S. Harborn (eds.). 2004. Molecular Analysis and Genome Discovery. John Wiley & Sons Ltd., Chichester.

Rasko, D.A., D.R. Webster, J.W. Sahl and B. Ali. 2011. Origins of the strain causing an outbreak of hemolytic–uremic syndrome in Germany. New England J. Medicine 365(8): 709–717.

Reddy, M.P., N. Sarla and E.A. Siddiq. 2002. Inter simple sequence repeat (ISSR) polymorphism and its application in plant breeding. Euphytica 128: 9–17.

Rehrig, E.M., H.M. Appel and J.C. Schultz. 2011. Measuring 'normalcy' in plant gene expression after herbivore attack. Mol. Ecol. Resources 11(2): 294–304.

Renart, J., J. Reiser and G.R. Stark. 1979. Transfer of proteins from gels to diazobenzyloxymethyl-paper and detection with antisera: a method for studying antibody specificity and antigen structure. Proc. Natl. Acad. Sci. USA 76(7): 3116–3120.

Restriction Fragment Length Polymorphism variation by site loss. 2008. Wikipedia Commons, Wikipedia.

Rogan, G.J. 1999. Immunodiagnostic methods for selection of 5-enolpyruvyl shikimate-3-phosphate synthase in Roundup Ready soybeans. Food Control 10: 407–414.

Rostoks, N., S. Mudie, L. Cardle, J. Russell, L. Ramsay, A. Booth, J.T. Svensson et al. 2005. Genome-wide SNP discovery and linkage analysis in barley based on genes responsive to abiotic stress. Mol. Genet. Genomics 274: 515–527.

Roy, D. 2000. Plant Breeding: Analysis and Exploitation of Variation. Alpha Science International Ltd. Pangbourne, UK, 703 pp.

Russel, P. 2001. iGenetics. Genetics Benjamin Cummings. New York 2001.

Sanger, F. and A.R. Coulson. 1975. A rapid method for determining sequences in DNA by primed synthesis with DNA polymerase. J. Mol. Biol. 94(3): 441–448.

Sanger, F., S. Nicklen and A.R. Coulson. 1977. DNA sequencing with chain-terminating inhibitors. Proc. Natl. Acad. Sci. USA 74(12): 5463–5467.

SanMiguel, P. and J.L. Bennetzen. 1998. Evidence that a recent increase in maize genome size was caused by the massive amplification of intergene retrotransposons. Annals of Botany 82(Suppl. A): 37–44.

Schlötterer, C. 2000. Evolutionary dynamics of microsatellite DNA. Nucleic Acids Res. 20: 211–215.

Schuster, S.C. 2008. Next-generation sequencing transforms today's biology. Nat. Methods 5(1): 16–18.

Selvaraj, D., J.-In. Park, Mi.-Y. Chung, Y.-Gu. Cho, S. Ramalingam and Ill.-Sup. Nou. 2013. Utility of DNA barcoding for plant biodiversity conservation. Pl. Breeding and Biotechnol. 1: 320–332.

Semagn, K., A. Bjørnstad and M.N. Ndjiondjop. 2006. An overview of molecular marker methods for plants. Afri. J. Biotech. 5(25): 2540–2568.

Shaked, H., K. Kashkush, H. Ozkan, M. Feldman and A.A. Levy. 2001. Sequence elimination and cytosine methylation are rapid and reproducible responses of the genome to wide hybridization and allopolyploidy in wheat. Plant Cell 13: 1749–1759.

Shimazu, K., C. Matsukura, M. et al. 2001. Characterization of a *DMC1* homologue *RiLIM15* in meiotic panicles, mitotic cultured cells, and mature leaves of rice (*Oryza sativa*). Theoretical and Applied Genetics 102: 1059–1163.

Shivaprasad, P.V., H.-M. Chen, K. Patel, D.M. Bond, B.A.C.M. Santos and D.C. Baulcombe. 2012. A microRNA superfamily regulates nucleotide bindings site-leucine-rich repeats and other mRNAs. The Plant Cell 24(3): 859–874.

Soltis, D.E., P.S. Soltis and J. Tate. 2004. Advances in the study of polyploidy since plant speciation. New Phytologist. 161: 173–191.

Song, K., P. Lu, K. Tang and T.C. Osborn. 1995. Rapid genome change in synthetic polyploids of *Brassica* and its implications for polyploid evolution. Proc. Natl. Acad. Sci. USA 92: 7719–7723.

Southern, E.M. 1975. Detection of specific sequences among DNA fragments separated by gel electrophoresis. J. Mol. Biol. 98(3): 503–517.

Staden, R. 1979. A strategy of DNA sequencing employing computer programs. Nucleic Acids Res. 6(7): 2601–2610.

Stam, M., C. Belele, W. Ramakrishna, J. Dorweiler, J.L. Bennetzen and V.L. Chandler. 2002. The regulatory regions required for B′ paramutation and expression are located far upstream of the maize b1 transcribed sequences. Genetics 162: 917–930.

Stamhuis, I.H., O.G. Meijer and E.J. Zevenhuizen. 1999. Hugo de Vries on heredity, 1889–1903. Statistics, Mendelian laws, pangenes, mutations. Isis; an international review devoted to the history of science and its cultural influences 90(2): 238–267.

Stoddart, D., G. Maglia, E. Mikhailova, A. Heron and H. Bayley. 2010. Multiple base-recognition sites in a biological nanopore: two heads are better than one. Angew. Chem. Intnl. Ed. Engl. 49(3): 556–559.

Streit, S., C.W. Michalski, M. Erkan, J. Kleef and H. Friess. 2009. Northern blot analysis for detection of RNA in pancreatic cancer cells and tissues. Nature Protocols 4(1): 37–43.

Sunkar, R., A. Kapoor and J.K. Zhu. 2006. Posttranscriptional induction of two Cu/Zn superoxide dismutase genes in *Arabidopsis* is mediated by down-regulation of miR398 and important for oxidative stress tolerance. Plant Cell 18: 2051–2065.

The Harvard Nanopore Group. 2009. Mcb.harvard.edu. http://mcb.harvard. edu/ branton/index.htm. Retrieved 2009-11-15.

Tikhonov, A.P., J.L. Bennetzen and Z. Avramova. 2000. Structural domains and matrix attachment regions along colinear chromosomal segments of maize and sorghum. Plant Cell 12: 249–264.

Towbin, H., T. Staehelin and J. Gordon. 1979. Electrophoretic transfer of proteins from polyacrylamide gels to nitrocellulose sheets: procedure and some applications. Proc. Natl. Acad. Sci. USA 76(9): 4350–4354.

Trayhurn, P. 1996. Northern Blotting. Proc. Nutrition Soc. 55: 583–589.

Valouev, A., J. Ichikawa, T. Tonthat et al. 2008. A high-resolution, nucleosome position map of *C. elegans* reveals a lack of universal sequence-dictated positioning. Genome Res. 18(7): 1051–1063.

van Vliet, A.H.M. 2010. Next generation sequencing of microbial transcriptomes: challenges and opportunities. FEMS Microbiology Letters 302(1): 1–7.

Visarada, K.B.R.S., K. Meena, C. Aruna, S. Srujana, N. Saikishore and N. Seetharama. 2009. Transgenic breeding: perspectives and prospects. Crop Sci. 49: 1555–1563.

Wacker, M.J. and M.P. Godard. 2005. Analysis of one-step and two-step real-time RT-PCR using SuperScript III. J Biomol Tech. Sep. 16(3): 266–271.

Wang, J. 2002. Cellular roles of DNA topoisomerases: a molecular perspective. Nat. Rev. Mol. Cell Biol. 3(6): 430–440.

Watson, J.D. 1968. The Double Helix: A Personal Account of the Discovery of the Structure of DNA. Atheneum. New York.

Weber, A.P., K.L. Weber, K. Carr, C. Wilkerson and J.B. Ohlrogge. 2007. Sampling the Arabidopsis transcriptome with massively parallel pyrosequencing. Plant Physiol. 144: 32–42.

Weiss, B., G. Davidkova and L.-W. Zhou. 1999. Antisense RNA gene therapy for studying and modulating biological processes. Cell. Mol. Life Sci. 55: 334–358.

Xu, M., D. Fujita and N. Hanagata. 2009. Perspectives and challenges of emerging single-molecule DNA. Sequencing technologies. Small 5(23): 2638–2649.

Xu, X., A.P. Hsia, L. Zhang, B.J. Nikolau and P.S. Schnable. 1995. Meiotic recombination break points resolve at high rates at the 5′ end of a maize coding sequence. The Plant Cell 7: 2151–2161.

Xu, Y. 2003. Developing marker-assisted selection strategies for breeding hybrid rice. Plant Breeding Rev. 23: 73–174.

Yakovchuk, P., E. Protozanova and M.D. Frank-Kamenetskii. 2006. Base-stacking and base-pairing contributions into thermal stability of the DNA double helix. Nucleic Acids Res. 34(2): 564–574.

Yang, H., J. McLeese, M. Weisbart, J.-L. Dionne, I. Lemaire and R.A. 1993. Simplified high throughput protocol for Northern hybridization. Nucleic Acids Res. 21(14): 3337–3338.

Yu, J., S. Hu, J. Wang, G.K.S. Wong, S. Li, B. Liu et al. 2002. Draft sequence of the rice genome (*Oryza sativa* L. ssp. *indica*). Science 296: 79–92.

Yu, J., J. Wang, W. Lin, S. Li, H. Li, J. Zhou et al. 2005. The genomes of *Oryza sativa*: a history of duplications. PloS Biology 32: e38.

Zhai, J., D.-H. Jeong, E. De Paoli1, S. Park et al. 2011. MicroRNAs as master regulators of the plant NB-LRR defense gene family via the production of phased, trans-acting siRNAs. Genes Dev. 25: 2540–2553.

Zhang, T., X. Zhang, S. Hu and J. Yu. 2011. An efficient procedure for plant organellar genome assembly based on whole genome data from 454 GS FLX sequencing platform. Plant Methods 7: 38.

Zhao, B., R. Liang, L. Ge, W. Li, H. Xiao et al. 2007. Identification of drought-induced microRNAs in rice. Biochem. Biophys. Res. Commun. 354: 585–590.

Zhu, Y.L., Q.J. Song, D.L. Hyten, C.P. Van Tassell, L.K. Matukumalli, D.R. Grimm, S.M. Hyatt et al. 2003. Single-nucleotide polymorphisms in soybean. Genetics 165: 1123–1134.

Z.S. Genetics. 2005. Systems and methods of analyzing nucleic acid polymers and related components. 14-07-2005.

CHAPTER 4

Transgenic Engineering

The term genetic engineering, also called **transgenic engineering**, was first coined by American science writer Jack Williamson in his science fiction novel *Dragon's Island*, published in 1951 [Stableford 2004], just a year before DNA's role in heredity was confirmed by Alfred Hershey and Martha Chase and two years before the double-helix structure of DNA was revealed by James Watson and Francis Crick. It took two decades before Paul Berg's group created, in 1972, the first recombinant DNA molecules by combining DNA from the monkey virus SV40 with that of the Lambda virus [Jackson et al. 1972]. The following year saw the creation of the first transgenic organism by inserting antibiotic resistance genes into the plasmid of *E. coli* bacterium by Herbert Boyer and Stanley Cohen [Arnold 2009; Cohen and Chang 1973]. They spliced the gene from one organism into the DNA of another to produce recombinant DNA which was then expressed normally, and this formed the basis of transgenic engineering.

In 1974, American biologists Beatrice Mintz and Rudolf Jaenisch created a transgenic mouse to study cancer and neurological diseases by introducing foreign DNA into its embryo, making it the world's first transgenic animal [Jaenisch and Mintz 1974]. In 1977, the San Francisco, CA-based biotech company Genentech produced a human protein, somatostatin, in *Eschericia coli* only to be followed a year later by the production of genetically engineered human insulin [Goeddel et al. 1979].

After these successes, the goal of plant molecular biologists and genetic engineers was to isolate one or more specific genes and introduce them into plants. In 1982, the first genetically modified plant was produced using an antibiotic-resistant tobacco plant [Fraley et al. 1982]. This was followed by the first field trials of genetically engineered plants in France and USA in 1986, when tobacco plants were engineered to make them resistant to herbicides [James and Krattiger 1996]. In 1987, a mutant ice-minus strain of *Pseudomonas syringe*—discovered by Stephen Lindow in 1978 to become the first genetically modified organism (GMO)—protected some of the strawberry and potato plants from frost at 25°F (–3.89°C) when sprayed on crops by Advanced Genetic Sciences, Inc. in California [Maugh 1987]. However, activist groups attacked and trashed a part of the test plants the night before the tests occurred. In 1992, China became the first country to commercialize transgenic plants, introducing a virus-resistant tobacco [James 1997], which was later followed by a virus-resistant tomato.

The watershed year in transgene engineering, however, was 1994 when the Davis, California-based Calgene produced world's first transgenetically engineered whole food tomato line 'Flavr Savr', and the European Union approved tobacco engineered to be resistant to the herbicide bromoxynil. The following year saw the U.S. Environmental Protection Agency approving Monsanto Company's insect-(Colorado potato beetle) resistant soil bacteria *Bacillus thuringiensis* (*Bt*) potato, 'New Leaf', making it the first pesticide producing crop to be approved in the USA [Genetically Altered Potato 1995]. The other transgenic crops that received marketing approval in the U.S. in 1995 included, a high-laurate rapeseed (the Canadian low-acid canola) variety Laurical with modified oil composition (Calgene), *Bt*-maize resistant to European corn borer (Ciba-Geigy), bromoxynil-resistant 'BXN' cotton (Calgene), *Bt* cotton (Monsanto), glyphosate-tolerant soyabean (Monsanto), and yellow mosaic and watermelon mosaic viruses-resistant squash 'ZW-20' (Asgrow). Since then, scores of transgenic lines incorporated with qualitative and quantitative traits have been developed in several crops around the world.

Recombinant DNA Technology

Recombinant DNA technology, the mainstay of transgenic engineering, is a laboratory gene-splicing procedure in which the DNA of the donor organism is cut into pieces using restriction enzymes followed by insertion of one of these fragments into the DNA of the host plant. Using recombinant DNA (rDNA) technology, we can isolate and clone a single copy of a gene or an rDNA fragment into an indefinite number of copies, all identical. This technology allows bringing together genetic material from multiple sources, creating sequences that would not otherwise be found in biological organisms. Most of the time, a bacterial or virus plasmid is used to insert the donor DNA.

Recombinant DNA is possible because DNA molecules from all organisms share the same chemical structure. But they differ only in the sequence of nucleotides within the identical overall structure. Consequently, when DNA from a foreign source is linked to host sequences that can drive replication and then introduced into a host organism, the foreign DNA is replicated along with the host DNA. The DNA sequences used in the construction of recombinant DNA molecules can originate from any species. Plant DNA may be joined to bacterial DNA. Furthermore, DNA sequences that do not occur in nature may be created by the chemical synthesis of DNA and incorporated into recombinant molecules. Using recombinant technology and synthetic DNA, literally any DNA sequence can be introduced and inserted into any of a wide variety of living organisms.

Proteins that result from the expression of recombinant DNA within living cells are termed **recombinant proteins**. When recombinant DNA encoding a protein is introduced into a host organism, the recombinant protein will not necessarily be produced. Expression of foreign proteins requires the use of specialized expression vectors and often necessitates significant restructuring of the foreign coding sequence. Recombinant DNA differs from genetic recombination in that the former results from artificial methods in the test tube, while the latter is a normal biological process that results in the remixing of existing DNA sequences in essentially all organisms. The recombinant technology relies on a very simple principle that almost all bacteria and virus as well as some fungi, protozoa, plants, and animals contain small, circular, double-stranded DNA fragments, first termed **plasmids** by Lederberg [Lederberg 1952].

Plasmids, Vectors

A plasmid is a small DNA molecule that is physically separate from, and can replicate independently of, chromosomal DNA within a cell. It can be opened with the same **restriction enzymes** as the DNA fragment of the donor (Fig. 4.1). A plasmid containing

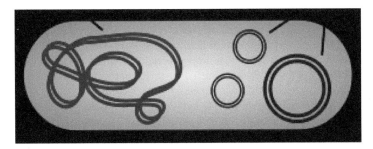

Fig. 4.1. The image showing a line drawing of a bacterium with its chromosomal DNA and several plasmids within it. The bacterium is drawn as a large oval. Within the bacterium, small to medium size circles illustrate the plasmids, and one long thin closed line that intersects itself repeatedly illustrates the chromosomal DNA [Spully 2007].

Color image of this figure appears in the color plate section at the end of the book.

DNA from the donor is called a **vector**. The recombinant vector can then be used to transform bacterial or virus cells.

As transformation vectors, plasmids are used to facilitate the generation of transgenic plants. They can replicate independently of chromosomal DNA within a cell [Lipps 2008]. Bacterial and virus plasmids provide extrachromosomal DNA that can be easily transferred from one organism to the other. Plasmids carry desired genes that may aid in conferring a variety of traits such as resistance to herbicides, plant pathogens, insects, abiotic stresses, etc. Plant biotechnologists use them as cloning vehicles, because they can be easily extracted from *E. coli* without contaminating genomic DNA and they replicate fast, creating identical copies (clones).

In the case of herbicide resistance, the gene to be replicated is inserted into copies of a plasmid containing genes that make cells resistant to particular herbicides and a multiple cloning site (MCS, or polylinker). MCS, extremely useful in transgenic engineering, is a short region of DNA containing several commonly used restriction enzyme recognition sites allowing easy insertion of DNA fragments at this location. In the next step, the plasmids are inserted into bacteria by a process known as transformation. The bacteria are then exposed to a particular herbicide. Those bacteria that take up copies of the plasmid survive, since the plasmid makes them resistant. The protecting genes used to make a protein are expressed, and the expressed protein breaks down the herbicide. In this way, the herbicide acts as a filter to select only the modified bacteria. These bacterial cells are plated and grown in large amounts, harvested, and lysed (using the alkaline lysis method) to isolate the plasmid of interest for insertion in a plant genome later.

A vector is a vehicle used to transfer the genetic material such as DNA sequences from the donor organism to the target cell of the recipient organism. It contains three key elements: a) plasmid selection, i.e., creating a custom circular strand of DNA; b) plasmid replication, so that it can be easily worked with; and c) T-DNA region, for inserting the DNA into the agrobacteria. The most commonly used vectors are termed binary vectors. The types of vectors currently used in transgenic engineering of plants include binary, super-binary, Gateway-based, co-integrate, and expression vectors.

Binary and Super-binary Vectors

The term 'binary vector' literally refers to the entire system that consists of two plasmids, one for the T-DNA and the other for the virulence genes (Fig. 4.2) [Patent Lens 2007], but the plasmid that carries the artificial T-DNA is frequently called a binary vector [Komori et al. 2007; Komori and Komari 2011]. The binary vector system has become the primary choice for the production of transgenic plants.

When the first generation of plant transformation binary vectors were introduced over two decades ago, they were rather simply designed, lacking cloning and expression versatility, and offered very little flexibility for their manipulation for specific research or application purposes. However, vector technology has evolved later with constant improvements, leading to introduction of several new and novel vectors, each suitable for performing various tasks for plant research and biotechnology [Tzfira et al. 2007]. The early binary vectors pBin19 [Bevan 1984] and pBI121 [Jefferson 1987], which were created by adding a reporter gene to pBin19, are still adequate for successful completion of plant transformation. However, the progress of DNA technology has made it possible to design binary vectors in a more sophisticated fashion [Komori and Komari 2011]. Currently, there is a plasmid for every task, including such relatively unique applications as activation tagging [Tzfira et al. 2007]. The commonly used binary and super-binary vectors include pBin19, pBI121, pPVP, pGA482, pPZP series, pCAMBIA series, pGreen series, pPCV001, pGA482, pCLEAN series, pORE O1, pSB11, pSB1, pBIBAC series, pYLTAC series [Komori and Komari 2011].

A binary vector consists of two elements [Komori et al. 2007]: a) T-DNA, the segment delimited by the homologous border sequences, the right border (RB) and the left border (LB), and it may contain multiple cloning site, a selectable marker gene for plants, a reporter gene, a promoter, and other genes of interest; and b) the vector backbone, carrying plasmid replication functions for *E. coli* and *Agrobacterium tumefaciens*, selectable marker genes for bacteria, and optimally a function for plasmid mobilization between the bacteria and other accessory components (Fig. 4.3) [Patent Lens 2007]. The imperfect RB and LB are integrated in binary vectors as DNA fragments cloned from well-known Ti plasmids of either octopine or nopaline type [Komori et al. 2007]. Because factors that enhance

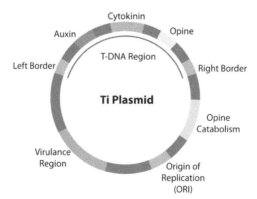

Fig. 4.2. Simplified two-vector system located in the same *Agrobacterium* strain having a T-region in one vector (A) and a *vir* region in another vector (B) [Patent Lens 2007].

Color image of this figure appears in the color plate section at the end of the book.

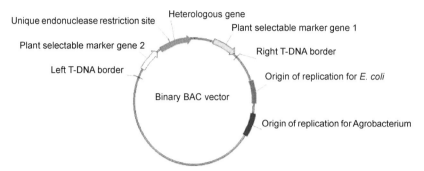

Fig. 4.3. Typical structure of a binary vector with key components [Patent Lens 2007].
Color image of this figure appears in the color plate section at the end of the book.

T-DNA transfer have been identified near the borders [Peralta et al. 1986; Wang et al. 1987], a few hundred bases of natural sequences adjacent to the T-DNA are retained by popular vectors, such as pBin19 [Bevan 1984], pPZP series [Hajdukiewicz et al. 1994], and pSB11 [Komari et al. 1996].

An improved version of a binary vector is the **super-binary vector** that carries additional virulence genes from a Ti plasmid and exhibits very high frequency of transformation, which is valuable for recalcitrant plants such as cereals. In this system, a 14.8-kb *Kpn*I DNA fragment that contains the *virB*, *virG*, and *virC* genes derived from pTiBo542 is introduced into a small T-DNA carrying plasmid [Komari 1990]. The strains of *A. tumefaciens* that carry pTiBo542 have wider host range and higher transformation efficiency than strains that carry Ti plasmids, such as pTiA6 and pTiT37 [Komari 1989]. The final step in the construction of super-binary vector is integration of an intermediate vector with an acceptor system in *A. tumefaciens*. The components of a super-binary vector [Komari et al. 2006] are grouped into two: intermediate vector components and acceptor vector components.

Gateway-based Binary Vectors

Binary vectors used for generation of transgenic cereal crops are cumbersome due to their large size and limited number of useful restriction sites. In order to circumvent the laborious preparation of constructs, Gateway Technology (invented and commercialized by the Carlsbad, CA-based Invitrogen) has developed binary vector sets for plant functional genomics, thereby allowing overexpression of large DNA fragments or knock-down of effector genes, expression of fusion proteins [Review by Earley et al. 2006], and transformation of multiple genes [Chen et al. 2006].

A fast, reliable, and high throughput restrictive-enzyme-free cloning methodology used to develop Gateway Binary Vectors (pGWBs) for plasmid construction is based on the site-specific reaction mediated by bacteriophage λ DNA fragments flanked by recombination sites (*att*). These sites can be transferred into vectors containing compatible recombination sites (*att*L *att*R or *att*B, *att*P) in a reaction mixture mediated by the Gateway clonase mix [Karimi et al. 2002; Magnani et al. 2006]. The plasmid pPZP200 is the backbone of all described Gateway-compatible plant transformation vectors [Hajdukiewicz et al. 1994].

Two recombination reactions, catalyzed by LR and BP recombinases (clonases) respectively, are used in Gateway cloning [Barampuram and Zhang 2011]. In the first step, catalyzed by LR, the gene of interest is inserted into the Gateway vector at the *att*L and *att*R sites. The resulting construct is called the entry clone. All entry clones have *att*Ls flanking the gene of interest. These are necessary in the Gateway system because these *att*L sites are cut to form sticky ends by the Gateway clonase. These sticky ends match with the sticky ends of the destination vector, which contains *att*R restriction sites. This process is called an LR reaction and is mediated by LR clonase mix, which contains the recombination proteins necessary for excision and incision. The product formed in the LR reaction is called the expression clone. It represents a subclone of the starting DNA sequence, correctly positioned in a new vector backbone.

The second BP recombinase reaction, the reverse of the LR reaction, the DNA insert, flanked by 25 bp *att*B sites, is transferred from the expression clone into a vector donated by a plasmid containing the *att*P sites. The final product is the destination clone, and it contains the transferred DNA sequence. Alternatively, these two sequential steps can be reversed to meet specific cloning needs. The BP reaction thus allows rapid, efficient, directional PCR cloning [Karimi et al. 2005, 2007].

Gateway's compatible binary vectors, including the pMDC series of binary vectors, are freely available for noncommercial use [Curtis and Grossniklaus 2003; Karimi et al. 2002]. These can be used for functional analysis of genes by constitutive or inducible ectopic expressions, antisense or RNAi expressions, promoter analyses, subcellular localizations, or complementation analyses [Curtis and Grossniklaus 2003]. The initial pGWB series (pGWBxx) include pGWB1, pGWB2, pGWB3 through pGWB45. This series was followed by Improved Gateway binary vectors (ImpGWBs) using pPZP as a backbone. This was followed by R4 series (R4pGWBs) and R4L1 series (R4L1pGWBs). As Gateway cloning is efficient, precise, flexible, and simple to use, its application will continue to grow in plant research [Tanaka et al. 2012].

Co-integrate Vectors

The co-integrate vector is a plant transformation vector containing both the T-DNA and the virulence genes in a single circular molecule. In this, the sequences to be transferred to the plant genome reside on the same plasmid as the *Vir* genes.

The difference between the binary vector system and the co-integrate vector system comes with the generation of the final plasmid. In the binary system, the plasmid has origins of replication for both *E. coli* and *A. tumefaciens*. It also has selectable markers for both systems and has the intact T-DNA with both left and right arms. The co-integrate vector system uses two plasmids with homologous regions that will allow recombination to generate a final plasmid with all of the necessary parts. Each of the two initial plasmids contains one origin of replication: one for *A. tumefaciens* and one for *E. coli*. When the plasmids recombine the final plasmid has both origins of replication. The same is true for the left and right borders of the T-DNA. In addition, the final T-DNA produced with the co-integrate system includes the homologous DNA sequence.

Co-integrate vector is constructed by recombining 'disarmed' Ti plasmid of *Agrobacterium*. It can also be constructed using a small vector plasmid, which is engineered to carry a gene of interest between a right and a left T-DNA border of the T-DNA region (engineered or modified T-DNA region). Recombination takes place through a single crossover event between homologous regions present in both plasmids. Although co-integrate vectors have become less popular in recent years due to some difficulties encountered in engineering them, they are still used to a certain extent when modified, for example, to allow site-specific recombination of the plasmids within the *Agrobacterium* genome [Patent Lens 2007].

Expression Vectors

An expression vector, also known as expression construct, is used to introduce a specific gene into a target cell. It has a vital role in the cell's mechanism for protein synthesis to produce the protein encoded by a gene. The vector is usually a plasmid or virus designed for protein expression in cells. The plasmid is engineered to contain regulatory sequences that act as enhancer and promoter regions and lead to efficient transcription of the gene carried on the expression vector. The goal of a well-designed expression vector is the production of significant amount of stable mRNA, and therefore proteins.

An expression vector has, like any vector, an origin of replication (a particular sequence in a genome at which replication is initiated), a selectable marker, and a suitable site for the insertion of a gene, multiple cloning site (polylinker). Cloning is generally done first using cloning vectors, and the cloned gene may then be transferred to an expression vector. The cloning process is normally performed in *E. coli*. The vectors used for protein expression in organisms other than *E. coli* would have, in addition to a suitable origin of replication for its propagation in *E. coli*, elements that allow them to be maintained in another organism, and these are called **shuttle vectors**. The elements of expression vectors include a strong promoter, the correct translation initiation sequence such as a ribosomal binding site and start codon, a strong termination codon, and a transcription termination sequence.

Many plant expression vectors are based on the Ti Plasmid of *A. tumefaciens*. It provides a mechanism for transformation, integration of T-DNA into the plant genome as well as promoter for the cloned genes. Another commonly used promoter in gene expression vectors in plant is CaMV 35S.

In the recent past, expression vectors have been used to introduce specific genes in plants to produce transgenic plants. For example, they have been used to introduce a vitamin A precursor, β-carotene, into rice plants and produced '**Golden Rice**' and *Bt* toxin into cotton plants to produce an insecticide that kills certain insects in cotton. Similarly, expression vectors have also been used to delay ripeness of tomatoes. Despite the consumer concerns about the safety of using expression vectors to modify food crops, they are still being used and heavily researched.

Choice of Vectors

Currently, a wide range of binary and super-binary vectors are available. The outcome of foreign gene expression in a vector-based transformation system is still somewhat unpredictable. Some genes are highly expressed but others are not at all for no obvious reason. Although there is no vector that is good for all purposes, there are many vectors which are versatile. Before choosing a suitable vector, the following criteria need to be considered: a) size and nature of DNA fragments, b) strains of *A. tumefaciens* to be employed, c) species of plants to be transformed, and d) purpose of the experiment.

Transgenesis

Transgenesis is the process of introducing an exogenous gene, called **transgene**, into a living organism so that the organism will exhibit a new property and transmit that property to its offspring. In this process, heterologous or homologous DNA is inserted into a plant nuclear genome, resulting in a stable integration and expression [Dixon et al. 2007]. Transgenes are genes that would never have been in their host plants but for biotechnology. Once they are inserted, they behave like normal plant genes.

This technology offers the opportunity to generate novel genetic variation because it allows the introduction of foreign genes from unrelated species and the down-regulation or up-regulation of genes. Over the past 30 years, the field of **transgenic engineering** has developed rapidly due to greater understanding of DNA. In this, the genetic makeup of a plant is altered by using techniques that remove heritable material or introduce DNA prepared outside the plant either directly into the host or into a cell that is then fused or hybridized with the host [The European Parliament 2001]. The resulting plant which now contains genetic material from another species is called "transgenic plant". Transgenic engineering does not normally include conventional plant breeding, *in vitro* fertilization, induction of polyploidy, mutagenesis, and cell fusion techniques that do not use recombinant nucleic acids or a genetically modified plant in the process [The European Parliament 2001]. However, it is expected to accelerate or complement conventional breeding efforts.

The primary benefit derived from transgenesis is an increase in the amount of genetic variability available for breeders to use beyond that which is accessible by conventional breeding methods. The process of transgenic engineering involves several biotechnology techniques, collectively known as **recombinant DNA** (rDNA) **technology**. The inserted genes may come from another plant of the same or a different species, or a completely unrelated kind of organism like bacteria as mentioned earlier. The gene being transferred may have its genetic code altered to modify its function, in addition to having different regulatory sequences spliced on to control how it is expressed (switched on or off) in the plant. The process of moving genes from one species to another is called **transformation**.

Transgenic plants are often referred to as genetically modified (GM) plants or genetically engineered (GE) plants. However, these terms are not preferred in a narrower sense because not all crop plants are genetically modified from their original wild state by domestication, selection, and controlled breeding over long periods of time.

Once a transgenic plant is created, the transgenes can be inherited along with the rest of the plant's genes through normal mating by pollination. The offspring are also transgenic when they acquire the transgenes this way. Because of this, plant breeders can take a transgenic plant made in the laboratory and use conventional breeding methods to develop different transgenic varieties of the crop that are adapted for specific uses.

Process of Transgenic Engineering

The process of transgenic engineering requires the successful completion of a series of steps: locating and identifying genes of interest, isolating and extracting DNA, cloning, designing, and constructing the gene of interest for plant infiltration, transformation, and finally testing and plant breed-back crossing.

Gene Identification

Locating and identifying genes for agriculturally important traits is currently the most vital step in the transgenic process. The traits include yield potential, resistance to herbicides, diseases, insects, and abiotic stress tolerance, salt tolerance, etc., as also higher nutrition and better plant characters. Once a gene of interest is identified, scientists need to understand how the gene is regulated, the effects it might have on other plant characters, and how it interacts with other genes active in the same biochemical pathway.

Isolation and Extraction of DNA

In this second step, a sample of an organism containing the gene of interest is taken through a series of steps to remove the DNA. In any isolation procedure, the initial stage is the disruption of the desired organism, which may be viral, bacterial, or plant cells, in order to extract the nucleic acid. The gene can be isolated using restriction enzymes to cut DNA into fragments and gel electrophoresis to separate them out according to length [Alberts et al. 2002]. Polymerase Chain Reaction may also be used to amplify the gene segment, which can be isolated through gel electrophoresis [Kaufman and Nixon 1996]. If the chosen gene or genome of donor organism is present in a genetic library it can be used. In case the DNA sequence is known, but no copies of the gene are available, it can be synthesized [Liang et al. 2011]. The nucleic acid is then extracted by precipitating it to form a thread-like pellets referring to the DNA.

Gene Design and Construction

In this step, the extracted gene is cloned. The cloned gene is mass-produced in a host cell to make thousands of copies. Once isolated and cloned, the gene of interest is designed and constructed to have it linked to pieces of DNA that will control how the gene will work once inside the host plant genome. Gene construct is an artificially constructed segment of nucleic acid that is going to be 'transplanted' into a target tissue or cell. Gene expression requires a promoter (start signal) and a terminator (stop signal). In most cases, additional sequences such as marker genes are initiated. The term 'construct' is used because the sequences normally do not exist in this combination, so it must be 'put together', i.e., constructed.

The gene of interest needs to undergo several construction stages (modifications) before it can be effectively inserted into a plant (Fig. 4.4). In the first construction stage, a promoter sequence is added for the gene to be correctly expressed (i.e., translation into protein). The **promoter** is an on/off switch that controls when and where in the plant the gene will be expressed.

Promoters allow differential expression of genes. Most promoters in transgenic crop varieties have been 'constitutive', i.e., causing gene expression throughout the life cycle of the plant in most tissues. The promoter region initiates transcription of the gene. The most commonly used constitutive promoter is CaMV 35S (vide Chapters 3 and 5) which generally results in a high degree of expression in plants. Other promoters are more specific and respond to cues in the plant's internal or external environment. One example is the light-inducible promoter from the 'CAB' gene, encoding the major chlorophyll a/b binding protein, localized in PS I and representing a system for balancing the excitation energy between two photosystems.

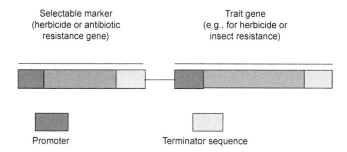

Selectable marker
(herbicide or antibiotic
resistance gene)

Trait gene
(e.g., for herbicide or
insect resistance)

Promoter

Terminator sequence

Fig. 4.4. Simplified presentation of a transgene construct that contains a gene (transgene), a promoter, and a terminator inserted into a plant's genome to perform the regulatory functions that lead to the production of a protein. The DNA sequence between the promoter and terminator determines the type of protein that is produced [Byrne 2014].

Color image of this figure appears in the color plate section at the end of the book.

The second stage involves modification of the cloned gene to enhance its expression in a plant. For example, *Bt* gene for insect resistance is of bacterial origin and has a higher degree of A-T nucleotide pairs compared to plants, which prefer G-C nucleotide pairs. In a clever modification, researchers substituted A-T nucleotide with G-C nucleotides in the *Bt* gene without significantly changing the amino acid sequence. The result enhances production of the gene product in plant cells. In the third stage, the terminator sequence signals to the cellular machinery that the end of the gene sequence has been reached.

In the final stage of construction, a **selectable marker gene** (vide 'Selectable Marker Genes' in this Chapter) is added to the gene construct in order to identify plant cells or tissues that have successfully integrated the transgene. This is necessary because achieving incorporation and expression of transgenes in plant cells is a rare event, occurring in just a few of the targeted tissues or cells. Selectable marker genes encode proteins that provide resistance to agents that are normally toxic to plants, such as antibiotics or herbicides. It is only the plant cells that have integrated the selectable marker gene which will survive when grown on a medium containing the appropriate antibiotic or herbicide. In regard to other inserted genes, marker genes also require promoter and termination sequences for proper function.

Transformation

Transformation, the fifth step in transgenic engineering, is the genetic alteration of a cell resulting from the uptake, incorporation, and expression of exogenous material (DNA) from its surroundings and taken up through the cell membrane(s). It occurs naturally in some species of bacteria, but it can also be effected by artificial means in other cells. Transformation is the heritable change in a cell or organism brought out by the uptake and establishment of introduced DNA. Once the gene of interest is packaged together with the promoter and the marker gene, it is then inserted into a bacterium to allow for the creation of many copies of the gene package.

Aside from transformation, three other terms, conjugation, transduction, and transfection, are used while referring to transferring foreign genetic material in host organism. **Conjugation** refers to direct transfer of genetic material between two bacterial cells. **Transduction** involves injecting foreign DNA by a bacteriophage virus into the host bacterium. In this, the desired genetic material is first inserted into a suitable virus, and then this modified virus is allowed to infect the plant. If the genetic material is DNA, it can recombine with the chromosomes to produce transformant cells. However, genomes of most plant viruses consist of single stranded RNA which replicates in the cytoplasm of infected cell. For such genomes, this method is a form of transfection and not a real transformation, since the inserted genes never reach the nucleus of the cell and do not integrate into the host genome. The progeny of the infected plants is virus-free and also free of the inserted gene.

Transfection is the process of deliberately introducing nucleic acids by non-viral methods into eukaryotic cells. Transfection can be two types, transient and stable. In transient transfection, the transferred DNA is not integrated into host chromosome. DNA is transferred into a recipient cell in order to obtain a temporary but high level of expression of the target gene. In stable transfection, also called permanent transfection, the transferred DNA is integrated (inserted) into chromosomal DNA, and consequently the genetics of recipient cells is permanently changed.

Since plants have millions of cells, it would be impossible to insert a copy of the transgene into every cell. Furthermore, only one percent of bacteria are naturally able to take up the foreign DNA. Therefore, tissue culture is used to propagate masses of undifferentiated plant cells called calluses (calli). These are the cells to which the new transgene will be added.

Tissue Culture

Plant cell and tissue transformation is the process through which the endogenous DNA is modified to enhance the genetic makeup of the plant. Using tissue culture, new plant or cell lines are generated. Recovery of transgenic plants requires transgenes to be targeted at tissues capable of a) receiving and integrating the introduced DNA, b) undergoing selection for the successful transgenic events, and c) regenerating to produce fertile, phenotypically normal plants under tissue culture conditions.

In plant tissue culture, various *in vitro* techniques are used to maintain or grow plant cells, tissues, or organs on a nutrient medium of known composition. It enables one type of tissue or organ to be initiated from another type. In this way, the whole plant can be subsequently regenerated. Tissue culture is widely used to produce clones of a plant in a method known as micropropagation. It has other wide range of applications in plant biology, including a) screening of cells rather than plants to induce herbicide resistance, b) growing plant cells in liquid cultures in bioreactors for production of, *inter alia*, recombinant proteins, c) crossing distantly related species by protoplast fusion to generate novel hybrids, and d) generating whole plants from plant cells that have been transgenetically modified.

For the majority of plant species, gene transfer is carried out using explants competent of regeneration to obtain complete, fertile plants. Cell division and callus formation, embryogenesis, and organogenesis can be induced using combinations of plant growth regulators such as a) auxins (2,4-D, IAA: indole-3-acetic acid, NAA: 1-naphthaleneacetic acid, dicamba, picloram, etc.), cytokinins, gibberellins, abscisic acid, etc. b) nutrients (macro elements such as nitrogen, potassium, calcium, magnesium, phosphorus, etc.); c) microelements (manganese, zinc, iodine, molybdenum, copper, etc.); d) organic supplements (thiamine and myoinositol); e) sucrose; and f) gelling agents like agar in the tissue culture media. As there are no universally applicable plant tissue culture methods, protocols need to be developed or modified for each cultivar, species, genus, and transformation method. The emerged shoots may be sliced off and rooted with auxin to produce plantlets which, when mature, can be transferred to potting soil for further growth in the greenhouse as normal plants.

Testing and Backcrossing

Testing is an extensive evaluation process to determine whether the inserted gene has been stably incorporated without detrimental effects to other functions, product quality, or the intended agroecosystem. This is generally carried out first in the greenhouse or screenhouse. Those that perform well are planted in the field for further testing. In the field, the plants are first grown in confined field trials to test whether the technology works and the plants express the desired traits in the open environment. If the technology works, then the plants are tested in multi-location field trials to establish whether the crop performs well in different environmental conditions.

If the transgenic crop passes all the tests, it may most likely not be used directly for commercial production. It will then be crossed with improved, elite varieties of the crop. This is because only a few varieties of a given crop can be efficiently transformed, and these generally do not possess all the producer and consumer qualities required of modern cultivars. The initial cross to the improved variety must be followed by several cycles of repeated crosses to the improved parent in a process known as **backcrossing** (vide 'Transgenic Breeding': Chapter 3). The goal is to cover as much of the improved parent's genome as possible, with the addition of the transgene from the transformed parent.

Once a new transgenic crop variety is in the process of development, it needs food and environmental safety assessment. This is carried out in conjunction with testing of plant performance. In this phase, the transgenic varieties need to be assessed for altered nutrient levels, known toxicants, new substances, antibiotic resistance markers, non-pathogenicity to animals and humans, toxicity to non-target organisms, stable integration of the introduced gene(s) in the plant's chromosomes, risk of creating new plant viruses, effects on plant biology and ecosystem, spread of the transgene to other crops and wild relatives, allerginicity, etc.

Transformation Methods

The chief objective of plant transformation is to produce transgenic plants with integrated transgenes from desirable backgrounds. Stable transformation methods consist of three steps: a) delivery of DNA into a single plant cell; b) integration of the DNA into plant cell genome; c) conversion of the transformed cell into a whole plant.

The common and vital feature in all transformation methods is that the foreign DNA first has to enter the plant cell by penetrating the cell wall and the plasma membrane and then must reach the nucleus and integrate into the resident chromosomes. The primary consideration before choosing a transformation method is the structure and complexity of transgenic loci they create. A transgenic plant is generated in a laboratory by altering its genetic makeup, by adding one or more genes to a plant's genome using different engineering methods. In order to achieve this, genes of interest are transported by using either of two major methods. One is vector-mediated, which is based on *Agrobacterium*, a bacterium whose plasmid is used to construct a vector which then incorporates genes that are to be transferred to the plant cells. A variant of vector-mediated approach is non-*Agrobacterium*-mediated transformation. The other method is based on the direct, vector-less, physical transfer of foreign genes into target plant cells. This method includes several approaches.

Each of these methods has varying transformation efficiency (TE), which reflects on the efficiency by which plant cells can take up extracellular DNA and express genes encoded by it. TE is calculated by the number of successful transformants from the amount of DNA used during a transformation process. Transformants are cells that have taken up DNA (foreign, artificial, or modified) and which can express genes on the introduced DNA.

Transformation efficiency is affected by a number of factors. These include a) method of transformation, b) type of plant species, c) forms of DNA, d) size of plasmid, e) type of explants used, f) genotype of cells, and g) growth of cells (Appendix: Table 3).

Vector-Mediated or Indirect Gene Transfer

Agrobacterium-Mediated Transformation

Among the various vectors employed in plant transformation, the tumor-(T) inducing, 'Ti', plasmid of *Agrobacterium tumefaciens* is being used for over 35 years. This soil-born, rod-shaped, Gram-negative, alphaproteobacterium is known as 'natural genetic engineer' of plants because it has the natural ability to transfer T-DNA of their plasmids into plant genome upon infection of cells at the wound site and cause an unorganized growth of a cell mass known as crown gall, a plant tumor (Fig. 4.5) [Clemson University 2002]. Once inside the plant, T-DNA is integrated into the host cell genome and stably maintained.

The *Agrobacterium*-mediated transformation is a highly complex and evolved process involving genetic determinants of both the bacterium and the host plant cell. The pathogenic genus *Agrobacterium* has been divided into a number of species, reflecting, for the most part, disease symptomology and host range. Aside from *A. tumefaciens*, other prominent species include the closely related *A. rhizogenes* (also called *Rhizobium rhizogenes*), which causes the hairy root tumors [Ganci 2012], and *A. radiobacter*, an 'avirulent' species, and *A. rubi* (causes cane gall disease). Another species includes *A. vitis* which causes galls

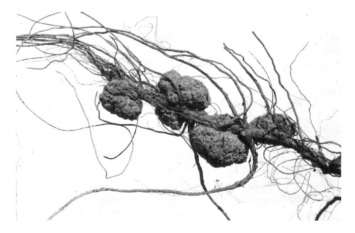

Fig. 4.5. *Agrobacterium tumefaciens*-induced galls at the root of *Carya illinoinensis* (pecan) [Clemson University-USDA Cooperative Extension Slide Series 2002].

Color image of this figure appears in the color plate section at the end of the book.

in grape and a few other plant species. Of these, *A. tumefaciens* is the most widely used species in plant biotechnology, followed by *A. rhizogenes*, the carrier of the distinct root (R)-inducing "Ri" plasmid.

Structure of Plasmid. Ti plasmids are used as gene vectors for delivering useful foreign genes into target plant cells and tissues. The foreign gene is cloned in the T-DNA region of Ti-plasmid (200–800 kb in size) in place of unwanted sequences. The T and R regions of Ti and Ri plasmids can be large enough to encode tens of genes. Both plasmids contain a segment of DNA, called T-DNA, which is integrated into the chromosomal DNA of host cells during transformation. The T-DNA contains two types of genes: oncogenic genes, encoding enzymes involved in the synthesis of auxins and cytokinins (causing tumor formation) and genes involved in opine production. The T-DNA segment is flanked by two 25-bp long direct repeat homologous sequences known as left and right borders (LB and RB). It is called the T-strand. It represents the substrate of DNA transfer to the host cell. Both oncogenic and opine catabolism genes are located inside the T-DNA of the Ti plasmid while the virulence (*Vir*) genes are situated outside the T-DNA of the Ti plasmid and bacterial chromosome. Many of the genes required for DNA transfer reside on a section of Ti (or Ri) plasmid known as the *Vir* region. The *Vir* genes are organized into several operons (*Vir*A, *Vir*B, *Vir*C, *Vir*D, *Vir*E, *Vir*F, *Vir*G, and *Vir*H) on the Ti plasmid and other operons (*chv*A, *chv*B, and *chv*F) that are chromosomal and essential for T-DNA transfer [Barampuram and Zhang 2011].

Wounded plant cells produce phenolic compounds such as acetosyringone. The compounds are detected by the protein product of one of the *Vir* genes, *Vir*A, which leads to the expression of other *Vir* genes. Together, the *Vir* proteins result in T-DNA transfer into the plant cell nucleus. The protein products of the *Vir* genes act in *trans*. This means that the T-DNA may be on a different plasmid from the *Vir* genes in the agrobacterial cell, and still be efficiently transferred into the plant genome, provided the *Vir* genes are intact and functional. This has led to specialized 'binary' cloning vectors (vide 'Binary Vectors', this chapter) becoming the most commonly used *Agrobacterium*-based transformation systems.

Mechanism of Transformation. The mechanism of gene transfer from *A. tumefaciens* to host plant cells involves several steps: bacterial colonization, induction of the bacterial virulence system, generation of the T-DNA transfer complex, T-DNA transfer, and integration of the T-DNA into the plant genome.

The T-DNA transfer process is initiated upon receipt of signals (in the form of phenolic compounds, monosaccharides, low pH, and low phosphate) received by *Vir*A from host cells, leading to activation of crown gall virulence machinery. This results in expression and activation of the virulence genes (*Vir*), grouped into six different loci, the largest locus being *Vir*B spanning 9.5 kb. During this process, *Vir*D1 and *Vir*D2 proteins nick both the left and right borders on the bottom strand of T-DNA molecule. The resulting single-stranded T-DNA molecule (T-strand), together with several *Vir* proteins, is then exported into the host cell cytoplasm, likely by a cellular motor-assisted mechanism, through a channel formed by the *Agrobacterium Vir*D4 and *Vir*B protein complex [Barampuram and Zhang 2011]. The T-complex then enters the cell nucleus by an activation mechanism mediated by the nuclear import machinery of the host cell. This facilitates integration of the T-strand into the host genome.

The integration and enhancement of gene expression in the plant genome greatly depends on the promoter that is fused with 5′ end of the gene of interest. The most widely used foreign regulatory elements include the CaMV and the transcriptional terminator of the *Agrobacterium* nopaline synthase (*nos*), which together promote high-level gene expression in transgenic plants [Lee and Gelvin 2008]. Another promoter is the *GUS* fusion gene found in maize and tobacco.

A brief protocol of this vector-based transformation process is as follows:

a) Leaf discs (in the case of dicots) or embryogenic calli (in the case of monocots) are collected and infected with *Agrobacterium* carrying recombinant disarmed Ti-plasmid vector.

b) The infected tissue is then cultured (co-cultivated) on shoot regeneration medium for 2–3 days during which time the transfer of T-DNA along with foreign genes takes place.

c) The transformed tissues (leaf discs/calli) are transferred onto selection cum plant regeneration medium supplemented with usually lethal concentration of an antibiotic to selectively eliminate non-transformed tissues.

d) After 3–5 weeks, the regenerated shoots (from leaf discs/calli) are transferred to root-inducing medium.

e) After another 3–4 weeks, complete plants are transferred to soil following the hardening (acclimatization) of regenerated plants.

f) Finally, the transgenic plants are assayed to detect the presence of foreign genes by using PCR and Southern hybridization techniques.

Host plant response to *Agrobacterium* takes place in two distinct stages, during the initial 12 hours after inoculation and subsequently. During the first stage, a 'general' response is observed with activation of many defense-related genes, akin to plant response to biotic stress. In the second stage, most of the defense-related genes are downgraded to their initial levels, and a second set of host genes involved in cell division and cell growth, important for the transformation process, is activated [Veena et al. 2003].

The target materials for *Agrobacterium*-mediated transformation include embryonic cultures, immature zygotic embryos, matured seed-derived calli, meristems, shoot apices, primary leaf nodes, excised leaf blades, roots, cotyledons, stem segments, and callus cultures which regenerate through either somatic embryogenesis or organogenesis regime.

The *Agrobacterium*-mediated transformation of higher plants is well established for dicot species, while the same for monocot species is on the rise. It is successfully employed in conjunction with organogenesis in tissue culture to transform dicot plants of commercial importance as well as monocotyledonous grain crops such as rice, wheat, maize, and barley, even though it is not a pathogen on cereals. The success rate is high when immature embryos (with or without pretreatment) as well as embryogenic calli derived from mature seeds are used as explants. In the case of rice, high-efficiency (100 percent) transformation system was obtained by co-cultivating calli with *Agrobacterium* on filter papers moistened with enriched media instead of using solid media [Ozawa 2013]. Enhancement of transformation from the normal range of 1–2 percent to greater than 7 percent was reported in sorghum by using immature embryos prior to *Agrobacterium* inoculation [Gurel et al. 2013].

A. tumefaciens carries an Ri plasmid that functions in a similar way to the Ti plasmid but induces a different response in the host plant cell. However, the *A. rhizogenes*-mediated root transformation has received considerably less attention than *A. tumefaciens* transformation. The main reason for this is the difficulty in regenerating plants from hairy roots transformed by *A. rhizogenes*. However, the hairy roots have the ability to grow in a hormone-free media.

The natural host range of *Agrobacterium* strains is very large, which includes most of the dicot and monocot species including rice maize, wheat, and barley, producing transgenic plants resistant to herbicides. In certain cases, *Agrobacterium* may persist in plant tissues after transformation. It has the ability to persist even in low populations in different plant organs, migrate inside of them, and survive diverse environments and soil conditions. Although addition of antibiotics in general to the culture media may lack the ability to eliminate *Agrobacterium*, the timentin, cefotaxime, and carbencillin have good antibacterial activity in suppressing this bacterium as has been found during transformation of wheat [Han et al. 2012] and rice [Priya et al. 2012].

Floral Dip Method

In floral-dip transformation, developing flowers (inflorescences) are immersed in a suspension of *A. tumefaciens*, and then forcing the bacteria into plant tissues by brief vacuum infiltration. This is followed by growing the plants to maturity, harvesting and germination of seeds, and the selection of seedlings.

Floral-dip method is the method of choice for rapeseed (canola) transformation. It allows plant transformation without the need for tissue culture, but the growth of the *A. tumefaciens* strain in liquid culture calls for pelleting the culture and resuspending in a buffered media, which takes one hour [Davis 2009]. However, selection for kanamycin- and hygromycin-B-resistance in the selection medium is a lengthy process, which may cause fungal contamination problem, resulting in depletion of the antibiotic present. This allows the non-transformants to remain green and crowd the seedlings for a much longer time on an agar plate [Li et al. 2010].

In an improved method [Li et al. 2010], the binary vectors used were pCAMBIA2200 and pCAMBIA1300, with the former conferring resistance to kanamycin via the selectable marker gene *npt*II and the latter to hygromycin B via the *hpt*. In this protocol, rapeseed plants were transformed at bud (initial blossom) stage by dipping the racemes briefly (15–20 min) into an *A. tumefaciens* culture (Silwet L-77, sucrose, and acetosyringone). The inflorescence of rapeseed was covered with paper bags to keep it moist for 24 hr. This was followed by incubating seeds in a selective medium {usually MS—Murashige and Skoog—plus kanamycin monosulphate (300 mg L^{-1}) or hygromycin B (100 mg L^{-1}} for 24–36 hr. The seeds were then planted in the soil and the growing seedlings examined after 2 weeks. Genomic DNA was extracted from the transformed plants by CTAB (cetyltrimethylammonium bromide) method and analyzed by PCR technique. The transformants had green, open, expanded chlorophyll-filled cotyledons, while the non-transformants failed to show these features.

Non-*Agrobacterium*-Mediated Transformation

Complex patent issues related to *Agrobacterium* as a tool for transgenic engineering and the general requirement of establishing transgenic plants can create obstacles in using this technology for speedy research and development and for agricultural improvements in many plant species [Chung et al. 2005].

Broothaerts et al. [2005] found three non-*Agrobacterium* species, *Rhizobium* sp. (NGR234), *Sinorhizobium meliloti*, and *Mesorhizobium loti*, being capable of genetically transforming different plant tissues and plant species. They introduced into these bacteria a disarmed Ti plasmid (pEHA105) from a hypervirulent *Agrobacterium* strain. To facilitate transfer of this large plasmid, the origin of transfer (*oriT*) of a broad host range IncP plasmid was integrated into the Ti plasmid of EHA105 at two different locations (pTiWB1, pTiWB3). The modified plasmids were then mobilized into: a) the *Rhizobium* sp. that has an exceptionally broad host range, capable of nodulating over 100 different plants [Pueppke and Broughton 1999]; b) the alfalfa-symbiont *Sinorhizobium meliloti*; and c) *Mesorhizobium loti*, a representative of a different family (Phyllobacteriaceae). To assay for gene transfer, three binary vectors were prepared. Of these, pCAMBIA1105.1R was introduced into *Rhizobia* bacteria, and either pCAMBIA1105.1 or pCAMBIA1405.1 into *Agrobacterium*. The plant transformation events were analyzed through *GUS* activity, Southern blotting, and PCR assays.

Although the transformation efficiency was considerably lower when compared with *Agrobacterium*-mediated transformation, all three non-virulent, non-*Agrobacterium* strains, belonging to two families of bacteria, were able to transform *Arabidopsis* and rice [Broothaerts et al. 2005]. Of these, *S. meliloti* was the most competent to transfer genes into both monocots and dicots and into a range of tissues, including leaf tissue, undifferentiated calli, and immature embryo ovules. T-DNA transformation appeared normal, even if it was a lower frequency.

Vector-less or Direct Gene Transfer

Direct gene transfer involves delivery of the foreign gene of interest into the host plant cell without the help of a vector. Molecules of the introduced naked DNA can interact via recombination and ligation reactions to form rearranged multimers. There are various vector-less transformation methods. The more prominent of them include biolistic, electroporation, bioactive beads-mediated, microinjection, liposome-mediated, chloroplast-mediated, chemical-mediated, pollen-based, and agrolistic methods. For full protocols, the reader may refer to other publications.

Biolistic Method

The biological ballistic (bioballistic, biolistic) method transforms almost any type of living cell. It is not limited to genetic material of the nucleus. It can also transform organelles, including plastids. Also known as the 'gene gun' or "article bombardment" method, this method is most commonly used in engineering plants with, *inter alia*, herbicide resistance in a crop. It involves physical impregnation of cells with nucleic acids or with other biological molecules. As it does not depend on bacteria, it has become a versatile, effective, and powerful transformation method for many species, particularly monocots including wheat, rice, maize, sorghum, barley, etc. and some dicot plants such as soyabean, cotton, papaya, pine, etc. [Wang and Ge 2006; Sujatha and Visarada 2013] for which transformation using *Agrobacterium tumefaciens* has been less successful.

A wide range of target tissues including cell suspension, meristems and embryogenic or organogenic cultures, or explants that give rise to these types of culture can be stably transformed by bombardment. Particle bombardment is also an effective method for embryogenic microspore and immature pollen transformation [Resch and Touraev 2011]. After transformation, transgenic homozygous diploid plants can be regenerated in a single step and these can be useful for creation of transgenic plants as well as for gene mapping, induction and selection of mutants, and plant breeding.

Mechanism of Transformation. The gene gun (22 caliber nail gun) that blasts microparticles coated with viral DNA into the plant can be a commercial or a home-made particle inflow gun. It has a small chamber connected to a cylinder powered by pressurized helium (900–1300 psi). A solenoid valve on the gas cylinder controls the release of helium, which passes into a vacuum chamber through a filter holding micron or submicron metal particles that carry nucleic acids.

The target of the biolistic transformation method is often the calli of undifferentiated plant cells growing on gel medium in a Petri dish. DNA is precipitated with additives such as spermidine, calcium chloride, or polyethylene glycol. The precipitated DNA is mixed with particles of tungsten or gold, and the DNA-coated particles (0.5–1.1 μm) are placed on the end of a large, plastic bullet. The bullet is then loaded into the gun barrel and positioned on the target tissue (target distance 7.5–10 cm) at the end of the barrel. Once the pressure that releases helium from the chamber reaches 7 in. of Hg, the gun is fired, accelerating the bullet to the end of the barrel where it is stopped by a perforated stopping plate (screen).

The small, DNA-coated slivers of metal pass through the perforation and hit the target tissue of living cells in Petri dish on the other side. Some particles pass through plant cell walls into cell nuclei, where they join the chromosome and transform the plant cell. However, some cells not obliterated in the impact, have successfully enveloped a DNA coated metal particle, whose DNA eventually migrates to and integrates into a plant chromosomal DNA. If the foreign DNA reaches the nucleus, then transient expression is likely to result and the transgene may become incorporated in a stable manner into host chromosome.

The transformed callus cells are re-collected and selected for successful integration and expression of new DNA for herbicide resistance using molecular techniques such as PCR and Southern and Northern hybridization analyses (vide Chapter 3). PCR is an effective method for rapid screening of transgenics. Southern blot hybridization analysis is used to confirm stable integration of a transgene in the plant genome. Expression levels of the transgene in individual plants can be detected by Northern blot hybridization analysis [Spangenberg et al. 1998].

Different methods are used to accelerate the particles. These include pneumatic devices; instruments utilizing a mechanical impulse or macroprojectile; centripetal, magnetic or electrostatic forces; spray or vaccination guns; and apparatus based on acceleration by shock wave, such as electric discharge [Christou and MaCabe 1992]. The cells that take up the desired DNA, identified by using a marker gene, β-glucuronidase (*GUS*), are then cultured to replicate the gene and possibly cloned.

The advantages of biolistic method are: a) it can be used to incorporate resistance to herbicides in plants, b) it is useful to transform even intracellular organelles, c) no binary vector is needed, d) no carrier DNA is required, e) both transient and stable transformation can be carried out, f) transformation protocol is relatively simple, g) diverse cell types can be targeted efficiently for foreign DNA delivery, and h) large intact DNA constructs can be introduced into the plant genome.

However, the procedure may cause serious damage to the cellular tissue, besides its high cost of equipment and microcarriers and difficulty in obtaining single copy of transgenic events. It also often results in the insertion of multiple gene copies and in complex rearrangements of transgenes compared to *Agrobacterium*-mediated transformation [Iyer et al. 2000]. These could adversely affect the stability of the transgenes. Furthermore, the transformation efficiency of this method is relatively low, but it depends on plant species.

Electroporation

Electroporation, also called electropermeability, involves introduction of foreign DNA into protoplasts suspension containing naked or recombinant plasmids by exposing it to pulses of transmembrane electric voltage (0.5–1.0 V) at each point on the lipid bilayers of cell membrane. The electric shocks induce the transient formation of large pores in the cell membrane. These transient pores allow diffusion of, and give a passage to, various classes of macromolecules including DNA, RNA, viral particles, antibodies, etc. They enter the protoplast and thus increase the transformation frequency. Electroporation is involved in the transfusion of DNA, frequently for stable transformation of various plant species.

Electroporation-based transformation system consists of a number of potentially important variables. These include method of protoplast preparation, strength and duration of electric pulse, concentration of ions, composition of the electroporation buffer, and purity of DNA. This method has been used for successful transformation of protoplasts of both monocot and dicots. The technique was originally developed for protoplast transformation, but has subsequently been used for intact tissues, pollen, and mature embryos as well. Electroporation of plant cells and tissues is very similar to that of protoplasts. This approach enabled the recovery of transgenic plants in barley [Salmenkallio-Marttila et al. 1995]. In sugarcane, a gene was transferred into intact meristematic tissue using electroporation-mediated transformation [Manickavasagam et al. 2004].

Using electroporation method, genes of choice have been successfully transferred into the protoplast of wheat, rice, sorghum, maize, tobacco, and petunia. In the recent past, transgenic trees have been created when a coat protein of the plum pox virus, which causes Sharka disease, has been introduced into apricot. The resulting transgenic tree shows a markedly reduced sensitivity to this virus.

With its transient expression, electroporation technique is rapid, allowing for reproducible detection of gene products within hours of the introduction of DNA [Fisk and Dandekar 2005]. It may also be used to rapidly demonstrate functionality of new transgene sequences before they are used to generate transformants by some other method of DNA introduction. Besides, this technique entails high frequency of transformation, versatility with nearly all cells and species, and simplicity of technique. Electroporation is convenient and the results are consistently duplicated as a daily routine. At times, it is more efficient than other methods including particle bombardment. Furthermore, it does not suffer from the host-range limitations posed by *Agrobacterium*-mediated method or the toxicity problems sometimes encountered using a PEG based procedure.

However, electroporation is not always as efficacious as biolistic method. Its limitation includes that cell suspensions employed for isolation of protoplast are tissue-dependent with a time-limited morphogenic potential competence and, hence, are prone to somaclonal variation. This method may not be suitable if the target cells are available in the form of multicellular explants with cell walls that could represent key physical barriers to electroporation. Furthermore, if the pulses are of wrong length and intensity, some pores may become too large or fail to close after membrane discharge, causing cell damage or rupture [Weaver 1995].

Bioactive Beads Method

In bioactive beads (BABs)-mediated transformation, the submicrometer-sized beads encapsulate DNA molecules and immobilize them into beads with the help of polyethylene glycol (PEG) solution. During immobilization within the beads, DNA molecules are physically stabilized and accumulated at a limited area on the cell surface.

Alginate, a hydrophilic polysaccharide that forms a gel in the presence of Ca^{2+} ions, is utilized as a barrier membrane to produce calcium alginate beads [Sone et al. 2002; Khemkladngoen et al. 2012] which are nontoxic to both plant and animal cells. The positive charge of calcium alginate allows an electrostatic interaction with the negative charge of cell membrane [Wada et al. 2011].

Mechanism of Transformation. In this procedure [Sone et al. 2002; Khemkladngoen et al. 2012; Wada et al. 2011], 100 µl of 0.5–2.0 percent sodium alginate solution is mixed with 900 µl of a water insoluble but partially hydrophilic solvent, isoamyl alcohol, using a sonicator (UR-20P; TomySeiko, Tokyo) for one minute to form a water and oil emulsion. In this emulsion, the alginate solution forms small micrometer-sized droplets. Then, 500 µl of a 100 mM $CaCl_2$ solution containing the plasmid DNA is added to the emulsion and the alginate is allowed to solidify. The bioactive beads (91 percent of them being spherical) are collected after centrifugation at 4000 rpm for 5 min and resuspended in 100 mM $CaCl_2$. The bioactive beads are washed four times in 100 mM $CaCl_2$ and then stored in $CaCl_2$.

In the next step, protoplast transformation with the BABs is performed in combination with PEG treatment [Sone et al. 2002; Higashi et al. 2004]. Protoplasts are isolated from a cell suspension. An aliquot of 500 µl of protoplast suspension is transferred to a 15 ml glass centrifugation tube. DNA-immobilized BABs are gently mixed with the protoplast suspension. Then 825 µl of 40 percent (w/v) PEG CMS6 solution is added, and the final concentration of PEG adjusted to 24 percent. After 10 min of PEG treatment, 815 µl of a 0.2 M $CaCl_2$ solution is added to dilute the PEG, and mixed to disperse the protoplasts. This procedure is repeated three times and the centrifugation tube filled with W5 solution [Menczel and Wolfe 1984]. The suspension is centrifuged for one min at 800 rpm, and the precipitated protoplasts washed again. The protoplasts are then suspended in protoplast culture medium. The transformation efficiency is estimated by the fluorescence of green fluorescence protein (GFP) in cells transformed with pUC18-sGFP. Efficiency is higher with beads of smaller size, with one to six µm diameter beads being more efficient [Khemkladngoen et al. 2012]. Smaller beads are more easily incorporated into the plant protoplasts because of the availability of a larger total surface area for the same volume of alginate solution being used.

The BAB method is simple, low-cost, and requires limited equipment. In addition, the BABs have the capability of stabilizing the large DNA fragments by encapsulating the DNA in the beads (as well as being stabilized on the surface of the beads) in suspension.

This method has been successfully used for transformation of protoplasts of various plant species (tobacco BY-2, tobacco SR-1, brinjal/eggplant, carrot, and rice) [Liu et al. 2004a,b; Wada et al. 2009]. Unlike *Agrobacterium*-mediated transformation and biolistic methods, the BAB method is more applicable for plant transformation with large DNA fragment as in the case of its successful use when 124 kb of YAC (yeast artificial chromosome) DNA was introduced into cultured tobacco BY-2 cells [Liu et al. 2004a]. Similarly, large BAC (bacterial artificial chromosome) DNA fragments (~100 kb) containing a set of *Aegilops tauschii* seed-hardness genes [Bhave and Morris 2008] has also been successfully used for transformation of rice (*Oryza sativa* L. ssp. *japonica* cv. Nipponbare) [Wada et al. 2009].

Bacterial beads method has also been successfully to introduce multiple genes into rice plants by producing stable transgenic plants that can pass introduced transgenes on to successive generations [Khemkladngoen et al. 2012]. Besides, the hardness gene, *puroindoline* b (*Pin*b), was functional when introduced in transgenic rice, suggesting that introduction of a genomic locus that controls a trait could be a good strategy for adding desirable traits to plants [Khemkladngoen et al. 2012].

The BAB method can also be applied to immobilize DNA-lipofectin complexes, instead of naked DNA, so the beads will have enhanced affinity to the protoplast surface [Murakawa et al. 2008] and to successfully entrap bovine serum albumin (BSA) protein. The positive charge on the lipofectin surface increases the electrostatic affinity for the negative charge on the surface of the protoplasts. The efficacy of BAB technique on protein immobilization can be further improved by treating the alginate solution with 1-ethyl-3-(3-dimethylaminopropyl) carbodiimide (EDC) and N-hydroxysulfosuccinimide (NHSS) to cross-link the BSA to the alginate carboxyl groups prior to solidification [Khemkladngoen et al. 2012]. Cross-linking beads provide high protein-retention ability for up to 2 wk after immobilization. Such improved protein-immobilizing beads with high retention capacity might have greater potential to be a preferred transformation technique.

Microinjection

Microinjection is a technique of delivering foreign DNA at a microscopic or borderline macroscopic level into a living cell through a glass micropipette with 0.5–1.0 µm diameter tip, resembling an injection needle. The desired contents are then injected into the desired sub-cellular compartment and the needle is removed. The process is frequently used as a vector in genetic engineering and transgenics to insert genetic material into a single cell. Microinjection can also be used in the cloning of organisms, and in the study of cell biology and viruses.

In order to manipulate the protoplast without damage, the protoplasts are cultured for 1 to 5 days before the injection is performed. The recipient protoplasts are immobilized on a solid support (cover slip, slide, etc.) and they are artificially bound or held by a pipette under suction. One end of a glass micropipette is heated until the glass becomes somewhat liquefied. It is quickly stretched to form a very fine tip at the heated end. The process of delivering foreign DNA is done under a specialized powerful optical microscope, called a micromanipulator. Cells to be microinjected are placed in a container. A holding pipette is placed in the field of view of the microscope. The pipette holds a target cell at the tip when gently sucked. The tip of the micropipette is injected through the membrane of the cell. Contents of the needle are delivered into the cytoplasm and the empty needle is taken out.

Employing needles with diameter greater than cell diameter, DNA has been injected into to the stem below the immature floral meristem of rye (*Secale cereale*) plant, so it can reach the sporogenous tissue [De la Pena et al. 1987]. This has led to the production of transgenic rye plants. However, this technique has not been successful with other cereals, making the validity of earlier experiments doubtful. Thus, microinjection process is considered as unreliable, with low success rate.

Liposome-Mediated Transformation

A liposome is an artificially-prepared vesicle composed of a lipid layer. It is an unilamellar, circular, bilayered, microscopic lipid molecule with an aqueous interior that can carry nucleic acids. Liposomes encapsulate the DNA fragments and then adhere to the cell membranes and fuse with them to transfer DNA fragments. They interact with many cells in such a way that plasmid will be transferred into plant genome. Thus, DNA enters the cell and then to the nucleus. Liposomes are constructed with PEG studding outside of the membrane. The PEG coating, which is inert in the body, allows for longer circulatory life for plant chemical delivery mechanism.

Liposome-mediated transformation is a very efficient technique used to transfer genes into genomes of a number of plant species. It generally uses a positively charged (cationic) lipid to form an aggregate with the negatively charged (anionic) genetic material [Felgner et al. 1987]. A net positive charge on this aggregate has been assumed to increase the effectiveness of transfection through the negatively charged phospholipid bilayer [Felgner et al. 1987]. This transfection technology performs the same tasks as other biochemical procedures utilizing polymers, diethylaminoethyl cellulose (DEAE-C) dextran, calcium phosphate, and electroporation.

The association of the lipofection reagent with nucleic acids results in a tight compaction and protection of the nucleic acids against nucleases present in the plant cell media. These cationic complexes are mainly internalized by endocytosis, a process by which cells absorb molecules such as proteins by engulfing them. Once inside the cells, the nucleic acids are released into the cytoplasm via two mechanisms. One mechanism relies on the endosomes (membrane-bound compartment inside the cell) buffering capacity of the polycationic residues, called 'proton sponge effect'. The other mechanism depends on the ability of negatively charged cellular lipids to neutralize the cationic residues of the transformation agent leading to destabilization of endosomal membranes. Finally, the cellular and molecular events leading to the nuclear uptake of DNA (not required for siRNA) following by gene expression remain highly speculative. However, the significance of cell division on transfection efficiency favors the assumption that nuclear membrane disruption during the mitosis process promotes DNA nuclear uptake. Nonetheless, transfections of primary cells (non-dividing) and *in vivo* are also achievable with lipofection, demonstrating that DNA can make its way to the nucleus where gene expression takes place.

The main advantages of liposome technique are its high efficiency, its ability to transfect all types of nucleic acids in a wide range of cell types, its ease of use, reproducibility, and low toxicity. Its other advantages include enhanced delivery of encapsulated DNA by membrane fusion, protection of nucleic acids from nuclease activity, targeting specific cells, delivery into a variety of cell types besides protoplasts by entry through plasmodesmata, and delivery of intact small organelles. In addition, this method is suitable for all transformation applications, including transient, stable, co-transfection, reverse, sequential, and multiple transfections. Successful transformation based on this system was reported for tobacco [Dekeyser et al. 1990], wheat [Zhu et al. 1993], and potato [Sawahel 2002]. However, the size of lipoplex (liposome/DNA complex) is a major factor determining lipofection efficiency, with large particles showing higher efficiency than small particles [Almofti et al. 2003].

Chloroplast-Mediated Transformation

Chloroplasts are specialized plant organelles, known to host photosynthesis, and harbor many other important biosynthetic pathways. During plant development, plastids arise by differentiation of proplastids, the precursors present in meristematic tissues. They also develop into many other forms, including chloroplasts in leaves, amyloplasts in roots, and chromoplasts in flowers and fruits. Plastid transformation can involve delivery of DNA into chloroplasts. Once stable transformation has been achieved, all plastid types within the plant contain the same transgenic plastome, the genome of a plastid. Thus, in flowering plants, which contain a variety of plastid developmental forms, the term **plastid transformation** is more accurate than chloroplast transformation [Day and Goldschmidt-Clermont 2011]. In this method, DNA has to be delivered in a non-lethal manner through cell wall and through at least three membranes (the plasma membrane and the double-membrane envelope), the principal barriers. Stable transformation of plastids in flowering plants is preceded by transient expression of genes. The biolistic, PEG, and glass-bead, microinjection, and protoplast-mediated methods give rise to stable chloroplast transformants. However, the biolistic technique allows the delivery of transforming DNA directly within the chloroplast, where it can integrate by homologous recombination in the plastid genome of the host. Integration of the transgene into the chloroplast occurs via homologous recombination of the flanking sequences used in the chloroplast vectors [Verma and Danielle 2007].

Chloroplast transformation, as opposed to nuclear transformation, offers several advantages: a) higher levels of foreign protein expression, b) expression of multiple genes as operons, suitable to introduce new metabolic pathways for the production of novel compounds, and c) containment of pollen-mediated foreign gene flow in most of the crop species. The target tissues in flowering plants include leaves (tobacco, cabbage, tomato, potato, lucerne, etc.), cotyledon petioles (rapeseed), leaf petioles (cabbage), protoplasts (lettuce, cauliflower), embryogenic calli (cotton, sugarcane), embryogenic tissues (soya bean/soybean), green stem segments (brinjal/eggplant: *Solanum melongena*), embryogenic cells (rice, carrot), immature embryos and immature influorescences (wheat) [Khan 2012].

Chloroplast transformation of monocot crops, particularly rice, wheat, and sugarcane, is challenging because of a couple of impediments. One impediment is their regeneration from non-green embryonic cells, containing proplastids, rather than plastids, while the other, especially in rice, being the low level of marker gene expression in non-green plastids in embryogenic cells [Khan 2012]. Another impediment is the availability of a single dominant marker gene, *aadA* that encodes aminoglycoside-3′-adenylyltransferase (AAD) and confers resistance to spectinomycin and streptomycin by adenylylation. The *aadA* has been used predominantly to transform plastids and the selection was carried out on spectinomycin in dicotyledonous plants (soyabean, lettuce, sugarbeet, cauliflower, oilseed rape, tobacco, potato, cabbage, eggplant/brinjal, etc.). Cells from cereal crops are naturally resistant to spectinomycin but sensitive to streptomycin. Hence, streptomycin can be used as a selection agent to recover transgenic clones on media, as demonstrated in rice [Khan and Maliga 1999]. This was confirmed by Lee et al. [2006] who transformed rice plastids using the selectable marker gene *aadA* and a reporter gene, *gfp*; both genes (*aadA* + *gfp*) were expressed under ribosomal RNA promoter in an operon. This suggested that adding a second marker for dual or stepwise selection on streptomycin and on a second antibiotic may facilitate selection and purification of transplastomic cells/shoots of cereals.

The inhibitory effects of herbicides are plant-specific and many of them take place in the plastid. Resistance to herbicides has been used for the design of chloroplast selectable marker genes. Several transgenes engineered through chloroplast transformation have conferred valuable agronomic traits in plants including herbicide, insect, and pathogen resistance as well as drought and salt tolerance. Advancement in chloroplast engineering has made it possible to use chloroplasts as bioreactors for the production of recombinant proteins. As plastid genes are maternally inherited, transgenes inherited into these plastids are not disseminated by pollen. Another advantage includes the ability to express several genes as a polycistronic unit, thereby potentially eliminating position effects and gene silencing [Kumar et al. 2004; Daniell et al. 2005]. All this makes chloroplast transformation an important technology.

Chemical-Mediated Transformation

The chemical-mediated transformation involves gene transfer using chemicals such as polyethylene glycol (PEG), DEAE dextran, calcium phosphate, and silicon carbide.

PEG-mediated Method. This technique is used to deliver DNA using protoplasts as explants. Lack of cell wall makes it easy for the passage of DNA into the cell. This method is similar to electroporation in that the DNA to be introduced is mixed with the protoplast, and uptake of DNA is then stimulated by the addition of PEG, rather than an electrical pulse. The PEG-mediated DNA transfer can be readily adapted to a wide range of plant species and tissue sources. Protoplasts can be derived from leaf mesophyll, roots, and cell suspensions. Cell suspensions provide an unlimited source of rapidly dividing protoplasts. A low autofluorescence of cell suspension and root-derived protoplasts is of particular importance, when light-emitting green fluorescence protein (GFP) or luciferase enzyme (luc) is being used as reporter protein in nondestructive *in vivo* gene expression assays. Root-derived protoplasts also feature a high division and regenerative capability. In leaf protoplasts, the red fluorescence of chloroplasts is a deterrent for effective monitoring of the GFP reporter gene activity.

In the case of plant cells, protoplasts may be regenerated into whole plants first by growing into a group of plant cells that develop calli and then by regeneration of shoots (caulogenesis) from the calli by using plant tissue culture methods. Growth of protoplasts into calli and regeneration of shoots requires the proper balance of plant growth regulators in the tissue culture medium that must be customized for each species of plant.

Although this PEG-mediated transformation is easy to handle and no specialized equipment is required, it is rarely used due to the low frequency of transformation and because many species cannot be generated into whole plants from protoplasts. Besides, fertility may be a concern because of somaclonal variation of the transgenic plants derived from protoplast cultures. However, this method has been used to develop transgenic maize and barley lines [Daveya et al. 2005]. Thus, protoplast transformation is feasible in cereals, even if fertility problems in the regenerants are often encountered. In cotton, transformation was achieved using polybrene-spermidine-based callus treatment [Sawahel 2001].

DEAE dextran-based Method. Another chemical method of transformation is with diethylaminoethyl-dextran (DEAE-dextran), a cationic polymer with high molecular weight. It links with nucleic acids and intake occurs presumably by endocytosis. This chemical binds and interacts with negatively charged DNA molecules via a largely unknown mechanism that brings about and uptake of nucleic acids by the cell. This DEAE-dextran-mediated transformation is generally used for transfecting animal cells with foreign DNA. Its utility in plant cells is far less common than in animal cells.

Calcium Phosphate-based Method. One of the cheapest chemical-based transformation methods is using calcium phosphate, discovered by Graham and van der Eb in 1973 [Graham and van der Eb 1973]. The HEPES-{4-(2-hydroxyethyl)-1-piperazineethanesulfonic acid} buffered saline solution (EeBS) containing phosphate ions is combined with a calcium chloride solution consisting of the DNA to be transfected. When the two are combined, a fine precipitate of the positively charged calcium and the negatively charged phosphate will form on its helix, binding the DNA to be transfected on its surface. The suspension of the precipitate is then added to the cells to be transfected. By a process not entirely understood, the cells take up some of the precipitate, and with it, the DNA.

Dendrimer-based Method. Another chemical-based method uses dendrimers, the highly branched organic compounds, which bind the DNA before entering the cell. The cationic dendrimers are used to traffic genes into cells without damaging and deactivating the DNA. The dendrimer/DNA complexes are encapsulated in a water soluble polymer, and then deposited on or sandwiched in functional polymer films with a fast degradation rate to mediate gene transfection. Based on this method, the G5 PAMAM-(generation 5 polyamidoamine) dendrimer/DNA complexes are used to encapsulate functional biodegradable polymer films for substrate-mediated gene delivery.

Silicon Carbide-based Method. In silicon carbide-based transformation (SCMT), small needle-type silicon carbide whiskers are mixed with plant cells and the gene of interest, and the mixture is then vortexed [Kaeppler et al. 1992]. In this process, the whiskers pierce the plant cells, permitting DNA entry into the cells. The fibers used are elongated, 10–80 mm long, 0.6 mm in diameter, and lacking elasticity. The efficiency of SCMT depends on fiber size, vortexing, the shape of vortex vessel, and plant material used for transformation. Stable transformants have been obtained by using SCMT technique in a variety of plants, including maize, rice, wheat, tobacco, cotton, etc. SCMT-mediated transgenic cotton plants showed significant improvement in salt tolerance [Asad et al. 2008]. Furthermore, silicon carbide fibers, which cause wounding of immature embryos, have been found to improve the efficiency of *Agrobacterium*-mediated transformation [Singh and Chawla 1999]. Although SCMT is simple, inexpensive, and effective on a variety of cells, its main disadvantages include low transformation efficiency and damage to cells. Inhalation of silicon fibers poses health hazard if the protocols are not properly performed.

Torney et al. [2007] circumvented this problem by using mesoporous silica nanoparticles (MSNs) of 100–200 nm in diameter to deliver DNA and chemicals via gene gun system into both plant cells and intact leaves. The pores of these fluorescein-doped MSNs were capped by surface-functionalized gold nanoparticles (10–15 nm in size) which not only acted as a biocompatible capping agent [Shukla et al. 2005], but more importantly added weight to each individual MSN to increase the density of the resulting complex material. This increase in MSN density improved transformation efficiency and the appearance of MSNs inside plant cells [Torney et al. 2007].

Pollen-based Transformation

Pollen grains and microspores are single, haploid cells that are available in large amounts as a synchronous population and allow production of transgenic plants by gametophytic and sporophytic routes [Resch and Touraev 2011]. Pollen transformation is divided into a) mature pollen-based transformation, b) microspore maturation-based transformation, and c) microspore and immature pollen embryogenesis-based transformation.

Mature pollen-based transformation involves the introduction of DNA into the generative or the sperm cell through the rigid pollen wall or through the germination pores. The sperm cell can deliver the DNA into the egg cell, after which integration might occur during or shortly after fertilization [Resch and Touraev 2011]. Over the years, different transformation methods have been used. These included *Agrobacterium* co-cultivation, pollen tube pathway, liposome-mediated delivery, biolistic, and electroporation. These methods employed dry pollen, incubated in a solution of DNA before pollination, but the major problem was that the nuclease activity of germinating pollen completely degraded DNA within 5 to 10 min of incubation, leading to low transformation frequency.

Microspore maturation-based transformation is based on male germ line transformation (MAGELITR) method, which uses biolistic method to deliver the foreign DNA into unicellular microspores. This technique was claimed to have 15 percent transformation efficiency, using *aph*IV (aminoglycoside phosphotransferase type IV) gene and *gus* gene for the recovery of transgenic plants [Aziz and Machray 2003]. It has also been used successfully to transform tobacco unicellular microspores with *dhfr* (dehydropholate reductase) gene [Aionesei et al. 2006].

Microspore and immature pollen embryogenesis-based transformation employs the microspores or immature pollen grains and the resulting transgenic embryogenic cells are divided to give rise to embryos and haploid plants. After transformation of microspores and immature pollen grains, transgenic homozygous diploid plants can be regenerated in a single step that can be used for the creation of transgenic plants. Of the various methods, biolistic technique is more effective. Transgenic plants have been produced by bombarding microspore culture using herbicide bialophos applied to the medium for 1–2 weeks in barley and to introduce DNA into immature tobacco pollen grains [Resch and Touraev 2011].

Agrolistic Method

Agrolistic method involves a novel technology in which the virulence genes from *Agrobacterium* that facilitate release of the T-DNA and contain nuclear targeting sequences are co-bombarded, using biolistic method, into the target tissue along with the selectable marker and gene of interest. This method combines the advantages of *Agrobacterium*-mediated transformation with the high efficiency of biolistic DNA delivery. In this approach, the virulence genes *Vir*D1 and *Vir*D2 essential for T-strand excision from Ti or Ri plasmids of *Agrobacterium* are co-delivered to the target material under the control of a constitutive promoter [Hansen and Chilton 1996]. These virulence genes have been shown to function *in planta* to produce the characteristic strand-specific nicking at the RB sequence, similar to that which leads to formation of the T-strand in *Agrobacterium*.

In tobacco, the calli were transformed after co-delivery of *Vir*D1 and *Vir*D2 with selectable marker genes flanked by RB and LB sequences [Hansen and Chilton 1996]. Examination of insertion sites revealed that some exhibited right junctions with the plant genomic DNA that corresponded precisely to the sequence expected for T-DNA insertion events. Such events were termed "agrolistic" inserts as opposed to biolistic insertion events. Around 20 percent of the transgenic lines obtained contained only agrolistic insertion events, while a further 20 percent contained both agrolistic and biolistic events. This method can be used with any target tissue susceptible to biolistic transformation. Its principal advantage is elimination of extraneous vector sequences from the insertion events and a reduction in copy number that will lead to increased stability of expression and inheritance of transgene.

Use of agrolistics has been shown to increase the number of transgenic plants receiving "clean" or precisely defined transgene inserts and to reduce the frequency of degraded transgene integrations [Hansen and Chilton 1996]. This technology requires further development but has the potential to address one of the major drawbacks of particle bombardment technology.

Delivery of Multiple Genes by Single Vector

Most of the plasmids and families of plasmids are limited to the expression of a single gene (excluding the selection marker, when present) [Dafny-Yelin and Tzfira 2007]. As many crop improvement traits are polygenic in nature, the use of such plasmids for the coordinated manipulation of multiple traits presents a unique challenge for the plant genetic engineer. There are several strategies to deliver multiple genes by using single-gene vectors. These include a) re-transformation by stacking of several transgenes by successive delivery of single genes into transgenic plants; b) co-transformation involving the combined delivery of several transgenes, using multiple plasmids or with a single plasmid, in a single transformation, and c) sexual crosses between transgenic plants carrying separate transgenes.

Re-transformation

Re-transformation by repeated transformation has been used in species as diverse as *Arabidopsis* and trees to successfully introduce unlimited number of transgenes or manipulate biochemical pathways controlling flower color or salinity tolerance, and involved in starch, fatty acids, and lignin production. However, this approach is severely limited by the fact that the introduced genes can segregate in subsequent generations as they are not linked, but sited at different, random loci in the plant's genome. Besides, re-transformation is a time-consuming and expensive process. It also demands distinct selectable marker gene for every round transformation. In the case of vegetatively propagated plants, selectable marker genes used during several rounds of re-transformation cannot be eliminated by outcrossing, resulting in their accumulation in transgenic plants. This may create a significant hurdle for regulatory approval and public acceptance. This problem can be circumvented by adopting techniques that enable the removal of selectable marker genes or that allow for transformation without the use of selectable markers.

Co-transformation

In co-transformation, multiple genes are introduced simultaneously followed by integration of the genes in the plant cell genome. The genes are either present on the same plasmid (single-plasmid co-transformation) or separate plasmids (multiple-plasmid co-transformation) [Xu 2010]. The advantage in this approach is that in a single transformation multiple transgenes are integrated. But the problem is the difficulty encountered in assembling complex plasmids with multiple gene cassettes [François et al. 2002]. Furthermore, co-transformation with separate multiple plasmids is a by-chance event, with the result that the inserted copy numbers and the relative arrangement among transgenes cannot be controlled [Lin et al. 2003]. This problem was circumvented via co-bombardment of several plasmids which integrated all transgenes at a single genomic locus [Agrawal et al. 2005; Zhu et al. 2008]. But this approach suffers from integration of multiple copies and complex integration patterns, thus complicating commercial use [Buntru et al. 2013].

However, the success of this technique depends on the frequency with which two or more independent transgenes are transferred to the plant cell and integrated into the cell genome (co-transformation efficiency) [Xu 2010]. Co-transformation has been used with success to deliver two different genes in aspen (*Populus* spp.) [Li et al. 2003], four transgene constructs in tobacco [Li et al. 2003], and four genes (*psy*, *crt*I, and *lyc* besides a selectable marker) from two independent constructs in rice to produce the high β–carotene containing "Golden Rice" [Ye et al. 2000; Datta et al. 2003] shuttled by two to four independent *Agrobacterium* strains. Agrawal et al. [2005] achieved multi-transgene co-transformation in rice using five multiple cassettes, with all transgenic plants generated containing at least two transgenes and 16 per cent containing all five.

Biolistic method is the most convenient method for multiple gene transfer to plants as it does not involve cloning, multiple *Agrobacterium* strains, or sequential crossing [Altepeter 2005]. This method was found to facilitate high levels of gene expression when nine transgenes were transferred to rice genome, with their expression being independent of each other [Wu et al. 2002], and a fifth gene encoding selectable marker, into rice, which simultaneously produced full-sized, multimeric proteins in transgenic plants.

Sexual Crossing

In sexual crossing, one gene is introduced in one parent and the other gene in the second parent, followed by crossing these two transgenic parents, with the resulting progeny carrying 25 percent of the transgenes. This approach is relatively simple as it involves transfer of pollen from one parent to the female reproductive organ of the other. Another advantage is that transgenic populations of each parent can be screened for optimal gene expression of each transgene, thus facilitating the combination of optimally expressed transgenes. But this approach suffers from its being time-consuming because two transgenes need to be combined by sequential crossing. Besides, this procedure is not suitable for crops with high levels of heterozygosity (e.g., potato) and vegetatively propagated crops (perennial fruit crops).

One significant success recorded with crossing-based approach was with Monsanto's multi-stacked maize, produced via conventional crossing of three inbred maize transgenic lines (vide Chapter 6}: MON863 (coleopteran-insect resistant), MON810 (lepidopteran-insect resistant), and NK603 (glyphosate-resistant) [Xu 2010]. The elements incorporated into the multi-stack included five loci, four carrying a synthetic gene linked to combinations of strong regulatory elements from viruses, bacteria, and unrelated plants. Expression of the first two synthetic genes produces an *epsps* that resembles the EPSPS from *E. coli* and is, unlike most plant versions, not activated by herbicides containing glyphosate. The third synthetic gene encodes the insecticidal *cry*3BP1 protein with activity against specific Coleoptera, whereas the fourth product, *cryaIAb*, provides tolerance against Lepidopteron insects. The fifth gene is kanamycin-resistance gene encoding neomycin phosphotransferase (*npt*II).

Modified Binary Vector Method

Majority of the binary vectors used in *Agrobacterium* mediated transformation systems have been limited to the cloning and transfer of just a single gene of interest, precluding introduction of multiple genes into transgenic plants. This problem can be overcome by using modified multi-transgene binary vectors and Gateway binary vectors.

Multi-transgene Binary Vectors. Multi-transgene binary vectors are engineered to carry an array of unique recognition sites for zinc finger nucleases (ZFNs) and homing endonucleases, and a family of modular satellite vectors [Zeevi et al. 2012]. ZFNs, also called "molecular scissors", are synthetic proteins consisting of an engineered zinc finger

DNA-binding domain fused to the cleavage domain of the bacterial FokI (*Flavobacterium okeanokoites*) restriction endonuclease (type IIS). ZFNs induce double-stranded breaks (DSBs) in specific DNA sequences and thereby promote site-specific homologous recombination and targeted manipulation of genomic loci in a variety of different cell types. These have been developed to target multiple genes in plant systems. By combining the use of designed ZFNs and commercial restriction enzymes, multiple plant expression cassettes were sequentially cloned into the acceptor binary vector [Zeevi et al. 2012]. These modified binary vectors carried up to nine genes while *Arabidopsis thaliana* protoplasts and plants were transiently and stably transformed, respectively, by several multigene constructs.

As ZFNs can be engineered to digest a wide variety of target sequences, Zeevi et al. [2012] suggested that it may be possible to reconstruct other types of binary vectors and adapt them for cloning on multigene vector systems in various binary plasmids. Thus ZFN-mediated gene targeting promises to be a powerful tool in the development of novel crop species possessing beneficial agronomic traits. ZFNs have been utilized successfully in both tobacco and *Arabidopsis* [Barampuram and Zhang 2011].

Modified Gateway Binary Vector. This approach is based on Gateway Binary vectors. In this, the MultiRound Gateway technology allows stacking multiple DNA fragments on a single vector. It is based on two different entry vectors which can be alternately used to deliver sequentially multiple DNA fragments into a Gateway-compatible destination vector [Chen et al. 2006]. This method was further improved by Buntru et al. [2013] who used a transformation-competent artificial chromosome (TAC)-based destination vector and a recombination-deficient strain of *Agrobacterium*. The system consisted of two *att*L-flanked entry vectors, which contained an *att*R cassette, and a transformation-competent artificial chromosome based destination vector. By alternate use of these two entry vectors, multiple transgenes can be delivered sequentially into the Gateway-compatible destination vector. Multigene constructs that carried up to seven transgenes corresponding to more than 26 kb were assembled by seven rounds of LR recombination. The constructs were successfully transformed into tobacco plants and stably inherited for at least two generations. Thus, this system represents a powerful, efficient tool for multigene plant transformation and may facilitate genetic engineering of agronomic traits or the assembly of genetic pathways for the production of biofuels as well as industrial or pharmaceutical compounds in plants.

Recombination-assisted Multifunctional DNA Assembly Platform (RMDAP)

An improved methodology that combined the advantages of existing transformation systems to design a 'One-stop Breeding' system that can easily provide the requirements for vector construction has been developed by Ma et al. [2011]. It was based on a platform called 'Recombination-assisted Multifunctional DNA Assembly Platform' (RMDAP) which incorporated 14 pairs of satellite vectors (pOSB series) and three kinds of recipient vectors. It was designed for the needs of transgenic research (e.g., overexpression, RNAi and marker-free vectors) and to resolve practical problems of plant genetic improvement. This platform incorporates three widely-used recombination systems, namely, Gateway technology, *in vivo* Cre/*lox*P, and recombineering into a highly efficient and reliable approach for gene assembly. RMDAP proposes a strategy for gene stacking and contains a wide range of flexible, modular vectors offering a series of functionally validated genetic elements to manipulate transgene overexpression or gene silencing involved in a metabolic pathway [Ma et al. 2011]. This versatile and easy-to-use multigene genetic transformation platform for the assembly of multiple-gene expression cassettes was used to successfully transfer several heterologous genes into the plant genome [Ma et al. 2011].

The coming years will witness a further refinement in these techniques and/or development of new methods for efficient, low cost, reproducible, and time-saving delivery of multiple genes using a single vector system.

Native-Gene Transfer-Mediated Transformation

Transfer of foreign genes from a wide spectrum of species, including viruses, bacteria, fungi, and animals into transgenic crops, has elicited perceived risks associated with this technology and opposition to their commercial deployment. Besides, concerns of consumers about the health safety and ethical justification of transgenic crops have led plant molecular biologists to employ alternative transformation methods. One of them is based on transfer of native genes (including regulatory elements), commonly known as 'intragenic' transformation.

Intragenic Transformation

In this, the inserted DNA can be a new combination of DNA fragments from the same species or from a cross-compatible species in a sense or antisense orientation. This method transforms crops with plant-derived transfer (P-) DNAs that consist of only native genetic elements, without affecting the overall structure of the plant's genome. These P-DNAs are introduced into plants by employing effective marker gene-free transformation systems or plant-derived marker genes. The genetic modification can be characterized molecularly so that any inadvertent transfer of undesirable DNA, as may be the case with traditional methods, is excluded.

In intragenic method, dormant benefit traits in a plant are activated by linking key genes in biosynthetic pathways to strong tissue-specific or near-constitutive promoters while eliminating the undesirable characteristics by employing silencing constructs [Rommens et al. 2011]. The undesirable genes are silenced or expression of desirable genes is enhanced when linked with beneficial genes.

Intragenic method has been used successfully in USA to improve the quality of a potato variety, 'Ranger Russet', which combined superior yield with disease resistance, adaptability, tuber uniformity, and high starch content but suffered from discoloration from impact-induced bruise as well as quality of fries and chips [Rommens et al. 2006]. These weaknesses were eliminated by down-regulating (silencing) its strengths, *Ppo* (polyphenol oxidase), *PhL* (phosphorylase-L), and the starch-associated *Rl* genes, resulting in black spot-resistant, sweetened, aroma-enhanced, and very high-starch tubers [Rommens et al. 2006]. This process has also been applied to increase the health-promoting flavonol, kaempferol, by activating phenylpropanoid pathway while partially-inhibiting anthocyanin biosynthesis by ~100-fold in a different potato variety 'Bintje' [Rommens 2008a). Traits such as tolerance to herbicides, diseases, salinity, drought, and frost that can be incorporated through intragenic modification are associated, directly or indirectly, with yield [Rommens 2008b).

Intragenic method was used in rapeseed (canola) which lacks genetic diversity to incorporate certain desirable traits through conventional breeding. It became a very useful tool to insert genes that enhance oleic acid, increase oil content, and incorporate shatter-resistance. The various elements regulating these traits in rapeseed include constitutive promoter (*fad*2D), seed-specific promoters that encode seed storage proteins such as oleosin, and cruciferin gene promoters, and terminator genes *cruT*, *E9T*, and *cabT* that terminate gene transcription in transgenic rapeseed [Rommens et al. 2011]. Employing *Brassica*-derived P-DNA vector yielded higher transformation frequencies for tobacco and rapeseed than were obtained by T-DNA vector [Rommens et al. 2011].

A variant of native-gene transfer method is 'cisgenic' transformation. In **cisgenesis**, the inserted gene is unchanged with its own introns and regulatory sequences in the normal sense orientation. In this method, the recipient plant is modified with a natural gene from a crossable (sexually compatible) plant [Schouten et al. 2006]. Such a gene includes its introns and is flanked by its native promoter and terminator in the normal-sense orientation. Cisgenic plants can harbor one or more cisgenes, but they do not contain any transgenes as in transgenesis.

Although both gene-transfer methods use the same genetic modification techniques— namely the introduction of one or more genes and their promoters into a plant—cisgenesis involves only genes from the plant itself or from a close relative, and these genes could also be transferred by traditional breeding techniques [Schouten et al. 2006] .

A broad variety of plant-derived marker genes associated with various traits has been identified for native-gene transfer-mediated transformation. This is discussed later in the Chapter.

Marker and Reporter Genes

Marker Genes

Marker genes, essential for the production of transgenic plants, are required to identify, to 'mark' the introduced genes, and finally to enable the selective growth of transformed cells. During genetic transformation, the new genes are inserted in only a fraction of the plant cells. They can be used to identify the tiny number of cells which have successfully acquired the new gene. The marker gene is transferred alongside the gene of interest. The desired gene and marker gene are inserted in the chromosome at the same location. Marker genes are also required to identify allele features easily besides being easy for detection, low in cost of development and genotyping, and lack pleiotropic effect on the expression of the target trait. They are derived from plant or bacterial sources. There are two types of marker genes: **selectable marker genes** and **reportable genes**.

The most commonly used marker genes at present are antibiotic or herbicide resistance genes. They all work by making the modified cells detoxify substances which would otherwise be fatal to them. For example, a herbicide resistance gene confers tolerance to the herbicide. If the cells come into contact with the corresponding substance following transformation (in the culture medium, for example), only those plants which have acquired the marker gene, and therefore the target gene as well, will survive. Transgenic plants are then grown from these cells.

Selectable Marker Genes

Selectable marker genes protect the organism from a selective agent that would normally kill or prevent its growth. Without them, the few plant cells that take up and stably integrate the foreign DNA would simply be lost in the presence of wild-type cells, which would overgrow these transformed cells in the absence of effective selection against them. All transformation systems have low efficiencies in the absence of selectable marker genes. Marker genes need to be polymorphic and duplicable, show co-dominance, and distribute evenly on the entire genome.

Selectable markers differ in various properties that confer advantages and drawbacks such as dominance, cell-autonomy, or portability. Some are dominant while others are recessive. Dominance is of particular relevance for transformation of the highly polyploidy plastome. Dominant markers increase the frequency because they have an effect at early stages during selection even though they may only be present in a minority of the plastomes. Conversely, recessive markers only confer resistance to herbicides if random segregation has produced a plastid that has enough transformed copies of the plastome for the selectable phenotype to emerge. As this is a rare event, recessive markers give lower transformation efficiencies than dominant ones. There are two major groups of selectable marker genes used during transformation: antibiotic-resistant genes and herbicide-resistant genes (vide Appendix: Table 4).

Antibiotic Resistance Genes. Antibiotic resistance genes act during pre-transformation as well as transformation process. The most frequently used selectable markers include kanamycin, hygromycin, paranomycin, gentamycin, etc. Exposure of plants to these antibiotics leads to an inhibition of chlorophyll synthesis and leaf bleaching. The most widely used selectable marker genes in cereal transformation are neomycin phosphotransferase (*npt*II), hygromycin phosphotransferase (*hpt*II), and phosphinothricin acetyltransferase (*bar*). These genes confer resistance to kanamycin and some related aminoglycosides (such as G418 and paranomycin), hygromycin, and phosphinothricin (PPT) respectively. Transformed cells in these systems are able to survive and non-transformed cells get killed by these selective marker genes. The transformed cells that express a resistance gene are able to survive by neutralizing the toxic effect of the selective agent, either by detoxification of the antibiotic through enzymatic modification or by evasion of the antibiotic through alteration of the target.

The utility of an antibiotic selectable marker gene is a function of both the properties of the resistance protein it encodes and the relative sensitivity of the target tissue to its corresponding selective agent. For example, the antibiotic kanamycin, used for transformation of dicot plants, is ineffective in the case of monocots, which then require a different selectable marker gene. There are concerns about the safety of their presence in transgenic crops or products destined for human and animal consumption (vide Chapter 9). This has made plant biotechnologists move away for marker-free transformation systems.

Metabolic markers are a potential alternative to the controversial antibiotic resistance genes. They enable the plants to grow on unfamiliar culture media or to produce metabolic products which allow only the transgenic cells to grow. However, this method is still in the early stages of development.

Herbicide Resistance Genes. A number of herbicide resistance genes are currently used as selective agents in crops. These have been developed by engineering resistance to herbicides that inhibit synthesis of phytochemicals. Both herbicides and antibiotics are used to select materials by adding them to culture media or by spraying plants.

Engineering for herbicide resistance in a normally sensitive-cereal crop is done by two methods. One is by introducing an herbicide-resistant variant of, for example, an amino acid biosynthetic enzyme gene, *als* (acetolactate synthase), for resistance to sulfometuron methyl. The other method is by directly introducing the herbicide-resistant gene, *bar*, which encodes phosphinothricin acyl transferase that inactivates bialophos. This gene, isolated from *Streptomyces hygroscopicus*, is used to develop transgenic lines in cereals including rice, maize, wheat, and barley which showed greater resistance to phosphinothricin-related herbicides. Another chloroplast-enzyme 5-enolpyruvylshikimate-3-phosphate synthase (ESPS) catalyzes aromatic amino acid biosynthesis in plants (vide Chapters 2 and 5). By inhibiting this plastid enzyme in resistant species, glyphosate prevents synthesis of chorismate-derived aromatic amino acids. Transformation by using the mutant *epsps* gene isolated from *Petunia hydrida* renders the normally sensitive plants resistant to this herbicide.

Once the gene is inserted, the plant tissue is transferred to a selective medium consisting of a herbicide or antibiotic, depending on the selectable marker used. During growth on selective medium, only plant tissues that have successfully integrated the transgene construct and express the selectable marker gene will survive. Once it is assumed that these plants possess the transgene of interest, the surviving plants will go through subsequent steps. The introduced DNA is covalently integrated within the host plant's genome.

The next step is to regenerate plants from the transformed cells. This can be done by using marker genes within the transgene and to select for their expression. They act by expressing an enzyme that inactivates the selective agent that causes detoxification and a resistant variant of a selective agent's target enzyme that confers tolerance. For example, the herbicide phosphinothricin, an analog of glutamine, acts by irreversibly inhibiting glutamine synthetase, a key enzyme for ammonium assimilation and the regulation of nitrogen assimilation in plants. The *bar* gene, allows growth of transformed cells in the presence of phosphinothricin or commercial glufosinate ammonium-based herbicides.

Plant-derived Marker Genes. Certain plant genes themselves can be used as transformation markers. For instance, *als* gene from *Arabidopsis* has been used to develop intragenic *Arabidopsis* plants displaying chlorsulfuron tolerance [Conner et al. 2007]. Another interesting native marker system is based on protoporphyrinogen I oxidase (Protox) genes. When plastidic Protox gene from *Arabidopsis* was overexpressed under the control of the CaMV 35S promoter in tobacco, the overproduction of Protox rendered plants resistant to the action of acifluorfen [Lermontova and Grimm 2000]. Maize transformants expressing a modified protoporphyrinogen oxidase were produced via butafenacil selection using a flexible light regime to increase selection pressure [Li et al. 2003]. Successful tobacco chloroplast transformation with a spinach betaine-aldehyde dehydrogenase gene [Daniell et al. 2001] suggests that native genes involved in the conversion of betaine aldehyde can also be used as markers for plant transformation. Several additional native markers function effectively but trigger cytokinin responses, which confer an undesirable phenotype to the transformed plant [Sun et al. 2003].

Reporter Genes

Reporter genes, also known as screenable marker genes, are used to confirm transformation, determine transformation efficiency, and monitor gene or protein activity. Visual observation of gene expression is achieved by using reporter genes that can represent important components for transient and stable expression studies. These genes are effective markers because their protein products are readily detectable in transgenic tissues but are absent in non-transgenic tissues.

Reporter gene generates a product that can be detected using a simple and quantitative assay. Its ability to form fusion genes at the transcriptional level can be used to assay the activity of regulatory elements. This property can be exploited, in the form of gene-trap vectors, to isolate and characterize new genes as part of a functional genomics strategy. The most commonly used reporter genes in plant transformation are β-glucuronidase, the fluorescent protein, luciferase, and β-galatcosidase.

The β-glucuronidase gene, **GUS**, is an excellent and most widely used reporter gene to visualize gene expression in transgenic plant cells. Its activity is visualized by adding 5-bromo-4-chloro-3-indolyl-β-D-glucuronide, the substrate which produces a blue pigment, following the cleavage of the gluguronic acid residue by the transgenic enzyme. *GUS* activity is generally regarded as a good indication of the strength or tissue-specific expression of the promoter to which the *GUS* gene is fused, although this is not universally so [Taylor 1997]. This *E. coli*-derived enzyme cleaves a group of sugars called β-glucuronides. It also cleaves a chemical that is added to the culture into an insoluble, visible blue precipitate at the site of enzyme activity. The level of gene expression can be measured by the intensity of the blue color produced. Many plants lack β-glucuronidase enzymes, so it is easy to determine whether the plant has been transformed besides identifying transformed cells and tissues. Enzyme activity can be easily assayed *in vitro*. Its drawback is that the cells are killed in the process. This is due to the toxicity associated with the ferricyanide and ferrocyanide in the *GUS* assay mix [Jefferson 1987] or the cleaved substrate. Once the problem is overcome with a better protocol, *GUS* remains an inexpensive and reliable tool for single point determinations of transgene expression in plants [Finer 2011].

Green fluorescent protein (GFP), isolated from jellyfish (*Aequorea victoria*), has a trio of amino acids that absorb blue light and fluoresce yellow-green light. GFP fluorescence is non-toxic and does not require addition of substrates or co-factors. Non-invasive detection allows monitoring of *gfp* gene expression in the same transgenic plant throughout its life cycle. It is detectable using fluorescence microscope simply by illuminating with blue light and observing the fluorescence. The main advantage of *gfp* is its ability to directly observe the presence of protein in living tissues. GFP can, therefore be used to monitor promoter activation, gene introduction, gene insertion (following recombination), promoter targeting/localization besides monitoring gene expression [Finer 2011]. This reporter gene has a relatively weak activity in transformed cells, but this can be modified by effecting mutations to increase signal intensity and shift excitation peak, alter codon usage for efficient translation, remove cryptic intron splice junctions to increase mRNA processing and stability, and inhibit thermosensitive protein misfielding [Xu 2010]. If its weaknesses are removed, *GFP*-monitoring would be a better monitoring system. Its activity in plant cells is visualized by adding 5-bromo-4-chloro-3-indolyl-β-D-glucuronide, the substrate which produces a blue pigment, following the cleavage of the gluguronic acid residue by the transgenic enzyme.

The luciferase gene, *luc*, derived from firefly (*Photnius pyralis*) or the luminescent bacterium *Vibrio*, catalyzes the oxidation of D(−)-luciferin in the presence of ATP to generate oxyluciferin and yellow green light. In transgenic plants, the protein product is detected after addition of luciferin substrate which, in the presence of O_2 and ATP, results in emission of green-yellow light. The activity of luciferase gene can be assayed in transformed cereal tissue without being destroyed. The problem of luciferin is its limited penetration in whole plant material. The detection equipment needed to monitor the expression of *luc* is relatively expensive. However, *luc* genes are widely used as an internal standard along with *gus* gene in transgenic plants.

The bacterial *lacZ* gene (LacZ), derived from *E. coli*, encodes a β-galactosidase, an intracellular enzyme that cleaves the disaccharide lactose into glucose and galactose. When medium containing certain galactosides (e.g., X-gal) is added, cells expressing the gene convert the X-gal to a blue product and can be seen with naked eye.

Removal of Selectable Marker Genes

Selectable marker genes which are indispensable during the process of plant transformation are dispensable once transgenic plants have been established. They serve no purpose once plants that are homologous for the transgene and transgenic cells have ben successfully identified. Their continuous expression may interfere with normal plant growth and development, besides possibly posing technological problems because they preclude re-transformation with the same marker systems. Besides, the fact that a selectable marker gene remains in the genomes of transgenic plants has raised concerns from both global regulatory agencies and consumers (vide Chapter 9). The increasing number of cases of bacteria that have evolved antibiotic resistance, such as methicillin-resistant *Staphylococcus aureus* (MRSA) has undoubtedly led to consumer concern [Yau and Stewart 2013]. Elimination of marker genes will appease some potential and consumer concerns, while removing technical barriers for plant genetic transformation.

Currently, a series of research projects are looking at ways of achieving gene transfer without marker genes, or the subsequent removal of the marker genes. Some strategies have been devised to make this happen. They can be removed by a) co-transformation, b) site-specific excision, and c) transposon-based elimination processes.

Co-transformation

In this, plant cells are co-transformed with two separate pieces of T-DNA, one with a selective marker gene and the other with genes of interest, and also to select marker-free progeny segregated from the co-transformants [Hohn et al. 2001]. Unlinked integrations of the two T-DNAs lead to the segregation of the marker gene from the gene of interest in the next generation.

Co-transformation is performed by using a separate transformation binary vector in a single *Agrobacterium* strain, or by using a mixture of *A. tumefaciens* strains, each harboring a binary vector. Although both methods are effective for transformation and marker gene removal, the one that uses a single strain yields higher co-transformation efficiency. These methods may be especially useful when using *Agrobacterium*-mediated transformation. However, this approach may not be suitable for species in which transformation efficiency is low. In order to improve co-transformation, two copies of T-DNA are placed on the same transformation vector. Both T-DNAs are located adjacent to each other using two extra T-DNA border regions, creating almost two contiguous T-DNAs.

In a different co-transformation method, the vector contains only one copy of T-DNA but carries two copies of its right border sequences, which flank a selectable marker gene [Lu et al. 2001]. The gene of interest is inserted next, and then one copy of the left border sequence is placed at the end. The vector enables two separate insertions: one develops from the first right border, which contains the selectable gene and the target gene, and the other comes from the second right border carrying only the target gene.

Binary Vector Method. In another co-transformation method to eliminate selective marker gene, two T-DNAs, belonging to the gene-of-interest and the marker gene, are placed on separate binary vectors. Contrary to the conventional vector design, the selectable marker gene is repositioned in the backbone of a regular binary vector and leaving only the gene of interest within the T-DNA region [Huang et al. 2004]. In this, the marker gene is transferred into the plant cell either along with the T-DNA that is initiated at the first border but failed to terminate at the second border, or as part of the T-DNA strand that is initiated at the second border. This shows that both the right and left borders can initiate and terminate T-strands. The co-transformation efficiency of this method in maize was much higher than other methods. It allows successful recovery of marker-free transgenic plants in subsequent generations.

The rationale of this approach is that after transformation, a portion of the resistant transgenic plants surviving antibiotic (or herbicide) selection should have also taken up the gene of interest-containing T-DNA cassette. Those transgenic plants with both selectable marker gene and gene of interest are allowed to set seeds for the next generation. By segregating away the marker gene which is, hopefully, unlinked to the gene of interest in the subsequent generation, plants with only the gene of interest can be obtained [Yau and Stewart 2013]. *Agrobacterium*-mediated transformation has been used more often than particle bombardment for co-transformation.

Super Binary Vector Method. In this method, two T-DNA constructs are produced on a single binary vector in such a way that they are separated with an intervening DNA fragment. The gene of interest and the selectable marker gene are transferred into plant cells independently through *Agrobacterium*-mediated transformation. Komari et al. [1996] developed a 50–55 kb 'super binary' transformation vector with two T-DNA regions that were separated by at least 15 kb. More than 50 percent progeny contained only the gene of interest, without the marker gene. Zhou et al. [2003] produced an intermediate-sized double T-DNA binary vector from two popular binary vectors (pBin19 and pCAMBIA2300) for co-transformation to produce SMG-free transgenic tobacco. Later, a standard binary vector (~11 kb) containing two independent T-DNA constructs (one with a SMG and one with the GOI) was used for co-transformation and successfully produced SMG-free transgenic sorghum [Lu et al. 2009].

Site-specific Excision

In this technique, the plant is transformed with a T-DNA vector carrying the gene of interest with two *lox* (locus of X) sites (34 bp repeats in direct orientation) flanking the selectable marker gene. In the second round, Cre (cyclization recombination) recombinase (derived from bacteriophage P1) is introduced to achieve precise excision of the marker gene. Cre enzyme specifically catalyzes the recombination between the *lox* repeat sequences, thereby eliminating the marker gene in the progeny [Wang et al. 2005; Jia et al. 2006]. This Cre-*lox* recombination system has been used in various plant species to generate marker-free transgenics.

In addition to Cre-*lox*, several other site-specific DNA excision systems derived from other organisms may be adapted for marker removal in transgenic plants. These include the R/*RS* system (R: R recombinase gene; RS: recombinase system) from *Zygosaccharomyces rouxii*, Flp/*frt* (Flp: enzyme flippase; *frt*: flippage recognition target) from *Saccharomyces cerevisiae*, and Gin/*gix* from bacteriophage Mu (Gin: G inversion; *gix*: the recombination sites) [Tian 2007].

Although site-specific excision is effective in marker gene removal, re-transformation is laborious, expensive, and time consuming, while outcrossing is not suitable for vegetatively propagated plants and woody tree species [Tian 2007]. Furthermore, the continuous and high level of expression of the *cre* recombinase gene may result in phenotypic aberrations in some plant species.

Transposon-based Elimination

Transposons are comprised by genetic elements that can 'jump' around in the genome of an organism. The best characterized transposons are those of the *Ac/Ds* family. *Ac* (activator) and *Ds* (dissociation) are two related elements. *Ac*, encodes the enzyme transposase, while the *Ds* is a deletion version of *Ac* element. Both share 11-bp terminal inverted repeat sequences (TIRs).

The maize *Ac/Ds* transposition system is used as a strategy to eliminate marker genes. In this approach, the marker gene is flanked by the inverted repeat sequences of the *Ds* mobile element. Subsequent to the transformation and T-DNA integration, excision of the *Ac* transposase from within the T-DNA results in excision of the gene of interest from the T-DNA insert containing the selectable marker gene. As a result, the gene of interest is transferred from the T-DNA site to another chromosomal location. Successful application of the system requires the activity of *Ac* transposase for the development of marker-free transgenic plants [Huang et al. 2004].

Thus, the transgene and the marker gene may be separated by transposon elements. The use of transposition to relocate the transgene to a new chromosomal locus is a useful approach for generating marker-free plants. However, genetic crossing or segregation is required to separate the gene of interest and the marker gene, and thus it is time consuming and entails a long process.

An alternative to the *Ac/Ds* transposable element system is a MAT (multiautotransformation) vector system that uses the *IPT* (isopentenyltransferase) gene and *Ac* element under the control of CaMV 35S promoter. The expression of *IPT* gene in the transformed plant generates an abnormal phenotype called the 'extreme shooty phenotype'. Subsequent to transformation, the *IPT* gene is removed using the *Ac* transposable element from the T-DNA, leaving only the gene of interest in the inserted copy of the T-DNA [Saelim et al. 2009]. This results in marker-free transgenic plants with a normal phenotype. However, there are several drawbacks using MAT system for marker gene removal.

Removal of Marker Genes from Chloroplasts

One approach to remove maker genes from transplastomic plants is by using Cre-*loxP* site-specific recombination system [Corneille et al. 2001; Jia et al. 2006; Lutz et al. 2006]. In this, *lox*P sites in the transformation vector flank the marker gene. The marker gene is stable in the absence of Cre recombinase that recombines a pair of short target sequences called the *lox* sequences. When excision of the marker gene is required, a plastid-targeted *cre* gene is expressed in the nucleus and the Cre protein is imported into chloroplasts where it excises the plastid marker gene. Although marker gene removal by this method is rapid, it encounters certain problems [Corneille et al. 2001; Tungsuchat et al. 2006]. The expression of Cre from a tobacco mosaic virus-(TMV) based vector could be used to eliminate selectable marker genes from transgenic tobacco plants produced by *Agrobacterium*-mediated transformation without sexual crossing and segregation, and this strategy could be extended to other TMV-infected plant species, and applied to other compatible virus-host plant systems [Jia et al. 2006].

Another approach for marker gene removal from plastids is via homologous DNA recombination. In this multiple-site specific excision method, several marker genes, *add*A (*Clostridium acetobutylicum* ATCC824), *bar*, and *gus*, are conventionally flanked by different recombination sites [Iamtham and Day 2000]. These are sequentially eliminated later via loop-out recombination. However, this system is difficult to regulate because of unpredictability of selection of homoplastic transformants.

A variant of this method is placing the marker gene outside of the flank regions for homologous recombination [Klaus et al. 2004]. Recombination via either left or right flanks resulted in co-integration of the vector, but a further round of recombination resulted in the loss of the marker gene. Thus, the marker gene was only transiently co-integrated into the genome. The marker gene expression for transplastomic selection was still provided effectively via this method [Klaus et al. 2004]. The advantage of this method is that marker-free plants can be generated directly in the first generation.

Selectable marker-removal techniques are labor-intensive and often too inefficient to allow their widespread use in commercial product development programs, especially in asexually reproducing or vegetatively propagated crops, and in cases where large numbers of primary transformation events are required. However, innovation for removal of selectable marker gene will continue by improving the existing techniques and designing new technologies such as transcription activation-like effector proteins (TALENs) produced by the plant pathogen *Xathomonas* [Yau and Stewart 2013]. Of particular importance is precision and robustness of removal without unintended consequences. Certainly, using the least amount of DNA possible is important for intellectual property and government regulatory concerns.

Transgene Integration, Expression, and Silencing

Integration and Expression

Integration of transgenes into the genomes of a plant is a random process. The precise integration of a transgene in a predetermined genomic location can reduce the variation in transgene expression [Day et al. 2000] and it can be achieved by the site-specific recombinase systems, such as Cre/*lox* and FLP/*frt* [Ow 2002]. Integration by homologous recombination would favor the establishment of a simple integration pattern and allow the insertion of a transgene into a known and stable region of the genome. Gene targeting has the potential to place foreign gene sequences in predetermined regions of the genome, thus potentially overcoming the so-called position effects on transgene expression [Xu 2010]. Transposons can be used to deliver recombination targets for subsequent site-specific integration.

Technologies that enhance the ability to create transgenic plants with the desired expression characteristics are of vital importance. One such technology involves the use of MARs (matrix attachment regions), the nuclear matrix of proteinaceous fibers that permeate the nucleus. MARs are DNA elements (sequences) that bind specifically to the nuclear matrix that often flank actively expressed genes and affect gene expression by influencing chromatin structure. Also known as 'SAR' (scaffold-attachment region), MARs enhance and stabilize transgene expression variably by reducing or eliminating some forms of gene silencing [Allen et al. 2000].

These MAR-matrix interactions organize chromatin into a series of independent loop domains. When MARs are positioned at 5'- and 3'-ends of a transgene, higher and more predictable expression of the transgene follows [Allen et al. 2004]. MARs are increasingly being applied to prevent or reduce some forms of unwanted transgene silencing, which is especially common when direct DNA transformation methods are used. Incorporation of MARS into transformation strategies can both improve transformation frequency and result in predictable, stable expression of the transgenic trait [Allen et al. 2004].

For successful integration and expression, the constructed transgene needs to have the following components: a) a tissue- or organ-specific promoter (e.g., CaMV 35S) that controls gene expression throughout the life cycle of the plant and in response to certain environmental changes; b) modification of gene (transgene) of interest (e.g., herbicide or insect resistant) to achieve greater expression in a plant; c) the termination sequence signal to the cellular machinery; and d) a selectable marker gene in the gene construct to identify with the integrated transgene.

Technologies such as Western blotting, Northern blotting, enzyme-linked immunosorbent assay (ELISA), and localization of mRNA transcripts are used for analysis of transgenic expression (proteins). However, these are not suitable for examination of the expression of specific genes in small amounts of tissues, nor do they have the precision and power to allow localization of the expression of a particular gene in specific cells or tissues [Page and Minocha 2004].

Introns

Introns stimulate gene expression in a wide range of organisms including plants. An intron is a nucleotide sequence within a gene that is removed from the corresponding RNA transcript of a gene by RNA splicing while the mature RNA product of a gene is being generated [Sambrook 1977; Alberts 2008].

Introns enhance gene expression by increasing the steady-state amount of mRNA in the cell [Callis et al. 1987] without affecting the stability of mRNA. They can elevate mRNA levels in two ways: first, by acting as transcriptional enhancers or alternative promoters located within the introns, and second, by a process termed intron-mediated enhancement (IME) [Rose 2008]. Although the mechanism of IME is largely unknown, it was hypothesized that the signals present in introns render the transcription machinery more processive, increasing the likelihood that full-length mRNAs will accumulate [Rose et al. 2008]. In the absence of these signals, the polymerase may tend to dissociate and produce truncated, rapidly degraded transcripts [Rose et al. 2008].

Introns are located within 1 kb of the transcription initiation site to stimulate gene expression. Introns proximal to promoters increase gene expression better than distal introns. Their effects vary depending on cell types, promoters or structural genes, and that a direct or indirect interaction between the introns and surrounding sequence elements was involved in the gene regulation. Therefore, when introns are used to strengthen gene expression, their influence on the pattern of the gene expression needs to be considered. Nevertheless, introns are very powerful tools for a higher gene expression.

The high expression capacity of promoters is usually due to an intron located within the 5' prime un-translated region (UTR), which contains elements for controlling gene expression by way of regulatory elements. It begins at the transcriptional start site and ends one nucleotide (nt) before the start codon (usually AUG) of the coding region. The 5' UTR has a median length of ~150 nucleotides in eukaryotes, but can be up to several thousand bases. Some viruses and cellular genes have unusually long and structured 5′ UTRs which may impact gene expression. On average, 3′ UTR tends to be twice as long as the 5′ UTR [Lodish et al. 2004].

Introns are used for optimizing transgene expression in higher plants. They sometimes enhance gene expression 10-fold or more, particularly in monocot species than dicot species. The insertion of the *ubiquitin* intron between CaMV 35S and a *gus* gene results in many-fold increase in the reporter gene expression. The rice *ubiquitin* intron also contributes to high degree of gene expression by increasing the steady-state level of mRNA as well as increasing the transcriptional and translational efficiencies. The maize polyubiquitin *Ubil* promoter has been extensively and successfully used to express chimeric genes in monocot transformation [Streatfield et al. 2004]. The rice RUBQ1 and RUBQ2 introns were found to be 782 and 962 bp, respectively [Wang et al. 2000]. When EPI, the first intron of rice *epsps* gene was inserted between CaMV 35S promoter and *GUS* gene, the expression level of the latter in transgenic tobacco increased 3 to 6 fold [Xu et al. 2003]. The 5′ introns of the two sugarcane polyubiquitin genes *ubi4* and *ubi9* were found to be 1360 and 1374 bp respectively [Wei et al. 2003], while *Ubil* gene has an intron of 1010 bp [Christensen et al. 1992].

Transgene Silencing and Inactivation

Gene silencing, an epigenetic process of gene regulation, is generally used to describe the 'switching off' of a gene by a mechanism other than genetic modification. That is, a gene which would be expressed ('turned on') under normal circumstances is switched off by machinery in the cell. Gene silencing occurs when RNA is unable to make a protein during translation.

Suppression of transgene expression through silencing is a relatively common phenomenon in transgenic plants, and it is exacerbated by the presence of multiple copies within the plant genome. Insertion of a transgene into the plant genome can trigger a partial or complete inactivation of homologous endogenous genes, transposable elements, and most frequently, transgenes. Transgene silencing occurs in untransformed plants where it reduces expression of endogenous genes. The silencing of transgenes is believed to be due to triggering of defense mechanisms, indicating that plants possess systems for controlling genome structure and gene expression [Matzke and Matzke 1998].

The introduced transgenes show varying levels and patterns of expression, even from the same gene construct, because they do not always behave as expected, leading to silencing. Introduction of foreign gene triggers a plant response that silences transgene in a way that is either integration-position specific or transgene-sequence specific [Graham et al. 2011]. Therefore, flanking sequences, transgene copy number, and rearrangements of the transgene and recipient genome vary between individual transformation events, and each of these can be a source of signals to trigger transgene silencing.

There are two kinds of transgene gene silencing: transcriptional gene silencing (TGS) and post-transcriptional gene silencing (PTGS). Both are linked to activation of RNAi (RNA interference) pathways [Graham et al. 2011]. TGS process, which blocks transcription, involves epigenetic changes, such as DNA methylation (the addition of a methyl group to the cytosine or adenine DNA nucleotides) during meiosis at the promoter regions [Park et al. 1996]. Besides, TGS is triggered by insertion near heterochromatin. In plants, multiple RNAi pathways play key roles in defense against viral infection and excessive transposon activity, controlling chromatin organization, regulating endogenous gene expression through transcriptional and translational repression, and DNA regulation.

The mechanism of RNAi-mediated gene silencing includes three classes of enzymes: RNA-dependent RNA polymerases (RdRP), dicer-like enzymes (DCL), and those belonging to Argonaute (AGO) protein family [Graham et al. 2011]. The RdRPs convert single-stranded RNA (ssRNA) templates to dsRNA, which is cleaved by DCL endonucleases into short dsRNA fragments of 19–24 nucleotides. Argonaute proteins select one strand from a short dsRNA complex and bind to it to form RNA-induced silencing complexes (RISCs), which then incorporate one strand of micro RNAs (miRNAs) or small interfering RNAs (siRNAs). Gene-silencing signals can be transmitted within plants by both short distance (cell-to-cell) and long distance pathways [Brosnan et al. 2007]. Therefore, silencing triggered in a number of cells might give rise to mosaic and sectored patterns, and ultimately systemic transgene silencing. TGS-mediated silencing of *trans*-inactivated loci can persist in subsequent generations.

PTGS can be triggered by transgenes, leading to the degradation of homologous RNA encoded by endogenous genes. Once activated, the RNA degradation machinery of PTGS becomes naturally efficient against endogenous gene or transgene RNA if it shares homology with the targeted transgene. The *de novo* methylation of transgene coding sequences is associated with PTGS [Park et al. 1996]. The mechanism of PTGS involves RNAi-mediated mRNA degradation. The siRNAs derived from transgene dsRNAs are the key specificity determinants in both TGS and PTGS, but the pathways of the latter process are different. PTGS is triggered by constructs designed to produce sense-strand RNA by co-suppression, antisense RNA, or hairpin RNA (hpRNA) [Graham et al. 2011].

Minimizing or Avoiding Transgene Inactivation and Silencing

Transgene silencing can be minimized and even avoided by several strategies [Graham et al. 2011; Xu 2010]. These include: a) using the recombinase systems such as Cre-*lox* (from bacteriophage 1) and FLP-*frp* for site-specific integration of transgenes into predetermined, suitable genomic locations; b) employing flanking S/MAR (scaffolding or matrix attachment regions that bind an internal nuclear network of nonhistone proteins) elements to, for example, *GUS* reporter constructs to increase transgene expression; c) eliminating silence triggers in the transgene sequence such as transgene expression cassette used for direct gene transfer and 'codon optimization' approach that improves transgene expression; d) inhibiting different steps in RNAi pathways by using RNA silencing suppressors; e) regulating transgene transcription rates and/or transcription structure to reduce excess antisense RNA-mediated turnover; to target transgenes into chromosomal regions that provide an optimum sequence environment for stable expression; f) induction of stress-mediated hyper-methylation in tissue culture using stress mimics such as propionic or butyric acid.

References

Agrawal, P., A. Kohli, R. Twyman and P. Christou. 2005. Transformation of plants with multiple cassettes generates simple transgene integration patterns and high expression levels. Mol. Breeding 16: 247–260.

Alberts, B. 2008. Molecular Biology of the Cell. Garland Science, New York.

Aionesei, T., J. Hosp, V. Voronin, E. Herbele-Voros and A. Touraev. 2006. Methotraxate and is a new selectable marker for tobacco immature pollen transformation. Plant Cell Rep. 25: 410–416.

Allen, G.C., S. Spiker and W.F. Thompson. 2000. Use of matrix attachment regions (MARs) to minimize transgene silencing. Plant Mol. Biol. 43: 361–376.

Allen, G.C., S. Spiker and W.F. Thompson. 2004. Transgene integration: use of matrix attachment regions. pp. 313–326. *In*: L. Peña (ed.). Transgenic Plants: Methods and Protocols. Methods Mol. Biol. 286.

Almofti, M.R., H. Hideyoshi, S. Yasuo, A. Almofti, W. Li and K. Hiroshi. 2003. Lipoplex size determines lipofection efficiency with or without serum. Mol. Membrane Biol. 20: 35–43.

Altpeter, F., N. Baisakh, R. Beachy, R. Bock, T. Capell, P. Christou et al. 2005. Particle bombardment and the genetic enhancement of crops: myths and realities. Mol. Breeding 15: 305–327.

Anuradha, T.S., S.K. Jami, R.S. Datla and P.B. Kirti. 2006. Genetic transformation of peanut (*Arachis hypogaea* L.) using cotyledonary node as explant and motorless *gus:npt*II fusion gene based vector. J. Biosci. 31: 235–246.

Arnold, P. 2009. History of Genetics: Genetic Engineering Timeline. http://www.brighthub.com/ science/ genetics/ articles/21983.aspx.

Asad, S., Z. Mukhtar, F. Nazir, J.A. Hashmi, S. Mansoor, Y. Zafar and M. Arshad. 2008. Silicon carbide whisker-mediated embryogenic callus transformation of cotton (*Gossypium hirsutum* L.) and regeneration of salt tolerant plants. Mol. Biotechnol. 40(2): 161–169.

Athmaram, T.N., G. Bali and K.M. Devaiah. 2006. Integration and expression of Bluetongue VP2 gene in somatic embryos of peanut through particle bombardment method. Vaccine 24: 2994–3000.

Australian Government Department of Agriculture. 2011. A background to genetically modified crops. Agriculture and Food. 4 July 2011.

Aziz, N. and A.N. Machray. 2003. Efficient male germ line transformation for transgenic tobacco production without selection. Plant Mol. Biol. 51: 203–211.

Baerson, S.R., D.J. Rodriguez, M. Tran, M. Feng, N.A. Biest and G.M. Dill. 2002. Glyphosate-resistant goosegrass identification of a mutation in the target enzyme 5-enolpyrvylshikimate-3-phosphate synthase. Plant Physiol. 129: 1265–1275.

Barampuram, S. and Z.J. Zhang. 2011. Recent advances in plant transformation. Methods Mol. Biol. 701: 1–35.

Bevan, M. 1984. Binary *Agrobacterium* vectors for plant transformation. Nucleic Acids Res. 12: 8711–8721.

Bhave, M. and C.F. Morris. 2008. Molecular genetics of puroindolines and related genes: allelic diversity in wheat and other grasses. Plant Mol. Biol. 66: 205–219.

Brãndo, L.B., N.P. Carneiro, A.C. de Oliveira, G.T.C.P. Coelho and A.A. Carneiro. 2012. Genetic transformation of immature sorghum influorescence via microprojectile bombardment. pp. 133–148. *In*: Yelda Özden Çiftçi (ed.). Transgenic Plants—Advances and Limitations. InTech Europe, Rijeka, Croatia, 478 pp.

Broothaerts, W., H.J. Mitchell, B. Weir, S. Kaines, L.M.A. Smith, W. Yang et al. 2005. Gene transfer to plants by diverse species of bacteria. Nature 433: 629–633.

Brosnan, C.A., N. Mitter, M. Christie, N.A. Smith, P.M. Waterhouse and B.J. Caroll. 2007. Nuclear gene silencing directs reception of long-distance mRNA silencing in *Arabidopsis thaliana*. Proc. Natl. Acad. Sci. USA 104: 14741–14746.

Buntru, M., S. Gärtner, L. Staib, F. Kreuzaler and N. Schlaich. 2013. Delivery of multiple transgenes to plant cells by an improved version of Multiround Gateway Technology. Transgenic Res. 22(1): 153–167.

Byrne, P. 2014. Department of Soil and Crop Science, Colorado State University. Personal Communication. Email of 24 April 2014.

Callis, J., M. Fromm and V. Walbot. 1987. Introns increase gene-expression in cultured maize cells. Genes Dev. 1: 1183–1200.

Charity, J.A., L. Holland, L.J. Grace and C. Walter. 2005. Consistent and stable expression of the nptII, uidA and bar genes in transgenic *Pinus radiata* after *Agrobacterium tumefaciens*-mediated transformation using nurse cultures. Plant Cell Rep. 23: 606–616.

Chen, Q.J., H.M. Zhou, J. Chen and X.C. Wang. 2006. A Gateway-based platform for multigene plant transformation. Plant Mol. Biol. 62: 927–936.

Cho, M.-J., H.W. Choi, D. Okamoto, S. Zhang and P.G. Lemaux. 2003. Expression of green fluorescent protein and its inheritance in transgenic oat plants generated from shoot meristematic cultures. Plant Cell Rep. 21: 467–474.

Christensen, A.H., R.A. Sharrock and P.H. Quail. 1992. Maize polyubiquitin genes—structure, thermal perturbation of expression and transcript splicing, and promoter activity following transfer to protoplasts by electroporation. Plant Mol. Biol. 18: 675–689.

Christou, P. and D. McCabe. 1992. Particle gun transformation of crop plants using electric discharge (ACCELL™ Technology). Agracetus Inc., Middleton, WI., USA.

Chung, S.-M., M. Vaidya and T. Tzfira et al. 2005. *Agrobacterium* is not alone: gene transfer to plants by viruses and other bacteria. Trends Pl. Sci. 20(20): 1–4.

Clemson University-USDA Cooperative Extension Slide Series, Bugwood.org. 2002. Agrobacterium tumefaciens gall at the root of *Carya illinoensis*. Wikipedia Commons: This file is licensed under the Creative Commons Attribution-Share Alike 3.0 United States license].

Cohen, S.N. and A.C.Y. Chang. 1973. Recircularization and Autonomous Replication of a Sheared R-Factor DNA Segment in *Escherichia coli* Transformants. Proc. Natl. Acad. Sci. USA 70(5): 1293–1297.

Conner, A.J., P.J. Barrell, S.J. Baldwin, A.S. Lockerse, P.A. Cooper, A.K. Erasmuson et al. 2007. Intragenic vectors for gene transfer without foreign DNA. Euphytica 154: 341–353.

Corneille, S. 2001. Efficient elimination of selectable marker genes from the plastid genome by the CRE-lox site-specific recombination system. Plant J. 27: 171–178.

Curtis, M.D. and U. Grossniklaus. 2003. A Gateway cloning vector set for high-throughput functional analysis of genes in planta. Plant Physiol. 133: 462–469.

Dafny-Yelin, M. and T. Tzfira. 2007. Delivery of multiple transgenes to plant cells. Plant Physiol. 145(4): 1118–1128.

Daniell, H., S. Chebolu, S. Kumar, M. Singleton and R. Falconer. 2005. Chloroplast derived vaccine antigens and other therapeutic proteins. Vaccine 23: 179–1783.

Daniell, H., B. Muthukumar and S.B. Lee. 2001. Marker free transgenic plants: engineering the chloroplast genome without the use of antibiotic selection. Current Genetics 39: 109–116.

Datta, K., N. Baisakh, N. Oliva, L. Torrizo, E. Abrigo, J. Tan et al. 2003. Bioengineered 'golden' *indica* rice cultivars with beta-carotene metabolism in the endosperm with hygromycin and mannose selection systems. Plant Biotechnol. J. 1: 81–90.

Daveya, M.R., P. Anthonya, J.B. Powera and K.C. Loweb. 2005. Plant protoplasts: status and biological perspectives. Biotech. Adv. 23: 131–171.

Davis, A.M., A. Hall, A.J. Millar, C. Darrah and S.J. Davis. 2009. Protocol: streamlined sub-protocols for floral-dip transformation and selection of transformants in *Arabidopsis thaliana*. Plant Methods 3: 5.

Day, A. and M. Goldschmidt-Clermont. 2011. The chloroplast transformation toolbox: selectable markers and marker removal. Plant Biotech. J. 9: 540–553.

Day, C.D., E. Lee, J. Kobayashi, L.D. Holappa, H. Albert and D.W. Ow. 2000. Transgene integration into the same chromosome location can produce alleles that express at a predictable level, or alleles that are differentially silenced. Genes and Development 14: 2869–2880.

Dekeyser, R.A., B. Claes, R.M.U. De Rycke, M.E. Habets, M.C. Van Montagu and A.B. Caplan. 1990. Transient gene expression in intact and organized rice tissues. Plant Cell 2: 591–602.

De La Peña, A., H.L. Örz and J. Schell. 1987. Transgenic rye plants obtained by injecting DNA into young floral tillers. Nature 325: 274–276.

De Padua, V.L.M., R.P. Ferreira, L. Meneses, L. Uchoa, M.-P. Marcia and E. Mansur. 2001. Transformation of Brazilian elite *Indica*-type rice (*Oryza sativa* L.) by electroporation of shoot apex explants. Plant Mol. Biol. Rep. 19: 55–64.

De Padua, V.L.M., M.C. Prestana, M. Margis-Pinheiro, D.E. De Oliviera and E. Mansur. 2000. Electroporation of intact embryonic leaflets of peanut: gene transfer and stimulation of regeneration of capacity. *In vitro* Cell Dev. Biol. Plant 36: 374–378.

Dixon, R.A., J.H. Bouton, B. Narasimhamoorthy, M. Saha, Z.-Y. Wang and G.D. May. 2007. Beyond structural genomics for plant science. Adv. Agron. 95: 77–161.

Earley, K.W., J.R. Haag, O. Pontes, K. Opper, T. Juehne, K. Song and C.S. Pikaard. 2006. GATEWAY-compatible vectors for plant functional genomics and proteomics. Plant J. 45: 616–629.

Felgner, P.L. et al. 1987. Lipofection: a highly efficient, lipid-mediated DNA-transfection procedure. Proc. Natl. Acad. Sci. USA 84: 7413–7417.

Finer, J.J. 2011. Visualizing transgene expression. pp.109–119. *In*: C.N. Stewart, A. Touraev, V. Citovsky and T. Tzfira (eds.). Plant Transformation Technologies. Wiley-Blackwell. Ames, Iowa, USA.

Fisk, H.J. and A.M. Dandekar. 2005. Electroporation: introduction and expression of transgenes in plant protoplasts. pp. 79–90. *In*: L. Pena (ed.). Transgenic Plants: Methods and Protocols. Human Press, Totowa, NJ.

Fraley, R.T., S.G. Rogers, R.B. Horsch, P.R. Sanders, J.S. Flick et al. 1983. Expression of bacterial genes in plant cells. Proc. Natl. Acad. Sci. USA 80: 4803–4807.

François, I., W. Broekaert and B. Commue. 2002. Different approaches for multi-transgene stacking in plants. Plant Science 163: 281–295.

Gao, X.R., G.K. Wang, Q. Su, Y. Wang and A.J. An. 2007. Phytase expression in transgenic soybeans: stable transformation with a vector-less construct. Biotechnol. Lett. 29: 1781–1788.

Genetically Altered Potato Ok'd For Crops. 1995. Lawrence Journal-World—6 May 1995.

Goeddel, D., D.G. Kleid, F. Bolivar, H.L. Heyneker, D.G. Yansura, R. Crea et al. 1979. Expression in *Escherichia coli* of chemically synthesized genes for human insulin. Proc. Natl. Acad. Sci. USA 76(1): 106–110.

Graham, F.L. and A.J. van der Eb. 1973. A new technique for the assay of infectivity of human adenovirus 5 DNA. Virology 52(2): 456–467.

Graham, M.W., S.R. Mudge, P.R. Sternes and R.G. Birch. 2011. Understanding and avoiding transgene silencing. pp. 171–196. *In*: C.N. Stewart, A. Touraev, V. Citovsky and T. Tzfira (eds.). Plant Transformation Technologies. Wiley-Blackwell. Ames, Iowa, USA.

Gurel, S., E. Gurel, R. Kaur, J. Wong, L. Meng, H.-Q. Tan and P. Lemaux. 2009. Efficient, reproducible *Agrobacterium*-mediated transformation of sorghum using heat treatment of immature embryos. Plant Cell Rep. 28: 429–444.

Gurel, S., E. Gurel, T.I. Miller and P.G. Lemaux. 2013. *Agrobacterium*-mediated transformation of sorghum bicolor using immature embryos. pp. 109–122. *In*: J.M. Dunwell and A.C. Wetten (eds.). Transgenic Plants: Methods and Protocols. Methods in Mol. Biol. 847.

Hajdukiewicz, P., Z. Svab and P. Maliga. 1994. The small, versatile pPZP family of Agrobacterium binary vectors for plant transformation. Plant Mol. Biol. 25: 989–994.

Han, S.-N., P.-R. Oh, H.-S. Kim, H.Y. Heo, J.C. Moon, S.-K. Lee et al. 2012. Effect of antibiotics on suppression of *Agrobacterium tumefaciens* and plant regeneration from wheat embryo. J. Crop Sci. Biotechnol. 10(2): 92–98.

Hansen, G. and M.D. Chilton. 1996. "Agrolistic" transformation of plant cells: integration of T-strands generated in planta. Proc. Natl. Acad. Sci. USA 93: 14978–14983.

Higashi, T., E. Nagamori, T. Sone, S. Matsunaga and K. Fukui. 2004. A novel transfection method for mammalian cells using calcium alginate microbeads. J. Biosci. Bioeng. 97: 191–195.

Hohn, B., A.A. Levy and H. Puchta. 2001. Elimination of selection markers from transgenic plants. Current Opinion in Biotech. 12: 139–143.

Huang, S.S., L.A. Gilbertson, T.H. Adams, K.P. Malloy, E.K. Reisenbigler, D.H. Birr et al. 2004. Generation of marker-free transgenic maize by regular two-border *Agrobacterium* transformation vectors. Transgenic Res. 13: 451–461.

Iamtham, S. and A. Day. 2000. Removal of antibiotic resistance genes from transgenic tobacco plastids. Nature Biotech. 18: 1172–1176.

Iyer, L.M., S.P. Kumpatla, M.B. Chandrasekharan and T.C. Hall. 2000. Transgene silencing in monocots. Plant Mol. Biol. 43: 323–346.

Jackson, D.A., R.H. Symons and P. Berg. 1972. Biochemical method for inserting new genetic information into DNA of Simian Virus 40: Circular SV40 DNA Molecules Containing Lambda Phage Genes and the Galactose Operon of *Escherichia coli*. Proc. Natl. Acad. Sci. USA 69(10): 2904–2909.

Jaenisch, R. and B. Mintz. 1974. Simian virus 40 DNA sequences in DNA of healthy adult mice derived from preimplantation blastocysts injected with viral DNA. Proc. Natl. Acad. USA 71(4): 1250–1254.

James, C. 1997. Global Status of Transgenic Crops in 1997. Briefs No. 5: 31. The International Service for the Acquisition of Agri-biotech Applications, Ithaca, NY.

James, C. and A.F. Krattinger. 1996. Global Review of the Field Testing and Commercialization of Transgenic Plants, 1986–1995: The First Decade Crop Biotechnology. The International Service for the Acquisition of Agri-biotech Applications, Ithaca, NY, 31 pp.

Jefferson, R.A. 1987. Assaying chimeric genes in plants: the GUS gene fusion system. Plant Mol. Biol. Rep. 5: 387–405.

Jia, H., Y. Pang, X. Chen and R. Fang. 2006. Removal of the selectable marker gene from transgenic tobacco plants by expression of Cre recombinase from a tobacco mosaic virus vector through agroinfection. Transgenic Res. 15: 375–84.

Kaeppler, H., D.A. Somers, H.W. Rines and A.F. Cockburn. 1992. Silicon carbide fiber-mediated stable transformation of plant cells. Theor. Appl. Genet. 84: 560–566.

Karimi, M., D. Inzé and A. Depicker. 2002. GATEWAY vectors for *Agrobacterium*-mediated plant transformation. Trends Plant Sci. 7: 193–195.

Karimi, M., D.M. Bjorn and P. Hilson. 2005. Modular cloning in plant cells. Trends Plant Sci. 10: 103–105.

Karimi, M., A. Depicker and P. Hilson. 2007. Recombinational cloning with plant gateway vectors. Plant Physiol. 145: 1144–1154.

Kaufman, R.I. and B.T. Nixon. 1996. Use of PCR to isolate genes encoding sigma54-dependent activators from diverse bacteria. J. Bacteriol. 178(13): 3967–3970.

Khan, A.S. 2012. Plastid genome engineering in plants: Present status and future trends. Mol. Plant Breed. 3(9). doi: 10.5376/mpb.2012.03.0009.

Khan, M.S. and P. Maliga. 1999. Fluorescent antibiotic resistance marker for tracking plastid transformation in higher plants. Nature Biotechnol. 17: 910–915.

Khemkladngoen, N., N. Wada, S. Tsuchimoto, J.A. Cartagena, S. Kajiyama and K. Fukui. 2012. Bioactive beads-mediated transformation of rice with large DNA fragments containing *Aegilops tauschii* genes, with special reference to bead-production methodology. pp. 117–132. *In*: Yelda Özden Çiftçi (ed.). Transgenic Plants—Advances and Limitations. InTech Europe, Rijeka, Croatia, 478 pp.

Kim, E.H., S.C. Such, B.S. Park, K.S. Shin, S.J. Keno, E.J. Han, S. Park et al. 2009. Chloroplast targeted expression of synthetic *cry1Ac* in transgenic rice as an alternative strategy for increased pest protection. Planta 230: 397–405.

Kim, T.-G., M.-Y. Back, E.-K. Lee, T.-H. Kwon and M.-S. Yang. 2008. Expression of human growth hormone in transgenic cell suspension culture. Plant Cell Rep. 27: 885–891.

Klaus, S.M.J. 2004. Generation of marker-free plastid transformants using a transiently cointegrated selection gene. Nature Biotechnol. 22: 225–229.

Komari, T. 1990. Transformation of cultured cells of *Chenopodium quinoa* by binary vectors that carry a fragment of DNA from the virulence region of pTiBo542. Plant Cell Rep. 9: 303–306.

Komari, T. 1989. Transformation of callus cultures of nine plant species mediated by Agrobacterium. Plant Sci. 60: 223–229.

Komari, T., T. Imayama, N. Kato, Y. Ishida, J. Ueki and T. Komari. 2007. Current status of binary vectors and superbinary vectors. Plant Physiol. 145: 1155–1160.

Komari, T., Y. Hiei, Y. Saito, N. Murai and T. Kumashiro. 1996. Vectors carrying two separate T-DNAs for co-transformation of higher plants mediated by *Agrobacterium tumefaciens* and segregation of transformants free from selection markers. Plant J. 10: 165–174.

Komari, T. and T. Komari. 2011. Current state and perspective of binary vectors and superbinary vectors. pp. 122–138. *In*: C.N. Stewart, A. Touraev, V. Citovsky and T. Tzfira (eds.). Vectors, Promoters, and Other Tools for Plant Transformation. Wiley-Blackwell, Ames, Iowa, USA.

Komari, T., Y. Takakura, J. Ueki, N. Kato, Y. Ishida and Y. Hiei. 2006. pp. 15–41. *In*: K. Wang (ed.). Binary Vectors and Super-binary Vectors. Methods in Molecular Biology 343: *Agrobacterium* Protocols, Ed. 2, Vol. 1. Humana Press, Totowa, NJ.

Kumar, S., A. Dhingra and H. Daniell. 2004. Stable transformation of the cotton plastid genome and maternal inheritance of transgenes. Plant Mol. Biol. 56: 203–216.

Lederberg, J. 1952. Cell genetics and hereditary symbiosis. Physiol. Rev. 32(4): 403–430.

Lee, L. and S.B. Gelvin. 2008. T-DNA binary vectors and systems. Plant Physiol. 146: 325–332.

Lee, S.M., K. Kang, H. Chung, S.H. Yoo, X.M. Xu, S.B. Lee et al. 2006. Plastid transformation in the monocotyledonous cereal crop rice (*Oryza sativa*) and transmission of transgenes to their progeny. Mol. Cells 21: 401–410.

Leelavathi, S., V.G. Sunnichan, R. Kumria, G.P. Vijayakanth, R.K. Bhatnagar and V.S. Reddy. 2004. A simple and rapid *Agrobacterium*-mediated transformation protocol for cotton (*Gossypium hirsutum* L.) embryogenic calli as a source to generate large numbers of transgenic plants. Plant Cell. Rep. 22: 465–470.

Lermontova, I. and B. Grimm. 2000. Overexpression of plastidic protoporphyrinogen IX oxidase leads to resistance to the diphenyl ether herbicide acifluorfen. Plant Physiol. 122: 75–84.

Liang, J., Y. Luo and H. Zhao. 2011. Synthetic biology: Putting synthesis into biology. Wiley Interdisciplinary Reviews: Systems Biology and Medicine 3: 7.

Li, H., H. Flachowsky, T.C. Fischer, M.-V. Hanke, G. Forkmann, D. Treutter et al. 2007. Maize Lc transcription factor enhances biosynthesis of anthocyanins, distinct proanthocyanidins and phenylpropanoids in apple (*Malus domestica* Borkh.). Planta 226: 1243–1254.

Li, J., X. Tan, F. Zhu and J. Guo. 2010. A rapid and simple method for *Brassica napa* floral-dip transformation and selection of transgenic plants. Int. Natl. Biol. 2(1): 126–131.

Li, L., Y. Zhou, X. Cheng, J. Sun, J.M. Marita, J. Ralph and V.L. Chiang. 2003. Combinatorial modification of multiple lignin traits in trees through multigene cotransformation. Proc. Natl. Acad. Sci. USA 100: 4939–4944.

Li, X., S.L. Volrath, D.B. Nicholl, C.E. Chilcott, M.A. Johnson et al. 2003. Development of protoporphyrinogen oxidase as an efficient selection marker for *Agrobacterium tumefaciens*-mediated transformation of maize. Plant Physiol. 133: 736–747.

Lin, L., Y.-G. Liu, X. Xu and B. Li. 2003. Efficient linking and transfer of multiple genes by a multiple gene assembly and transformation vector system. Proc. Natl. Acad. Sci. USA 100(10): 5962–5967.

Lipps, G. 2008. Plasmids: Current Research and Future Trends. Caister Academic Press, Norfolk, England.

Liu, C.-W., C.-C. Lin, J.-C. Yiu, J.J.W. Chen and M.-J. Tseng. 2008. Expression of a *Bacillus thuringiensis* toxin (*cry1Ab*) gene in cabbage (*Brassica oleracea* L. var. *capitata* L.) chloroplasts confers high insecticidal eYcacy against *Plutella xylostella*. Theor. Appl. Genet. 117: 75–88.

Liu, H., A. Kawabe, S. Matsunaga, T. Murakawa, A. Mizukami, M. Yanagisawa et al. 2004a. Obtaining transgenic plants using the bio-active beads method. J. of Plant Res. 117: 95–99.

Liu, H., A. Kawabe, S. Matsunaga, A. Kobayashi, S. Harashima, S. Uchiyama et al. 2004b. Application of the bio-active beads mediated in rice transformation. Plant Biotechnol. 21: 303–306.

Lodish, H., A. Berk, P. Matsudaira, C.A. Kaiser, M.P. Scott, L. Zipursky and J. Darnell. 2004. Molecular Cell Biology (5th ed.). W.H. Freeman and Company, New York, 113 pp.

Lowe, B.A., N.S. Prakash, W. Melissa, M.T. Mann, T.M. Spencer and R.S. Boddupalli. 2009. Enhanced single copy integration events in corn via particle bombardment using low quantities of DNA. Transgenic Res. 18: 831–840.

Lu, L., X. Wu, X. Yin, J. Morrand, X. Chen, W.R. Folk and Z.J. Zhang. 2009. Development of marker-free transgenic sorghum [*Sorghum bicolor* (L.) Moench] using standard binary vectors with *bar* as a selectable marker. Plant Cell Tissue Organ Cult. 99: 97–108.

Lu, H.J., X.R. Zhou, Z.X. Gong and N.M. Upadhyaya. 2001. Generation of selectable marker-free transgenic rice using double right-border (DRB) binary vectors. Australian J. Pl. Physiol. 28: 241–248.

Lutz, K.A., M.H. Bosachi and P. Maliga. 2006. Plastid marker-gene excision by transiently expressed CRE recombinase. Plant J. 45: 447–456.

Ma, L., J. Dong, Y. Jin, M. Chen, X. Shen and T. Wang. 2011. RMDAP: A Versatile, Ready-To-Use Toolbox for Multigene Genetic Transformation. PLoS One 6(5): e19883 (Published Online 13 May 2011. doi: 10.1371/ journal.pone.0019883.

Magnani, E., L. Bartling and S. Hake. 2006. From Gateway to MultiSite Gateway in one recombination event. BMC Mol. Biol. 7: 46.

Manickvasagam, M., A. Ganapathi, V.R. Ambuzhagan, B. Sudhakar, N. Selvaraj, A. Vasudevan and S. Kasthurirengan. 2004. Agrobacterium-mediated genetic transformation and development of herbicide-resistant sugarcane (*Saccharum* species hybrids) using axillary buds. Plant Cell Rep. 23: 134–143.

Matzke, M.A. and A.J.M. Matzke. 1998. Epigenetic silencing of plant transgenes as a consequence of diverse cellular defense responses. Cell. Mol. Life Sci. 54: 94–103.

Maugh II, T.H. 1987. The Los Angeles Times. 9 June 1987. Altered Bacterium Does Its Job: Frost Failed to Damage Sprayed Test Crop.

eyer, P. 1998. Stabilities and instabilities in transgene expression. pp. 263–275. *In*: K. Lindsey (ed.). Transgenic Plant Research. Harwood Academic Publishers, Switzerland.

Murakawa, T., S. Kajiyama, T. Ikeuchi, S. Kawakami and K. Fukui. 2008. Improvement of transformation efficiency by bioactive beads mediated gene transfer using DNA-lipofection complex as entrapped genetic materials. J. Biosci. Bioengi. 105: 77–80.

Ganci, M. 2012. *Rhizobium rhizogenes=Agrobacterium rhizogenes*. Pathogen profile. College of Agri. Life Sci., N.C. State University, Raliegh, NC, USA.

Nicholson, L., P. Gozalez-Melendi, C. van Dolleweerd, H. Tuck, Y. Perren, J.K.-C. Ma, R. Fischer et al. 2005. A recombinant multimeric recombinant immunoglobulin expressed in rice shows assembly-dependent subcellular localization in endosperm cells. Plant Biotechnol. J. 3: 115–127.

Ow, D.W. 2002. Recombinase-directed plant transformation for the post-genomic era. Plant Mol. Biol. 48: 183–200.

Ozawa, K. 2013. A High-Efficiency *Agrobacterium*-mediated transformation system of rice (*Oryza sativa* L.). pp. 51–57. *In*: J.M. Dunwell and A.C. Wetten (eds.). Transgenic Plants: Methods and Protocols. Methods Mol. Biol. 847 pp.

Park, Y.D., I. Papp, E.A. Mascone, V.A. Iglesias, H. Vaucheret, A.J.K. Matzke et al. 1996. Gene silencing mediated by promoter homology occurs at the level of transcription and results in meiotically heritable alterations in methylation and gene activity. Plant J. 9: 183–194.

Patent Lens. 2007. *Agrobacterium*-mediated transformation—Overview: Binary Vectors. www.patentslens. net.

Peralta, E.G., R. Hellmiss and W. Ream. 1986. Overdrive, a T-DNA transmission enhancer on the *A. tumefaciens* tumour-inducing plasmid. EMBO J. 5: 1137–1142.

Priya, A.M., S.K. Pandian and R. Manikandan. 2012. The effect of different antibiotics on the elimination of *Agrobacterium* and high frequency of *Agrobacterium*-mediated transformation of India rice (*Oryza sativa* L.). Czech. J. Genet. Plant Breed. 48: 120–130.

Pueppke, S.G. and W.J. Broughton. 1999. *Rhizobium* sp. strain NGR234 and *R. fredii* USDA257 share exceptionally broad, nested host ranges. Mol. Plant Microbe Interact. 12: 293–318.

Resch, T. and A. Touraev. 2011. Pollen transformation technologies. pp. 83–91. *In*: C.N. Stewart, A. Touraev, V. Citovsky and T. Tzfira (eds.). Plant Transformation Technologies. Wiley-Blackwell. Ames, Iowa, USA.

Rommens, C.M. 2004. All-native DNA transformation: a new approach to plant engineering. Trends Plant Sci. 9: 457–464.

Rommens, C.M., O. Bougri, H. Yan, J.M. Humara, J. Owen, K. Swords and J. Ye. 2005. Plant-derived transfer DNAs. Plant Physiol. 139: 1338–1349.

Rommens, C.M., A. Conner, H. Yan and Z. Hanley. 2011. Intragenic vectors and marker-free transformation: tools for a green biotechnology. pp. 93–107. *In*: C.N. Stewart, A. Touraev, V. Citovsky and T. Tzfira (eds.). Plant Transformation Technologies. Wiley-Blackwell, Ames, Iowa, USA.

Rommens, C.M., C.M. Richael, H. Yan, D.A. Navarre, J. Ye, M. Krucker and K. Swords. 2008a. Engineered native pathways for high kaempferol and caffeoylquinate production in potato. Plant Biotechnol. J. 6: 870–886.

Rommens, C.M., H. Yan, K. Swords, C. Richael and J. Ye. 2008b. Low-acrylamide French fries and potato chips. Plant Biotechnol. J. 6: 843–853.

Rommens, C.M., J. Ye, C.M. Richael and K. Swords. 2006. Improving potato storage and processing characteristics through all-native DNA transformation. J. Agri. Food Chem. 54: 9882–9887.

Rose, A.B. 2008. Intron-mediated regulation of gene expression. pp. 277–290. *In*: A.S.N. Reddy, M. Golovkin and M.B. Heidelberg (eds.). Current Topics in Microbiology and Immunol. 326: Springer-Verlag. Berlin Heidelberg, Germany.

Rose, A.B., T. Elfersi, G. Parra and I. Korf. 2008. Promoter-proximal introns in *Arabidopsis thaliana* are enriched in dispersed signals that elevate gene expression. Plant Cell 20: 543–551.

Saelim, L., S. Phanisiri, M. Suksangpanomrung, S. Netrphan and J. Narangajavana. 2009. Evaluation of a morphological marker selection and excision system to generate marker-free transgenic cassava plants. Plant Cell Rep. 28: 445–455.

Salmenkallio-Marttila, M., K. Aspegren, S. Akerman, U. Kurten, L. Mannonen et al. 1995. Transgenic barley (*Hordeum vulgare* L.) by electroporation of protoplasts. Plant Cell Rep. 15: 301–304.

Sambrook, J. 1977. Adenovirus amazes at Cold Spring Harbour. Nature 268: 101–104.

Satyavathi, V.V., V. Prasad, A. Khandelwal, M.S. Shaila and G. Lakshmi Sita. 2003. Expression of hemagglutinum protein of Rinderpest virus in transgenic pea [*Cajanus cajan* (L.) Millsp.] plants. Plant Cell Rep. 21: 651–658.

Sawahel, W.A. 2002. The production of transgenic potato plants expressing human Ü-interferon using lipofection-mediated transformation. Cell. Mol. Biol. Lett. 7: 19–29.

Sawahel, W.A. 2001. Stable genetic transformation of cotton plants using polybrene spermidine treatment. Plant Mol. Biol. Rep. 19: 377a–377f.

Schouten, H.J., F.A. Krens and E. Jacobsen. 2006. Cisgenic plants are similar to traditionally bred lines: International regulations for genetically modified organisms should be altered to exempt cisgenesis. EMBO Reports 7: 750–753.

Seema, G., H.P. Pande, J. Lal and V.K. Madan. 2001. Plantlet regeneration of sugarcane varieties and transient GUS expression in calli by electroporation. Sugar Tech. 3: 27–33.

Shukla, R., V. Bansal, M. Chaudhary, A. Basu, R.R. Bhonde and M. Sastry. 2005. Bioincompatibility of gold nanoparticles and their endocytotic fate inside the cellular compartment: a microscopic overview. Langmuir. 21(23): 10644–10654.

Singh, N. and H.S. Chawla. 1999. Use of silicon carbide fibers for *Agrobacterium*-mediated transformation in wheat. Current Sci. 76: 1483–1485.

Sone, T., E. Nagamori, T. Ikeuchi, A. Mizukami, Y. Takakura et al. 2002. A novel gene delivery system in plants with calcium alginate micro-beads. J. Biosci. Bioengi. 94: 87–91.

Spangenberg, G., Z.-Y. Wang and I. Potrykus. 1998. Biotechnology in forage and turf grass improvement. Springer, Berlin Heidelberg, New York.

Spokevicius, A.V., K.V. Beveren, M.A. Leitch and G. Bossinger. 2005. *Agrobacterium* mediated *in vitro* transformation of wood-producing stem segments in eucalypts. Plant Cell Rep. 23: 617–624.

Spully. 2007. Plasmid. Spaully. Wikimedia Commons.

Stableford, B.M. 2004. Historical dictionary of science fiction literature. The Rowman & Littlefield Publishing Group, Inc. Lanham, Maryland, USA, p. 133.

Streatfield, S.J., M.E. Magallanes-Lundback, K.K. Beifuss, C.A. Brooks, R.L. Harkey, R.T. Love et al. 2004. Analysis of the maize polyubiquitin-1 promoter heat shock elements and generation of promoter variants with modified expression characteristics. Transgenic Res. 13: 299–312.

Sujatha, M. and K.B.R.S. Sarada. 2013. Transformation of nuclear DNA in meristematic and embryogenic tissues. pp. 27–44. *In*: S. Sudowe and A.B. Reske-Kunz (eds.). Biolistic DNA Delivery: Methods and Protocols. Methods Mol. Biol. 978-1-62703-110-3 (Online).

Sun, J., Q.W. Niu, P. Tarkowski, B. Zheng, D. Tarkowska, G. Sandberg et al. 2003. Arabidopsis AtIPT8/PGA22 gene encodes as isopentenyl transferase that is involved in *de novo* cytokinin biosynthesis. Plant Physiol. 131: 167–176.

Tanaka, Y., T. Kimura, K. Hikino, S. Goto, M. Nishimura, S. Mano and T. Nakagawa. 2012. Gateway vectors for plant genetic engineering: Overview of plant vectors, application for biomolecular fluorescence complementation (BiFC) and multigene construction. pp. 35–58. *In*: H.A. Barrera-Saldana (ed.). Genetic Engineering–Basics, New Applications and Responsibilities. InTech, Rijeka, Croatia, 256 pp.

Taylor, C.B. 1997. Promoter fusion analysis: An insufficient measure of gene expression. Plant Cell 9: 273–275.

The European Parliament and the Council of the European Union. 12 March 2001. Directive on the release of genetically modified organisms (GMOs) Directive 2001/18/EC ANNEX I A. Official Journal of the European Communities. p. 17.

Tian, L. 2007. Selectable marker gene excision from transgenic plants. ISB News Report (July 2007) 1–3.

Torney, E., B.G. Trewyn, T. Shimada and T. Kabayashi. 2007. Mesoporous silica nanoparticles deliver DNA and chemicals into plants. Nature Nanotech. 2: 295–300.

Tungsuchat, T., H. Kuroda, J. Narangajavana and P. Maliga. 2006. Gene activation in plastids by the CRE site-specific recombinase. Plant Mol. Biol. 61: 711–718.

Tzfira, T., S.V. Kozlovsky and V. Citovsky. 2007. Advanced expression vector systems: New weapons for plant research and biotechnology. Plant Physiol. 145: 1087–1089.

Um, M.K., T.I. Park, Y.J. Kim, H.Y. Seo, J.G. Kim, S.Y. Kwon et al. 2007. Particle bombardment mediated transformation of barley with Arabidopsis NDPK2 cDNA. Plant Biotech. Rep. 1: 71–77.

Veena, J.H., R.W. Doerge and S.B. Gelvin. 2003. Transfer of T-DNA and Vir proteins to plant cells by *Agrobacterium tumefaciens* induces expression of host genes involved in mediating transformation and suppresses host defense gene expression. Plant J. 35: 219–236.

Vega, J., W. Yu, A.M. Kennon, X. Chen and Z.J. Zhang. 2008. Improvement of *Agrobacterium*-mediated transformation in Hi-II maize (*Zea mays* L.) using standard binary vectors. Plant Cell Rep. 27: 297–305.

Verma, D. and H. Daniell. 2007. Chloroplast vector systems for biotechnology applications. Plant Physiol. 145: 1129–1143.

Wada, N., S. Kajiyama, Y. Akiyama, S. Kawakami, D. No, S. Uchiyama et al. 2009. Bioactive beads-mediated transformation of rice with large DNA fragments containing *Aegilops tauschii* genes. Plant Cell Rep. 28(5): 759–768.

Wada, N., S. Kajiyama, N. Khemkladngoen and K. Fukui. 2011. A novel gene delivery system in plants with calcium alginate micro-beads. pp. 73–81. *In*: C.N. Stewart, A. Touraev, V. Citovsky and T. Tzfira (eds.). Plant Transformation Technologies. Wiley-Blackwell. Ames, Iowa, USA.

Wang, J.L., J.D. Jiang and J.H. Oard. 2000. Structure, expression and promoter activity of two sugarcane polyubiquitin genes from rice (*Oryza sativa* L.). Plant Sci. 156: 201–211.

Wang, Y., B. Chen, Y. Hu, J. Li and Z. Lin. 2005. Inducible excision of selectable marker gene from transgenic plants by the Cre/lox site-specific recombination system. Transgenic Res. 14: 605–614.

Wang, K., C. Genetello, V.M. Montagu and P.C. Zambryski. 1987. Sequence context of the T-DNA border repeat element determines its relative activity during T-DNA transfer to plant cells. Mol. Gen. Genet. 210: 338–346.

Wang, Z.-Y. and Y. Ge. 2006. Recent advances in genetic transformation of forage and turf grasses. *In vitro* Cell Dev. Biol. Plant 42: 1–18.

Weaver, J.C. 1995. Electroporation theory: concepts and mechanisms. pp. 1–26. *In*: J.A. Nickoloff (ed.). Electroporation Protocols for Microorganisms. Humana Press, Totowa, New Jersey.

Wei, H.R., M..L. Wang, P.H. Moore and H.H. Albert. 2003. Comparative expression analysis of two sugarcane poly-ubiquitin promoters and flanking sequences in transgenic plants. J. Plant Physiol. 160: 1241–1251.

Wheeler. 2007. A diagram of the main components of an electroporator with cuvette loaded. En.User:Zephyris. Wikipedia Commons.

Wu, H., A. Doherty and H.D. Jones. 2008. Efficient and rapid *Agrobacterium*-mediated genetic transformation of durum wheat (*Triticum turgidum* L. var. durum) using additional virulence genes. Transgenic Res. 17: 425–436.

Wu, L., S. Nandi, L. hen, R.L. Rodriguez and N. Huang. 2002. Expression and inheritance of nine transgenes in rice. Transgenic Res. 11: 533–541.

Xu, J., D. Feng, G. Song, X. Wei, L. Chen, X. Wu et al. 2003. The first intron of rice EPSP synthase enhances expression of foreign gene. Sci. China C. Life Sci. 46(6): 561–569.

Xu, Y. 2010. Molecular Plant Breeding. CABI Head Office, Oxfordshire, UK, 734 pp.

Ye, X., S. Al-Babili, A. Kloti, J. Zhang, P. Lucca, P. Beyer and I. Potrykus. 2000. Engineering the provitamin A (beta-carotene) biosynthetic pathway into (carotenoid-free) rice endosperm. Science 287: 303–305.

Yau, Y.-Y. and C.N. Stewart. 2013. Less is more: strategies to remove marker genes from transgenic plants. BMC Biotechnology 13: 36.

Zeevi, V., Z. Liang, U. Arieli and T. Tzfira. 2012. Zinc finger nuclease and homing endonuclease-mediated assembly of multigene plant transformation vectors. Plant Physiol. 158(1): 132–144.

Zeng, P., D. Vadnais, Z. Zhang and J. Polacco. 2004. Refined glufosinate selection in *Agrobacterium*-mediated transformation of soybean [*Glycine max* (L.) Merr.]. Plant Cell Rep. 22: 478–482.

Zhou, H.Y., S.B. Chen, X.G. Li, G.F. Xiao, X.L. Wei and Z. Zhu. 2003. Generating marker-free transgenic tobacco plants by *Agrobacterium*-mediated transformation with double T-DNA binary vector. Acta Botanica Sinica 45: 1103–1108.

Zhu, C., S. Naqvi, J. Breitenbach, G. Sandmann, P. Christou and T. Capell. 2008. Combinatorial genetic transformation generates a library of metabolic phenotypes for the carotenoid pathway in maize. Proc. Natl. Acad. Sci. USA 105: 18232–18237.

Zhu, Z., B. Sun, C. Liu, G. Xiao and X. Li. 1993. Transformation of wheat protoplasts mediated by cationic liposome and regeneration of transgenic plantlets. China. J. Biotechnol. 9: 257–261.

CHAPTER 5

Transgenes in Herbicide and Insect Resistance

The backbone of a transgenic line is the gene construct which must contain all appropriate elements critical for gene expression. The elements of a gene construct include the transgene, a promoter gene, a reporter gene, and an intron. This chapter deals with transgenes responsible for herbicide and insect resistance besides gene promoters, the essential components of gene expression in transgenic engineering. The other genetic elements required during genetic transformation have been discussed in Chapters 3 and 4.

Herbicide-Resistant Transgenes

Glyphosate

There are several glyphosate resistance genes: *EPSPS*, *aro*A, *gat, gox* depending on the source of their isolation. All of them are used in developing transgenic crops.

EPSPS Gene

Glyphosate inhibits 5-enolpyruvylshikimate-3-phosphate synthase (EPSPS), the sixth enzyme involved in the shikimate pathway of biosynthesis of aromatic amino acids (tryptophan, tyrosine, and phenylalanine) and subsequently phenolics, liginins, tannins, and other phenylpropanoids. Another aromatic amino acid, tyrosine, can be synthesized from phenylalanine. Shikimate pathway, through which 20 percent of the carbon fixed by plants is routed, is also used by bacteria and fungi.

EPSPS enzyme belongs to the family of transferases specific to those transferring aryl or alkyl groups other than methyl groups. EPSPS, also called by its systematic name, phosphoenolpyruvate:3-phosphoshikimate 5-O-(1-carboxyvinyl)-transferase, catalyzes the chemical reaction (Fig. 5.1).

The monomeric EPSPS is composed of two domains, joined by a protein strand. This strand acts as a hinge, and can bring the two protein domains closer together. When a substrate binds to the enzyme, ligand bonding causes the two parts of the enzyme to clamp down around the substrate in the active site.

As a competitive inhibitor of the EPSPS enzyme, glyphosate resembles the transition state that transforms the reactants into products in the reaction that is catalyzed by EPSP synthase. As a transition state analog, glyphosate binds more tightly to EPSP synthase than

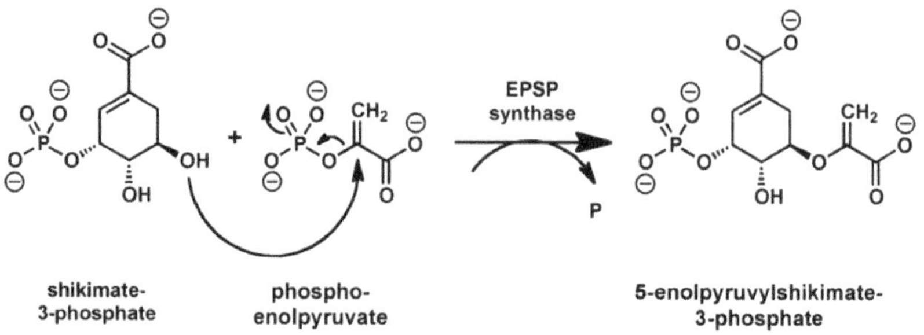

Fig. 5.1. Conversion of shikimate-3-phosphate (S3P) plus phosphophenol pyruvate (PEP) to 5-enolpyruvylshikimate-3- phosphate (EPSP) by 5-enolpyruvyl shikimate-3-phosphate synthase (EPSPS) [Boghog, Wikipedia Commons 2011].

its natural substrate and thereby preventing binding of substrate to the enzyme [Schönbrunn et al. 2001]. This binding leads to the inhibition of the enzyme, and consequently shuts down the entire shikimate pathway. This also means that glyphosate is not toxic to animals and humans since they are not dependent on the shikimate pathway for the synthesis of tryptophan, tyrosine, and phenylalanine which they obtain from their diet.

Resistance to glyphosate has been engineered into numerous crops since 1985 by introducing glyphosate-resistant variant EPSPS enzymes [Comai et al. 1985] or genetic constructions for the overproduction of EPSPS [Shah et al. 1986]. EPSPS proteins are universally present in plants and microorganisms. Although their sequences are variable, their chemical function is highly specific and conserved [CERA 2010].

Initial efforts in using EPSPS enzyme to develop transgenic crop lines were focused on identifying insensitive mutant versions of *EPSPS* gene from mutagenesis screens in bacteria. Genes encoding the chloroplast-localized EPSPS have been cloned from *Arabidopsis* [Klee et al. 1987], tomato [Gasser et al. 1988], petunia [Gasser et al. 1988], and tobacco [Wang et al. 1991]. Two distinct complementary DNAs (cDNAs) encoding EPSPS have been identified in tobacco cell cultures. One of these genes (*EPSPS-1*) was amplified in glyphosate-tolerant cultures [Wang et al. 1991]. Expression of the *Arabidopsis EPSPS* gene under the direction of the CaMV 35S promoter resulted in overproduction of *EPSPS,* leading to glyphosate resistance in transformed *Arabidopsis* calli and plants [Klee et al, 1987].

A variant of *EPSPS* is the *cp4 EPSPS* gene, derived from *Agrobacterium tumefaciens* strain CP4 [Barry et al. 1992a]. This gene is highly resistant to inhibition by glyphosate. It encodes a 47.6 kDa EPSPS protein consisting of a single polypeptide of 455 amino acids. The CP4 EPSPS protein expressed in transgenic varieties is functionally equivalent to endogenous plant EPSPS enzymes with the exception of its reduced affinity for the glyphosate molecule [CERA 2010].

The other variants of *epsps* gene include *mepsps*, *2mepsps*, *zm-epsps*, and *zm-2epsps* genes. The *mepsps* gene is a modified EPSPS enzyme which confers tolerance to glyphosate. The *2mepsps*, a double mutant version of the *epsps* gene, encodes 2mEPSPS protein. It is significantly insensitive to glyphosate inhibition (due to lower binding affinity for EPSPS substrates), but still retains its function in the shikimate pathway to continue aromatic amino acid synthesis as well as secondary metabolites. The *zm-epsps* gene is derived from maize (*Zea mays*) while *zm-2epsps* differs from the naturally occurring maize *epsps* gene by two amino acid substitutions, resulting in a protein with greater than 99 percent identity to that of the maize protein. Another gene, *epsps* (*Ag*), is derived from *Athrobacter globiformis*.

AroA Gene

Aromatic amino acid (AroA) is a key enzyme involved in aromatic amino acid biosynthesis. It catalyzes an unusual reaction between shikimate-3-phosphate (S3P) and phosphoenolpyruvate (PEP), with transfer of the carboxyvinyl moiety from PEP to S3P to form EPSP and inorganic phosphate [Herrmann and Weaver 1999]. Glyphosate inhibits AroA in a slowly reversible reaction, which is competitive versus PEP and noncompetitive versus S3P [Boocock and Coggins 1983; Steinrucken and Amrhein 1984]. Glyphosate binds to the binding site for PEP and forms a stable ternary complex with the enzyme and S3P (AroA-S3P-glyphosate) [Schönbrunn et al. 2001]. This ternary complex represents the actual enzyme-bound form of glyphosate, which is responsible for its herbicide activity [Sikorski and Gruys 1997].

There are two classes of AroA enzymes. The Class I AroA enzymes, derived from *E. coli*, *Aeromonas salmonicida*, *Petunia petunia*, and *A. thaliana*, are naturally sensitive to glyphosate. Their mutants with tolerance to glyphosate can be generated by different methods [Comai et al. 1983; He et al. 2001; Padgette et al. 1991]. For example, the highly glyphosate-tolerant AroA has been produced by a G96A substitution in the *aroA* gene of *E. coli* [Padgette et al. 1991]. However, glyphosate tolerance in class I enzymes is often paralleled by a decrease in the affinity of AroA synthase for PEP [Barry et al. 1992a; Sikorski and Gruys 1997].

Class II AroA enzymes, which share less than 30 percent amino acid identity with class I AroA enzymes, have been identified from *A. tumefaciens* CP4, *Bacillus subtilis*, and *Pseudomonas* sp. strain PG2982. In contrast to class I enzymes, class II enzymes usually have not only high affinity for PEP but also natural tolerance to glyphosate [Barry et al. 1992b].

The bacterial *aroA* gene, which codes for EPSPS, influences the virulence of a number of pathogenic microorganisms. Pathogenic bacteria with either defective or without *aroA* genes (i.e., *aroA⁻* mutants) are unable to produce aromatic intermediates and therefore are autotrophic, i.e., they are dependent upon the supply of aromatic substrates such as para-aminobenzoic acid. As humans and animals do not produce aromatic precursors, *aroA⁻* mutants of pathogens are unable to multiply in their bodies. In *E. coli*, the *aroA* genes are located in the chromosome region and distributed throughout the genome. Its amplification results in tolerance to glyphosate. In an amplified form, it can serve as a selectable glyphosate-resistance marker.

Mutants defective in EPSP synthase, are located on the standard *E. coli* map and require amino acids for growth. The use of *aroA* mutants allows specific isolation of the EPSP synthase gene by genetic complementation. The *EPSPS* gene is expressed constitutively. Therefore, cells that carry *aroA* gene in a multi-copy plasmid will over-synthesize EPSP synthase. If this enzyme is the target of glyphosate, the cells will be more tolerant. Cells that contain a multi-copy plasmid carrying the *EPSPS* gene overproduce the enzyme 5- to 17-fold and exhibit at least an eight-fold increased tolerance to glyphosate [Rogers et al. 1983]. EPSP synthase is a major site for glyphosate action in *E. coli* and it can, in an amplified form, serve as a selectable glyphosate resistance marker [Rogers et al. 1983].

GAT Gene

A different approach to inactivate glyphosate is by derivatization in which the herbicide is transformed into a product of similar chemical structure. This is achieved by *N*-acetylation of glyphosate, and the acetylated non-herbicidal product, *N*-acetylglyphosate, is stable not metabolized in plants. This strategy has been used to successfully insert phosphinothricin- or bialaphos-resistant gene (*pat*, *bar*) in transgenic maize, rice, cotton, soyabean, sugarbeet, rapeseed (canola), chicory (*Cichorium intybus*) [Castle and Lassner 2004].

Castle et al. [2004] looked for a similar enzyme capable of carrying *N*-acetylation of glyphosate to produce *N*-acetylglyphosate and found a strain of the common benign, Gram-positive, saprophytic *Bacillus licheniformis* having three alleles of a glyphosate *N*-acetyltransferase (*gat*) gene. However, this bacterium detoxified at rates too low to be of value if transferred to a transgenic crop. The native enzymes exhibit poor kinetic properties. In order to improve the kinetic efficiency of *gat* and to create improved enzymes that exhibit higher turnover rates and increased specificity to glyphosate, DNA shuffling was done to optimize kinetics of corresponding GAT proteins to acetylate the herbicide active ingredient glyphosate [Castle and Lassner 2004]. DNA shuffling [Stemmer 1994; Crameri et al. 1998; Ness et al. 2002] is a process that recombines genetic diversity from parental genes to create libraries of gene variants that are screened to identify those progeny with improved properties.

Shuffling is an effective technique for producing proteins with altered properties such as improved kinetics, altered substrate specificity, changes in temperature or pH optima, ligand binding, and solubility. At each shuffling stage, the most active genes were identified by cloning all variants in *E. coli* and assaying the recombinant colonies for GAT activity. The most active genes were then used as substrates for the next round of shuffling. After 11 rounds, a gene specifying a GAT with 10,000 times the activity of the enzyme present in the original *B. licheniformis* strain was obtained [Brown 2010]. The *gat* genes were introduced into the nuclear genomes of *Arabidopsis*, tobacco, and maize for the expression of proteins in the cytosol [Castle and Lassner 2004]. The resultant transgenic plants became resistant to even 6 to 20 times higher than the commercial application rate without any reduction in plant productivity. A variant of *gat* gene is the *gat4601* derived from the sequences of three weakly active *N*-acetyltransferase isozymes from *B. licheniformis* [Castle et al. 2004].

Another *gat* gene was derived from the same DNA shuffling process that was used to produce *gat4601* is *gat4621*. It was also isolated from *B. licheniformis*. The GAT4621 protein encoded by *gat4621* gene is 147 amino acids long with a molecular weight of 17 kDa. It is 75–78 percent identical and 90–91 percent similar at amino acid level to each of the three native GAT enzymes from which it was derived. It also has higher catalytic efficiency for glyphosate than GAT4601. Both GAT4601 and GAT4621 are 91 percent identical and 96 percent similar at the amino acid level.

GOX Gene

Another approach to inactivate glyphosate is accomplished by glyphosate oxidase (GOX), an enzyme which can actively break down glyphosate into two nontoxic compounds, aminomethylphosphonic acid (AMPA), the main metabolite, and glyoxalate. AMPA is mildly toxic, and under some conditions the AMPA accumulating in glyphosate-resistant soyabean correlates with glyphosate-caused phytotoxicity. The *gox* gene produces GOX enzyme. Neither an isolated plant GOX enzyme nor a gene for it has yet been reported in plants. A variant of the *gox* gene is *goxv247* (glyphosate oxidoreductase variant 247).

The *goxv247* gene encodes a single polypeptide of 431 amino acids with a molecular mass of 46.1 kDa. This gene has improved affinity for glyphosate and therefore degrades the herbicide more efficiently. The *goxv247* gene varies from the *gox* gene by only five nucleotides and the variant GOXv247 protein is 99 percent identical to the native GOX enzyme, differing by 3 amino acids out of 400 [Woodward et al. 1994].

A *GOX* gene isolated *Ochrobactrum anthropi* strain LBAA [Barry et al. 1992b] has been used as a selectable marker gene in tobacco, *Arabidopsis*, potato, and sugarbeet [Barry and Kishore 1995]. It is not employed alone but rather used in combination with *cp4 EPSPS*, and when both are targeted to the plastids, the effectiveness of conferred tolerance is higher [Zhou et al. 1995; Mannerlöf et al. 1997]. A combination of these two genes has been used to transform wheat [Zhou et al. 1995] and sugarbeet [Mannerlöf et al. 1997]. A bacterial *GOX* is used in glyphosate-resistant rapeseed, and an altered bacterial glyphosate *N*-acetyltransferase is planned for a new generation of glyphosate-resistant crops.

Glufosinate/Phosphinothricin

A natural herbicide bialaphos (L-Alanyl-L-alanyl-phosphinothricin), also known as phosphinothricin tripeptide (PTT), is an antibiotic product of some strains of the genus *Streptomyces* (Actinobacteria: Actinomycetales). This unique tripeptide, composed of two L-alanine residues and one glutamic acid analogue moiety (L-phosphinothricin having C-P-C bond which is rare in natural compounds) as shown in Fig. 5.2, is also called glufosinate {2-amino-4-(hydroxymethylphosphinyl) butanoate}.

Phosphinothricin (PPT) is the active ingredient in bialaphos which inhibits glutamine synthetase (GS), the key enzyme in the nitrogen assimilation pathway (vide Chapter 2), by competitively binding in place of the normal substrate, glutamate (glutamic acid). This prevents the synthesis of L-glutamine, which is not only an important chemical precursor for the synthesis of nucleic acids and proteins, but serves as the mechanism of ammonia incorporation for plants.

PPT becomes active after removal of the alanine residues by intracellular peptidases. The remaining glufosinate {4-(hydroxy(methyl) phosphonoyl) butanoic acid}, the structural analog of glutamate, inhibits GS, leading to rapid accumulation of toxic levels of ammonia and a concomitant depletion of glutamine and several other amino acids in both bacteria and plant cells. These effects in plants are accompanied by a rapid decline of photosynthetic CO_2-fixation, followed by chlorosis and desiccation.

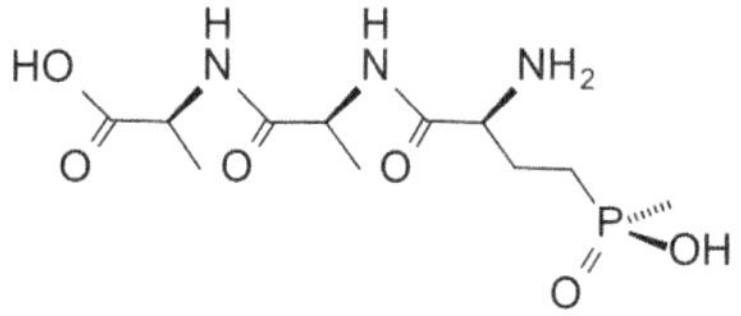

Fig. 5.2. Structure of Phosphinothricin-alanyl-alanine (Bialaphos). [Hennix, Wikipedia Commons 2007].

BAR, PAT Genes

Some microorganisms can detoxify glufosinate by producing an enzyme that causes acetylation of the amino group. A *bar* (bialaphos resistance) gene encodes a phosphinothricin acetyl transferase (PAT) enzyme. This *bar* gene was isolated from *Streptomyces hygroscopicus*, an organism which produces the tripeptide bialaphos as a secondary metabolite [Thompson et al. 1987], as well as from *Streptomyces viridochromogenes* [Wohlleben et al. 1988]. A *bar* gene has also been isolated from *Alcaligenes faecalis*.

PAT is encoded by either the *bar* [Thompson et al. 1987] gene or *pat* [Strauch et al. 1988] gene, and both detoxify PPT by acetylation. The PAT enzymes encoded by these two genes are functionally identical and show 85 percent identity at the amino acid level and 87 percent similarity as the nucleotide level, and both require some codon optimization for expression in plants [Wohlleben et al. 1988; Wehrmann et al. 1996]. The only recorded differences in activity between the two proteins are minor differences in the optimal pH, and a significantly different affinity for acetyl-coA (a co-substrate); these differences are not expected to be meaningful *in planta* [Wehrmann et al. 1996].

PPT resistant crops have been obtained by expressing chimeric *bar* or *pat* genes in the cytoplasm from nuclear genes. Both genes, which code for polypeptides of 183 amino acids, are widely used in plant transgenic engineering as selectable marker genes, reporter genes in chimeric gene constructs, and dominant genes for engineering herbicide-resistance in crop species. There are no differences in phenotype or properties of gene products in plants expressing these *pat* and *bar* genes [Wohlleben et al. 1988; Wehrmann et al. 1996]. Several crop plants have been transgenically modified for resistance to glufosinate over the past two decades.

Bromoxynil

Bromoxynil (3,5-dibromo-4-hydroxybenzonitrile) is a benzonitrile contact herbicide used for Post-emergence broad-leaf weed control in a variety of crops including wheat, barley, maize, sorghum, onion, garlic, and certain varieties of lucerne (alfalfa).

This PS II-inhibiting contact herbicide, affecting the electron transport complex (vide Chapter 2), is formulated as an octanoate ester. In resistant crops, bromoxynil is detoxified by hydrolysis of the nitrile group, first to 3,5-dibromo-4-hydroxybenzamide and then to 3,5-dibromo-4-hydroxybenzoic acid [Buckland et al. 1973]. These metabolites can be further degraded by other enzymes that cleave bromine atoms and subsequently attack the benzoic acid ring structure. The two metabolites are at least 100-fold less toxic to plant cells than bromoxynil. Therefore, conversion of bromoxynil to either of these metabolites should inactivate the compound and result in bromoxynil-resistant phenotypes.

In the soil, bromoxynil is rapidly degraded via pathways similar to those in resistant plants. The degradation bypasses the amide derivative of bromoxynil and proceeds directly to form 3,5-dibromo-4-hydroxybenzoic acid in a single step [Rosenbrock et al. 2004; Hsu and Camper 1975]. This is due to the activity of the soil bacterium *Klebsiella pneumoniae* subsp. *ozaenae* which detoxifies bromoxynil via hydrolysis of nitrile group. This reaction is catalyzed by a bromoxynil-specific enzyme 'nitrilase' which does not release bromoxynil from the catalytic site during conversion to 3,5-dibromo-4-hydroxybenzoic acid [McBride et al. 1986; Stalker et al. 1988a].

Nitrilase acts on the cyno moiety of nitrile-containing compounds. It detoxifies bromoxynil in the cytosol even before it reaches its site of action in the chloroplast. Although nitrilase enzymes are common in many organisms, they are present only in certain plants belonging to Gramineae, Brassicaceae, and Musaceae families. This explains why weeds belonging to these families are resistant to bromoxynil. The dicot crops such as tobacco, cotton, and potato do not have nitrilase activity and this makes them sensitive to this herbicide.

BXN Gene

The *bxn* gene from *Klebsiella ozaenae*, which encodes the bromoxynil-specific nitrilase has been cloned [Stalker et al. 1988b]. This gene encoding 37,000-mol wt polypeptide was cloned and expressed in *E. coli* [Stalker et al. 1988a]. This chimeric bacterial gene has been constructed for expression in plants by using a plant promoter derived from a tobacco light-inducible tissue-specific ribulose biphosphate (RuBP) small subunit (SSU-mRNA) which expresses only in photosynthetic tissue [Stalker et al. 1988b]. Transgenic tobacco plants expressing the bromoxynil-specific nitrilase in photosynthetic tissues were resistant up to eight times the commercial application rate, and the resistance trait was stably inherited in the succeeding generations [Stalker et al. 1988b].

ALS-(AHAS) Inhibitors

Acetolactate synthase (ALS), also called acetohydroxyacid synthase (AHAS), is the primary target site for at least five structurally distinct classes of herbicides including sulfonylureas, imidazolinones, triazolopyrimidines, sulfonyl-aminocarbonyl triazolinones, and pyrimidinyl thiobenzoates. ALS, to which these herbicides bind, catalyzes the first common step in the branched-chain amino acid biosynthesis in chloroplasts leading to leucine, isoleucine, and valine (vide Chapter 2). Sulfonylurea herbicides are broad-spectrum selective, post-emergence herbicides, with high potency at very low rates, making them environmentally safe.

The imidazolinones (IMIs) are a group of broad spectrum pre-emergence and post-emergence herbicides. These include the non-selective imazapyr and imazapic and the selective imazaquin, imazethapyr, imazamethabenz, and imazamox. All IMIs, which have an imidazole moiety in their molecular structure, are divided into three groups based on the second cyclic structure of their molecules excluding the imidazole ring. Imazaquin and imazamethabenz have quinoline moiety and benzene ring respectively. The other four IMIs (imazapyr, imazapic, imazethapyr, imazamox) have the pyridine ring, but differ only at position five (having hydrogen, methyl, ethyl, and methoxymethyl functional groups respectively), and these are the ones used to evolve IMI resistant transgenic crops [Tan et al. 2005].

ALS is a nuclear-encoded enzyme that moves to its active site in the chloroplast with the help of a chloroplast-transit peptide. This enzyme has two substrates, 2-ketobutyrate and pyruvate, which are used to produce the essential branched-chain amino acids valine, leucine, and isoleucine. Binding of ALS blocks substrate access to a deeply buried active site [McCourt et al. 2006].

ALS and *csr*1-1 Genes

Resistance to ALS-inhibitors is conferred by a single, dominant, or at least partially dominant, nuclear-encoded gene, *ALS*. This gene codes for an enzyme of 670 amino acids with a molecular weight of about 73 kDa (vide Chapter 2). Plant *ALS* genes encoding the catalytic (large) subunits were first isolated from *Arabidopsis thaliana* and tobacco utilizing the yeast ALS gene as a heterologous hybridization probe [Mazur et al. 1987]. Since then, some plant *ALS* genes encoding catalytic subunits have been cloned and characterized [Bernasconi et al. 1995; Fang et al. 1992; Grula et al. 1995; Rutledge et al. 1991]. The plant ALS regulatory small subunit enhances the catalytic activity of the large subunit and confers sensitivity to feedback inhibition by branched-chain amino acids [Lee and Duggleby 2001]. *ALS* genes encoding catalytic small subunits have been cloned from some of these plants, and their sequences were found to possess single or double mutations. These mutated *ALS* (*mALS*) genes confer resistance to ALS-inhibiting herbicides [Shimizu et al. 2011].

The most commonly occurring mutations that confer resistance to these herbicides occur at positions of Ala122, Pro197, Ala205, Trp574, and Ser653 (vide Chapter 2) [Tan et al. 2005; Heap 2014]. However, it is the mutation at Ser653 that confers resistance to IMIs, not to other ALS-inhibiting herbicides. In contrast, mutation at Pro197 generally confers good resistance to sulfonylureas. The mutation at Trp574 is generally cross-tolerant to different ALS-inhibiting herbicide families, while mutation at Ala205 exhibit acceptable tolerance to IMI-resistant crops. The majority of commercialized IMI-resistant crops are currently developed from either one or a combination of Ala205, Trp574, and Ser653 mutations [Tan et al. 2005].

The target tissue used to develop IMI-resistant line depends on the crop. In the case of maize, cell cultures and pollen were employed for mutagenesis, while microspores were used in rapeseed. By comparison, IMI-resistant rice and wheat varieties were developed from the mutagenesis of seeds (vide Chapter 6). Unlike these crops, IMI-resistant sunflower lines were obtained by selecting naturally occurring resistant mutants in wild sunflower and transferring the trait to cultivated types. Because the IMI resistance was achieved without inserting foreign DNA, all commercialized IMI-resistant crops are non-transgenic, and marketed as non-transgenic crop varieties [Tan et al. 2005].

Sulfonylureas inhibit ALS as a competitive inhibitor [Chaleff and Mauvais 1984]. A mutant ALS gene, *csr*1-1, (chromosome-segregation and RNAi deficient), was isolated from sulfonylurea-resistant mutant line GH50 of *A. thaliana* [Haughn and Somerville 1986]. The mutant was caused by a single base conversion from C to T at nucleotide 870 relative to the initiation codon in the ALS coding sequence and leads to proline and serine substitution in a conserved region of the ALS amino acid sequence [Haughn et al. 1988; Lee et al. 1988]. The mutant enzyme, which has much lower affinity to sulfonylureas in *Arabidopsis*, has been used to transform tobacco and rapeseed (canola) and to confer sulfonylurea resistance to heterologous dicotyledon cells [Charest et al. 1990; Gabard et al. 1989; Miki et al. 1990; Odell et al. 1990]. The mutant *ALS* gene is efficiently expressed in transgenic rice cells when placed under the control of CaMV 35S promoter, thus conferring upon transformed cells, at least 200-fold greater resistance to chlorsulfuron than non-transformed control cells [Li et al. 1992]. This suggested that the mutant *ALS* gene of *A. thaliana* is an alternative selectable marker gene to the hygromycin resistance gene for rice transformation and fertile transgenic rice production [Li et al. 1992].

Although the *csr1-1* gene has been widely used to generate sulfonylurea-resistant crops, some weeds resistant to this herbicide group have appeared in the field. This increased resistance is due to occurrence of mutations in *csr1-1*, the *ALS* gene [Cui et al. 2008]. The rice *CYP81A6* gene, which mediates sulfonylurea tolerance through a different mechanism from that of the mutant *ALS*, appears to be a viable alternative to produce transgenic crop plants resistant to some sulfonylureas.

GM-ALS Gene

The *ALS* gene is isolated from numerous organisms. The one derived from *Glycine max* (soyabean) is designated '*GM-ALS*' gene which expresses GM-ALS protein. The *gm-als* gene is used as terminator while producing transgenic crop varieties.

GM-HRA Gene

The *gm-hra* gene, which encodes GM-HRA protein (*Glycine max* herbicide-resistant acetolactate synthase) is a modified *als* gene that was produced by site-specific mutagenesis of two amino acids of the endogenous soyabean *gm-als* gene. The GM-HRA protein is > 99 percent homologous to the native GM-ALS protein from which it was derived. Compared to the GM-ALS I protein, the GM-HRA protein sequence contains two amino acid substitutions important for tolerance to the ALS inhibiting class of herbicides, including chlorsulfuron, and five additional N-terminal amino acids derived from the translation of 15 nucleotides of the *gm-als* 5' untranslated region [USFDA 2007]. While soya bean expressing the GM-HRA protein is considered to be transgenic, it would actually be more accurate to consider it cisgenic since the *gm-hra* gene originated from soyabean.

The *in planta* expression of the *gm-hra* gene is controlled by the promoter derived from the S-adenosyl-L-methionine synthetase (*SAMS*) gene from soya bean, and an intron that interrupts the *SAMS* 5' untranslated region (5'-UTR). Termination of transcription of the *gm-hra* gene is under the control of the native soyabean *gm-als* terminator [USFDA 2007].

Sethoxydim

Sethoxydim, belonging to cyclohexanedione herbicide group, is applied early post-emergence to control perennial and annual grasses, while being effective against certain broadleaf weeds. The affected plants cease to grow within 2–3 days after application, with growing tissue in the nodes and buds becoming necrotic. Young leaves are affected first, turning yellow (1–3 wk) and then brown. Susceptible plants are typically dead within 2–4 weeks and those not completely dead may show excessive tillering.

Cyclohexanediones, like aryloxyphenoxypropionics (AOPP) and phenylpyrazoline (PPZ), inhibit acetyl-CoA carboxylase (ACCase), a key enzyme in fatty synthesis in plants (vide Chapter 2). It blocks the production of phospholipids, the products of fatty acid biosynthesis, used in building new membranes required for cell growth. Broadleaf weeds are resistant to these herbicides because of an insensitive ACCase. Similarly, natural tolerance of some grass species appears to be due to a less sensitive enzyme [Stoltenberg 1989].

ACCase I, ACCase II Genes

Two distinct isoforms of *ACCase* gene, *ACCase* I and *ACCase* II, have been found in maize. *ACCase* I occurs in the plastids of mesophyll cells, and is sensitive to inhibition by sethoxydim and haloxyfop, while *ACCase* II, not occurring in plastids, is relatively insensitive to, or less inhibited by them [Tan and Bowe 2012]. Besides, two very similar partial cDNAs coding for the multifunctional ACCase (pA3 and pA4) enzyme have been sequenced [Ashton et al. 1994].

The DNA analysis of sethoxydim- and haloxyfop-resistant maize DK592 showed that the *ACCase* gene, presumably the *ACCase* I gene, has a single nucleotide substitution at codon 1781 in reference to blackgrass (*Juncus gerardii*) or at codon 1769 in reference to wheat [Tan and Bowe 2012]. Codon nucleotides T-T-A replaced A-T-A and, as a result, the encoded ACCase enzyme has the I1781L (Ile-1781-Leu) amino acid substitution. The alteration of the *ACCase* prevents some AOPP and CD herbicides from binding the enzyme and makes the plant resistant to the herbicides [Tan and Bowe 2012]. Sethoxydim-resistant maize, the first ACCase-resistant crop, was not transgenic but cisgenic as it had an altered ACCase enzyme that was generated from tissue culture selection [Somers 1996].

2,4-D

Auxinic herbicides (vide Chapter 2) act rapidly at multiple receptors and compete with an essential plant hormone pathway, making crops resistant by modifying the site of auxin action difficult. In addition, these receptors respond differently to different auxin herbicide classes.

Axr1 Genes

The Auxinic 2-4-D acts by promoting the degradation of transcriptional regulators called Aux/IAA proteins. The mechanism of resistance to 2,4-D is considered to be due to mutations caused by certain dominant and recessive genes. Resistance of *Arabidopsis thaliana* to 2,4-D was due to a recessive mutation, caused by a single gene, *axr1*; the mutant producing bushy and short plants, with small, thin roots and smaller leaves [Estelle and Somerville 1987]. The *axr1* (auxin-resistance 1) gene may encode for a different auxin receptor, and resistance may be due to an alteration that has a greater effect on the affinity of this receptor for 2,4-D and IAA. Of the 20 *axr1* mutants found in *A. thaliana*, at least five different *axr1* alleles had similar phenotypes, and the extent of auxin resistance of each mutant line could be directly correlated with severity of the morphological alterations [Lincoln et al. 1990].

The mutant *axr4-2* disrupts auxin transport into cells, thereby making a plant resistant to 2,4-D [Gleason et al. 2011]. Of the five auxin-signaling F-box receptor mutants, *tir1-1*, *afb1*, *afb2*, *afb3*, and *afb5*, four of them (*tir1-1*, *afb1*, *afb2*, *afb3*) showed resistance to 2,4-D. The F box protein TIR1 (Transport Inhibitor Response) is required for Aux/IAA degradation. Besides, three additional F box proteins (AFBs), called AFB1, AFB2, and AFB3, also regulate auxin response [Dharmasiri et al. 2005]. Like TIR1, these proteins interact with the Aux/IAA proteins in an auxin-independent manner. Plants deficient in all four proteins are auxin insensitive and exhibit a severe embryonic phenotype similar to the *mp/arf5* and *bdl/iaa12* mutants [Dharmasiri et al. 2005]. Correspondingly, all TIR1/AFB proteins interact with Aux/IAA BODENLOS (BDL), and BDL is stabilized in triple mutant plants. TIR1 and the AFB proteins collectively mediate auxin responses throughout plant development [Dharmasiri et al. 2005].

*Tfd*A Gene

Numerous strains of soil bacteria contain plasmid-borne genes that confer the ability to catabolize 2,4-D. One such gene, *tfd*A (2,4-dichlorophenoxyacetate/α-ketoglutarate dioxygenase), encodes 2,4-D monooxygenase which converts 2,4-D into 2,4-dichlorophenol (DCP) and glyoxylate by cleavage of the aliphatic side chain [Streber et al. 1987]. In *Alcaligenes eutrophus*, 2,4-D degradation pathway comprises six specialized enzymes, encoded by a 75 kb plasmid designated pJP4. These enzymes convert 2,4-D to chloromaleylacetic acid, a compound subsequently metabolized to succinic acid; the genes for these enzymes include *tfd*A, *tfd*B, *tfd*C, *tfd*D, *tfd*E, and *tfd*F [Don et al. 1985].

The chimeric *tfd*A genes were cloned directly into binary vector pGA470 [An et al. 1985] or pBin19 [Bevan 1984] which was conjugally transferred into disarmed strains of *Agrobacterium tumefaciens*, or cloned indirectly first into an intermediate vector that was transformed to *Agrobacterium* and recombined into disarmed Ti plasmid. The binary plasmids are self-replicating and carry right and left T-DNA borders that direct integration into the plant genome, and a kanamycin resistance gene that allows the recovery of transformed plant cells [Llewellyn and Last 1996].

The transfer of modified versions of this gene into tobacco conferred resistance to 2,4-D [Lyon et al. 1989; Streber and Willmitzer 1989]. Subsequent biochemical analysis indicated that this *tfd*A gene product is a ferrous iron-dependent dioxygenase that uses α-ketoglutarate as a co-substrate [Fukumori and Hausinger 1993]. The α-ketoglutarate is converted to succinate and carbon dioxide concomitant with the conversion of 2,4-D to DCP and glyoxylate.

Aad Genes

A family of *aad* genes that code for aryloxyalkanoate dioxygenase (AAD) provides resistance to certain auxin herbicides [Müller et al. 2006; Schleinitz et al. 2004]. These genes have low homology with *tfdA* genes. The *aad* gene isolated from a Gram-negative soil bacteria, *Sphingobium herbicidovorans,* codes for a Fe(II) and 2-ketoglutarate-dependent dioxygenase that degrades the alkanoate side chain of the synthetic auxin herbicide 2,4-D to a hydroxyl [Fukumori and Hausinger 1993; Wright et al. 2005]. Various *aad* gene sequences code 2-ketoglutarate–dependent dioxygenases that differentially inactivate the two herbicide types. Some are active on phenoxyalkanoate auxins and aryloxyphenoxypropionate (AOPP), with others being inactive on both. One gene, called *aad-12*, isolated from *Delftia acidovorans* (previously identified as *Pseudomonas acidovorans and Comamonas acidovorans*), codes for a 2-ketoglutarate-dependent dioxygenase that inactivates phenoxyacetate auxins such as 2,4-D and pyridinyloxyacetate auxins such as triclopyr and fluoroxypyr, but not AOPPs [Wright et al. 2007, 2010].

A second gene known as *aad-1,* derived from *Sphingomonas herbicidovorans*, inactivates auxins and ACCase-inhibiting herbicides by side-chain degradation in some AOPP herbicides (e.g., fluazifop) [Wright et al. 2010]. This trait, DHT1, is being developed in maize. It degrades the alkanoate side chains to a hydroxyl of AOPP herbicides (e.g., quizalofop) [Wright et al. 2010]. DHT1 maize reportedly tolerates post-emergence applications of quizalofop of up to 184 g ha^{-1} with no adverse agronomic effects. Both DHT1 and DHT2 (Dow AgroSciences Herbicide Tolerance 1 and 2), which provide resistance to high rates of 2,4-D, are combined with glyphosate and glufosinate resistance traits in maize and soya bean [Simpson et al. 2008].

HPPD Inhibitors (Isoxaflutole, Mesotrione)

The HPPD-inhibiting herbicides are, in some ways, ideal to complement glyphosate. Many HPPD (4-hydroxyphenyl-pyruvatedioxygenase) herbicides have soil residual activity and control key broadleaf weeds that have already evolved resistance to glyphosate. Resistance mechanisms for HPPD herbicides include a less sensitive target site, overexpression of the enzyme, alternate pathway, increasing flux in the pathway, and metabolic inactivation [Matringe et al. 2005; Green and Castle 2010].

Isoxaflutole is a selective preplanting and preemergence herbicide belonging to the isoxazole family. It inhibits the action of 4-HPPD, an Fe(II)-dependent nonheme oxygenase enzyme, which converts 4-hydroxyphenyl pyruvate to homogentisic acid, a key step involved in degradation of tyrosine and biosynthesis of tocopherols and plastoquinones [Pallet et al. 1998; Meazza et al. 2002]. This is a key step in plastoquinone biosynthesis and its inhibition of HPPD results in depletion of carotenoids. Biosynthesis of plastoquinone, a cofactor of phytoene desaturase, is a key factor in the synthesis of carotenoid pigments which protect a plant's chlorophyll from decomposition by sunlight. Inhibition of HPPD enzyme causes disruption of the biosynthesis of carotenoids leading to destabilization of photosynthesis followed by bleaching of the foliage and death of the plant.

HppdPfW336, Ahvppd-03 Genes

The coding sequence of the 4-HPPD protein was isolated from the *Pseudomonas fluorescens* strain A32. The primers in the amplification were based on amino acid sequence of the HPPD protein present in *Pseudomonas* strain P.J. 874. The isolated DNA sequence was modified using site-directed mutagenesis to improve resistance against HPPD inhibitors by the replacement of amino acid glycine with tryptophan at position 336 [Boudec et al. 2001]. The modified protein designated as HPPDW336 is encoded by *hppdPfW336* gene. This 40 kDa protein consists of 358 amino acids. Soyabean is the first crop to engineer for isoxaflutole-resistance when the *hppdPfW336* gene was incorporated into the glyphosate-isoxaflutole stacked FG72 variety in 2009 (vide Chapter 6).

Like isoxaflutole, mesotrione, a triketone herbicide, also interferes with normal plant metabolism by tightly inhibiting plant HPPD. But in the case of mesotrione, it is the *ahvppd-03* gene that encodes HPPD enzyme, AvHPPD-03. In March 2013, Syngenta produced a double-herbicide (glufosinate-cum-HPPD-inhibitor mesotrione) stacked soya bean variety SYHT0H2 (vide Chapter 6).

Phenmedipham

Phenmedipham of phenylcarbamate group is a selective postemergence herbicide, with translocation primarily in the apoplast. It is applied in sugar beet crop preferably at 2-leaf stage of weeds for best results. It is effective against a broad-range of weeds, mainly broadleaf species besides a few annual grasses (*Setaria viridis, Setaria glauca*).

PED Gene

Phenylcarbamate herbicides, which inhibit electron transport complex in PS-II (vide Chapter 2), are hydrolyzed in the soil by bacteria. In the case of phenmedipham, the major metabolite is methyl-N-(3-hydroxyphenyl) carbamate (MHPC) followed by 3-aminophenol. Hydrolysis reaction is catalyzed by phenylcarbamate hydrolase (PMPH) enzyme, a monomer with a molecular weight of 55 kDa. This enzyme detoxifies phenmedipham through hydrolytic cleavage of the carbamate bond between the benzene rings [Pohlenz et al. 1992]. The gene responsible for this degradation of phenmedipham is *pcd*. This gene is isolated from the bacterium *Arthrobacter oxydans* strain P52.

Pohlenz et al. [1992] determined two partial amino acid sequences that allowed the localization of the coding sequence of the *pcd* gene on a 3.3-kb PstI restriction fragment within pHP52 DNA by hybridization with synthetic oligonucleotides. The *p*-Cresol methylhydroxylase (PCMH) enzyme was functionally expressed in *E. coli* under control of the *lacZ* promoter after the 3.3-kb PstI fragment was sub-cloned into the vector pUC19. Sequence analysis revealed an open reading frame of 1,479 bases containing the amino acid partial sequences. Sequence comparisons revealed significant homology between the *pcd* gene product and the amino acid sequences of esterases of eukaryotic origin. Subsequently, it was demonstrated that the esterase substrate p-nitrophenylbutyrate is hydrolyzed by phenmedipham hydrolase.

Paraquat

Paraquat, belonging to bipyridilium group, is a nonselective contact herbicide. Being a divalent cation, paraquat short-circuits the light-driven photosynthetic electron transport in PS I by accepting electrons that normally flow to ferredoxin (vide Chapter 2). Paraquat, also called methyl viologen (mv), is a powerful propagator of highly toxic superoxide radicals, catalyzed by the enzyme superoxide dismutase.

MvrA, MvrC, EmrE, PqrA Genes

Paraquat resistant genes such as *mvr*A (*mv*-resistant A) and *mvr*C (*mv*-resistant C) have been isolated from a paraquat-sensitive *E. coli* mutant [Morimyo 1988; Morimyo et al. 1992]. The *mvr*A gene encodes an enzyme protein which plays a role in reducing the toxic species produced by paraquat [Morimyo 1988], and the *mvr*C gene encodes a membrane protein which prevents the incorporation of paraquat into cells, and they thus allow the host to exhibit specific resistance to paraquat [Morimyo et al. 1992]. Besides, *emr*E gene (previously called also *mvr*C gene) is a small (110 residues) SMR (Small Multidrug Resistance) transporter isolated from *E. coli*; it confers resistance to paraquat by actively removing it during proton exchange [Yerushalmi et al. 1995]. The *emr*E gene was identified and cloned from the genome of *E. coli* on the basis of its ability to confer resistance to ethidium [Purewal 1991] and to methyl viologen (paraquat) [Morimyo et al. 1992].

   Another gene, *pqr*A, confers paraquat resistance to the heterologous host *E. coli*. Won et al. [2001] isolated, cloned, and analyzed this gene from a chromosomal DNA library of the paraquat-resistant *Ochrobactrum anthropic* JW2 from orchard soil. It was identified as a Gram-negative, aerobic, non-fermentative, and non-fastidious bacillus that was positive for oxidase and urease production [Won et al. 2001; Kern et al. 1993]. Cells of *E. coli* transformed with a plasmid carrying the *pqr*A gene have elevated resistance to paraquat, but not to hydrogen peroxide [Won et al. 2001]. Resistance to paraquat was by reducing the permeability or enhancing the efflux of paraquat, not by detoxifying superoxide radicals while the PqrA protein is a membrane protein that plays an important role in protecting cells against paraquat toxicity [Won et al. 2001].

Dalapon

Dalapon (2,2-dichloropropionic acid), a chlorinated aliphatic acid, is a systemic herbicide that controls both annual and perennial grasses. It is selective at lower rates and non-selective at higher rates. Transgenic resistance to dalapon may be induced by a) modifying the target site in the plant and render it insensitive to the herbicide and by b) introducing an enzyme or enzyme system into the plant to detoxify the herbicide before it reaches target site [Botterman and Leemans 1988].

DehI Gene

Dehalogenases are key enzymes in the metabolism of halo-alkanoic herbicides. Many soil bacteria utilize dalapon as a source of carbon and energy and this suggests that dehalogenase genes are well distributed across microbial genera [Wong and Huyop 2012]. One such microbe is *Pseudomonas putida* from which a dehalogenase gene, *deh*I, was identified [Beeching et al. 1983]. Strain PP3 of this bacterium produces two hydrolytic dehalogenases encoded by *deh*I and *deh*II, which are new members of different *deh* gene families. Of the two, *deh*I is most active against dalapon. The 9.74-kb DEH transposon containing *deh*I and its cognate regulatory gene, *deh*R$_1$, was isolated from strain PP3 by using the TOL plasmid pWW0 [Weightman et al. 2002]. The *deh*I gene, which degrades dalapon by detaching the chlorine ions from the herbicide structure, has also been expressed in *E. coli*.

The dehalogenase gene together with a kanamycin-resistance gene, under the control of CaMV 35S and *nos* promoters respectively, were cloned into a plant transformation vector to give plasmid pVW291 [Buchanan-Wollaston et al. 1992]. When this plasmid was transferred into *Nicotiana plumbagimfolia* via *Agrobacterium*-mediated transformation, the subsequent transgenic calli and plants were resistant to dalapon toxicity, with some individual plants showing resistance to a level of 110 kg ha^{-1} of the herbicide [Buchanan-Wollaston et al. 1992].

Dicamba

Dicamba (3,6-dichloro-2-methoxybenzoic acid) is a widely used benzoic acid herbicide, applied at pre-emergence and post-emergence, to control annual and perennial broadleaf weeds in maize, sorghum, small-grain crops, sugarcane, etc. Maize is naturally tolerant to dicamba, but the dicots soya bean, cotton, tobacco, potato, sunflower, etc. are sensitive. It is readily degraded by soil microbes as a source of carbon. The first chemical step in the mineralization of dicamba by *Stenotrophomonas* (formerly *Pseudomonas*) *maltophilia*, strain DI-6 is catalyzed by the O-demethylating enzyme dicamba monooxygenase, which oxidizes dicamba to the herbicidally-inactive compounds 3,6-dichlorosalicylic acid (DCSA) and formaldehyde (Fig. 5.3) [Behrens et al. 2007; Dumitru et al. 2009].

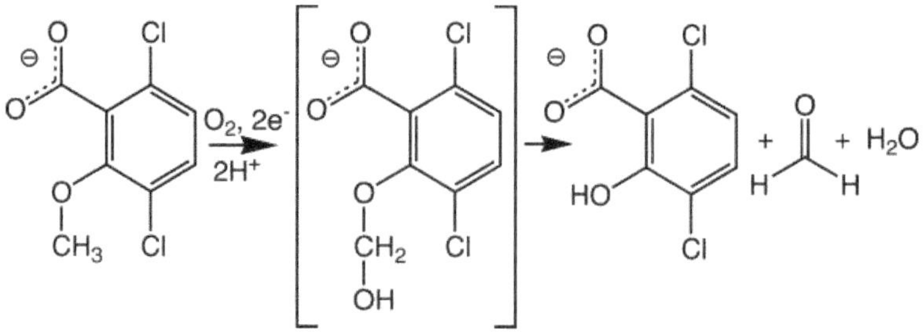

Fig. 5.3. Dicamba catalyzd by dicamba *O*-demethylase (DMO) enzyme to 3,6-dichlorosalicylic acid through an intermediate [Dumitru et al. 2009].

Dmo Gene

The enzyme responsible for the first breakdown step in a three-component system is called dicamba *O*-demethylase (DMO). Therefore, this demethylase gene, *dmo*, is a potential resource for producing an enzyme capable of inactivating dicamba in transgenic plants. Genetic engineering of *dmo* for expression in higher plants has allowed production of crop plants that are resistant to treatment with dicamba [Behrens et al. 2007]. Resistance is due to the ability of *dmo* to efficiently convert dicamba into DCSA in transgenic plants before the applied herbicide can build to toxic levels within plant cells.

This enzyme (*O*-demethylase), also known as dicamba monooxygenase, is a terminal Rieske nonheme oxygenase of a three-component enzyme system consisting of a monooxygenase, a ferredoxin, and a reductase. It catalyzes the NADH-dependent oxidative demethylation of dicamba. Electrons are shuttled from the reduced NADH through the reductase to the ferredoxin, and finally to the terminal component DMO [Herman et al. 2005; Chakraborty et al. 2005; Wang et al. 1997]. The ferredoxin component of dicamba closely resembles that found in plant chloroplasts. DMO is effective when expressed from the nucleus with a chloroplast transit peptide (CTP). Expression of DMO is better with a chloroplast transit peptide and best when present in the chloroplast genome where the monooxygenase would have a steady stream of electrons from reduced ferredoxin produced by photosynthesis and where transgene proteins can be expressed at higher levels.

Atrazine

AtzA Gene

In plants, atrazine, like other triazine herbicides, affects Photosystem II by inhibiting electron transport from Q_A to Q_B. The D1-protein is encoded by a chloroplast gene called *psbA*. Atrazine chlorohydrolase (AtzA) is an Fe(II) metalloenzyme belonging to the amidohydrolase superfamily. In the soil, it is the first enzyme in a six-step pathway leading to degradation of atrazine in Gram-negative bacteria. AtzA enzyme catalyzes the hydrolytic dechlorination and detoxification of atrazine to hydroxyatrazine [Wang et al. 2005]. The gene that encodes the atrazine chlorohydrolase enzyme is "*atzA*" which rapidly transforms atrazine to CO_2 and NH_3. The soil bacterium *Pseudomonas* sp. strain ADP uses atrazine as the sole source of nitrogen via hydroxyatrazine for growth and liberates the ring carbon atoms as carbon dioxide [Mandelbaum et al. 1995]. The genetic potential of this bacterium was attributed to a 1.9-kb *Ava*I DNA fragment from strain ADP [De Souza et al. 1995]. This and many other Gram-negative bacteria [De Souza et al. 1998b; Sadowsky and Wackett 2000; Wackett et al. 2002] initiate atrazine catabolism by a plasmid-borne *atzA* gene [De Souza et al. 1996; Martinez et al. 2001], encoding hydrolytic dechlorination of atrazine, resulting in the production of hydroxyatrazine as the first intermediate in the degradation pathway. This non-herbicidal and biodegradable hydroxyatrazine is more strongly sorbed to soils than atrazine [Clay and Koskinen 1990]. This 1,419 nucleotides-long *atzA* gene encodes a polypeptide of 473 amino acids with a predicted molecular weight of 52,421 [De Souza et al. 1996].

The wild-type *atzA* gene which doesn't express *in planta* to degrade atrazine in lucerne (alfalfa) does so when *p-atz*A is modified by changing 359 nucleotides (representing changes to 312 codons) [Sadowsky et al. 2005; Wang et al. 2005]. Uefuji et al. [2005] studied if transgenic plants expressing bacterial *atzA* transformed atrazine to hydroxyatrazine *in planta*. The wild type bacterial *atzA* gene was modified for plant codon usage, and its translation initiation site to produce *p-atzA*. The *p-atzA* was inserted between the CsVMV (cassava vein mosaic virus) promoter and the *nos* terminator in the binary vector pILTAB381. The resultant gene construct was used for *Agrobacterium*-mediated plant transformation of tobacco, lucerne, and *Arabidopsis*. All transgenic plant species expressed *p-atzA* with tobacco, alfalfa, and *Arabidopsis* being resistant to 15, 10 and 5 μg ml^{-1} atrazine, respectively while wild type plants died when atrazine exceeded 0.4, 0.2, and 0.1 μg ml^{-1} respectively. In hydroponic assays, transgenic tobacco and alfalfa plants absorbed 88 percent and 56 percent of the applied atrazine, and dechlorinated 100 percent and 97 percent of this triazine herbicide into hydroxyatrazine, and a small amount of other metabolites, respectively [Uefuji et al. 2005].

Bentazon, Bensulfuron

CYP81A6, CYP76B1, CYP71A10 Genes

Bentazon and bensulfuron, the benzothiadiazinone and sulfonylurea herbicides respectively, are widely used in rice against annual broadleaf weeds, sedges, and some grasses in rice fields. The tolerance for these two classes of herbicides is controlled by the *Bel* locus in rice [Zhang et al. 2002]. The *bel* gene encodes a cytochrome P450 protein, CYP81A6, which presumably catalyzes the hydroxylation of bentazon and bensulfuron as a way of their detoxification [Pan et al. 2006; Zhang et al. 2007].

Cytochrome P450, a heme-protein present in all organisms (vide Chapter 7), activates the oxygen molecule to produce a water molecule and oxygenated substrate using electrons removed from NADPH. P450s are involved in the detoxification of herbicides in plants via hydroxylation, dealkylation, deamination, decarboxylation, isomerization, dimerization, epoxidation, dehydration, and aryl-migration of the substrates. Genes encoding P450s have been used to generate herbicide-resistant crops. For example, *CYP76B1* gene from *Helianthus tuberosus* (Jerusalem artichoke) encodes a dealkylase that catalyzes the metabolism of phenylurea herbicides and the transgenic tobacco and *Arabidopsis* plants expressing this gene enhanced resistance to these herbicides, including chlortoluron, linuron, and diuron [Didierjean et al. 2002; Robineau et al. 1998]. Soyabean gene *CYP71A10* catalyzes the oxidative *N*-demethylation or ring-methyl hydroxylation of a variety of phenylurea herbicides, and its expression in tobacco leads to enhanced tolerance to them [Siminszky et al. 1999; Siminszky et al. 2003].

Liu et al. [2012] introduced the rice *CYP81A6* gene into *Arabidopsis* and tobacco plants to test the possibility of engineering tolerance to bentazon and bensulfuron in other susceptible plants. The full-length open reading frame of *CYP81A6* amplified from rice was introduced into pCAMBIA3301 by replacing the *GUS* gene to generate the expression plasmid pCYP86A1 in which *CYP81A6* was driven by the CaMV35S promoter. Following transformation, 4-week plants, derived from homozygous *CYP81A6*-transgenic *Arabidopsis* T$_3$ lines, were tested for bentazon tolerance. The transgenic seedlings survived up to 0.12 g L^{-1} concentration while wild type plants died even at 0.03 g L^{-1} within 1–3 d of treatment. Similarly, the transgenic plants grew well at 1×10^{-5} g L^{-1} of bensulfuron-methyl whereas wild type *Arabidopsis* plants succumbed, suggesting that expression of the rice *CYP81A6* gene conferred tolerance to this sulfonylurea herbicide as well.

Liu et al. [2012] also introduced *CYP81A6* gene into tobacco plants through *Agrobacterium*-mediated transformation. Fifteen glufosinate-tolerant tobacco transformants were tested for bentazon tolerance. They all exhibited normal growth at 0.12 g L^{-1} rate, but wild type plants turned yellow and died gradually 3 days after treatment. The tolerance of transgenic tobacco plants was also tested at 1×10^{-5} g L^{-1} of bensulfuron-methyl. Wild type tobacco seedlings showed pale-green coloration, and their growth was suppressed by this sulfonylurea herbicide, while the transgenic tobacco plants exhibited strong tolerance to it. Besides, producing transgenic plants resistant to both these herbicides, the *CYP81A6* gene can also be used as a novel plant-derived selection marker in plant transformation [Liu et al. 2012].

PPO-Inhibitors

PPX Genes

Protoporphyrinogen oxidase (PPO), a key enzyme in the chlorophyll/heme biosynthetic pathway, catalyzes the oxidation of protoporphyrinogen IX (Protogen IX) to protoporphyrin IX (Protox IX). PPO is inhibited by the PPO family of herbicides which include diphenylethers, oxadiazoles, phenylpyrazoles, pyrimidinediones, thiadiazoles, and triazolinones (vide Chapter 2).

There are two nuclear genes that encode PPO isozymes in the plastid and mitochondria. These are called *PPX1* and *PPX2* for the plastid and mitochondria respectively. Protogen IX accumulates in sensitive plants treated with PPO inhibitors. When exported to the cytoplasm, protogen IX is converted to protox IX which in the presence of light causes the formation of singlet oxygen ($^1O_2$) leading to membrane damage and plant death. One gene from the resistant biotype, designated *PPX2L*, contained a codon deletion (G210) [Patzoldt et al. 2005]. *PPX2L* is predicted to encode both plastid- and mitochondria-targeted PPO isoforms, allowing a mutation in a single gene to confer resistance to two herbicide target sites. Resistant biotypes of *Amaranthus tuberculatus* have robust resistance to most PPO-inhibiting herbicides (lactofen, sulfentrazone, flumioxazin). Deletion of a codon rather than substitution is a unique formation of target site resistance to these herbicides.

Jung et al. [2004] introduced the protox gene derived from *Myxococcus xanthus* into rice genome and studied its effect on its targeting site and the effect on herbicide resistance to PPO-inhibitors. Scutellum-derived rice calli of cultivar 'Dongjin' were co-cultured with *A. tumefaciens* LBA4404 harboring pGA1611:MP binary vector before transforming them via *Agrobacterium*-mediated method. The transgenic rice lines had 10–18 times greater protox activity in both chloroplasts and mitochondria than in the wild-type line. Seeds of transgenic lines M4 and M7 germinated at 500 μm of oxyfluorfen while wild-type seeds failed to do so even at 1.0 μm. Four-week old transgenic plants treated with foliar application of oxyfluorfen at 50 μm for 3 days exhibited normal growth whereas plants of wild-type line were found severely bleached and necrotized [Jung et al. 2004]. Resistance of transgenic rice plants was due to the accumulation of photodynamic Protox IX in cytosol. Higher chloroplastic and mitochondrial protox activity in oxyfluorfen-treated transgenic lines metabolized Protogen IX to chlorophyll and heme [Jung et al. 2004].

Dinitroanilines

Alpha2-Tubulin Mutant Gene

Microtubules are important for maintaining cell structure, providing platforms for intracellular transport, forming mitotic spindle, as well as other cellular processes. They are polymers constructed from α-β-tubulin heterodimers. These structures are rapidly assembled and disassembled to create essential components of plant cells, such as spindles and flagella. Dinitroaniline herbicides disrupt the microtubules of plants (vide Chapter 2).

The α-*tubulin* gene causes Thr239 to be replaced by Ile239 in the resistant biotype of *Eleusine indica* (goosegrass) [Yamamoto et al. 1998]. The *α2-tubulin* mutant gene is a good candidate for creating new herbicide-resistant germplasm in cereals. Attempts to improve maize using this mutant gene have been performed through classical breeding [Landi et al. 1999] and through genetic engineering [Anthony and Hussey 1999].

Carotenoid-Inhibitors

PDS, Crtl Genes

Phytoene desaturase (PDS) is the first and key enzyme in the carotenoid biosynthetic pathway to convert the colorless phytoene to colored carotenoids which constitute the essential photosynthetic components in plants. It catalyzes two sequential steps of dehydrogenation of phytoene to produce ζ-carotene via phytofluene as an intermediate. PDS has been implicated as a rate-limiting step in the pathway in a number of plant species [Chamovitz et al. 1993; Lopez et al. 2007]. In addition, PDS is also a target site for herbicides, such as norflurazon (vide Chapter 2). PDS-inhibiting herbicides prevent the formation of carotenoids, resulting in the degradation of chlorophyll and the destruction of chloroplast membranes, which is characterized by the photobleaching of green tissues [Böeger and Sandmann 1998].

Besides PDS, *lycopene β-cyclase* is another crucial enzyme in carotenoid biosynthesis pathway in plants. Ripening in tomato fruit is associated with vivid changes in color. The change in fruit color from green to orange, pink, and then red is accompanied by shift in carotenoid profile from β-carotene at breaker stage to lycopene at red ripe stage. These changes are brought about by transcriptional upregulation of *Phytoene Synthase (PSY1)* and *PDS* genes [Corona et al. 1996; Alba et al. 2005] besides the down regulation of *lycopene β-cyclase (LCY-B)* and *lycopene ε-cyclase (CRTL-e)* genes [Alba et al. 2005; Ronen et al. 1999, 2000].

Three natural somatic mutations at codon 304 of the *pds* gene of *Hydrilla verticillata* (L.f. Royle) have been reported to provide resistance to the herbicide fluridone [Arias et al. 2006]. This gene, first isolated from the norflurazon-resistant mutants of *Synechococcus* sp. PCC7942 [Chamovitz et al. 1990], has three putative start codons: ATG1, ATG2, and ATG3 [Arias et al. 2006]. The Thr304 *Hydrilla pds* gene mutant proved to be an excellent marker for the selection of transgenic plants. Seedlings harboring Thr304 *pds* had a maximum resistance to sensitivity (R/S) ratio of 57 and 14 times higher than that of the wild-type for treatments with norflurazon and fluridone, respectively. These plants exhibited normal growth and development, even after long-term exposure to herbicide. As Thr304 *pds* is of plant origin, it could become more acceptable than other selectable markers for use in genetically modified food [Arias et al. 2006].

The lycopene ε-cyclase enzyme-encoding gene, *crtl*, derived from the phytopathogenic bacterium *Erwina ueredovora*, is resistant to bleaching herbicides. This gene was fused to the transit peptide nucleotide sequence of pea Rubisco small subunit, and expressed under the control of CaMV 35S promoter in tobacco plants. Transformation of tobacco with this gene yielded transgenic plants which, with enhanced β–carotene synthesis, acquired elevated resistance to norflurazon [Misawa et al. 1993]. The chimeric gene product, Crtl, was targeted into chloroplasts and processed in the transgenic plants [Misawa et al. 1993].

Significant increase in carotenoid content was achieved by genetic engineering of carotenoid biosynthesis pathway in rapeseed (canola), rice, potato and maize [Aluru et al.

2008; Ducreux et al. 2005; Paine et al. 2005]. In contrast to these transgenic crops, only limited success has been achieved in increasing the carotenoid levels in tomato [Rosati et al. 2000].

Insect-Resistant Transgenes

Along with herbicide-tolerant crops, transgenic insect-resistant crops have been one of the major commercial successes in agriculture. At the present time, transgenic cotton, resistant to lepidopteran larvae (moths, butterflies, etc.), and transgenic maize, resistant to both lepidopteran and coleopteran larvae (rootworms) are being widely grown in global agriculture.

The source of the insecticidal toxins produced in transgenic plants is the soil bacterium, *Bacillus thuringiensis* (*Bt*). The first commercial insecticide based on *Bt* spore-based formulations called "sporine" was produced in France in 1938 and used primarily to kill flour moths. In the United States, *Bt* was first manufactured commercially in 1958 and, by 1961, *Bt*-based bioinsecticides were being registered by the US Environmental Protection Agency. Since 1996, insect-resistant transgenic crops, known as *Bt* crops, have expanded around the globe and are proving to be quite efficient and helpful in reducing the use of chemical insecticides [Qaim and Zilberman 2003; Kleter et al. 2007].

Bacillus thuringiensis

Strains of *Bt* which show differing specificities of insecticidal activity towards insects constitute a large reservoir of genes encoding insecticidal proteins. The *Bt* proteins are toxic to various herbivorous insects. The protein, known as *Bt* toxin, is produced in an inactive, crystalline form. The larvicidal activity of this Gram positive, entamopathogenic bacterium is attributed to its parasporal crystals it produces. As the structure of these toxic proteins is crystalline, they are called crystal (Cry) proteins. The genes encoding for the crystal proteins are named '*cry*' genes, and their common characteristic is the expression of the genes during the stationary phase.

The crystal *Bt* toxin are aggregates of a large protein (130–140 kDa), called protoxin, which must be activated before it has any effect. As the crystal protein is highly insoluble in normal conditions, it is safe to humans, higher animals, and most insects. However, when it is solubilized to a high pH (> 9.5), the condition commonly found in the mid-gut of lepidopteran larvae, *Bt* toxin becomes a highly specific insecticidal agent [Deacon 2001].

Once it has been solubilized in the insect gut, the protoxin is cleaved by a gut protease to produce an active toxin of about 60 kDa. This toxin is termed delta (δ)-endotoxin. It binds to the mid-gut epithelial cells, creating pores in the cell membranes and leading to equilibration of ions. As a result, the gut is rapidly immobilized, the epithelial cells lyse, the larva stops feeding, and the gut pH is lowered by equilibration with the blood pH. This lower pH enables the bacterial spores to germinate, and the bacterium can then invade the host, causing a lethal septicaemia [Deacon 2001].

The δ-endotoxin has three domains. Domain I is a bundle of seven alpha-helices, some or all of which can insert into the gut cell membrane, creating a pore through which ions can pass freely. Domain II consists of three antiparallel beta-sheets, similar to the antigen-binding regions of immunoglobulins, suggesting that this domain binds to receptors in the gut. Domain III is a tightly packed β-sandwich which is thought to protect the exposed end (C-terminus) of the active toxin, preventing further cleavage by gut proteases [Deacon 2001].

Crystal Proteins and Genes

The Cry proteins are classified into different categories based on the homology of their amino acid sequences. Over 300 *cry* genes have been classified into 47 groups. Additionally, certain strains of cytolysin (Cyt) protein are found in the crystal inclusions of the strains active against dipterous strains [Smith and Couch 1991]. The 22 cytolytic (*cyt*) genes so far discovered have been divided into two classes. The number and type of δ-endotoxins produced determine the bioactivity of a *Bt* strain [Crickmore et al. 1998; Schnepf et al. 1998; Hofte and Whiteley 1989]. The morphology, size, and number of parasporal inclusions vary among different *B. thuringiensis* strains.

The specificity of *Bt* Cry toxins towards target insect species is a major advantage in agriculture because the effects on non-target insects and other organisms in the ecosystem are minimized [Gatehouse 2008]. However, deployment of transgenic crops expressing a single specific *Bt* toxin can lead to problems in the field where secondary insect species are not affected, and these can cause significant damage to the crop [Gatehouse 2008]. *Bt* preparations are commonly used in organic agriculture to control insects because *Bt* toxins occurs naturally and these are completely safe for humans.

The Cry1, Cry2, Cry8, Cry9, and Cry20 groups exhibit strongest activity to Lepidoptera while the Cry3, Cry7, and Cry8 groups, which have dual activity, are most toxic to Coleoptera. The Cry4 and Cry11 groups are highly active to Diptera. Some Cry proteins were reported to display toxicity to more than one insect order. For example, the Cry1I protein that is toxic to insects of both Lepidoptera and Coleoptera [Tailor et al. 1992] while the Cry1B protein is active against Lepidoptera, Coleoptera, and Diptera [Zhong et al. 2000]. The Cry5, Cry12, Cry13, Cry14, Cry21, and Cry55 proteins are nematicidal [Crickmore et al. 1998; Guo et al. 2008].

The genes predominantly distributed in natural environments are *cry1*-allied genes encoding Lepidoptera-specific Cry1 proteins. Strains containing *cry1*-type genes are the most abundant, representing 237 of the 310 *B. thuringiensis* isolates (76.5 percent) [Wang et al. 2003]. Within the Cry1 protein group, there are around ten different subclasses, and each subclass has a specific range of activity against different lepidopteran insects. Many of the *cry* genes such as *cry1Aa*, *cry1Ab*, *cry1Ac*, *cry1Ba*, *cry1Ca*, *cry1H*, and *cry1F* have been engineered into plants. Most Cry proteins, even within the Cry1A subfamily have a distinct insecticidal spectrum. Notwithstanding their classification, Cry proteins have similar mechanism of action.

Vegetative and Proteins and Genes

Apart from Cry proteins, some *B. thuringiensis* and *B. cereus* strains also secrete insecticidal proteins into the culture medium. These proteins have been referred to as vegetative insecticidal proteins (Vips) since they are expressed during the vegetative stage [Estruch et al. 1996]. Unlike *Bt* toxins whose expression is restricted to sporulation, Vip insecticidal proteins, encoded by *Vip* genes, are expressed in the vegetative growth stage starting at mid-log phase as well as during sporulation. Ingestion of the single-chain Vip proteins causes swelling and disruption of the midgut epithelial cells by osmotic lysis in the target insects.

One group of Vip toxins consists of binary toxins made of two components, Vip1 and Vip2. The combination of Vip1 and Vip2 is highly insecticidal to an agronomically important insect like the Western maize rootworm (*Diabrotica virgifera*), but does not show any insecticidal activity to lepidopteran insects. The other group consists of Vip3 toxins, which share no sequence similarity to Vip1 and Vip2 toxins. The first identified Vip3 toxin, Vip3Aal, is highly insecticidal to several major lepidopteran insects of maize and cotton [Estruch et al. 1996; Fang et al. 2007]. According to Agricultural Biotechnology Annual Report for Australia [Crothers 2006], genetically modified Vip cotton has the license for only limited and controlled release. Another group of binary toxins includes Cry34/35 proteins, active against maize the coleopteran rootworm, and have been expressed in transgenic maize [Moellenbeck et al. 2001].

Multiple Genes

Development of transgenic crops expressing a single specific *Bt* toxin can lead to problems in the field because the unaffected secondary insect species can cause significant damage to the crop. Introduction of additional *Bt cry* genes into the crop can afford protection against a wider range of insects. Transgenic cotton expressing both Cry1Ac and Cry2Ab proteins are more toxic to both the target bollworm (*Helicoverpa zea*) and the two secondary target species of armyworm (*Spodoptera frugiperda* and *Spodoptera exigua*) than cotton expressing either of them alone as was evident in MON15985 maize variety approved by USA in 2002, followed thereafter by many countries including Canada, Australia, Brazil, China, India, Mexico, South Africa, etc.

However, simultaneous exposure of insects to plants carrying either of the two or both *Bt* genes actually increased resistance breakdown in the two-gene plants [Zhao et al. 2005], suggesting that the multiple-toxin approach is not a cure-all solution. Other results have also shown that insects can acquire resistance to multiple toxins as well. For example, a strain of the lepidopteran cotton insect *Heliothis virescens* has simultaneous resistance to *cry1Ac* and *cry2Aa* genes, with a different genetic basis of resistance to each toxin [Gahan et al. 2005].

Improvements in plant transformation methods, such as extending the species-range of *Agrobacterium*-mediated gene transfer methods to monocots and using plasmid vectors containing multiple gene constructs to allow introduction of multiple transgenes at a single genetic locus, have enabled the expression of multiple toxins in transgenic plant varieties [Gatehouse 2008]. One example is the development in 2010 of a quadruple transgenic maize hybrid containing six insect resistance genes active against maize coleopteran rootworm (*cry34Ab1*, *cry35Ab1*, modified *cry3Bp1*) and lepidopteran insects (*cry1Fa2*, *cry1A.105*, *cry2Ab*) and two genes conferring tolerance to herbicides glyphosate and glufosinate (vide Chapter 6). Further to this is the development of quintuple- and hextuple-stack herbicide cum insect-resistant maize hybrids (vide Chapter 6).

Gene Promoters

A promoter is a segment of DNA that initiates transcription of a particular gene (transgene). It is usually located near the genes they transcribe, on the same strand on the DNA (towards the 3′ region of the anti-sense strand, also called template strand and non-coding strand). It acts as a controlling element in the expression of the gene by serving as a catalyst to enhance the rate of gene activity (vide Chapter 3).

Cauliflower Mosaic Virus

Cauliflower mosaic virus (CaMV) is a member of genus Caulimovirus, one of the six genera in the Caulimoviridae family. Belonging to the group of pararetroviruses that infect plants [Pringle 1999], CaMV replicates through reverse transcription just like retroviruses, but the viral particles contain DNA instead of RNA [Rothnie et al. 1994]. A detailed discussion on this constitutive promoter gene is presented in Chapter 3.

CaMV 35S is used in most transgenic crops to activate foreign genes which have been artificially inserted into the host plant. Although the 35S promoter and its derivatives can drive high levels of transgene expression in dicotyledonous plants [Battraw and Hall 1990; Benfey et al. 1990], their activities are substantially lower in monocotyledonous plants [Christensen et al. 1992; Gupta et al. 2001; Weeks et al. 1993]. Conversely, the activity of monocot-derived promoters is higher in monocots than in dicots [Cornejo et al. 1993] necessitating the development of both monocot and dicot promoters. CaMV 35S is inserted into transgenic plants in a form which is different to that found when it is present in its natural *Brassica* plant hosts. This enables it to operate in a wide range of host-organism environments which would otherwise not be possible. CaMV 35S is the promoter of choice for plant genetic engineering as it is a strong and constitutive promoter.

Cassava Vein Mosaic Virus

The cassava vein mosaic virus (CsVMV) is a double stranded DNA (dsDNA) pararetrovirus which infects cassava (*Manihot esculenta*) plants in the northeastern region of Brazil. The CsVMV has spherical particles of 50–60 nm in diameter occurring in cytoplasmic inclusion bodies. It contains a circular dsDNA genome of 8158 nucleotides. The molecular characterization of the CsVMV genome organized in five open reading frames is different from other characterized plant pararetroviruses.

Several transcriptional promoters capable of causing high levels of gene expression in transgenic plants have been isolated pararetrovirus genomes. However, the speciality of CsVMV is the arrangement of genes in the genome, which is distinct from that of caulimoviruses and badnaviruses. It was proposed that CsVMV be the type species of the genus *Cavemovirus*, a genus of the plant pararetrovirus family *Caulimoviridae* [Hull et al. 2005]. As in the genomes of other pararetroviruses, CsVMV contains a sequence that is complementary to a plant cytoplasmic tRNA. The numbering of the CsVMV genome starts at the 5′-end of this putative primer-binding site.

ORF1 potentially encodes a protein of 1372 amino acids with a molecular mass of 186 kDa. ORF 2 overlaps ORF 1 by 14 nucleotides [Marmey and Fauquet 2006]. The putative encoded protein of 71 amino acids, with a molecular mass of 8.8 kDa, has no homology with other proteins and its role is unknown. The putative protein encoded by ORF 3 is 652 amino acids long with a molecular mass of 77 kDa. It contains sequences similar to protease, reverse transcriptase, and RNase H consensus regions. The presence of such proteins makes CsVMV similar to other plant pararetroviruses. The other two ORFs, ORF 4 and ORF 5, which overlap the preceding ORF, encode putative proteins of 201 and 220 amino acids, with molecular masses of 24 kDa and 26 kDa respectively [Marmey and Fauquet 2006]. There is no similarity with known proteins encoded by caulimoviruses or badnaviruses.

CsVMV promoters are useful in the production of transgenic plants. Desired phenotypes are produced in transgenic plants as a result of transformation of plant cells by a DNA construct containing heterologous DNA sequence operably linked to a CsVMV promoter. The CsVMV promoter can be used to link with numerous transgenes including herbicide tolerance, insect resistance, and disease resistance. The CsVMV promoter fragment is made up of different regions that confer distinct tissue-specific expression of the gene [Verdaguer et al. 1998]. The region encompassing nucleotides −222 to −173 contains *cis* elements that control promoter expression in green tissues and root tips. In vascular elements, expression from the CsVMV promoter is directed by the region encompassing nucleotides −178 to −63. However, elements located between nucleotides −149 and −63 are also required to activate promoter expression in green tissues suggesting a combinatorial mode of regulation [Verdaguer et al. 1998].

Figwort Mosaic Virus

Figwort mosaic virus (FMV), a caulimovirus of the pararetrovirus group, is very similar to cauliflower mosaic virus. FMV, with a genome of approximately 8 kb, has eight ORFs, seven of which correspond in size and location to loci in CaMV. Like CaMV, FMV employs methionine t-RNA as initiator of replication [Richins et al. 1987]. Plant expression vectors have been developed using FMV [Maiti et al. 1997]. FMV has many of the same advantages as CaMV for genetic manipulation by molecular cloning procedures. Its relatively small double-stranded DNA can be easily isolated from plants and analyzed by restriction enzyme cleavage.

The FMV promoter is characterized to be constitutive in most of the plant tissues. Its strength is comparable or even somewhat stronger than CaMV promoter and much stronger than nopaline promoter [Sanger et al. 1990; Matai et al. 1997]. The FMV promoter is a useful alternative to CaMV promoter in crop genetic engineering.

The sub-genic transcript (Sgt) FMV promoter is linked to heterologous coding sequences to form a chimeric gene construct. The 5′-3′-boundaries required for maximal activity and involvement of *cis*-sequences for optimal expression in plants are defined by 5′-, 3′-end deletion and internal deletion analysis of FMV Sgt promoter fragments coupled with a β-glucuronidase reporter gene in both transient protoplast expression experiments and in transgenic plants. A 301 bp FMV Sgt promoter fragment (sequence −270 to +31 from the transcription start site: TSS) provided maximum promoter activity [Bhattacharya et al. 2002]. An activator domain located upstream of TATA box (also called Goldberg-Hogness box) [Lifton et al. 1978] at −70 to −100 TSS is absolutely required for promoter activity and its function is critically position-dependent with respect to TATA box. The FMV Sgt promoter is less active in monocot cells. Its expression level was about 27.5-fold higher in tobacco cells compared to that in maize cells. The FMV Sgt promoter is about two-fold stronger than the CaMV 35S promoter [Bhattacharya et al. 2002].

Cestrum Yellow Leaf Curling Virus

Cestrum yellow leaf curling virus (CmYLCV) is another member of Caulimoviridae family first isolated from *Cestrum paraqui*(called green cestrum, green poison berry, Chilean Jessamine, native to central and South America) belonging to Solanaceae family. Like other members of the Caulimoviridae, CmYLCV possesses a dsDNA genome replicated via a pre-genomic/polycistronic RNA [Rothnie et al. 1994] that is transcribed under the control of the genomic full-length transcript promoter.

The CmYLCV promoter is a novel tool with broad utility for regulating the constitutive expression of transgenes at high levels in both monocot and dicot crops. The constitutive activity of the CmYLCV promoter is equal to or higher than the level obtained with any of the most commonly used constitutive promoters [Stavolone et al. 2003]. It is unlikely that the different activity of the promoters tested was the result of a transgene positional effect.

The CmYLCV promoter has high activity in meristematic cells and callus. This indicates that, in addition to being useful to direct high levels of gene expression in most cell and tissue types of a wide selection of crop plants, the CmYLCV promoter may also be useful for driving expression of a selectable marker gene for transgenic plant production. Furthermore, the level of expression is comparable to, or higher than that from the CaMV 35S, the 'super-promoter', or the maize ubiquitin-1 promoters, the frequently used promoters in agricultural biotechnology. The heritable, strong and constitutive activity in both monocotyledonous and dicotyledonous plants, combined with the extremely narrow CmYLCV host range, makes the CmYLCV promoter an attractive tool for regulating transgene expression in a wide variety of plant species [Stavolone et al. 2003].

Peanut Chlorotic Streak Caulimovirus

Peanut chlorotic streak virus, a unique caulimovirus, has been first reported to naturally infect groundnut (peanut) in India [Reddy et al. 1993]. It has a broad host range compared to other caulimoviruses. The circular double-stranded genome of PClSV is 8174 bp. The genetic map of PClSV contains a set of three open reading frames A, B, and C located between genes 1 and IV and it is not similar to other caulimoviruses. However, with some minor differences, the genomic organization and transcription of this virus is similar to CaMV and FMV [Richins 1993; Mushegian et al. 1995].

The outer double circle represents the double-stranded genomic DNA. The position and the orientation of the major open reading frames (ORFs) as they occur around the circular PClSV DNA are indicated by arrows on the double circle. The ORFs are designated (I, A, B, C, IV, V, VI and VII) based on observed relationships with spatially homologous ORFs occurring in other caulimovirus genomes. The tRNAMet primer binding site (PBS), and the position (indicated by small stippled circle) of two promoters (pregenomic transcript promoter (P_{FLt}) and subgenomic transcript promoter (P_{Sgt}) of the PClSV are shown. The two major viral transcripts, Sgt RNA (equivalent to CaMV 19S RNA) and FLt RNA (equivalent to CaMV 35S RNA) are shown by thin line as arrows along their respective coding region. These RNAs share the same 3' ends (arrow heads). In addition, the position of several selected restriction endonuclease sites is shown.

Tobacco Etch Virus

Tobacco etch virus (TEV) is a plant pathogenic virus belonging to the genus *Potyvirus* and family Potyviridae. Like other members of the *Potyvirus* genus, TEV is a monopartite strand of positive-sense, single-stranded RNA surrounded by a capsid made for a single viral encoded protein. The virus is a filamentous particle of about 730 nm long. It is transmissible in a non-persistent manner by more than 10 species of aphids including *Myzus persicae*. It is also easily transmitted by mechanical means but is not known to be transmitted by seeds. Like other Potyviruses, TEV RNA codes for at least seven proteins including protease. The TEV protease is a highly site-specific cysteine protease.

Tobacco etch virus RNA has a 3' terminal poly(A) sequence, ranging from about 200 to less than 33 residues but with two size classes, averaging about 150 and 30 residues [Hari 1981]. A genome-linked protein, estimated MW 6000, has been detected but apparently is not required for infectivity [Hari 1981]. Tobacco etch virus RNA is translated efficiently *in vitro* in rabbit reticulocyte lysates [Dougherty and Hiebert 1980a, 1980b, 1980c].

References

Alba, R., P. Payton, Z. Fei, R. McQuinn, P. Debbie, G.B. Martin et al. 2005. Transcriptome and selected metabolite analyses reveal multiple points of ethylene control during tomato fruit development. Plant Cell 17: 2954–2965.

Aluru, M., Y. Xu, R. Guo, Z. Wang, S. Li, W. White et al. 2008. Generation of transgenic maize with enhanced provitamin A content. J. Exp. Bot. 59(13): 3551–3562.

An, G., B. Watson, S. Stachel, M. Gordon and E. Nester. 1985. New cloning vehicle for transformation of higher plants. EMBO J. 4: 277–284.

Angus, T.A. 1953. Progress Report. Vol. 9. Forest Biology Division, Canada Department of Science Service; 1953. Studies of *Bacillus* spp. pathogenic for silkworm. p. 6.

Anthony, R.G. and P. Hussey. 1999. Double mutation in *Eleusine indica* α-tubulin increase the resistance of transgenic maize calli to dinitroaniline and phosphorothioamidate herbicides. Plant J. 18: 669–674.

Anthony, R.G., S Reichelt and P.J. Hussey. 1999. Dinitroaniline herbicide-resistant transgenic tobacco plants generated by co-overexpression of a mutant alpha-tubulin and a beta-tubulin. Nat. Biotechnol. 17(7): 712–716.

Arias, R.S., F.E. Dayan, A. Michel, J.L. Howell and B.E. Scheffler. 2006. Characterization of a higher plant herbicide-resistant phytoene desaturase and its use as a selectable marker. Plant Biotechnol. J. 4(2): 263–273.

Ashton, A.R., C.L. Jenkins and P.R. Whitfeld. 1994. Molecular cloning of two different cDNAs for maize acetyl CoA carboxylase. Plant Mol. Biol. 24(1): 35–49.

Barry, G.F. and G.A. Kishore. 1995. Glyphosate tolerant plants. US Patent 5,463,175.

Barry, G.F., G.M. Kishore and S.R. Padgette. 1992a. DNA encoding class II 5′-enolpyruvyl shikimate-3-phosphate synthase–for producing plants and bacteria tolerant to glyphosate herbicides. World Patent WO 9. 204: 449.

Barry, G.F., G.M. Kishore, S.R. Padgette, M. Taylor, K. Kolacz, M. Weldon et al. 1992b. Inhibitors of amino acid biosynthesis: strategies for imparting glyphosate tolerance to crop plants. pp. 139–145. *In*: B.K. Singh, H.E. Flores and J.C. Shannon (eds.). Biosynthesis and Molecular Regulation of Amino Acids in Plants. Amer. Soc. Plant Physiol. Rockville, MD, USA.

Battraw, M.J. and T.C. Hall. 1990. Histochemical analysis of CaMV 35S promoter-β-glucuronidase gene expression in transgenic rice plants. Plant Mol. Biol. 15: 527–538.

Beeching, J.R., A.J. Weightman and J.H. Slater. 1983. The formation of an R-prime carrying the fraction I dehalogenase gene from *Pseudomonas putida* PP3 using the IncP plasmid R.6844. J. Gen. Microbiol. 129: 2071–2078.

Behrens, M.R., N. Mutlu, S. Chakraborty, R. Dumitru, W.Z. Jiang, B.J. Lavallee et al. 2007. Dicamba resistance: enlarging and preserving biotechnology-based weed management strategies. Science 316: 1185–1188.

Benfey, P.N., L. Ren and N.H. Chua. 1990. Tissue-specific expression from CaMV 35S enhancer subdomains in early stages of plant development. EMBO J. 9: 1677–1684.

Bernasconi, P., A.R. Woodworth, B.A. Rosen, M.V. Subramanian and D.L. Siehl. 1995. A naturally occurring point mutation confers broad range tolerance to herbicides that target acetolactate synthase. J. Biol. Chem. 270: 17381–17385.

Bevan, M. 1984. Binary *Agrobacterium* vectors for plant transformation. Nucl. Acids Res. 12: 8711.

Bhattacharyya, S., N. Dey and I.B. Maiti. 2002. Analysis of cis-sequence of subgenomic transcript promoter from the Figwort mosaic virus and comparison of promoter activity with the cauliflower mosaic virus promoters in monocot and dicot cells. Virus Res. 90: 47–62.

Bhatti, M., P.C.C. Feng, J. Pitkin and S.-W. Hoi. 2008. Methods for improving dicamba monooxygenase transgene encoding plant resistance to stress and disease by application of dicamba herbicide and its metabolites. World Intellectual property Organization Patent Application WO/2008/048964 A2, 1-50, filed 24 April 2008.

Boeger, P. and G. Sandmann. 1998. Carotenoid biosynthesis inhibitor herbicides—mode of action and resistance mechanisms. Pestic. Outlook 9: 29–35.

Boocock, M.R. and J.R. Coggins. 1983. Kinetics of 5-enolpyruvylshikimate-3-phosphate synthase inhibition by glyphosate. FEBS Lett. 154: 127–133.

Botterman, J. and J. Leemans. 1988. Engineering of herbicide resistance in plants. Biotechnol. Genet. Eng. Rev. 6: 321–340.

Boudec, P., M. Rodgers, F. Dumas, A. Sailland and H. Bourdon. 2001. Mutated hydroxyphenylpyruvate dioxygenase, DNA sequence and isolation of plants which contain such a gene and which are tolerant to herbicides. US Patent US6245968B1(12-JUN-2001). Aventis CropScience S.A. (FR).

Brown, T.A. 2010. Gene Cloning and DNA Analysis: An Introduction (6th edition). John & Wiley & Sons Ltd., Chichester, West Sussex, UK.

Buchanan-Wollaston, V., A. Snape and F. Cannon. 1992. A plant selectable marker gene based on the detoxification of the herbicide dalapon. Plant Cell Rep. 11: 627–631.

Buckland, J., R. Collins and E. Pullin. 1973. Metabolism of bromoxynil octanoate in growing wheat. Pestic. Sci. 4: 149–162.

Castle, L.A. and M.W. Lassner. 2004. A New Strategy for Glyphosate Tolerant Crop Plants. linda.castle@ verdiainc.com; http://www.verdiainc.com, http://www.pioneer.com.

Castle, L.A., D.L. Siehl, R. Gorton, P.A. Patten, Y.H. Chen, S. Bertain et al. 2004. Discovery and directed evolution of a glyphosate tolerance gene. Science 304: 1151–1154.

Center for Environmental Risk Assessment. 2010. A Review of the Environmental Safety of the CP4 EPSPS Protein. ILSI Research Foundation. 26 May 2010.

Chakraborty, S., M. Behrens, P.L. Herman, A.F. Arendsen, W.R. Hagen, D.L. Carlson et al. 2005. A three-component dicamba O-demethylase from *Pseudomonas maltophilia*, strain DI-6: purification and characterization. Arch. Biochem. Biophys. 437: 20–28.

Chaleff, R.S. and C.J. Mauvais. 1984. Acetolactate synthase is the site of action of two sulfonylurea herbicides in higher plants. Science 224(4656): 1443–1446.

Chamovitz, D., I. Pecker, G. Sandmann, P. Böeger and J. Hirschberg. 1990. Cloning a gene for norflurazon resistance in cyanobacteria. Z Naturforsch. 45c: 482–486.

Charest, P.J., J. Hattori, J. DeMoor, V.N. Iyer and B.L. Miki. 1990. *In vitro* study of transgenic tobacco expressing *Arabidopsis* wild type and mutant acetohydroxyacid synthase genes. Plant Cell. Rep. 8: 643–646.

Christensen, A.H., R.A. Sharrock and P.H. Quail. 1992. Maize polyubiquitin genes: structure, thermal perturbation of expression and transcript splicing, and promoter activity following transfer to protoplasts by electroporation. Plant Mol. Biol. 18: 675–689.

Clay, S.A. and W.C. Koskinen. 1990. Characterization of alachlor and atrazine desorption from soils. Weed Sci. 38: 74–80.

Cornejo, M.J., D. Luth, K.M. Blankenship, O.D. Anderson and A.E. Blechl. 1993. Activity of a maize ubiquitin promoter in transgenic rice. Plant Mol. Biol. 23: 567–581.

Corona, V., B. Aracci, G. Kosturkova, G.E. Bartley, L. Pitto, L. Giorgetti et al. 1996. Regulation of a carotenoid biosynthesis gene promoter during plant development. The Plant J. 9: 505–512.

Crameri, A., S.-A. Raillard, E. Bermudez and W.P.C. Stemmer. 1998. DNA shuffling of a family of genes from diverse species accelerates directed evolution. Nature 391: 288–291.

Crickmore, N., D.R. Zeigler, J. Feitelson, E. Schnepf, J. Vanrie, D. Lereclus et al. 1998. Revision of the nomenaclature for the *Bacillus thuringiensis* pesticidal crystal proteins. Microbiol. Rev. 62: 807–813.

Crothers, L. 2006. Agricultural biotechnology annual report 2006. GAIN Report. http://www.fas.usda. gov/gainfiles /200606/146198091.doc.

Cui, H.L., C.X. Zhang, H.J. Zhang, X. Liu, Y. Liu, G.Q. Wang et al. 2008. Confirmation of flixweed (*Descurainia sophia*) resistance to tribenuron in China. Weed Sci. 56: 775–779.

Deacon, J. 2001. The Microbial World: *Bacillus thuringiensis*. University of Edinburgh, UK. http://www. biology.ed.ac.uk/archive/jdeacon/microbes/bt.htm.

De Souza, M.L., M.J. Sadowsky, J. Seffernick, B. Martinez and L.P. Wackett. 1998. The atzABC genes encoding atrazine catabolism are widespread and highly conserved. J. Bacteriol. 180: 1951–1954.

De Souza, M.L., M.J. Sadowsky and L.P. Wackett. 1996. Atrazine chlorohydrase from *Pseudomonas* sp. strain ADP: Gene sequence, enzyme purification, and protein characterization. J. Bacteriol. 178(16): 4894–4900.

De Souza, M.L., L.P. Wackett, K.L. Boundy-Mills, R.T. Mandelbaum and M.J. Sadowsky. 1995. Cloning, characterization, and expression of a gene region from *Pseudomonas* sp. strain ADP involved in the dechlorination of atrazine. Appl. Environ. Microbiol. 61: 3373–3378.

Dharmasiri, N., S. Dharmasiri and M. Estelle. 2005. The F-box protein TIR1 is an auxin receptor. Nature 435: 441–445.

Didierjean, L., L. Gondet, R. Perkins, S.M. Lau, H. Schaller, D.P. O'Keefe et al. 2002. Engineering herbicide metabolism in tobacco and *Arabidopsis* with CYP76B1, a cytochrome P450 enzyme from Jerusalem artichoke. Plant Physiol. 130: 179–189.

Don, R.H., A.J. Weightman, H.-J. Knackmuss and K.N. Timmis. 1985. Transposon mutagenesis and cloning analysis of the pathways for degradation of 2,4-dichlorophenoxyacetic acid and 3-chlorbenzoate in *Alcaligenes eutrophus* JMP134(pJP4). J. Bacteriol. 161: 85–90.

Dougherty, W.G. and E. Hiebert. 1980a. Translation of potyvirus RNA in a rabbit reticulocyte lysate: reaction conditions and identification of capsid protein as one of the products of *in vitro* translation of tobacco etch and pepper mottle viral RNAs. Virology 101: 466–474.

Dougherty, W.G. and E. Hiebert. 1980b. Translation of potyvirus RNA in a rabbit reticulocyte lysate: identification of nuclear inclusion proteins as products of tobacco etch virus RNA translation and cylindrical inclusion protein as a product of the potyvirus genome.Virology 104(1): 174–182.

Dougherty, W.G. and E. Hiebert. 1980c. Translation of potyvirus RNA in a rabbit reticulocyte lysate: Cell-free translation strategy and a genetic map of the potyviral genome. Virology 104(1): 183–194.

Ducreux, L.J.M., W.L. Morris, P.E. Hedley, T. Shepherd, H.V. Davies, S. Millam et al. 2005. Metabolic engineering of high carotenoid potato tubers containing enhanced levels of beta-carotene and lutein. J. Exp. Bot. 56: 81–89.

Dumitru, R., W.Z. Jiang, D. Weeks and A. Wilson. 2009. Crystal structure of dicamba monooxygenase: a Rieske nonheme oxygenase that catalyzes oxidative demethylation. J. Mol. Biol. 392(2): 498–510.

Estelle, M.A. and C. Somerville. 1987. Auxin-resistant mutants of *Arabidopsis thaliana* with an altered morphology. Mol. Gen. Genet. 206: 200–206.

Estruch, J.J., G.W. Warren, M.A. Mullins, G.J. Nye, J.A. Craig and M.G. Koziel. 1996. Vip3A, a novel *Bacillus thuringiensis* vegetative insecticidal protein with a wide spectrum of activities against lepidopteran insects. Proc. Natl. Acad. Sci. USA 93: 5389–5394.

Fang, J., X. Xu, P. Wang, J.Z. Zhao, A.M. Shelton, J. Cheng et al. 2007. Characterization of chimeric *Bacillus thuringiensis* Vip3 toxins. Appl. Environ. Microbiol. 73: 956–961.

Fang, L.Y., P.R. Gross, C.H. Chen and M. Lillis. 1992. Sequence of two acetohydroxyacid synthase genes from *Zea mays*. Plant Mol. Biol. 18: 1185–1187.

Fukumori, F. and R.P. Hausinger. 1993. Purification and characterization of 2,4-dichlorophenoxyacetate/α-ketoglutarate dioxygenase. J. Biol. Chem. 268: 24311–24317.

Gabard, J.M., P.J. Charest, V.N. Iyer and B.L. Miki. 1989. Cross-resistance to short residual sulfonylurea herbicides in transgenic tobacco plants. Plant Physiol. 91: 574–580.

Gahan, L.J., Y.T. Ma, M.L.M. Coble, F. Gould, W.J. Moar and D.G. Heckel. 2005. Genetic basis of resistance to Cry1Ac and Cry2Aa in *Heliothis virescens* (Lepidoptera: Noctuidae). J. Econ. Entomol. 98: 1357–1368.

Gasser, C.S., J.A. Winte, C.M. Hironaka and D.M. Shah. 1988. Structure, expression, and evolution of the 5-enolpyruvylshikimate-3-phosphate synthase genes of petunia and tomato. J. Biol. Chem. 263: 4280–4287.

Gatehouse, J.A. 2008. Biotechnological prospects for Engineering Insect-Resistant Plants. Plant Physiol. 146(3): 881–887.

Gleason, C., R.C. Foley and K.B. Singh. 2011. Mutant analysis in *Arabidopsis* provides insight into the molecular mode of action of the auxinic herbicide dicamba. PLoS One 6(3): e17245. doi:10.1371/journal.pone.0017245.

Grula, J.W., R.L. Hudspeth, S.L. Hobbs and D.M. Anderson. 1995. Organization, inheritance and expression of acetohydroxyacid synthase genes in the cotton allotetraploid *Gossypium hirsutum*. Plant Mol. Biol. 28: 837–846.

Guo, S., M. Liu, D. Peng, S. Ji, P. Wang et al. 2008. New strategy for isolating novel nematicidal crystal protein genes from *Bacillus thuringiensis* strain YBT-1518. Appl. Environ. Microbiol. 74: 6997–7001.

Gupta, P., S. Raghuvanshi and A.K. Tyagi. 2001. Assessment of the efficiency of various gene promoters via biolistics in leaf and regenerating seed callus of millets, *Eleusine coracana* and *Echinochloa crus-galli*. Plant Biotechnol. 18: 275–282.

Hannay, C.L. 1953. Crystalline inclusions in aerobic sporeforming bacteria. Nature 172: 1004.

Hannay, C.L. 1955. Fitz-James P. The protein crystals of *Bacillus thuringiensis* Berliner. Can. J. Microbiol. 1: 694–710.

Hari, V. 1981. The RNA of tobacco etech virus: further characterization and detection of protein linked to RNA. Virology 112: 391–399.

Haughn, G.W., J. Smith, B. Mazur and C. Somerville. 1988. Transformation with a mutant Arabidopsis acetolactate gene renders tobacco resistant to sulfonylurea herbicides. Mol. Gen. Genet. 211: 266–271.

Haughn, G.W. and C. Somerville. 1986. Sulfonylurea resistant mutants of *Arabidopsis thaliana*. Mol. Gen. Genet. 204: 430–434.

He, M., Z.Y. Yang, Y.F. Nie, J. Wang and P. Xu. 2001. A new type of class I bacterial 5-enopyruvylshikimate-3-phosphate synthase mutants with enhanced tolerance to glyphosate. Biochim. Biophys. Acta 1568: 1–6.

Herman, P.L., M. Behrens, S. Chakraborty, B.M. Chrastil, J. Barycki and D.P. Weeks. 2005. A three-component dicamba O-demethylase from *Pseudomonas maltophilia*, strain DI-6: gene isolation, characterization, and heterologous expression. J. Biol. Chem. 280: 24759–24767.

Herrmann, K.M. and L.M. Weaver. 1999. The shikimate pathway. Annu. Rev. Plant Physiol. Plant Mol. Biol. 50: 473–503.

Hofte, H. and H.R. Whiteley. 1989. Insecticidal crystal protein of *Bacillus thuringiensis*. Microbiol. Rev. 53: 242–255.

Hsu, J.C. and N.D. Camper. 1975. Degradation of ioxynil and bromoxynil as measured by a modified spectrophotometric method. Canadian J. Microbiol. 21(12): 2008–2012.

Hull, R., A. Geering, G. Harper, B.E. Lockhart and J.E. Schoelz. 2005. Caulimoviridae. pp. 385–396. *In*: C.M. Fauquet, M.A. Mayo, J. Maniloff, U. Desselberger and L.A. Ball (eds.). Virus Taxonomy: Eighth Report of the International Committee on Taxonomy of Viruses. Elsevier Academic Press, San Diego, CA.

Jung, S., Y. Lee, K. Yang, S.B. Lee, S.M. Jang, S.B. Ha et al. 2004. Dual targeting of *Myxococcus xanthus* protoporphyrinogen oxidase into chloroplasts and mitochondria and high level oxyfluorfen resistance. Plant Cell Environ. 27: 1436–1446.

Kern, W.V., A. Oethinger, A. Kaufhold, E. Rozdzinski and R. Marre. 1993. *Ochrobactrum anthropi* bacteria: Report of four cases and short review. Infection 21: 306–310.

Klee, H.J., Y.M. Muskopf and C.S. Gasser. 1987. Cloning of an *Arabidopsis thaliana* gene encoding 5-enolpyruvylshikimate-3-phosphate synthase: sequence analysis and manipulation to obtain glyphosate-tolerant plants. Mol. Gen. Genet. 10: 437–442.

Kleter, G.A., Ad A.C.M. Peijnenburg and H.J.M. Arts. 2005. Health considerations regarding horizontal transfer of microbial transgenes present in genetically modified crops. J. Biomed. Biotechnol. 2005(4): 326–352.

Landi, P., E. Frascaroli and M.M. Guilani. 1999. Genetic variability for resistance to trifluralin in *Zea mays*. Weed Sci. 47: 369–374.

Lee, Y.T. and R.G. Duggleby. 2001. Identification of the regulatory subunit of *Arabidopsis thaliana* acetohydroxyacid synthase and reconstitution with its catalytic subunit. Biochemistry 40: 6836–6844.

Lee, K.Y., J. Townsend, J. Tepperman, M. Black, C.F. Chui, B. Mazur et al. 1988. The molecular basis of sulfonylurea herbicide resistance in tobacco. EMBO J. 7: 1241–1248.

Li, Z., A. Hayashimoto and N. Murai. 1992. A sulfonylurea herbicide resistance gene from *Arabidopsis thaliana* as a new selectable marker for production of fertile transgenic rice plants. Plant Physiol. 100: 662–668.

Lifton, R.P., M.L. Goldberg, R.W. Karp and D.S. Hogness. 1978. "The organization of the histone genes in Drosophila melanogaster: functional and evolutionary implications". Cold Spring Harb. Symp. Quant. Biol. 42: 1047–1051.

Lincoln, C., J.H. Britton and M. Estelle. 1990. Growth and development of the axr1 mutants of *Arabidopsis*. The Plant Cell 2: 1071–1080.

Liu, C., S. Liu, F. Wang, Y. Wang and K. Liu. 2012. Expression of a rice CYP81A6 gene confers tolerance to bentazon and sulfonylurea herbicides in both *Arabidopsis* and tobacco. Plant Cell Tissue Organ Cult. 109: 419–428.

Llewellyn, D. and D. Last. 1996. Genetic engineering of crops for tolerance to 2,4-D. pp. 159–174. *In*: S.O. Duke (ed.). Herbicide-Resistant Crops. CRC Press, Boca Raton, Florida, USA.

Lyon, B.R., D.J. Llewellyn, J.L. Huppatz, E.S. Dennis and W.J. Peacock. 1989. Expression of a bacterial gene in transgenic tobacco plants confers resistance to the herbicide 2,4-dichlorophenoxyacetic acid. Plant Mol. Biol. 13(5): 533–540.

Maiti, I., S. Gowda, J. Kiernan, S. Ghosh and R. Shepherd. 1997. Promoter/leader deletion analysis and plant expression vectors with the figwort mosaic virus full length transcript promoter containing single or double enhancer domains. Transgenic Res. 6: 143–156.

Mandelbaum, R.T., D.L. Allan and L.P. Wackett. 1995. Isolation and characterization of a *Pseudomonas* sp. that mineralizes the s-triazine herbicide atrazine. Appl. Environ. Microbiol. 61: 1451–1457.

Mannerlof, M., S. Tuvesson, P. Steen and P. Tenning. 1997. Transgenic sugar beet tolerant to glyphosate. Euphytica 94: 83–91.

Marmey, P. and C.M. Fauquet. 2006. Description of Plant Viruses: Cassava vein mosaic virus. Amer. Assoc. Biologists 413.

Martinez, B., J. Tomkins, L.P. Wackett, R. Wing and M.J. Sadowsky. 2001. Complete nucleotide sequence and organization of atrazine catabolic plasmid pADP-1 from *Pseudomonas* sp. strain ADP. J. Bacteriol. 183: 5684–5697.

Matringe, M., A. Sailland, B. Pelissier, A. Roland and O. Zind. 2005. *p*-Hydroxyphenylpyruvate dioxygenase inhibitor-resistant plants. Pest Manag. Sci. 61: 269–276.

Mattes, O. 1927. Parasitare krankheiten der mehimottenlarven und versuche uber ihre verwendbarkeit als biologisches bekamfungsmittel. Ges Naturw Marburg Schrift. 1927: 381. (Ita).

Mazur, B.J., C.-F. Chui and J.K. Smith. 1987. Isolation and characterization of plant genes coding for acetolactate synthase, the target enzyme for two classes of herbicides. Plant Physiol. 85: 1110–1117.

McCourt, J.A., S.S. Pang, J. King-Scott, L.W. Guddat and R.G. Duggleby. 2006. Herbicide-binding sites revealed in the structure of plant acetohydroxyacid synthase. Proc. Natl. Acad. Sci. USA 103: 568–573.

Meazza, G., B.E. Scheffler, M.R. Tellez, A.M. Rimando, J.G. Romagni, S.O. Duke et al. 2002. The inhibitory activity of natural products on plant p-hydroxyphenylpyruvate dioxygenase. Phytochemistry 59: 218–288.

Miki, B.L., H. Labbe, J. Hattori, T. Ouellet, J. Gabard, G. Sunohara et al. 1990. Transformation of *Brassica napus* canola cultivars with *Arabidopsis thaliana* acetohydroxyacid synthase genes and analysis of herbicide resistance. Theor. Appl. Genet. 80: 449–458.

Misawa, N., S. Yamano, H. Linden, M.R. de Filipe, M. Lucas, H. Ikenga et al. 1993. Functional expression of *Erwina ueredovora* carotenoid biosynthesis gene *crtl* in transgenic plants showing an increase in β-carotene biosynthesis activity and resistance to the bleaching herbicide norflurazon. The Plant J. 4(5): 833–840.

Moellenbeck, D.J., M.L. Peters, J.W. Bing, J.R. Rouse, L.S. Higgins, L. Sims et al. 2001. Insecticidal proteins from *Bacillus thuringiensis* protect corn from corn rootworms. Nat. Biotechnol. 19: 668–672.

Morimyo, M. 1988. Isolation and characterization of methyl viologen-sensitive mutants of *Escherichia coli* K-12. J. Bacteriol. 170: 2136–2142.

Morimyo, M., E. Hongo, H. Hama-lnaba and I. Machida. 1992. Cloning and characterization of the *mcrV* gene of *Escherichia coli* K-122 which confers resistance against methyl viologen toxicity. Nucleic Acids Res. 20: 3159–3165.

Müller, T.A., T. Fleischmann, J.R. van der Meer and H.-P.E. Kohler. 2006. Purification and characterization of two enantioselective alpha-ketoglutarate-dependent oxygenases, RdpA and SdpA from *Sphingomonas herbicideovorans* MH. Appl. Environ. Microbiol. 72: 4853–4861.

Mushegian, A.R., J.A. Wolff, R.D. Richins and R.J. Shepherd. 1995. Molecular analysis of the essential and nonessential genetic elements in the genome of the peanut chlorotic streak caulimovirus. Virology 206: 823–834.

Ness, J.E., S. Kim, A. Gottman, R. Pak, A. Krebber, T.V. Borchert, S. Govindarajan et al. 2002. Synthetic shuffling expands functional protein diversity by allowing amino acids to recombine independently. Nat. Biotechnol. 20: 1251–1255.

Odell, J.T., P.G. Caimi, N.S. Yadav and C.J. Mauvais. 1990. Comparison of increased expression of wild type and herbicide resistant acetolactate synthase genes in transgenic plants and indication of post-transcriptional limitation of enzyme activities. Plant Physiol. 94: 1647–1654.

Padgette, S.R., D.B. Re, C.S. Gasser, D.A. Eichholtz, R.B. Frazier, C.M. Hironaka et al. 1991. Site-directed mutagenesis of a conserved region of the 5-enolpyruvylshikimate-3-phosphate synthase active site. J. Biol. Chem. 266: 22364–22369.

Paine, J.A., C.A. Shipton, S. Chaggar, R.M. Howells, M.J. Kennedy, G. Vernon et al. 2005. Improving the nutritional value of golden rice through increased pro-vitamin A content. Nature Biotechnol. 23: 482–487.

Pallett, K.E., J.P. Little, M. Sheeky and P. Verasekaran. 1998. The mode of action of isoxaflutole. I. Physiological effects, metabolism and selectivity. Pestic. Biochem. Physiol. 62: 113–124.

Pan, G., X. Zhang, K. Liu, J. Zhang, X. Wu, J. Zhu and J. Tu. 2006. Map-based cloning of a novel rice cytochrome P450 gene CYP81A6 that confers resistance to two different classes of herbicides. Plant Mol. Biol. 61: 933–943.

Patzoldt, W.L., A.G. Hager, J.S. McCormick and P.J. Tranel. 2005. A codon deletion confers resistance to herbicides inhibiting protoporphyrinogen oxidase. Proc. Natl. Acad. Sci. USA 103: 12329–12334.

Pohlenz, H.-D., W. Boidol, I. Schuttke and W.R. Streber. 1992. Purification and properties of an *Arthrobacter oxydans* P52 carbamate hydrolase specific for the herbicide phenmedipham and nucleotide sequence of the corresponding gene. J. Bacteriol. 174(20): 6600–6607.

Pringle, C.R. 1999. Virus taxonomy—1999. The universal system of virus taxonomy, updated to include the new proposals ratified by the International Committee on Taxonomy of Viruses during 1998. Arch. Virol. 144(2): 421–429.

Purewal, A.S. 1991. Nucleotide sequence of the ethidium efflux gene from *Escherichia coli*. FEMS Microbiol. Lett. 82: 229–232.

Qaim, M. and D. Zilberman. 2003. Yield effects of genetically modified crops in developing countries. Science 299: 900–902.

Reddy, D.V.R., R. Richins, N. Lizuka, S.K. Manohar and R.J. Shepherd. 1993. Peanut chlorotic streak virus—a new caulimovirus infecting peanuts (*Arachis hypogea* L.) in India. Phytopathology 83: 129–133.

Reinbothe, S., A. Nelles and B. Parthier. 1991. N-(phosphonomethyl)glycine (glyphosate) tolerance in *Euglena gracilis* acquired by either overproduced or resistant 5-enolpyruvylshikimate-3-phosphate synthase. Eur. J. Biochem. 198: 365–373.

Richins, R.D. 1993. Organization and expression of the peanut chlorotic streak virus genome. Ph.D. Dissertation, Univ. of Kentucky, Lexington, KY, USA, p. 153.

Richins, R., H. Scholtof and R. Shepherd. 1987. Sequence of figwort mosaic virus. Nucl. Acids Res. 15: 8451–8466.

Robineau, T., Y. Batard, S. Nedelkina, F. Cabello-Hurtado, M. LeRet, O. Sorokine et al. 1998. The chemically inducible plant cytochrome P450 CYP76B1 actively metabolizes phenylureas and other xenobiotics. Plant Physiol. 118: 1049–1056.

Rogers, S.G., L.A. Band, S.B. Holder, E.S. Sharps and M.J. Brackin. 1983. Amplification of the aroA gene from *Escherichia coli* results in tolerance to the herbicide glyphosate. Appl. Environ. Microbiol. 46(1): 37–43.

Ronen, G., L. Carmel-Goren, D. Zamir and J. Hirschberg. 2000. An alternative pathway to β-carotene formation in plant chromoplasts discovered by map-based cloning of Beta and old-gold color mutations in tomato. Proc. Natl. Acad. Sci. USA 97: 11102–11107.

Ronen, G., M. Cohen, D. Zamir and J. Hirschberg. 1999. Regulation of carotenoid biosynthesis during tomato fruit development: expression of the gene for lycopene epsilon-cyclase is down-regulated during ripening and is elevated in the mutant Delta. The Plant J. 17: 341–351.

Rosati, C., R. Aquilani, S. Dharmapuri, P. Pallara, C. Marusic, R. Tavazza et al. 2000. Metabolic engineering of beta-carotene and lycopene content in tomato fruit. The Plant J. 24: 413–419.

Rutledge, R.G., T. Quellet, J. Hattori and B.L. Miki. 1991. Molecular characterization and genetic origin of the *Brassica napus* acetohydroxyacid synthase multigene family. Mol. Gen. Genet. 229: 31–40.

Rothnie, H.M., Y. Chapdelaine and T. Hohn. 1994. Pararetroviruses and retroviruses: a comparative review of viral structure and gene expression strategies. Adv. Virus Res. 44: 1–67.

Sadowsky, M.J., D.A. Samac, C.P. Vance and L.P. Wackett. 2005. Transgenic Plants for Bioremediation of Atrazine and Related Herbicides. USEPA. 2005 Annual Report.

Sadowsky, M.J. and L.P. Wackett. 2000. Genetics of atrazine and s-triazine degradation by *Pseudomonas* sp. strain ADP and other bacteria. pp. 268–282. *In*: J.C. Hall, R.E. Hoagland and R.M. Zablotowic (eds.). Pesticide Biotransformations in Plants and Microorganisms. ACS Symp. Ser. 777. Washington, DC: Oxford Uni. Press.

Sanger, M., S. Daubert and R. Goodman. 1990. Characteristics of a strong promoter from figwort mosaic virus: comparison with the analougous 35s promoter from cauliflower mosaic virus and the reulated mannopine synthetase promoter. Plant Mol. Biol. 14: 433–443.

Schlenitz, K.M., S. Kleinsteuber, T. Vallaeys and W. Babel. 2004. Localization and characterization of two novel genes encoding for stereospecific dioxygenases catalyzing 2(2,4-dichlorophenoxy)propionate cleavage in *Delftia acidovorans* MC 1. Appl. Environ. Microbiol. 70: 5351–5365.

Schnepf, E., N. Crickmore, J. Vanrie, D. Lereclus, J. Baum, J. Feitelson et al. 1998. *Bacillus thuringiensis* and its pesticidal crystal proteins. Microbiol. Molec. Bio. Rev. 62: 775–806.

Schonbrunn, E., S. Eschenburg, W.A. Shuttleworth, J.V. Schloss, N. Amrhein, J.N. Evans and W. Kabsch. 2001. Interaction of the herbicide glyphosate with its target enzyme 5-enolpyruvylshikimate 3-phosphate synthase in atomic detail. Proc. Natl. Acad. Sci. USA 98: 1376–1380.

Shimizu, M., K. Kawai, K. Kaku, T. Shmizu and H. Kobayashi. 2011. Application of Mutated Acetolactate Synthase Genes to Herbicide Resistance and Plant Improvement. InTech Europe. Rijeka, Croatia. www.intechopen.com.

Siminszky, B., F.T. Corbin, E.R. Ward, T.J. Fleischmann and R.E. Dewey. 1999. Expression of a soybean cytochrome P450 monooxygenase cDNA in yeast and tobacco enhances the metabolism of phenylurea herbicides. Proc Natl. Acad. Sci. USA 96: 1750–1755.

Siminszky, B., A.M. Freytag, B.S. Sheldon and R.E. Dewey. 2003. Co-expression of a NADPH:P450 reductase enhances CYP71A10-dependent phenylurea metabolism in tobacco. Pestic. Biochem. Physiol. 77: 35–43.

Sikorski, J.A. and K.J. Gruys. 1997. Understanding glyphosate's molecular mode of action with EPSP synthase: evidence favoring an allosteric inhibitor model. Acc. Chem. Res. 30: 2–8.

Simpson, D.M., T.R. Wright, R.S. Chambers, M.A. Petersom, C. Cui, A.E. Robinson et al. 2008. Introduction to Dow AgroSciences herbicide tolerance traits. Abstr. Weed Sci. Soc. Amer. 48: 115.

Smith, R.A and G.A. Couch. 1991. The phylloplane as a source of *Bacillus thuringiensis*. Appl. Environ. Microbiol. 57: 311–315.

Somers, D.A. 1996. Aryloxyphenoxypropionate- and cyclohexanedione-resistant crops. pp. 175–188. *In*: S.O. Duke (ed.). Herbicide-Resistant Crops: Agricultural, Environmental, Economic, Regulatory, and Technical Aspects. CRC and Lewis, Boca Raton, FL.

Stalker, D.M., L. Malyj and K.E. McBride. 1988a. Purification and properties of a nitrilase specific for the herbicide bromoxynil and corresponding nucleotide sequence analysis of the *bxn* gene. J. Biol. Chem. 263: 6310–6314.

Stalker, D.M., K.E. McBride and L. Malyj. 1988b. Herbicide resistance in transgenic plants expressing a bacterial detoxification gene. Science 242: 419–423.

Stavolone, L., M. Kononova, S. Pauli, A. Ragozzino, P. de Haan, S. Milligan, K. Lawton and T. Hohn. 2003. Cestrum yellow leaf curling virus (CmYLCV) promoter: a new strong constitutive promoter for heterologous gene expression in a wide variety of crops. Plant Mol. Biol. 53: 703–713.

Steinrucken, H.C. and N. Amrhein. 1984. 5-enolpyruvylshikimate-3-phosphate synthase of *Klebsiella pneumoniae*. 1. Purification and properties. Eur. J. Biochem. 143: 341–349.

Stemmer, W.P.C. 1994. DNA shuffling by random fragmentation and reassembly: *In vitro* recombination for molecular evolution. Proc. Natl. Acad. Sci. USA 91: 10747–10751.

Stoltenbrerg, D.E., J.W. Gronwald, D.L. Wyse, J.D. Burton, D.A. Somers and B.G. Gengenbach. 1989. Effect of sethoxydim and haloxyfop on acetyl-CoA carboxylase activity in festuca species. Weed Sci. 37: 512–517.

Strauch, E., W. Wohlleben and A. Pühler. 1988. Cloning of a phosphinothricin N-acetyltransferase gene from Streptomyces viridochromogenes Tü 494 and its expression in Streptomyces lividans and *Escherichia coli*. Gene 63: 65–74.

Streber, W.R., K.N. Timmis and M.H. Zenk. 1987. Analysis, cloning, and high level expression of 2,4-dichlorophenoxyacetate monooxygenase gene of *Alcaligenes eutrophus* JMP134. J. Bacteriol. 169: 2950–2955.

Streber, W.R. and L. Willmitzer. 1989. Transgenic tobacco plants expressing a bacterial detoxifying enzyme are resistant to 2,4-D. Biotechnology 7: 811–816.

Tailor, R., J. Tippett, G. Gibb, S. Pells, D. Pike, L. Jordan and S. Ely. 1992. Identification and characterization of a novel *Bacillus thuringiensis* δ-endotoxin entomocidal to coleopteran and lepidopteran larvae. Mol. Microbiol. 6: 1211–1217.

Tan, S. and S.J. Bowe. 2012. Herbicide-tolerant crops developed from mutations. pp. 423–436. *In*: Q.Y. Shu, B.P. Forster and H. Nakagawa (eds.). Plant Mutation Breeding and Biotechnology. CABI, Cambridge, MA, USA.

Tan, S.Y., R.R. Evans, M.L. Dahmer, B.K. Singh and D.L. Shaner. 2005. Imidazolinone-tolerant crops: history, current status and future. Pest Manag. Sci. 61: 246–257.

Thompson, C.J., N.R. Movva, R. Tizard, R. Crameri, J.E. Davies, M. Lauwereys and J. Botterman. 1987. Characterization of the herbicide-resistance gene bar from *Streptomyces hygroscopicus*. EMBO J. 6: 2519–2523.

Uefuji, H., L. Wang, D.A. Samac, N. Shapir, L.P. Wackett, C.P. Vance et al. 2005. Development of transgenic plantsexpressing a bacterial gene to remove and degrade an environmentally persistant herbicide, atrazine. USDA-ARS: Plant Biology Annual Meeting. Abstract No. 111.

U.S. Food and Drug Administration. 2007. Biotechnology Consultation Note to the File BNF No. 000108. Biotechnology Notification File (BNF) BNF 000108; GAT4601 and GM-HRA proteins; Soybean Transformation Event 356043. 21 September 2007.

Verdaguer, B., A. De Kochko, C.I. Fux, R.N. Beachy and C. Fauquet. 1998. Functional organization of the cassava vein mosaic virus (CsVMV) promoter. Plant Mol. Biol. 37: 1055–1067.

Wackett, L.P., M.J. Sadowsky, B. Martinez and N. Shapir. 2002. Biodegradation of atrazine and related triazine compounds: from enzymes to field studies. Appl. Microbiol. Biotechnol. 58: 39–45.

Wang, J.A. Boets, J.V. Rie and G. Ren. 2003. Characterization of cry1, cry2, and cry9 genes in *Bacillus thuringiensis* isolates from China. J. Invertebrate Pathol. 82(1): 63–71.

Wang, Y.X., J.D. Jones, S.C. Weller and P.B. Goldsbrough. 1991 Expression and stability of amplified genes encoding 5-enolpyruvylshikimate-3-phosphate synthase in glyphosate-tolerant tobacco cells. Plant Mol. Biol. 17: 1127–1138.

Wang, X., B. Li, P.L. Herman and D.P. Weeks. 1997. A Three-Component Enzyme System Catalyzes the O Demethylation of the Herbicide Dicamba in *Pseudomonas maltophilia* DI-6. Appl. Environ. Microbiol. 3: 1623–1626.

Weeks, J.T., O.D. Anderson and A.E. Blechl. 1993. Rapid production of multiple independent lines of fertile transgenic wheat (*Triticum aestivum*). Plant Physiol. 102: 1077–1084.

Wehrmann, A., A.V. Vilet, C. Opsomer, J. Botterman and A. Schulz. 1996. The similarities of bar and pat gene products make them equally applicable for plant engineers. Nature Biotechnol. 14: 1274–1278.

Weightman, A.J., A.W. Topping and A.W. Thomas. 2002. Transposition of *DEH*, a Broad-Host-Range Transposon Flanked by IS*Ppu12*, in *Pesudomonas putida* is associated with Genomic Rearrangements and Dehalogenase Gene Silencing. J. Bacteriol. 184(23): 8581–8591.

Wikimedia Commons. 2007. Bialaphos. File: Bialaphos. svg. 21 December.

Wikipedia/Wikimedia Commons. 2011. Glyphosate. EPSP reaction II.tif. 11 December.

Wohlleben, W., W. Arnold, I. Broer, D. Hilleman, E. Strauch and A. Puhler. 1988. Nucleotide sequence of the phosphinothricin N-acetyltransferase gene from *Streptomyces viridochromogenes* Tü494 and its expression in *Nicotiana tabacum*. Gene 70: 25–37.

Won, S.H., B.H. Lee, H.S. Lee et al. 2001. An *Ochrobacterium anthropi* gene conferring paraquat resistance to heterologous host *Escherichia coli*. Biochem. Biophys. Res. Commun. 285(4): 885–890.

Wong, W.-Y. and F. Huyop. 2012. Molecular identification and characterization of Dalapon-2,2-dichloropropionate (2,2DCP)-degrading bacteria from a Rubber Estate Agricultural area. Afric. J. Microbiol. Res. 6(7): 1520–1526.

Woodward, H.D., G.F. Barry, R. Forgey, M.L. Taylor, S.R. Padgette, M.H. Marino and G.M. Kishore. 1994. Isolation and characterisation of a variant of the enzyme glyphosate oxidoreductase with improved kinetic properties. MSL-13246.

Wright, T.R., J.M. Lira, T.A. Walsh, D.M. Merlo, N. Arnold, J. Ponsamuel et al. 2010. Improving and preserving high-performance weed control in herbicide tolerant crops: development of a new family of herbicide tolerant traits. Abstracts, 239th Meeting of the Amer. Chem. Soc. Washington, DC 78: 202.

Wright, T.R., J.M. Lira, T.A. Walsh, D.J. Merlo, P.S. Jayakumar and G. Lin. 2007. Dow Agrosciences, assignee. Novel herbicide resistance genes World Intellectual Property Organization patent WO/2007/053482. Oct 5. 105 inventors.

Wright, T.R., J.M. Lira, D.J. Merlo and N. Hopkins. 2005. Dow Agrosciences assignee Novel herbicide resistance genes World Intellectual Property Organization patent WO/2005/107437. Nov 17. 1117 inventors.

Yamamoto, E., L. Zeng and W.V. Baird. 1998. α-tubulin missense mutations correlate with antimicrotubule drug resistance in *Eleusine indica*. Plant Cell 10: 297–308.

Yerushalmi, H., M. Lebendiker and S. Schuldiner. 1995. EmrE, an *Escherichia coli* 12-kDa multidrug transporter, exchanges toxiccations and H⁺ and is soluble in organic solvents. J. Biol. Chem. 270: 6856–6863.

Zhang, J., Y. Xu, X. Wu and L. Zhu. 2002. A bentazon and sulfonylurea sensitive mutant: breeding, genetics and potential application in seed production of hybrid rice. Theor. Appl. Genet. 105: 16–22.

Zhang, L., Q. Lu, H. Chen, G. Pan, S. Xiao, Y. Dai et al. 2007. Identification of a cytochrome P450 hydroxylase, CYP81A6, as the candidate for the bentazon and sulfonylurea herbicide resistance gene, *Bel*, in rice. Mol. Breed. 19: 59–68.

Zhao, J.Z., J. Cao, H.L. Collins, S.L. Bates, R.T. Roush, E.D. Earle et al. 2005. Concurrent use of transgenic plants expressing a single and two *Bacillus thuringiensis* genes speeds insect adaptation to pyramided plants. Proc. Natl. Acad. Sci. USA 102: 8426–8430.

Zhong, C.H., D.J. Ellar, A. Bishop, C. Johnson, S.S. Lin and E.R. Hart. 2000. Characterization of a *Bacillus thuringiensis* δ-endotoxin which is toxic to insects in three orders. J. Invertebr. Pathol. 76: 131–139.

Zhou, H., J.W. Arrowsmith, M.E. Fromm, C.M. Hironaka, M.L. Taylor, D. Rodriguez et al. 1995. Glyphosate-tolerant CP4 and GOX genes as a selectable marker in wheat transformation. Plant Cell. Rep. 15: 159–163.

CHAPTER 6

Herbicide-Resistant Transgenic Crops

Herbicide resistance is one of the first traits engineered into plants. Generally, there are two approaches in transgenic engineering for herbicide resistance. One is the modification of a plant enzyme or other sensitive biochemical target of herbicide action to render it insensitive to the herbicide, or to induce the overproduction of the unmodified target protein permitting normal metabolism to occur. The other approach is the introduction of an enzyme or enzyme system that degrades or detoxifies the compound in the plant before the herbicide reaches the site of action. Plants modified by both approaches may be obtained either by selection for resistance against a specific herbicide or by applying gene transfer techniques utilizing genes encoding herbicide resistance determinants.

The first step in transgenic engineering is extraction and construction of the DNA sequence to be inserted into the plant. The sequence should include the gene(s) conferring resistance to a herbicide, a selectable marker gene, a promoter, and a terminator. The constructed DNA sequence is isolated and transferred into the plant cell by using a vector-mediated or vector-less transformation procedure (vide Chapter 4). Once inside the plant cell, the DNA sequence is integrated into the chromosome of the plant by a process known as recombination, which is an exchange of nucleotides between two molecules of RNA or DNA. Recombination plays an important role during meiosis in creating genetic diversity.

During the next crucial step, the successfully transformed cells are selected. Selection is mediated by a selectable marker gene, an antibiotic resistance gene, which is co-introduced into the plant cell along with the functional gene to be inserted. Following selection, the transformed transgenic tissues such as immature embryos are grown successively under controlled environmental conditions in a series of media containing different concentrations of nutrients and growth hormone in a process of tissue culture (vide Chapter 4). This process ends in the regeneration of a whole plant from a single transformed cell. This key regeneration step in producing transformed plants has been a stumbling block in producing transgenic plants in many species, but specific varieties of most crops can now be transformed and regenerated.

Once whole plants are generated and seeds produced, evaluation of the progeny begins. Intrinsic to the production of transgenic plants is an extensive evaluation process to verify whether the inserted gene has been stably incorporated into the plant genome without detrimental effects to other plant functions, product quality, or the agroecosystem. Initial evaluation includes attention to the: a) activity of the introduced gene, b) stable inheritance of the gene, and c) unintended effects on plant growth, yield, and quality.

If a plant passes these tests, most likely it will not be used directly for crop production, but will be crossed with improved elite varieties of the crop (vide 'Transgenic Breeding', Chapter 3). This is because only a few varieties of a given crop can be efficiently transformed, and these generally do not possess all the producer and consumer qualities required of modern cultivars. Crossing with elite lines is done to combine the desired traits of elite parents and the transgene into a single line. The initial cross to the improved variety must be followed by several cycles of repeated crosses to the improved parent through a process called backcross breeding. The offspring are repeatedly crossed back to the elite line to obtain a high yielding transgenic line. The result will be a plant with a yield potential close to current hybrids that expresses the trait encoded by the new transgene. This is followed by measuring and quantifying the expression of the inserted transgene(s) in the the transgenic line (vide 'Gene Expression', Chapter 3).

The next step in the process is multi-location and multi-year evaluation trials in greenhouse and field environments to test the effects of the transgene and overall performance of the transgenic line. This phase also includes evaluation of environmental effects and food safety.

The first herbicide-resistant transgenic crop variety to be released commercially was the BXN cotton line developed in 1994 by CalGene and Rhône-Poulenc. Since then, scores of transgenically modified herbicide-resistant crops continued to be released. Genes used to transfer resistance traits belong to glyphosate, glufosinate, bromoxynil, imidazolinones, sulfonylureas, sethoxydim, 2,4-D, dalapon, dicamba, atrazine, phenmedipham, paraquat, isoxafutole, mesotrione, etc. Initially, transgenic engineering was used to generate plants that made greater than normal amounts of herbicide genes, with the expectation that they would withstand higher doses of herbicides than non-targeted plants. Later, some of these lines were modified to introduce higher tolerance level to the same herbicide or resistance to a second and third herbicide, or to provide farmers more flexibility and options in weed management. Transgenic varieties were also developed to simultaneously introduce insect- and disease-resistant genes and make them resistant to a herbicide as well.

Event, Gene/Trait Stacking

Transformation Event

The transgenic line which underwent all these stages is released for commercial purpose as a 'transformation event'. The term 'event' in this context refers to the original transformed plant and its progeny that contain a heterologous DNA construct as also the regenerated population resulting from the insertion of the transgene into the genome of the plant. It also refers to the progeny produced by sexual backcrosses between a donor inbred line (the original transformed line and the progeny) comprising the insert DNA and the adjacent flanking genomic sequences and a recipient inbred line (or recurrent line) that does not contains the said insert DNA. After repeated back-crossing, the insert DNA is present in the recipient line at the same locus in the genome as in the donor line.

Transgenic events are created by transforming constructs carrying one or two trait genes individually into plants. Products with single or double traits provide flexibility in trait combinations for each region. However, the effort, which required to bring a large number of trait loci into multiple cultivars even in a single crop, quickly becomes unmanageable if each transgenic locus carries more than one or two traits. Hence, it is very desirable, from a scientific and breeding point of view, to deliver several traits simultaneously into a single locus. This is made possible by stacking of genes/traits.

Gene/Trait Stacking (Pyramiding)

Gene stacking refers to the process of combining two or more genes of interest into a single plant [ISAAA 2013a]. Gene pyramiding and multigene transfer are other monikers in the scientific literature referring to the same process. The combined traits resulting from this process are called stacked traits. A biotech crop variety that bears stacked traits is called a 'biotech stack' or simply 'stack'. The stacks are also called 'gene stack', 'stacked events', or 'pyramided stacks'. An example of a stack is a plant transformed with two or more genes that code for proteins having different modes of action. For example, it is a hybrid plant expressing both herbicide resistance and insect resistance genes derived from two parent plants [ISAAA 2013a]. Resistance to insects conferred by a single *Bt* gene has the potential to break down as the target insect pest mutates and adapts to defeat the *Bt* trait.

Gene stacking also broadens weed control efficiency by stacking up genes resistant to herbicides with different modes (sites) of action. For example, this is done by combining glyphosate resistance gene *epsps* with the *pat* gene to confer resistance to glufosinate and/ or with the *dmo* gene to confer resistance to dicamba. Biotech stacks are also engineered to have better chances of overcoming myriad other problems in the field such as diseases and abiotic stresses by stacking the concerned genes with those that confer resistance to herbicides and insect-pests, so that farmers can increase crop productivity.

The easiest and quickest way to stack up genes into a plant is to make crosses between parental plants that have different biotech traits, an approach known as hybrid stacking. Most of the commercially available biotech stacks are products of serial hybrid stacking which is widely adapted and accepted [ISAAA 2013a]. Another method of gene stacking, known as molecular stacking, involves the introduction of gene constructs simultaneously or sequentially into the single locus of the target plant by standard delivery systems such *Agrobacterium*-mediated and biolistic methods [Halpin 2005; Que et al. 2010]. In some stacks, molecular stacking has been done with conventional breeding approaches to put together the desirable traits [ISAAA 2013a].

Gene Stacking Methods

Methods of gene stacking are: a) hybrid stacking, b) co-transformation, c) re-transformation, and d) linked genes or multigene cassette transformation. As mentioned above, **hybrid stacking** involves cross-hybridization of a plant that harbors one or more transgenes with another plant containing other transgenes. Multi-stack hybrids are developed via iterative hybridization [ISAAA 2013a). An example of hybrid stack is the quadruple gene stack BT11 x MIR62 x TC1507 x GA21 in maize.

In **co-transformation**, a plant is transformed with two or more independent transgenes. The transgenes of interest are in separate gene constructs and delivered to the plant simultaneously. The **re-transformation** method involves transformation of a plant harboring a transgene with additional transgenes. In **linked genes** or **multi-gene cassette** transformation, a plant is transformed with a single gene construct that harbors two or more linked transgenes [ISAAA 2013a).

Following the commercial success of MON802 maize stack referred earlier, biotech companies sought to stack up more biotech traits into their crop portfolio to create multi-stack varieties and hybrids. In addition to the herbicide tolerance and insect resistance traits, there are several quantitative and quality traits that the agricultural biotech industry is currently developing. The increasing number of biotech traits in recent stacks has set the trend for the next generation of biotech crops. It is estimated that by 2015, 24 mono-trait events may have been commercialized for maize alone. This potentially translates to over 2,000 triple-stacks and over 12,000 quadruple-stacks that may enter the pipeline [Stein and Rodriguez-Cerezo 2009].

In the following part of the chapter, a number of transgenic crop events and stacked varieties/hybrids are described with a brief procedure of transformation for each of them. As it is impossible to include all the events and stacks developed in the world thus far, only a few selected ones are included for discussion. For more detailed information on the transgenic events and stacks discussed here, the reader may refer to other sources. The various events and stacked lines and hybrids of some of the major global crops discussed in this part of the chapter are generally transgenic. In the case of non-transgenic events and stacks, the word 'non-transgenic' is mentioned in parenthesis.

Maize (Corn)

The transgenically modified maize is one of the 25 crops grown worldwide, with majority of the area under varieties engineered for herbicide resistance. About 75 percent of the transgenic varieties were stacked with either herbicide resistance trait alone or herbicide-insecticide traits in a combination. Lately, maize farmers have shown increased preference for adopting varieties stacked with herbicide-insecticide resistance traits.

Mono-Trait Events: Herbicides

Glyphosate

GA21. The GA21 maize line (Agrisure GT) was developed by Monsanto and DeKalb through biolistic transformation of embryogenic cells with DNA-coated gold particles and regeneration of plants by tissue culture on selective medium. This first transgenic maize, commercialized as 'Roundup Ready Corn GA21', was derived from the inbred 'AT' corn variety by insertion of an additional copy of maize EPSPS encoding gene. This modified *epsps* (*zm-2mepsps*) gene, containing the EPSPS coding sequence from *E. coli*, was completely sequenced. It encodes a 47.7 kDa protein consisting of 445 amino acids. It differs from wild type maize EPSPS by two amino acid substitutions. This results in a protein with greater than 99.3 percent sequence identity to that of the maize protein [Monsanto 1997].

The transformation vector used was plasmid pDPG434. A purified 3.4 kb DNA restriction fragment (*Not*I digested pDPG434) was used in the transformation. The *Not*I fragments contain the *mepsps* gene, originally isolated from maize, under the regulation of the rice *actin I* promoter and rice *actin* intron besides the 3′ polyadenylation signal *nos* gene terminator from *A. tumefaciens*. Post-translational targeting of this gene product to the chloroplast organelles was accomplished by fusion of the 5′-terminal coding sequence with the chloroplast transit peptide (CTP) DNA sequences derived from *RuBisCo* genes isolated from maize and sunflower. The *mepsps* coding sequence was fused to the optimized transit peptide (OTP) coding sequence from sunflower. The OTP transports the EPSPS protein expressed by the *mepsps* gene to the chloroplast, the site of shikimate pathway as well as glyphosate site of action. Transit peptides are typically cleaved form the "mature" protein following delivery to the plasmids [Della-Cioppa et al. 1986].

Data from five generations of backcrossing and one generation of self-pollination demonstrated stable inheritance of the novel trait [Health Canada 1999d]. GA21, which received nonregulated status from APHIS in November 1997 [US Federal Register 1997] and commercialized in the U.S. in 1998, was embroiled in legal dispute over infringement of intellectual property rights between Monsanto and Syngenta, which named it as 'Agrisure GT.' Canada approved this transgenic line in May 1999, followed by Australia and New Zealand in November 2000, Korea in 2002 and ten other countries by 2008, with the European Union giving its approval in 2006. The GA21(R) technology has largely been phased out in the marketplace in favor of an improved glyphosate-tolerant event called 'NK603'.

NK603. NK603 of Monsanto was produced by using an inbred maize cell culture line 'AW x CW'. It was developed by introducing two *cp4 epsps* coding sequences through direct DNA transformation by microparticle bombardment of plant embryogenic cells with DNA-coated gold particles and regeneration of plants by selective culture medium. The plasmid vector used was a *Mlu*I fragment of the bacterial plasmid vector PV-ZMGT32. It contained two adjacent plant gene expression cassettes, each containing a single copy of *cp4 epsps* gene. The vector also contains an *npt*II gene bacterial selectable marker gene encoding kanamycin resistance allowing selection of bacteria containing the plasmid, and an origin of replication (ori) necessary for replicating the plasmid in *E. coli* [Monsanto 2000].

In one gene cassette, the *cp4 epsps* coding sequence was under the regulation of the rice *actin* 1 promoter gene and rice *actin* intron, and it contains the *nos*3′ polyadenylation sequence for termination of transcription as described in GA21. In the second gene cassette, the *cp4 epsps* coding sequence was under the regulation of the CaMV e35S with an enhanced duplicator region, maize *hsp70* intron, and the *nos*3′ polyadenylation sequence [CERA 2009c]. In both cassettes, the *cp4 epsps* coding sequences were fused to chloroplast transit peptide (CTP2) sequences, based on sequences isolated from *A. thaliana* EPSPS. The CTP targets the CP4 EPSPS proteins to the chloroplast, the location of EPSPS in plants and the site of aromatic amino acid biosynthesis [Health Canada 2001; Monsanto 2002]. Expression of the first *ctp2-cp4 epsps* cassette is regulated by the rice acting promoter and a rice intron sequence introduced upstream of the ctp2 sequence. Expression of the second *ctp2-cp4 epsps* cassette is regulated by the *35S* promoter from CaMV with a double enhancer region (*e35S*) and a maize intron derived from a *hsp70* gene encoding a heat shock protein. Each cassette is linked to the transcription terminator, the non-translated DNA sequence nopaline synthase (*nos 3′*) sequence from A. *tumefaciens* [Europa 2005].

Segregation and Southern blot analyses across nine generations demonstrated the stable inheritance of the novel trait in the transgenic maize line NK603. Event NK603 was approved by USA in 2000 and Canada in 2001, followed by Australia and New Zealand in 2002, and the European Union in 2004. Twelve other countries also approved this glyphosate-resistant transgenic line (Roundup Ready Corn NK603) by 2009.

MON832. In 1996, Monsanto developed another 'Roundup Ready' maize line Mon832. A commercial variety was transformed by biolistic transformation of embryonic maize cells using a mixture of DNA solution containing two plasmids, PV-ZMBK07 and PV-ZMGT10. Plasmid PV-ZMBK07 contained the synthetic *cr1Ab* gene regulated by the enhanced duplicated cauliflower mosaic virus 35S promoter (CaMV E35S) and the maize hsp70 heat shock protein intron. The polyadenylation signal was from the *Agrobacterium tumefaciens* nopaline synthase (*nos*) gene.

The second plasmid, PV-ZMGT10, contained genes encoding the CP4 EPSPS enzyme from the common soil bacterium, *A. tumefaciens* sp. CP4, and glyphosate oxidoreductase (*goxv247*) from *Ochrobactrum anthropi*. Constitutive expression of these genes in plant cells was under the control of the CaMV E35S promoter, the hsp70 intron, and *nos 3'* terminator. Post-translational targeting of the CP4 EPSPS and glyphosate oxidoreductase enzymes to the chloroplast was accomplished by fusion of the 5'-terminal coding sequences with the chloroplast transit peptide DNA sequences from the *Arabidopsis thaliana* EPSPS gene (CTP2) and from *A. thaliana* SSU1A gene (CTP1), respectively. Plasmid PV-ZMGT10 also contained sequences from the *lacZ* operon, ori-pUC and the *neo* gene. The *neo* gene on both plasmids was included as a selectable marker to identify bacteria transformed with recombinant plasmid DNAs. Expression of the *neo* gene was regulated by a bacterial promoter and therefore is not functional in plants.

Southern blot analysis of genomic DNA from MON832 (Roundup Ready) confirmed that sequences from plasmid PV-ZMBK07, containing the *cry1Ab* gene, were not incorporated, but the CP4 EPSPS encoding sequences along with three copies of the *gox* gene and two NPTII/ori-pUC sequences (one complete and one rearranged) were integrated as a single 16 kb insert. Canada approved this transgenic maize line in 1997 while it was not available for commercial use in other countries including USA [CERA 2001b].

MON87427. Conventional maize hybrid seed production has historically relied upon detasseling using either manual methods or semi-automated processes to ensure the purity of the hybrid cross. Monsanto has developed biotechnology-derived MON87427 (Roundup Ready) maize with tissue-selective glyphosate tolerance to facilitate the production of hybrid maize seed. This male-sterile maize event was developed in 2012 by introducing the *cp4 epsps* gene to confer tolerance to glyphosate in a tissue-specific manner. After infection through *Agrobacterium*-mediated transformation, the immature embryos of conventional cultivar 'LH198 x Hill' inbred were placed in a selection medium containing glyphosate to select for positive transformed cells, as well as in a carbenicillin disodium salt to limit the growth of *Agrobacterium*. Event MON87427 was selected as the lead event based on superior phenotypic characteristics and its comprehensive molecular profile.

The plasmid PV-ZMAP1043 of approximately 8.9 kb contains *cp4 epsps* gene within the T-DNA that is delineated by Left and Right border sequences. The expression cassette of T-DNA also contained the CaMV *e35S* promoter, the *hsp70* intron, the *CTP2* targeting sequence, and the *nos 3'*-nontranslated region of nopaline synthase. The backbone region of PV-ZMAP1043, located outside of the T-DNA, contains two origins of replication for maintenance of the plasmid vector in bacteria (*ori V*, *ori-pBR322*), a bacterial selectable marker (kanamycin-resistance) gene (*aadA*), and a coding sequence for repressor of primer protein for maintenance of plasmid vector copy number in *E. coli* (*rop*).

This specific promoter and intron combination drives CP4 EPSPS protein expression in female reproductive tissues, conferring resistance to glyphosate in the leaves, stalk and root tissues, and tissues that develop into seed or grain and silks but not in two key male reproductive tissues: a) pollen microspores, which develop into pollen grains and b) tapetum cells that supply nutrients to the pollen. Thus, in this male sterile MON87427, male reproductive tissues, critical for male gametophyte development, are not resistant to glyphosate, and application of herbicide during tassel development stage will produce a male sterile phenotype through tissue-selective glyphosate resistance. This eliminates or greatly reduces the need for detasseling during the production of hybrid maize seed. Tissue-specific expression of CP4 EPSPS protein in this transgenic maize enables an extension of the use of glyphosate-resistant maize as a tool in hybrid maize seed production [USFDA 2012; Canadian Food Inspection Agency 2012a].

This event is compositionally equivalent to a near-isogenic conventional comparator, suggesting that hybrid maize seed production is possible without affecting crop composition [Venkatesh et al. 2014]. This maize event was deregulated by APHIS of USDA in September 2013 [ISAAA 2013a].

HCEM485. Stine Seed Farm produced the glyphosate-resistant event HCEM485 using a particle acceleration transformation system with a gel-isolated *Not*I DNA restriction fragment of plasmid vector pDGP434 containing the modified EPSPS encoding gene (as in the case of GA21) by introducing a 6 kb maize fragment derived from the inbred line 'B73', containing a modified form of the endogenous maize EPSPS encoding gene [Held et al. 2006]. The fragment, flanked by unique *Cla*I and *Eco*RV restriction endonuclease sites, contained an *epsps* 5' regulatory sequence (before position 1868), an EPSP synthase coding region (positions 1868–5146) comprised of 8 exons and 7 introns, and a 3' untranslated region (after position 5146). The EPSP synthase coding region also contained sequences encoding an endogenous N-terminal chloroplast transit peptide (position 1868–2041) [Stine 2009].

This maize genome fragment was cloned into the *Cla*I and *Eco*RV sites of pBlueScript vector and subjected to site-directed mutagenesis using the QuickChange Site-Directed Mutagenesis Kit (Stratagene). Two mutations were introduced into the EPSPS coding sequence: a cytosine to thymine substitution at position 2886 and a second cytosine to thymine substitution at position 2897. These two point-mutations resulted in two amino acid changes within the sequence of the mature EPSPS protein, a Thr-102→Ile and Pro-106→Ser substitution [Stine 2009]. These mutations result in functional tolerance to glyphosate-containing herbicides [Spencer et al. 2000; Lebrun et al. 2003)]. They were the same mutations as those introduced into the modified EPSPS synthase encoding gene in GA21 line. The pBlueScript vector containing the genome fragment bearing double-mutated EPSPS-encoding gene is designated pHCEM.

The pHCEM plasmid was subjected to agarose gel electrophoresis, and the purified maize DNA fragment was introduced into immature maize embryos derived from inbred line 'Stine 963' by aerosol beam injection [Held et al. 2004]. Upon culturing for 5 days in nonselective medium, embryos were transferred to medium containing glyphosate (100 ml L^{-1}). After 14 days, embryos were transferred to a medium containing successive greater glyphosate concentrations, up to 540 mg L^{-1}, before regeneration.

The EPSPS (2mEPSPS) enzyme expressed in maize line HCEM485 was identical to the native wild-type maize EPSPS sequence. It did not contain any heterologous DNA sequences, either coding or non-coding, from any other species [Stine 2009].

Event HCEM485 is similar to GA21. In 2013, the US regulatory agency, APHIS extended the non-regulated status of GA21 to HCEM485 whereby this maize line was no longer subject to the regulations governing the introduction of certain genetically engineered organisms [APHIS-USDA 2013].

VCO-01981-5. In 2012, Genective SA, France, has genetically engineered VCO-01981-5 maize line by disarmed *Agrobacterium*-mediated transformation to contain the *epsps grg23ace5* gene, derived from *Arthrobacter globiformis* expressing the EPSPS ACE5 protein, an improved EPSPS enzyme which confers resistance to glyphosate [USFDA 2013]. Two separate dent-type maize partially inbred lines 'Hi-IIA' and 'Hi-IIB' (selected out of a cross between A188 and B73) were crossed. The resulting embryos of 'Hi-II' were used as target tissue for transformation.

The vector pAG3541 was used to transfer the *epsps grg23ace5* expression cassette within the T-DNA region. The gene construct in this event comprises a DNA molecule of a sugarcane ubiquitin-4 promoter and intron, operably linked to the DNA molecule coding for the maize aceto-hydroxyacid synthase (AHAS) chloroplast transit peptide which was linked to a DNA molecule coding for the EPSPS GRG23ACE5. The gene construct also comprises an upstream terminator CaMV 35S. The promoter::gene::terminator string was mobilized into *A. tumefaciens* strain LBA4404 which also harbors the plasmid pSBl, using triparental mating and plating on media containing spectinomycin, streptomycin, tetracycline, and rifampicin to form a final plasmid, pAG3541. This glyphosate-resistant VCO-01981-5 maize event was deregulated by APHIS, USDA in 2013 for commercialization [ISAAA 2013e].

Glufosinate

DLL25 (B16). The novel genetic material contained in event DLL25 (synonym B16; DKB -89790-5) of DeKalb Genetics was inserted into type II callus line cells of the maize genotype 'A188 x B73', a cross between two inbred lines. The original transformant was then crossed with elite inbred lines.

The vector pDGP165, derived from pUC19 plasmid from *Agrobacterium*, was used for transformation by biolistic method. The gene cassette contained the *bar* gene, CaMV 35S promoter, and a partial copy of the reporter gene, *bla*, of bacterial origin (from plasmid pBR232 of *E. coli*) for gene expression [CERA 2001f]. The *bar* gene was modified from GTG to ATG to optimize its expression in plants without altering the amino acid sequence it encodes.

Genetic segregation and Southern blot analyses of genomic DNA indicated that the *bar* gene was integrated at a single site of insertion into maize line DLL25. The insert also contained single incomplete copies of the 35S promoter and β-lactamase gene which, however, was truncated in the 3' region [Biosafety Clearing House 2006]. The truncated β-lactamase did not produce any protein detectable on a Western blot. This DLL25 event was approved by USA and Canada in 1996, Japan in 1999, Taiwan and Philippines in 2003, and Korea in 2004 [CERA 2001f].

T14, T25. The glufosinate-resistant transgenic T14 and T25 maize lines developed by Bayer CropScience (Aventis/AgrEvo) were derived by direct gene uptake by protoplast cultures, obtained from a yellow dent maize line 'He/89', with additional genetic material [APHIS-USDA 1995]. In this chemically mediated transformation using plasmid p35S/AC (derived from pDH51), the introduced DNA is a synthetic version of the *pat* gene which encodes the enzyme PAT. The introduced *pat* gene was modified to optimize its expression in plants without altering the amino acid sequence of the PAT enzyme. Expression of the *pat* gene was regulated by CaMV 35S promoter. The transformed lines also contained the *bla* gene as a selectable marker to identify transformed bacterial cell colonies during the initial stages of cloning the recombinant *pat* gene. The *bla* gene which codes for β-lactam antibiotics, including the moderate-spectrum penicillin and ampicillin, was not functional in the modified maize lines, as its promoter was only active in bacteria.

The genetic construct was introduced via direct uptake of the plasmid DNA by maize protoplasts. Protoplasts and plasmid DNA were mixed together in a buffered solution containing polyethylene glycol. After mixing and incubation at room temperature, the protoplasts were pelleted, washed, and suspended in a protoplast culture medium. The putatively transformed protoplasts were cultivated under various conditions until microcolonies of 20–50 plant cells were formed. These microcolonies were transferred to solid medium and grown for several weeks before selection of transformants on medium containing L-PPT (L-isomer of phosphinothricin).

Molecular analyses confirmed that T14 contained three copies of the *pat* gene, but no intact copies of the *ampR* (*bla*) gene while T25 had a single copy of the *pat* gene but no intact copy of the *ampR* (*bla*) gene. The glufosinate-resistant maize events T 14 and T25 were approved in the U.S. in 1995 and Canada in 1997. T25 was approved by the European Union and several countries between 1998 and 2012 [CERA 2006f].

676, 678, 680. Pioneer Hi-Bred developed maize lines 676, 678, and 680 which were genetically engineered to express male sterility and resistance to glufosinate [CERA 2001d]. The male-sterile trait was introduced by inserting the *dam* gene derived from *E. coli* encoding the enzyme DNA adenine methylase (DAM). The *dam* gene expresses DAM enzyme in specific tissue, which results in the inability of the transformed plants to produce anthers or pollen.

These lines also contained the *pat* gene, isolated from *Streptomyces viridochromogenes*, to confer resistance to glufosinate and to serve as a selectable marker to identify transformed plants during tissue culture regeneration. Insertion of this *dam* gene in specific plant tissues results in the inability of the transformed plants to produce anthers or pollen, leading to a male-sterile plant. Under field conditions, plants that were not male-sterile could be eliminated by application of glufosinate. This novel hybrid system provides an efficient and effective way to identify male-sterile plants for use in hybrid seed production. Both genes were inserted as a single copy locus in the genome. However, a partial copy of the *pat* gene insertion was also detected at another locus in the genome.

These transgenic maize lines were produced using biolistic technique to introduce the genes of interest, followed by plant regeneration in tissue culture on glufosinate-containing medium to select transformed plants. The plasmid DNA, PHP6710, used in the transformation contained three genes: a) DNA adenine methylase encoding *dam* gene; b) the *pat* gene; and c) the β-lactamase encoding *bla* gene, which conferred resistance to the antibiotic ampicillin and was used to select bacterial colonies transformed with plasmid DNA. Tissue-specific expression of the *dam* gene achieved included sequences from the maize anther-specific 512del promoter, and the constitutive expression of the *pat* gene was regulated by CaMV 35S promoter. Prior to biolistic transformation, PHP6710 plasmid DNA was subjected to restriction enzyme digestion to isolate a 4.5 kb DNA fragment containing the *dam* and *pat* genes, but not the *bla* gene. It was this DNA fragment, not the entire plasmid that was used for the transformation procedure [CERA 2001d]. These glufosinate-resistant male sterile mazie lines were approved by the U.S. in 1998. Auxins (2,4-D) and Aryloxyphenoxypropionates (AOPPs).

DAS-40278-9. The transgenic DAS-40278-9 maize event was developed in 2012 by Dow AgroSciences to provide increased tolerance to auxin and the ACCase-inhibiting resistance to AOPP herbicides. The process involves direct Whiskers-mediated transformation to stably incorporate the *aad-1* gene from the soil bacterium, *Sphingobium herbicidovorans*, into maize. The *aad-1* gene encodes the aryloxyalkanoate dioxygenase (AAD-1) enzyme which, when expressed in plants, degrades the 2,4-D into herbicidally-inactive 2,4-dichlorophenol (DCP). Additionally, plants expressing AAD-1 converted certain AOPP herbicides into their corresponding inactive phenols [Dow 2012]. Therefore, plants that contain this enzyme are tolerant to these herbicides.

The *aad-1* gene sequence was adapted for expression in maize, with many changes in synonymous codons (coding for the same amino acids) and inserted into a plant expression cassette to make a plasmid pDAS1740. The final transformation fragment was a 6236 bp DNA which contained the matrix attachment region (MAR) from *Nicotiana tobacum*, maize *ZmUbi1* promoter particularly in monocots, synthetic plant optimized *aad-1* gene, and ZmPer5 3' UTR regulatory sequence from maize DNA that promotes the efficient transcription and translation of a maize gene. The linear DNA fragment, without the remainder of the plasmid, was used to transform embryogenic cell suspensions of maize line 'Hi-II' using silicon carbide whisker fibers for direct DNA insertion. Both the *ZmUbi1* promoter and the ZmPer5 3' UTR genetic sequences served only as a regulatory role, and not translated with the AAD-1 protein. Selection of transformation events was based on resistance to AOPP herbicide R-haloxyfop and regeneration of *aad-1* maize plants [Dow 2012].

The maize plants containing the inserted genetic material was initially isolated by growing the transformed plant tissue on a medium containing R-haloxyfop inhibiting ACCase enzyme. Whole plants were regenerated from the surviving tissue and the selected plants were bred with desirable maize lines to generate elite varieties and hybrids containing the *aad-1* gene, leading to selection of the transgenic DAS-40278-9 event which had the inserted DNA stably integrated over several breeding generations [Dow 2012]. Analyses to evaluate 82 different nutrient compositions indicated that it is substantially equivalent to its non-transgenic counterparts derived by traditional breeding methods [Herman et al. 2010].

Mono-Trait Events: Herbicides (Non-Transgenic)

Imidazolinones (Selection-Based Approach)

In selection-based approach, embryonic cell culture of maize were subjected to sublethal doses of imidazolinone (IMI) herbicide and sectors of rapidly growing tissue were subsequently sub-cultured. These subcultures were then treated in successive selection cycles of increasing herbicide concentrations. The resistant cell lines were selected and plants regenerated in the presence of IMI herbicides [Anderson and Georgeson 1989].

Of the several IMI-resistant maize lines (XA17, XI12, QJ22, XS40, Za54, UV18, AC17, and QT15) following tissue culture selection of cell calli of the maize hybrid 'A188' x 'B73', 'XA17' and 'XI12' showed a high degree of resistance to IMIs, leading to their subsequent marketing in 1992 [Tan et al. 2005]. Plants from nonsegregating resistant progenies from the XA17 line were crossed with susceptible inbred lines. The F_1 hybrids, which showed uniform herbicide resistance, were self-pollinated and crossed back to the susceptible parent. When treated with IMI herbicide, selfed progenies segregated in a 3 (resistant):1 (susceptible) ratio, while the crossed progenies segregated in 1:1 ratio, suggesting that resistance is conferred by a single dominant nuclear allele [Shaner and Singh 1997].

The IMI-resistant lines XA17 and XI12 contain the alleling mutations of *als2*, a locus of *ALS* gene, while QJ22 and XS40 apparently have the mutations of *als1* [Tan et al. 2005; Shaner et al. 1996; Dietrich 1998]. The XI12 mutation has a single nucleotide substitution at codon 653 where Ser was substituted by Asn of the ALS protein [Dietrich 1998]. Although QJ22 and XI12 were located at the same loci, QJ22 had the same mutation as XI12 (Ser653→Asn653) [Dietrich 1998]. Similarly, mutant 2 selected from pollen mutagenesis also had the same mutation as XI12 and QJ22 [Dietrich 1998; Bright et al. 1992]. In contrast, the XA17 mutation had a single nucleotide substitution at codon 574. As a result, Trp was replaced with Leu at position 574 of the ALS protein. RSC or IMR maize obtained through tissue culture selection and tolerant to AHAS inhibitors was believed to have the same mutation as XA17 based on the sensitivity of its AHAS enzyme to imazethapyr, nicosulfuron, and primisulfuron [Tan et al. 2005; Currie et al. 1995].

Mutation Ser653→Asn653, occurring in XI12, QJ22, and mutant 2, confers tolerance only to IMIs [Tan et al. 2005]. In contrast, mutation Trp574→Leu574 of XA17 confers tolerance to IMIs besides other ALS-inhibiting pyrimidinylthiobenzoate triazolopyrimidine, and sulfonylurea herbicide families. Mutation Ala155→Thr155 confers tolerance to IMIs and pyrimidinylthio-benzoates but not to sulfonylureas or triazolopyrimidines. Mutation Ala122→Thr122 confers tolerance only to IMIs.

3417IR. The imidazolinone-tolerant maize 3417IR was developed by Pioneer-Hi-Bred by tissue culturing somatic embryos on imidazolinone-enriched media. The regenerated somaclonal variant, XA17, was subsequently crossed with the inbred maize line B73 and the resulting progeny were subsequently backcrossed into each of two Pioneer hybrids, which were then crossed to produce the hybrid 3417IR. The tolerance to imidazolinone resulted from the selection of a mutation within the ALS encoding gene. Besides, the ALS enzyme in 3417IR accumulated normal levels of valine, leucine, and isoleucine. The progeny of XA17 crosses with inbred maize lines showed regular Mendelian segregation of the novel trait. No novel proteins were produced, while the activity site of the ALS enzyme did not otherwise affect its activity except to provide resistance to imidazolinone. Agronomic, disease, and insect characteristics of 3417IR were comparable to its unmodified counterpart. This modified IMI-resistant novel maize hybrid was approved by Canada in 1994 [Health Canada 1999a].

Imidazolinones (Mutagenesis-Based Approach)

In the mutagenesis-based approach, IMI-resistance is induced by mutation achieved through chemical mutagenesis. In this, the pollen of a maize line is exposed to chemical mutagens followed by employing the mutagenized pollen to fertilize the parent line and screening the progeny for IMI herbicide tolerance.

Exp1910IT. The imazethapyr resistant trait in line EXP1910IT was selected by Zeneca Seeds following chemical mutagenesis subsequent to exposing pollen to ethyl-methane sulfonate [Health Canada 1999b]. Mutagenized pollen was then used to fertilize the inbred line, UE95, and progeny plants were screened for tolerance to imazethapyr. EMS, a commonly used chemical mutagen, affects DNA by chemically altering base pairs. The resistance to imazethapyr resulted from a single nucleotide substitution within the ALS encoding gene. This substitution resulted in a single amino acid change (Ser21→Asp21) in the sequence of the enzyme, which prevented the binding of imazethapyr to the active site, thus maintaining normal enzyme activity. This novel trait was stably inherited after several generations of backcrossing, while no novel proteins were produced in EXP1910IT variety. The single amino acid change within the active site of the ALS enzyme did not otherwise affect its activity except to provide resistance to imazethapyr at up to 500 uM. Other than resistance to this IMI herbicide, the disease, insect pest, and other agronomic characteristics of EXP1910IT were comparable to unmodified UE95 maize. This imazethapyr-resistant maize event was approved by Canada in 1996.

Sethoxydim

Two regenerable, friable, embryogenic callus cultures have been selected from a maize tissue culture of 'A188' x 'B73' cross in a medium containing sethoxydim, a cyclehexanedione herbicide [Parker 1990; Tan and Bowe 2012]. These sethoxydim-tolerant callus culture lines, S1 and S2, exhibited 100- and >100-fold increases in sethoxydim resistance, respectively, compared to the unselected control callus lines. Later, sethoxydim-resistant inbred lines were developed by backcrossing public inbred lines with plants regenerated from S2 callus cultures. These plants were heterozygous for a single, partially dominant allele that conferred tolerance to sethoxydim [Parker et al. 1990]. They were more than 40 times more tolerant to sethoxydim than plants regenerated from tissue cultures not exposed to sethoxydim (Parker et al. 1990). ACCase activity from S1 and S2 was inhibited 50 percent by sethoxydim concentrations that were 4-fold and 40-fold higher than concentrations required for 50 percent inhibition in wild type ACCase activity [Parker et al. 1990].

   In 1990, these materials were used to develop sethoxydim-resistant maize hybrids which were first marketed on a limited basis in 1996 [Owen 1998; Tan and Bowe 2012]. This technology was positioned as a strategy to control specific weeds such as *Eriochloa villosa* (woolly cupgrass), *Panicum miliaceum* (wild proso millet), *Agropyron repens* (*Elytrigia repens* subsp. *repens*; *Elymus repens*; quackgrass, couchgrass). Although this technology has been good, it is often not as effective, given the biological characteristics of the target weed, to meet farmer expectations. Although using sethoxydim-resistant maize hybrids and sethoxydim constitutes an important component in a *Eriochloa villosaa* management program, they can not be the real answer [Owen 1998].

DK412SR, DK404SR. The sethoxydim resistance trait was introduced by BASF Canada into the registered maize hybrids 'DK412' and 'DK381' via tissue culture by a phenomenon known as somaclonal variation [Health Canada 1997; CERA 2001c]. Somatic embryos of these maize hybrids were grown on sethoxydim-enriched culture media. The original sethoxydim tolerant mutant lines, which produced an altered ACCase enzyme while retaining its original catalytic properties, were selected from somaclonal variants from maize embryo tissue grown under sethoxydim selection pressure. From the somatic embryos that survived, the somaclonal variant cell line S2 was selected and subsequently regenerated. The regenerated plants were backcrossed at least six times with both parental lines of the hybrid DK412SR and DK404SR to transfer the sethoxydim-resistant trait. There was no new genetic material introduced into the genomes of these sethoxydim-tolerant lines as a result of the modification. Performance factors to measure the growth and development of DK412SR and DK404SR maize lines were comparable to the performance factors for unmodified maize lines and were within the normal ranges for the characteristics tested. These non-transgenic maize lines, primarily intended for animal feeding, were made available for commercial use in Canada in 1997 [Health Canada 1997].

Double-Trait Stacks: Herbicides

Glyphosate and ALS-Inhibitor

Event 98140. The 98140 maize double-herbicide-resistant line, a transgenic product developed by Pioneer Hi-Bred in 2007, provides resistance to two different classes of herbicides: glyphosate and ALS-inhibiting herbicides. The inbred 'PHWVZ' (derived from two inbreds 'PH09B' and 'Hi-II') was genetically engineered to express the GAT4621 and ZM-HRA (the modified version of ALS) proteins. The GAT4621 protein was encoded by the *gat4621* gene. The ZM-HRA protein, encoded by the *zm-hra* gene, confers tolerance to the ALS-inhibiting herbicides. The *zm-hra* gene was derived by isolating the herbicide-sensitive maize *als* gene and introducing two specific amino acid changes known to confer herbicide resistance to tobacco ALS [CERA 2012f; Pioneer 2007]. Expression of the *zm-hra* gene is driven by the maize ALS promoter.

This transgenic line was produced by *A. tumefaciens* (strain LBA4404)-mediated transformation of immature maize embryos (removed from the developing caryopsis 9–10 d after pollination) using the plasmid vector PHP2479 with a 7440 bp T-DNA region containing *gat4621* and *zm-hra* gene cassettes. The *gat4621* cassette comprised of the *ubiZM1* promoter, the *ubiZM1* intron, the *gat4621* gene, and the *pin*II terminator from *Solanum tuberosum*. The *zm-hra* cassette consisted of the maize *als* promoter, the *zm-hra* gene, and the *pin*II terminator. Located between the two cassettes are three copies of the CaMV 35S enhancer element to provide transcription enhancement to both cassettes [CERA 2012f; Pioneer 2007].

After 6 d, the embryos were transferred to selective medium that contained glyphosate for selection of cells expressing the *gat4621* transgene. The medium also contained carbenicillin to kill any remaining *Agrobacterium*. After 2 weeks, healthy, growing calli which showed resistance to glyphosate were identified. The putative transgenic calli were continuously transferred to fresh selection medium for further growth until the start of the regeneration process. The presence of both transgenes was confirmed by PCR analysis. The regenerated whole transgenic plants (T$_0$) were transferred to the greenhouse for evaluation for glyphosate and ALS-inhibitor herbicide tolerance. The stability of the *gat4621* and *zm-hra* genes across four generations of event 98140 maize (T0S3, BC0S2, BC1, and BC1S1) was determined by Southern hybridization analysis of *Eco*RV digests of genomic DNA [Pioneer 2007]. Both transgenes were stably integrated at a single site in the 98140 maize genome.

This dual herbicide resistant transgenic maize stack, marketed in the U.S. under the brand name "Optimum", was expected to enable farmers to choose an optimal combination of glyphosate and ALS-herbicide to best manage a broad spectrum of weeds while delaying the population shifts of troublesome weeds or the development of resistance. USA, Canada, and Korea approved this transgenic maize variety in 2008, 2009, and 2010 respectively.

Double-Trait/Event Stacks: Herbicide-cum-Insect Resistance

Glyphosate-cum-Lepidopteran Insects

MON802. This novel MON802 maize glyphosate-cum insect stack was transformed to be resistant to the European corn borer, a major lepidopteran insect pest, and glyphosate. This 'Insect-Protected Roundup Ready' maize produces Cry1Ab, the insecticidal protein sensitive to lepidopteran insects (vide Chapter 5). A synthetic *cry1Ab* gene was developed for maximum expression in maize. Besides, two glyphosate-resistant genes, *cp4 epsps* and *gox247*, were also introduced into this transgenic line.

MON802 line (YieldGard) was produced by biolistic transformation with a mixture of DNAs from two plasmids, PV-ZMBK15 and PV-ZMGT03. Plasmid PV-ZMBK15 contained the synthetic *cry1Ab* gene regulated by the CaMV 35S promoter while plasmid PV-ZMGT03 contained *cp4 epsps* and the *goxv247* genes. The expression of the *cp4 epsps* gene was regulated via the CaMV 35S promoter and the *hsp70* intron, while *gox247* gene expression was under the control of the CaMV 35S promoter only. Both genes were terminated with sequences from the 3′ polyadenylation signal from *nos* gene from *A. tumefaciens*. Post-translational targeting of CP4 EPSPS and GOX enzymes to the chloroplast was accomplished by fusion of the 5′-terminal coding sequences with chloroplast transit peptide (CTP) DNA sequences from *A.thaliana epsps* gene (*CTP2*) and from *A. thaliana SSU1A* gene (*CTP1*), respectively. The plasmids also contained sequences encoding the enzyme NPTII from the *Tn5* transposon of *E. coli*, strain K12. Expression of the *neo* gene was regulated by a bacterial promoter and used as a selectable marker to identify bacteria transformed with recombinant plasmid DNAs, but it is not functional in plants. The expressions of CP4 EPSPS and GOX activity were used as selectable traits to screen transformed plants for the presence of the *cry1Ab* gene and to confer resistance to glyphosate [USDA-APHIS 1997; Health Canada 1999f].

Both glyphosate-resistance and insect-tolerance traits in MON802 event have remained stable for six generations. This transgenic line was approved by USA in 1996, Canada and Japan in 1997 [CERA 2005e], and Australia and New Zealand in 2000 [ANFZA 2000].

Glyphosate-cum-Coleopteran Insects

MON88017. The glyphosate-resistant and the coleopteran rootworm protected maize stack MON88017 (YieldGard VT Rootworm) of Monsanto was produced by *Agrobacterium*-mediated transformation of the hybrid maize line 'LH198' [CERA 2009a]. The T-DNA segment of the vector plasmid PV-ZMIR39 contained sequences corresponding to a synthetic variant of the *cry*3Bb1 gene and the *cp4 epsps* gene. Transcription of the *cry3Bb1* gene was directed by the 35S promoter with a duplicated enhancer region (*P-e35S*) from CaMV to control the constitutive expression of the synthetic variant of the *Bt* gene; the 5′ untranslated leader from the wheat chlorophyll a/b-binding protein; and the rice *actin* gene first intron sequence (*I-ract1*) to stabilize levels of gene transcription. The terminator and polyadenylation sequences were derived from the 3′ UTR of the wheat 17.3 kDA heat shock protein (tahsp17 3′).

The *cp4 epsps* gene was fused to a*ctp2* gene, isolated from *A. thaliana* to target the expression of the novel protein to the chloroplasts, the site of aromatic amino acid synthesis. The constitutive expression of the *cp4 epsps* gene was regulated by the rice *actin* gene promoter and enhanced by the rice *actin* gene first intron promoter (*P-ract1*) and the rice actin 1 intron sequence (*I-ract1*) for gene transcript stability. Terminator and polyadenylation sequences were derived from the 3′ untranslated region of the nopaline synthase (NOS) sequence. The left border sequence (from octopine Ti plasmid pTi15955) and right border (from nopaline Ti plasmid pTiT37) contained non-coding sequences essential for the transfer of the T-DNA segment.

Both novel genes, *cry3Bb1* and *cp4 epsps*, along with their respective promoter, enhancer and terminator sequences, were stably integrated into the genome of MON88017 across 10 generations. This transgenic glyphosate-cum-insect resistant maize line was approved in the U.S. in 2005, Canada, Australia, New Zealand, Japan, Korea, and Taiwan in 2006, China in 2007, the European Union in 2009, and Colombia in 2011 [CERA 2009a].

Glufosinate-cum-Lepidopteran Insects

CBH351. This glufosinate-cum-lepidopteran insect resistant CBH351 maize stack was produced by Aventis CropScience by using biolistic transformation of the backcrossed hybrid maize line (PA91 x H99) x H99 with two pUC19 based plasmids, PRSVA9909 and pDE110. Both plasmids contained the modified *cry9C* gene (derived from *B. thuringiensis* subsp. tolworthi strain BTS02618A) and the *bar* gene, respectively, engineered for enhanced expression in plants [CERA 2001e]. The *cry9C* and *bar* genes were fused to non-coding regulatory sequences that enabled them to be expressed at high levels, constitutively throughout most of the plant. Specifically, the expression of the modified *cry9C* gene was regulated by the promoter and terminator sequences from the CaMV 35S, along with the leader sequence of the *cab22L* gene from petunia. The expression of the *bar* gene was also directed by the CaMV 35S promoter along with the 3' untranslated region from the *nos* gene derived from *A. tumefaciens*. Additional genetic elements present on the transforming plasmids included the ampicillin resistance gene β-*lactamase* (*bla*) and the origin of replication (ori) both from *E. coli*. Both were introduced into CBH-351 maize. However, these elements are non-functional in plants. The *bla* gene was present on the plasmids only as a selectable marker to detect transformed *E. coli* host bacteria [CERA 2001e]. This maize line CBH-351, under the commercial name 'StarLink' has only been approved by USA in 1998 for livestock feed use [CERA 2001e].

TC1507. The glufosinate-cum-lepidopteran insect resistant maize stack TC1507 was developed by Mycogen (parent Dow Ago Sciences) and Pioneer Hi-Bred (later DuPont Pioneer) from a hybrid line 'Hi-II' using biolistic method of transformation by introducing a DNA construct into plant cells [Health Canada 2002; USFDA 2001]. The gene construct consisted of a linear portion of DNA containing the truncated *cry1F* gene, the *pat* gene, and regulatory sequences necessary for expression of both genes. The protein encoded by the truncated *cry1F2* gene (derived from *B. thuringiensis* var. *aizawai* strain PS811) is nearly identical to the first 605 amino acids of the Cry1F protein protoxin produced by *Bt* bacterium, which is effective in controlling certain lepidopteran larval pests common to maize, such as European corn borer and black cutworm. In order to optimize expression of the Cry1F protein, the nucleotide sequence of the *cry1Fa2* gene was modified via *in vitro* mutagenesis to contain plant-preferred codons.

The *pat* gene was used as the selectable marker. Transcription of the *cr1Fa2* gene was directed by the promoter and a 5' untranslated region from the maize ubiquitin (*ubi*) gene including the first exon and intron. The 3' termination/polyadenylation sequences were derived from the *Agrobacterium tumefaciens* open reading frame 25 (ORF25 PolyA). The *ubi* exon and intron included in this construct (PHI8999) have no effect on the structure of the Cry1F product, but only on the expression of the gene. Located next to the *cry1F2* gene, the *pat* gene is controlled by a different promoter and trailing regulatory element. The presence of PAT protein in the transformed maize cells enables selection of these cells for further development and makes line 1507 resistant to glufosinate. The gene construct also has CaMV 35S as promoter. The *cry1F* and *pat* genes were stably integrated in the plant genome and they were also inherited as dominant genes. This glufosinate-cum-insect resistant transgenic stack TC1507 (Herculex 1) became commercially available in USA in 2001 and Canada in 2002.

DAS-51922. DAS-51922 maize stack of Dow AgroSciences is another glufosinate-cum-coleopteran insect resistant maize line. Maize line 'HI-II', derivative of 'A188' and 'B73' inbred lines, was modified by *Agrobacterium*-mediated method transforming plasmid PHP17662 carrying a T-DNA vector containing three genes *cry34Ab1*, *cry35Ab1* (conferring resistance to rootworm insect), and *pat* gene for insertion into the plant genome. The *cry34Ab1* and *cry35Ab1* genes were isolated from the common *B. thuringiensis* (*Bt*) strain PS149B1.

The plasmid vector plasmid PHP17662 contained the spectinomycin and tetracycline resistance genes in the backbone and an origin of replication for *Agrobacterium*. The T-DNA region contained the elements in the following order: Right T-DNA border, corn ubiquitin promoter, *cry34Ab1* gene, *PIN*II terminator, tobacco anionic (TA) peroxidase promoter, *cry35Ab1*, *PIN*II terminator, CaMV 35S promoter, *pat* gene, CaMV 35S terminator, Left T-DNA border [Dow 2004].

Immature maize embryos, removed from the developing caryopsis, were co-cultivated with *A. tumefaciens* strain LBA4404 on solid culture medium followed by their subsequent transfer to fresh culture medium that contained antibiotics and glufosinate-ammonium. The calli that survived the herbicide proliferated and produced the genetically transformed embryonic tissue, which was then manipulated to regenerate whole transgenic plants before transferring to the greenhouse. Positive plants were crossed with an inbred line to obtain seed from the initially transformed plants. A number of lines were field-evaluated resulting in the selection of line 59122, based on its good agronomic characteristics and resistance to corn rootworm. The transgenes were stably inserted and integrated at the genome level of DAS-59122 in different environments. This transgenic insect-cum-glufosinate resistant maize variety was approved in USA and Mexico in 2004 and various other countries between 2005 and 2011.

DP004114. This maize stack, referred to as "4114 maize" or '4114' combines genes and traits from two maize lines previously approved for commercial use: event TC1507 which expresses the lepidopteran-insect protected Cry 1F and event DAS-59122 which expresses the coleopteran-tolerant Cry 34Ab1, Cry 35Ab1 proteins. Both events contain the glufosinate-resistant PAT proteins. It was developed by Pioneer Hi-Bredusing *Agrobacterium*-mediated transformation of the inbred line PHWWE. The disarmed *A. tumefaciens* strain LBA4404 carried a binary plasmid vector PHP27118. The 11,978 bp T-DNA carried four gene expression cassettes: *cry1F*, *cry34Ab1*, *cry35Ab1* and *pat* [APHIS-USDA 2012]. These cassettes consisted of the following elements [APHIS-USDA 2012].

a) *cry1F*: promoter, 5' untranslated region (UTR) and intron from the maize polyubiquitin gene, a truncated version of the gene from *B. thuringiensis* var. *aizawai*, and a terminator sequence from the *A. tumefaciens* pTi15955 ORF 25; b) *cry34Ab1*: the promoter as in *cry1F*, codon-optimized version of the gene from *B. thuringiensis* strain PS149B1, and the *pin*II gene as terminator; c) *cry35Ab1*: promoter from *Triticum aestivum* peroxidase including leader, codon-optimized version of the *cry35Ab1* gene encoding a 44 kDa δ-endotoxin, and the *pin*II gene; and d) *pat*: promoter from CaMV 35S, codon-optimized *pat* gene, and CaMV 35S terminator. Besides, the inserted T-DNA also carried polylinkers containing restriction enzyme recognition sites [APHIS-USDA 2012]. The regenerated plants from the transformation event contained a single copy of intact T-DNA which was stably inserted as evident by analysis over five generations.

DBT418. The glufosinate-resistant and the lepidopteran insect-protected DBT418 (DKB-89614-9) maize stack line was originally produced by DeKalb Genetics via biolistic transformation of the inbred line 'DBT418' with DNA from three different plasmid vectors, respectively containing the genes *cry1Ac* {from *B. thuringiensis* subsp. *kurstaki* strain HD-73 (B.t.k.)}, *bar* (from *S. hygroscopicus*), and *pin*II (proteinase inhibitor from potato: *S. tuberosum*). The synthetic *cry1Ac* gene introduced into line DBT418 encodes the first 613 amino acids of the native Cry1Ac protein. It was modified to reflect plant-preferred codon frequencies in order to maximize translation in maize cells [CERA 2005g].

Constitutive expression of the *cry1Ac* gene was under the control of the CaMV 35S promoter and two copies of the octopine synthase (OCS) enhancer from *A. tumefaciens*. The OCS enhancer promotes the expression of genes in most vegetative tissues. An intron from a maize alcohol dehydrogenase gene (*adhI* intron VI) was positioned downstream from the promoter to enhance gene expression. The 3'-terminal sequences from the potato *pin*II gene were used as a termination signal for the *cry1Ac* gene. Expression of the *bar* gene was regulated using the CaMV 35S promoter and termination sequences from the Ttr 7 transcript 7 polyadenylation region of the *A. tumefaciens* T-DNA. The third gene, *pin*II was under the control of the CaMV 35S promoter and *adhI* intron I but was not completely expressed [CERA 2005g]. Each plasmid also included the *bla* gene, derived from *E. coli* and expression was under the control of its own bacterial promoter. It was included as a selectable marker to identify bacteria transformed with recombinant plasmid DNAs, but is not functional in plants [CERA 2005g].

The co-transformed maize stack DBT418 has two intact copies of the *cry1Ac* gene, one intact copy of the *bar* gene and one rearranged copy of the *bar* gene stably integrated into its genome. The *bla* gene was not expressed. The non-functional *pin*II gene too was not expressed. This glufosinate- and insect-resistant transgenic maize stack DBT418 (commercial name: 'Bt Extra') of Monsanto (purchaser of Dekalb in 1998) was approved by USA and Canada in 1997 and other countries later [CERA 2005g].

SYN-BT11 (BT11). The lepidopteran insect-protected glufosinate-resistant BT11 maize stack was developed by Syngenta through direct transformation of plant protoplasts of inbred line 'H8540' and regeneration on selective medium. The plasmid pZO1502 contained a truncated synthetic *cry1Ab* gene (derived from *B. thuringiensis* subsp. *kurstaki*, strain HD-1) encoding Cry1Ab endotoxin and a synthetic *pat* gene to express PAT enzyme. Prior to transformation, the plasmid vector was treated with the restriction endonuclease *Not*I in order to remove the *bla* gene from the DNA fragment containing the *cry1Ab* and *pat* genes. Both genes were introduced in one T-DNA insert.

Constitutive expression of the *cry1Ab* gene was controlled by the CaMV 35S promoter modulated by the IVS6 intron (from maize alcohol dehydrogenase 1S gene), and the 3'*nos* gene from *A. tumefaciens*. The *pat* gene, present as a selectable marker, was under the control of CaMV 35S promoter, the IVS2 intron, and NOS 3' terminator. The plasmid also contained the *bla* gene that confers resistance to β-lactum antibiotics, as selectable marker to screen transformed bacterial cells. The *bla* gene was excised from the plasmid vector prior to transformation of the maize tissue [CERA 2005d]. Commercialized as 'Agrisure CB/LL', this maize stack line BT11was approved by USA, Canada, and Japan in 1996 followed by the European Union in 1998. Other countries approved it between 2002 and 2004.

Double-Event/Trait Stack Hybrids: Herbicide-cum-Insect Resistance

Glufosinate-cum-Lepidopteran Insects

SYN-BT011-1 x SYN-IR604-5 (BT11 x MIR604). BT11 x MIR604 stacked line developed by Syngenta is an F_1 transgenic maize hybrid resulting from the hybridization of the lepidopteran-resistant cum glufosinate-tolerant maize parental line BT11 and the coleopteran-tolerant maize parent MIR604 [CERA 2007b]. This stacked hybrid expresses four novel proteins: the δ-endotoxin Cry1Ab which confers resistance to the European corn borer and other lepidopteran insects; the PAT protein which confers tolerance to the herbicide glufosinate; the δ-endotoxin mCry3A which confers resistance to corn rootworm; and the PMI protein which allows growth on mannose as a carbon source and is used as a selectable marker. The genes inserted include the *bar* (*pat*) synthetic gene from the native *S. viridochromogenes* sequence, modified *cry1Ab* gene from *B. thuringiensis*, modified *cry3A* gene, and *pmi* (manA) from *E. coli* encoding phosphomannose isomerase.

The novel traits of each parental line have been combined, through traditional plant breeding, to produce this new hybrid which was approved in 2007 by several countries such as Mexico, Korea, Japan, Colombia, and Philippines [CERA 2007b] under the commercial name 'Agrisure CB/LL/RW'.

Double-Event/Triple-Trait Stack Hybrids: Herbicide-cum-Insect Resistance

Glyphosate and Glufosinate-cum-Lepidoteran and Coleopteran Insects

BT11 x GA21. Syngenta developed another F_1 hybrid resulting from the hybridization of the lepidopteran-resistant and herbicide-resistant maize line BT11 and dual herbicide (glufosinate and glyphosate)-resistant maize line GA21. This double-event (triple-trait) stack expresses three novel proteins: the δ-endotoxin Cry1Ab which confers resistance to the European corn borer and other lepidopteran insects; the PAT protein which confers resistance to glufosinate; and a modified EPSPS protein to confer resistance to glyphosate. The modified EPSPS protein was produced by the *mepsps* gene from GA21, the insecticidal protein Cry1Ab by the *cry1Ab* gene, and PAT by the *pat* gene, both from BT11. The novel traits of each parental line have been combined, through traditional plant breeding, to produce this new hybrid [CERA 2012g].

For a full description of each parental line please refer to the individual product descriptions in the crop database for BT11 and GA21. This double-event stack maize hybrid (Agrisure GT/CB/LL) was approved by several countries between 2007 and 2012, but not USA [CERA 2012g].

Triple-Event/Trait Stack Hybrids: Herbicide-cum-Insect Resistance

Glyphosate and Glufosinate-cum-Lepidoptern-Coleopteran Insects

BT11 x MIR604 x GA21. The maize hybrid stack BT11 x MIR604 x GA21 was obtained by means of traditional cross breeding of three genetically modified parental lines: BT11, MIR604, and GA21. It was produced by Syngenta to protect the plant from lepidopteran (Pyralidae and Sesamia) and coleopteran (diabrotica) insect attacks, and for resistance to glyphosate and glufosinate herbicides. This hybrid has inherited all the transgenes present in the transgenic maize progenitors: the *cry1Ab* and the *pat* genes from line Bt11; a copy of the modified *cry3A* (*mcry3A*) gene from line MIR604; and the modified *epsps* (*zm-*

2mepsps) gene from line GA21. As a consequence, this triple-event (four-gene) stacked maize produces the Cry1Ab proteins, active against lepidopteran insects, the modified Cry3A proteins active against coleopteran insects, and the PAT and EPSPS enzymes for resistance to glufosinate and glyphosate. This transgenic maize hybrid was approved by Canada, Japan, Philippines, and Korea in 2007 [CERA 2007a] for commercial use as 'Agrisure 3000GT'. Colombia approved in 2012.

BT11 x GA21 x MIR162. Developed by Syngenta by cross breeding three genetically modified lines BT11, MIR161, and GA21, the maize hybrid stack has inherited all the four transgenes present in the its progenitors: the *cry1Ab* and *pat* genes from BT11; the *vip3Aa* and *pmi* genes from maize MIR162; and the *mepsps* gene from GA21. As a consequence, this triple-event stacked maize produces the Cry1Ab δ-toxin, the Vip3Aa vegetative insecticidal protein, active against a broad-range of lepidopteran insects, and the EPSPS and PAT enzymes for resistance to glyphosate and glufosinate. The insecticidal *cry1Ab* and *vip3Aa* genes were derived from *B. thuringiensis* subsp. *kurstaki* and *B. thuringiensis* strain ABBB respectively, while the herbicidal *pat* and *mepsps* genes were sourced from *S. viridochromogenes* strain Tu494 and *Zea mays* respectively. The *pmi* gene, an environment friendly selectable marker, which encodes mannose-6-phosphate (phosphomannose) isomerase was derived from *E. coli*. This stacked maize hybrid was approved in 2011 by Brazil and in 2012 by Colombia under the commercial name 'Agrisure Viptera 3111' [CERA 2012d].

MON89034 x TC1507 x NK603. This triple event maize line was produced by means of conventional cross breeding of maize parental lines MON89034, TC1507 and NK603. The stacked maize hybrid has inherited the *cry1A* and the *cry2Ab* transgenes and resistance to lepidopteran insects from the event MON89034. It has inherited the *cry1F* gene, which also confers resistance to many lepidopteran insects, and the *pat* and *cp4 epsps* genes for resistance to glufosinate and glyphosate respectively from TC1507 and NK603.

The molecular analysis confirmed the integrity and stability of the inserted DNA in these three events. The patterns of hybridization (for the individual events and for the stacked hybrid) and inheritance of all the new genes were analyzed, confirming that they were inherited in a predictable pattern according to the principle of Mendelian genetics [SENASA 2012]. This five-gene transgenic maize hybrid developed by Monsanto and Dow AgroSciences has been approved for commercialization by Brazil, Argentina, and Paraguay under its commercial name 'Power Core' (vide Chapter 8).

Quadruple-Event Stack Hybrids: Herbicide-cum-Insect Resistance

Glyphosate and Glufosinate-cum-Lepidopteran and Coleopteran Insects

MON89034 x TC1507 x MON88017 x DAS-59122-7. This multi-gene (eight) quadruple-event stacked maize hybrid, commercialized under the name 'SmartStax' by Monsanto, produces six lepidopteran- and coleopteran-active *B. thuringiensis* (*Bt*) proteins, as well as CP4 EPSPS and PAT proteins to confer resistance to glyphosate and glufosinate respectively. This hybrid is a product of traditional plant breeding between four transgenic events: Mon89034 (lepidopteran insect resistant); TC1507 (glufosinate and lepidopteran insect resistant); MON88017 (glyphosate and coleopteran insect resistant); and DAS59122-7 (glufosinate and coleopteran insect resistant) [CERA 2010a]. The gene elements introduced in this transgenic hybrid are presented in Table 6.1.

Table 6.1. The introduced genetic elements in MON89034 x TC1507 x MON88017 x DAS-59122-7 quadruple-event stacked maize hybrid (SmartStax) [CERA 2010a].

Gene	Protein/toxin and Gene Source	Promoter, etc.	Terminator
pat	Phosphinothricin N-acetyltransferase *Streptomyces viridochromogenes*)	CaMV 35S	CaMV 35S 3′ poly-adenylation signal
cp4 epsps	5-enolpyruvyl shikimate-3-phosphate synthase (*A.tumefaciens CP4*)	Rice *actin* 1 promoter and intron sequences chloroplast transit peptide from *A. thaliana*	*A. tumefaciens* nopaline synthase (*nos*) 3′-UTR
cry35Ab1	Cry35Ab1 δ-endotoxin (*Bacillus thuringiensis* strain PS149B1)	*Triticum aestivum* peroxidase gene root-preferred promoter	*S. tuberosum* proteinase inhibitor II (PINII)
cry1A.105	Chimeric cry1 delta-endotoxin (*B. thuringiensis*)	CaMV 35S 5′ untranslated leader from wheat chlorophyll a/b-binding protein	3′ UTR of wheat heat shock protein 17.3
cry2Ab	Cry2Ab δ-endotoxin (*Bacillus thuringiensis*)	FMV-35S—promoter from Figwort Mosaic Virus; *hsp70* intron from maize heat shock protein gene	*A. tumefaciens* nopaline synthase (*nos*) 3′-untranslated region
cry3Bb1	Cry3Bb1 δ-endotoxin (*Bacillus thuringiensis* subsp. *kumamotoensis* strain EG4691)	CaMV 35S promoter with duplicated enhancer region 5′ UTR from wheat chlorophyll a/b-binding protein; rice *actin* gene first intron	3′ UTR from wheat heat shock protein (tahsp173′)
cry34Ab1	Cry34Ab1 δ-endotoxin (*B. thuringiensis* strain PS149B1)	*Zea mays* ubiquitin gene promoter, intron and 5′ UTR	*Solanum tuberosum* PINII
cry1Fa2	Cry1F δ-endotoxin (*Bacillus thuringiensis* var. *aizawai*)	Ubiquitin (*ubi*) ZM (*Zea mays*) promoter and the first exon and intron	3′ poly-adenylation signal from ORF25 (*A. tumefaciens*)

Hybridization patterns for the combined trait product were identical to those of the parental lines with *cry1F, cry34Ab1, cry35Ab1,* and *pat* gene probes indicating that the TC1507 and DAS-59122-7 insertions were unaffected by combining with MON89034 and MON88017 through conventional breeding. Despite concerns about the safety of the accumulated toxins to the consumers and environment, this multi-transgene maize line was approved by USA, Japan, Korea, and Taiwan in 2009 followed by Colombia and Philippines in 2010 [CERA 2010a].

This SmartStax maize hybrid stack builds towards Monsanto's promise of doubling yields by 2030 on the same or less land [Genuity 2013; Smartstax 2013].

BT11 x MIR162 x MIR604 x GA21. This multi-gene stacked maize, with a commercial name 'Agrisure Viptera 3111', was developed by Syngenta by means of traditional cross breeding of four genetically modified maize lines: BT11, MIR162, MIR604, and GA21. It has inherited all the transgenes present in the transgenic maize progenitors: the *cry1Ab* {Cry1Ab δ-endotoxin (*Btk* HD-1)} and the *pat* genes from BT11; the *vip3Aa* (vegetative insecticidal protein from *B. thuringiensis* strain AB88) and the *pmi* (mannose-6-phosphate isomerase from *E. coli*) genes from MIR162; the *mcry3A* (Cry3A-δ-endotoxin from *B. thuringiensis* subsp. *tenebrionis*) and the *pmi* genes from MIR604; and the *mepsps* gene from GA21. Consequently, the maize hybrid produces Cry1Ab and the Vip3Aa proteins, active against lepidopteran insects, the Cry3A proteins, active against coleopteran insects, and the PAT and the EPSPS enzymes for resistance to glyphosate and glufosinate herbicides. This multi-gene hybrid maize was approved by Colombia in 2012 [CERA 2012c].

Quintuple-Event Stack Hybrids: Herbicide-cum-Insect Resistance

Glyphosate and Glufosinate-cum-Lepidopteran and Coleopteran Insects

BT11 x DAS59122 x MIR604 x TC1507 x GA21. This quintuple-event stack maize hybrid of Syngenta was obtained by conventional breeding of five genetically modified maize lines: BT11, 59122, MIR604, TC1507, and GA21. This maize hybrid has inherited all the transgenes present in these five events (Table 6.2): the *cry1Ab* and the *pat* genes from maize line BT11; the *Cry34Ab1*, the *Cry35Ab1* and the *pat* genes from maize line 59122; the *mcry3A* and the *pmi* genes from MIR604; the *cry1F* and the *pat* genes from maize line TC1507; and the *mepsps* gene from GA21 [ISAAA 2013c; Biosafety Scanner 2013]. As a consequence, the stacked GM plant produces the Cry1Ab and Cry1F proteins active against certain lepidopteran insects, different Cry3A proteins active against coleopteran insects, and the PAT and EPSPS enzymes which allow the plant to tolerate glufosinate-ammonium and glyphosate based herbicides. The *pmi* gene serves as a selection marker/reporter. This maize is commercialized under the name 'Agrisure 3122' [ISAAA 2013c; Biosafety Scanner 2013].

SYN5307 x BT11 x MIR604 x TC1507 x GA21. Syngenta developed another quintuple-event stack maize hybrid (Agrisure Duracade 5122). This was produced through conventional cross hybridization and selection involving five parental lines (SYN 5307, BT11, MIR604, TC1507 and GA21) which contributed four genes that confer resistance to lepidopteran and coleopteran insects in addition to the two genes conferring tolerance to glyphosate and glufosinate. There is only one difference between this hybrid and the other quintuple-event hybrid (Agrisure 3122) described earlier. The maize Event DAS59122 (contributor of coleopteran insect-resistant *cry34Ab1* and *cry35Ab1* genes) in Agrisure 3122 was replaced with SYN5307 in this hybrid. The SYN5307 line, which confers tolerance to both coleopteran and lepidopteran insects, contains *ecry3.1Ab* gene, derived from synthetic form of *cry3A* gene and *cry1Ab* gene from *B. thuringiensis* [ISAAA 2013d; CERA 2013].

This unique *ecry3.1Ab* gene encodes eCry3.1Ab protein for control of certain coleopteran insects like *Diabrotica virgiferavirgifera* LeConte (Western corn rootworm: WCR) and related species. The coleopteran-active eCry3.1Ab protein was generated following variable-region exchange of *B. thuringiensis* lepidopteran-active protein, Cry1Ab, with a Cry3A region [Walters et al. 2010]. This variable-region exchange is

Table 6.2. The introduced genes, gene sources, gene products and resistance targets in quintuple-stacked herbicide-cum-insect resistance maize hybrid (Agrisure 3122) [ISAAA 2013c; Biosafety Scanner 2013].

Event	Gene	Gene Product	Resistance
BT11	*cry1Ab*	Cry1Ab δ-endotoxin	Lepidopteran insects
TC1507	*cry1Fa2*	Modified Cry1F protein	Lepidopteran insects
BT11; DAS59122	*pat*	PAT enzyme	Glufosinate
GA21	*mepsps*	Modified EPSPS enzyme	Glyphosate
MR604	*mcry3A*	Modified Cry3A δ-endotoxin	Coleopteran insects
DAS59122	*cry34Ab1*	Cry34Ab1 δ-endotoxin	Coleopteran insects
DAS59122	*cry35Ab1*	Cry35Ab1 δ-endotoxin	Coleopteran insects
MR604	*pmi* (Selection Marker/Reporter)	Phosphomannose isomerase (PMI)	Metabolizes mannose for selection of transformants

responsible for imparting strong bioactivity against the larvae of WCR, an insect pest species which is not susceptible to either parent protein sequence.

Hextuple-Event Stack Hybrids: Herbicide-cum-Insect Resistance

Glyphosate and Glufosinate-cum-Lepidopteran and Coleopteran Insects

SYN5307 x MIR604 x BT11 x TC1507 x GA21 x MIR162. This multi-transgene line stacked F$_1$ hybrid with six maize lines was developed by Syngenta through conventional cross hybridization and selection involving multi-transgenic donors, which conferred resistance to glufosinate and glyphosate herbicides as well as tolerance to lepidopteran and coleopteran insects. It inherited seven genes from six parental transgenic lines (Table 6.3) [ISAAA 2012].

This hextuple-event stacked transgenic maize hybrid was approved by USA in August 2012 for marketing under the commercial name 'Agrisure-Duracade 5222' [ISAAA 2012].

Table 6.3. The introduced genes, gene sources, gene products and resistance targets in hextuple-event stacked herbicide-cum-insect resistance maize hybrid ('Agrisure-Duracade 5222') [CERA 2010; ISAAA 2012].

Event	Gene	Gene Source	Gene Product	Resistance
SYN 5307	*ecry3.1Ab*	Synthetic form of *cry3A* gene & *cry1Ab* gene from *B. thuringiensis*	Chimeric (Cry3A-Cry1Ab) δ-endotoxin	Coleopteran and lepidopteran insects
MIR604	*mcry3A*	Synthetic form of *cry3A* gene from *B. thuringiensis* subsp. *tenebrionis*	Modified Cry3A δ-endotoxin	Coleopteran (corn rootworm) insects
BT11, TC1507	*cry1Ab*	*B. thuringiensis* subsp. *kurstaki*	Cry1Ab δ-endotoxin	Lepidopteran insects
BT11	*pat*	*S. viridochromogenes* strain Tu494	PAT enzyme	Glufosinate
TC1507	*cry1Fa2*	Synthetic form of *cry1F* gene from *B. thuringiensis* var. *aizawai*	Modified Cry1F protein	Lepidopteran insects
GA21	*mepsps*	*Zea mays*	modified EPSPS enzyme	Glyphosate
MIR162	*vip3Aa20*	*B. thuringiensis* strain AB88	Variant of vegetative insecticidal protein (vp3Aa)	Lepidopteran insects
MIR162	*pmi* (Selection marker/ Reporter)	*E. coli*	Phosphomannose Isomerase (PMI) enzyme	Metabolizes mannose for selection of transformed plants

Soya Bean (Soybean)

Mono-Trait Events: Herbicides

Glyphosate

MON40302-6 (GTS 40-30-2). Soya bean was the first crop to be engineered for glyphosate resistance when *cp4 EPSPS* gene was introduced by Monsanto. Event MON40302-6 was produced by biolistic transformation of plant cells from a high-yielding commercial soya bean cultivar 'A5403' with DNA-coated gold particles. The plasmid PV-GMGT04 used for transformation contained the genes coding for glyphosate tolerance and for production of *gus* (β-glucuronidase) gene, a selectable marker. Expression of the *CP4 EPSPS* gene was regulated by an enhanced 35S promoter (e35S) of CaMV, a chloroplast transit peptide (CTP4) coding sequence from *Petunia hybrida*, and a *nos* 3' transcriptional termination element from *A. tumefaciens*. The CTP4 facilitated the translocation of newly translated EPSP synthase into chloroplasts, the site of aromatic amino acid biosynthesis and glyphosate site of action [CERA 2009e].

Southern blot analysis of genomic DNA from MON40302-6 (Roundup Ready) demonstrated that there were two sites of integration: one containing a functional copy of *CP4 EPSPS* gene and the other containing a non-functional segment of the *CP4 EPSPS* gene. The *gus* gene was not integrated into the host genome and there were no antibiotic resistance marker genes introduced into this soya bean event. The *epsps* gene was stably integrated into the plant genome [CERA 2009e].

DNA analyses over six generations showed that the insertion was stable. This transgenic line was approved by USA in 1994 followed by 19 other countries by 2005 [CERA 2009e]. This glufosinate-tolerant was grown on approximately 87 percent of the U.S. and 60 percent [USDA-NASS 2005a] of the global soya bean acreage in 2005.

MON89788. In 2006, Monsanto produced a second generation glyphosate-resistant soya bean, MON89788, to provide farmers flexibility, simplicity and cost-effective weed control options. This was developed by *Agrobacterium*-mediated transformation of meristematic tissue from soya bean cultivar 'A3244'. The plasmid PV-GMGOX20 used for transformation contained the *aroA* gene from *A tumefaciens* strain CP4 for glyphosate tolerance. Expression of this gene was regulated by a chimeric promoter combining the enhancer sequences from the FMV 35S promoter and the promoter from the *Tsf1* gene from *A. thaliana*, which codes for the elongation factor, EF-1 alpha. A chloroplast transit peptide (CTP2) coding sequence from the *ShkG* (Shank girth) gene of *A. thaliana* facilitated the translocation of newly translated EPSP synthase into chloroplasts, the site of aromatic amino acid biosynthesis and glyphosate site of action. The 3' non-translated sequence from *RbcS2* (the ribulose-1, 5-bisphosphate carboxylase small subunit) *E9* gene of pea provided the transcriptional termination sequences. This cassette, coding for the expression of the CP4 EPSPS enzyme, was contained between the Right and Left Border sequences of the T-DNA region in PV-GMGOX20 [CERA 2009h].

Subsequent analyses showed that the *cp4 epsps* gene was stably inserted and integrated at a single locus within the soya bean genome while the glyphosate-resistant trait was stably inherited across multiple generations. USA, Canada, Japan, Taiwan, and Philippines approved MON89788 (Genuity Roundup Ready 2 Yield) in 2007 [CERA 2009h] followed by other countries by 2010.

Glufosinate

A2704-12; *A5547-127*. The transgenically modified soyabean events A2704-12 (ACS-GM005-3) and A5547-127 (ACS-GM006-4) resistant to glufosinate ammonium were derived from two different cultivars 'A2704' and 'A5547' respectively [CERA 2009d]. Both transgenic lines were produced via biolistic transformation of soya bean cells by Aventis CropScience (later Bayer CropScience) using the circular plasmid pB2/35Sack, a derivative of the vector pUC19 which contains the RB fragment from the *A. tumefaciens* Ti plasmid pTiAch5 and the synthetic *pat* gene fused to CaMV 35S promoter and terminator.

Vector pUC19 was digested with a restriction enzyme *Pvu*I at Nt positions 337 and 3456 to disrupt the region coding for the selectable marker gene *bla* (β-lactamase). The inserted DNA sequence consists of the bacterial *pat* gene (derived from *S. viridochromogenes* strain Tü 494), CaMV 35S promoter, and a partial copy of the *bla* gene (originally isolated from pBR322 a plasmid of *E. coli* to confer resistance against β-lactam antibiotics, ampicillin penicillin, etc.), a reporter gene of bacterial origin (from plasmid pBR232 of *E. coli*) for gene expression. The *pat* gene has been altered with plant codons for increased expression without altering the amino acid profile of the protein [CERA 2009d].

The transformed plants showed that the glufosinate-resistant A2704-12 soya bean contained two copies of the *pat* gene in the genome, while A5547-127 soya bean contained one copy of the *pat* gene. In both, the *pat* gene was joined by one copy of the 3'*bla* sequences and one copy of the 5'*bla* sequences. Both of the integrated parts of the *bla* gene did not constitute an intact, functional *bla* gene as the 5' *bla* sequences were integrated in an inverted orientation. These two glufosinate-resistant soya bean lines (Liberty Link soybean) were approved by USA in 1998, Canada in 2000, Japan in 2002, Australia in 2005, the European Union in 2008 [CERA 2009d], Brazil in 2010, and Argentina in 2011.

W62, W98, GU262. AgrEvo (Bayer CropScience) also developed two other transgenic lines 'W62' and 'W98' via biolistic transformation of soya bean cells with a pUC19 based plasmid containing a modified form of the *bar* gene (originally isolated from *S. hygroscopicus*, unlike the *pat* gene in the case of A2704-12 and A5547-127) placed under the control of CaMV 35S promoter [CERA 2005c]. In addition, these lines also express a selectable marker gene *gus*, coding for β-D-glucuronidase, isolated from *E. coli*. The glufosinate-resistant transgenic soya bean lines W62 and W98 (Liberty Link soybean), field tested in the U.S. during 1990–1993, were approved in 1998.

Another glufosinate-resistant soya bean event developed by AgrEvo (Bayer CropScience) is GU262 (ACS-GM003-9). It expresses two copies of the synthetic *pat* gene isolated from a soil bacterium *S. viridochromogenes*. In addition, GU262 also contains partial copies of the bacterial selection marker gene *bla* that is not expressed. This event was produced via biolistic transformation of a soya bean line with a pUC19 based plasmid vector, pB2/35SacK, containing the *pat* gene and sequences encoding the β-lactamase enzyme that confers resistance to β-lactam antibiotics such as ampicillin, peniciliin, etc. Expression of the *pat* gene was regulated by CaMV 35S promoter and terminator. The *bla* gene was employed as a selection method during the development process in order to identify bacterial colonies that had been transformed with recombinant plasmids. This transgenic line (Liberty Link) was approved by USA in 1998 [CERA 2005f].

Dicamba

MON87708. Soya bean event MON87708 has been genetically engineered in 2010 to contain *dmo* gene from the bacteria *Stenotrophomonas maltophilia* Strain DI-6 that expresses a monooxygenase enzyme, DMO, which rapidly demethylates dicamba rendering it inactive, thereby conferring resistance to the herbicide. This transgenic line was developed through *Agrobacterium*-mediated transformation of meristem tissue of soya bean cultivar 'A3525' using the 2T-DNA plasmid vector PV-GMHT4355. This vector contains two T-DNAs. The first T-DNA, designated T-DNA I, contains the *dmo* expression cassette and this is maintained in MON 87708. The second T-DNA, T-DNA II, contains the glyphosate-tolerant *cp4 epsps* gene expression cassette that was used for early event selection, and it was segregated away from T-DNA I by conventional breeding (self-pollination) [Monsanto 2011b].

These gene constructs of PV-GMHT4355 intended for insertion into the soybean genome comprised between the T-DNA I borders are, from the RB region, the *PC1SV* promoter (P-*PC1SV*: peanut chlorotic streak caulimovirus that directs transcription in plant cells), the *TEV* leader (L-*TEV*: 5' non-translated region of the tobacco mosaic virus which regulates gene expression), the *RbcS* targeting sequence (TS-*RbcS*: sequences encoding transit peptide of the mature protein of the *RbcS* gene from *Pisum sativum*), the *dmo* coding sequence (CS-*dmo*), and the *E9* 3' non-translated region (T-*E9*). These elements together constitute the *dmo* expression cassette [Monsanto 2011b].

The molecular analysis showed that MON 87708 contained a single copy of the *dmo* expression cassette stably integrated into a single locus of the soya bean genome [Monsanto 2011b]. MON87708 facilitates a wider window of application of dicamba in soya bean, allowing pre-emergence application up to the day of crop emergence and post-emergence application through the early crop reproductive stage [Monsanto 2011b]. Its commercial name 'Genuity Roundup Ready2 Xtend' suggests that this dicamba-resistant line extends the utility of transgenic glyphosate-resistant soya bean discussed earlier.

Mono-Trait Events: Herbicides (Non-Transgenic)

Imidazolinones

BPS-CV127. The event CV127 of BASF was modified to express an AtAHASL (altered acetohydroxy-acid synthase large unit) protein of 670 amino acids encoded by the *csr1-2* gene from *A. thaliana* to confer resistance to IMI herbicides. This protein is structurally and functionally identical to the native AtAHASL, except for substitution of serine for asparagine at residue 653 (Ser653→Asn653). The *csr1-2* gene in this soya bean line also contains a second mutation, in which arginine at 272 is replaced by lysine (Arg272→Lys272) [CERA 2010b].

A purified, 6.2 kb linear *Pvu*II fragment derived from plasmid pAC321 containing the *csr1-2* gene cassette was used in the transformation of the embryonic axis tissue obtained from the apical meristem of a single commercial soya bean Brazilian cultivar 'Conquista'. Transformation was done by biolistic method, without using a carrier DNA. The *csr1-2* coding sequence is 2013 bp long and includes the S653N point mutation which confers tolerance to IMIs. Besides, a second mutation was discovered in the *ahas* (of *Arabidopsis*) gene coding sequence integrated in the CV127 soya bean genome. This second mutation, in which Arg at position 272 of the AtAHAS protein is replaced by Lys, does not impact the function of the AHAS enzyme or its herbicide-resistance properties.

The *Pvu*II fragment derived from plasmid pAC321 was integrated at a single locus as one single copy in the soya bean genome. The complete CV127 insert sequence was 4758 bp long. The *csr1-2* gene, stably integrated into the nuclear soya bean genome, has made CV127 resistant to IMIs [CERA 2010b]. The transformed T_0 generation plants were advanced to the fourth generation via four cycles of self-pollination. Some T_4 plants were then backcrossed to Conquista to reduce the number of copies of the transgenic insert to a single locus. The progeny of the first filial generation were advanced to F_8 by successive rounds of self-pollination and selection. The homozygous transgenic CV127 of F_8 generation was approved by Brazil in 2009 [CERA 2010b] and Colombia in 2012.

Double-Trait Stacks: Herbicides

Glyphosate and Isoxaflutole

FG72. In 2009, Bayer CropScience has developed this dual herbicide resistant soya bean in response to the epidemic glyphosate-resistant (GR) weeds fostered by widespread use of transgenic soya bean varieties. This transgenic double-trait stack has been field tested under APHIS regulations since 2001. The company engineered soya bean variety 'Jack' to make it resistant to both glyphosate and isoxaflutole so that the latter will help manage non-GR weeds.

Isoxaflutole, an isoxazole group herbicide, inhibits 4-hydroxyphenyl-pyruvate-dioxygenase (4-HPPD) (vide Chapters 2 and 5). This selective preplanting and preemergence herbicide inhibits the biosynthesis of carotenoid pigments. The wild type *hppd* gene isolated from *Pseudomonas fluorescens* was mutated by site-directed mutagenesis, resulting in *hppdPfW336* gene. Event FG72 of soya bean, the first crop engineered for resistance to isoxaflutole, was developed by direct transfer of *2mepsps* and *hppdPfW336* genes, which encode 2mEPSPS and HPPD proteins respectively [APHIS-USDA 2011]. These two genes were introduced into the soya bean genome by using biolistic transformation process.

The *2mepsps* gene expression cassette, borne by plasmid pSF10, is represented by "Ph4a748-intron1 h3At-TpotpC::2*mepsps*::3'histonAt" string. The wild type *epsps* gene isolated from maize was mutated by adding a methionine codon to the N-terminal of the 2mEPSPS protein sequence in order to restore the cleavage site of the optimized plastid transit peptide. The Ph4a748 promoter sequence, which controls expression of *2mepsps* gene, was derived from H4 of *A.thaliana*. The optimized transit peptide (TpotpC) contains sequences from the RuBisCO small subunit genes of maize and sunflower, while 3'histonAt is the 3'untranslated region of the histone from *A. thaliana*. The *hppdPfW336* gene expression cassette is represented by "Ph4a748 ABBC-5'tev-TpotpY::*hppdPfW336*::3'nos" string. Ph4a748 promoter was duplicated to increase the promoter activity by fusing with the leader sequence of 5'tev (tobacco etch virus) whereas 3' *nos*, the 3'untranslated region of napoline synthase from *A. tumefaciens*, is a polyadenylation signal [APHIS-USDA 2011].

The HPPD and 2mEPSPS proteins expressed in Event FG72 soya bean are nearly identical to the native proteins produced by *P. fluorescens* and maize respectively [APHIS-USDA 2011]. This transgenic glyphosate- and isoxaflutole-resistant soybean line FG72 has received non-regulated status in USA in 2013 [USDA-APHIS 2013].

Glyphosate and Chlorsulfuron

DP 356043. DuPont Pioneer developed soya bean DP 356043 which can withstand glyphosate and chlorsulfuron, an ALS-inhibitr. When planting this double-trait transgenic line, a farmer can spray the crop with glyphosate and/or ALS inhibitor, killing only the weeds and leaving soya bean [USFDA 2007]. Soya bean cultivar 'Jack' was used for transformation via microprojectile bombardment.

Plasmid PHP20163 was constructed with two expression cassettes and an antibiotic (hygromycin) resistance marker gene. From this plasmid, a linear fragment was excised using two restriction enzymes *Asc* I and *Not* I. This linear transformation fragment, PHP20163, contained the glyphosate *gat*4601 gene cassette and the chlorsulfuron *gm-hra* gene cassette [USFDA 2007].

The *gat*4601 gene has a high degree of similarity to the amino acid sequence of the native GAT protein. Its *in planta* expression was under the control of two regulatory elements: a) SCP1 promoter a synthetic constitutive promoter, containing a portion of the CaMV 35S promoter and the Rsyn7-Syn II Core synthetic consensus promoter, and b) TMV omega 5′-UTR (the omega 5′ untranslated leader of the Tobacco Mosaic Virus). The SCP1 promoter drives the transcription of the *gat*4601 gene. The TMV omega 5′-UTR, located downstream from the SCP1 promoter, enhances translation. Termination of transcription of the *gat*4601 gene was under the control of the *pin*II terminator gene [USFDA 2007].

The *gm-hra* gene cassette contained the coding sequence of the *gm-hra* gene, which encodes the GM-HRA protein. It is a modified version of the endogenous *gm-als* gene that encodes the GM-ALS I protein. Compared to the GM-ALS I protein, the GM-HRA protein sequence contains two amino acid substitutions important for tolerance to the ALS-inhibiting herbicides, and five additional N-terminal amino acids derived from the translation of 15 nucleotides of the *gm-als* 5′ untranslated region. The *in planta* expression of the *gm-hra* gene was controlled by the promoter derived from the S-adenosyl-L-methionine synthetase (SAMS) gene from soya bean, and an intron that interrupts the SAMS 5′-UTR). Termination of transcription of the *gm-hra* gene was under the control of the native *gm-als* terminator [USFDA 2007].

Soya bean embryonic somatic cultures, derived from explants from small, immature soya bean seeds were transformed by biolistic method using microscopic gold particles coated with purified PHP20163A DNA fragment. Following transformation, the soya bean tissue was transferred to liquid culture flasks and allowed to recover for 7 days before being transferred to a selective medium supplemented with chlorsulfuron. The chlorsulfuron-resistant green embryogenic tissue was excised after several weeks and subsequently regenerated. Southern blot assay showed that 356043 soya bean had one insert with a single, intact copy of the transformation fragment PHP20163A containing the *gat*4601 and *gm-hra* gene cassettes [USFDA 2007]. This 356043 soyabean (Optimum GAT) was approved by USA and Mexico in 2008 followed by other countries later.

2,4-D and Glufosinate

DAS 68416-4. Soya bean DAS 68416-4 was developed by Dow AgroSciences to confer resistance to 2,4-D and glufosinate. Cotyledonary node explants of cultivar 'Maverick' were transformed via *Agrobacterium*-mediated method by using disarmed *A. tumefaciens* strain EHA 101, carrying the binary vector pDAB4468. The T-DNA contains two gene expression cassettes for insertion into soya bean [Dow 2010].

The *aad-12* cassette contains the plant-optimized aryloxyalkanoate dioxygenase (*aad-12*) gene that encodes the AAD-12 protein. The *aad-12* gene, which catalyzes the side chain degradation of 2,4-D and confers resistance to soya bean, was isolated from *Delftia acidovorans*. The synthetic version of the gene was optimized to modify the G+C codon bias to a level more typical for plant expression. This gene encodes a protein of 293 amino acids that has a molecular weight of approximately 32 kDa. Expression of this *aad-12* was controlled by the AtUbi10 promoter from *A. thaliana* and AtuORF23 3′ UTR sequence from *A. tumefaciens* plasmid pTi15955. Besides, a matrix attachment region (MAR) of RB7 from *Nicotiana tabacum* was also included at the 5′ end of the aad-12 PTU (plant transcriptional unit, includes promoter, gene, and terminator sequences) to potentially facilitate expression of the *aad-12* gene in the plant [Dow 2010].

The *pat* expression cassette contained *pat* gene to express the PAT protein. The synthetic version isolated from the *S. viridochromogenes* gene was optimized to modify the G+C codon bias to a level more typical for plant expression. The insertion of the *pat* gene into soya bean genome confers tolerance to glufosinate and it was used as a selectable marker during the soya bean transformation. The *pat* gene encodes a protein of 183 amino acids that has a molecular weight of approximately 21 kDa. Expression of the *pat* gene was controlled by CsVMV (cassava vein mosaic virus) promoter and AtuORF1 3' UTR sequence from *A. tumefaciens* plasmid pTi15955 [Dow 2010].

The hybridization patterns across four sample sets representing three generations of DAS-68416-4 soya bean were identical, indicating that the insertion is stably integrated in the soya bean genome [Dow 2010]. This double-herbicide trait soya bean stack (Enlist) was approved by USA in 2012.

Glufosinate and Mesotrione

SYHT0H2. In 2013, Syngenta and Bayer CropScience produced soya bean line SYHT0H2 to confer resistance to glufosinate and the HPPD-inhibiting herbicide mesotrione. Immature embryos of seeds of soya bean cultivar 'Jack' were transformed by *A. tumefaciens* strain EHA101 using the binary vector pSYN15954 for insertion of three-gene expression cassettes into the soybean genome [Cogem 2013].

The *avhppd-03* gene which encodes HPPD (AvHPPD-03) enzyme was derived from seedlings of the common oat *Avena sativa*. This gene is regulated by the FMV, CaMV 35S, and TMV enhancer sequences, and the *nos* terminator sequence. The other two gene cassettes express the *pat-03-01* gene and *pat-03-02* gene (derived from *S. viridochromogenes*) encoding PAT enzymes. The *pat-03-01* gene was regulated by the FMV, CaMV 35S promoter sequence, and *nos* terminator sequence, while the *pat-03-02* gene was regulated by the CMP (cestrum yellow leaf curling virus) promoter sequence, TMV enhancer sequence, and *nos* terminator sequence. Both versions of *pat* genes encode the identical PAT protein sequence. Expression of *pat* confers a glufosinate-tolerance phenotype [Cogem 2013].

This dual-herbicide transgenic soya bean line contained two inverted and truncated copies of T-DNA at a single integration locus [Cogem 2013]. The stably inserted T-DNA consisted of a single copy of *avhppd-03*, four copies of *pat*, a single copy of the *avhppd-03* enhancer complex sequence, two copies of the CaMV 35S promoter, two copies of the CMP promoter, two copies of the TMV enhancer, and five copies of the *nos* terminator. It did not contain any extraneous DNA fragments of these functional elements elsewhere in the SYHT0H2 soya bean genome, and it also did not contain the FMV enhancer or plasmid backbone sequence from pSYN15954. The inserted DNA did not disrupt any known endogenous soya bean gene [Cogem 2013].

Triple-Trait Stacks: Herbicides

2,4-D, Glufosinate, and Glyphosate

DAS-44406-6. This tri-herbicide transgenic soya bean line of Dow provides resistance to 2,4-D, glufosinate, and glyphosate so the farmers can have a new management tool that allows control of a broad-spectrum weeds comprising of annual and perennial grasses and broadleaf weeds. DAS-44406-06 was developed through *Agrobacterium*-mediated transformation of cotyledonary node plants of soybean cultivar 'Maverick'. The disarmed *A. tumefaciens* strain EHA101 binary vector pDPAB8264 was used to carry the *pat, aad-*

12, and *2mepsps* gene expression cassettes within the T-DNA region [Dow 2011]. The synthetic versions of *pat* and *aad-12* genes were optimized to modify the G+C codon bias to a level more typical for plant expression [Dow 2011].

Expression of *pat* gene was controlled by CsVMV promoter, AtuORF1 3' UTR sequence from *A. tumefaciens* plasmid pTi15955 while that of *aad-12* gene is controlled by the AtUbi10 promoter from *A. thaliana* and AtuORF23 3' UTR sequence from *A. tumefaciens* plasmid pTi15955. Expression of *2mepsps* gene was controlled by the Histone H4A748 promoter from *A. thaliana* and Histone H4A748 3' UTR sequence from *A. thaliana* [Dow 2011]. Molecular characterization of the transformed DAS-44406-6 line confirmed that a single, intact DNA insert containing the three gene expression cassettes was stably integrated into the soya bean genome without the plasmid backbone. The integrity of the inserted DNA was demonstrated in five different breeding generations [Dow 2011].

Rapeseed (Canola)

Early rapeseed (mustard) cultivars had high levels of erucic acid in the oil and glucosinolates in the meal and their presence was a health concern. In 1959, Canada produced 'Liho', a variety which contained low levels of erucic acid. This trait was subsequently transferred to agronomically adapted rapeseed cultivars. This led to the first low erucic acid cultivar of *B. napus*, 'Oro', in 1968 [Przybylski et al. 2005].

However, the hydrolyzed products of the undesirable glucosinolates present in low erucic acid varieties interfere with the uptake of iodine by the thyroid gland, contribute to liver disease, and reduce growth and weight gain in animals. In 1977, the National Research Council of Canada released the first low erucic acid and low glucosinolate cultivar of *B. rapa*, 'Candle' [Przybylski et al. 2005].

In 1978, Western Canadian Oilseed Crushers changed the name of rapeseed to 'canola', which was approved the Canola Council of Canada in 1980. The name canola at that time included those cultivars which contained less than five percent erucic acid in the oil and 3 mg g^{-1} aliphatic glucosinolates in the meal. However, the definition of **canola** (**Can**adian **o**il, **l**ow **a**cid) was amended in 1986 to *B. napus* and *B. rapa* lines with less than two percent erucic acid in the oil and less than 30 μmol g^{-1} glucosinolates in the air-dried, oil-free meal [Przybylski et al. 2005].

Mono-Trait Events: Herbicides

Glyphosate

GT73. Canola line GT73 (synonym RT73) containing two glyphosate-resistant genes was developed by Monsanto by transforming the cultivar 'Westar'. The disarmed *A. tumefaciens* plant transformation system was used to produce this 'Roundup Ready' canola variety. The transformation process used plasmid PV-BNGT04, a double-border vector.

Constitutive expression of the *cp4epsps* gene was regulated using the FMV 35S promoter and the 3' end of the pea rbcS E9 gene (E9 3'). The *cp4 epsps* gene was fused to a chloroplast transit peptide (CTP 2) sequence derived from the *Arabidopsis thaliana* to allow targeting of the newly translated EPSPS enzyme into the chloroplast, the site of aromatic amino acid biosynthesis and endogenous EPSPS and GOX activity. Similarly, expression of *goxv247* gene was under control of the 35S promoter from a modified FMV and the 3' end of the pea rbcS E9 gene (E9 3'). Expression of the *goxv247* gene was targeted to the chloroplast by the action of the N-terminal of the small subunit 1A (SSU1A) of the ribulose 1,5-bisphosphate carboxylase chloroplast transit peptide (Arab-SSU1A/CTP1) of *A. thaliana*. Both genes, placed under the control of FMV 35S promoter from a modified figwort mosaic virus (P-CMOVb), were optimized for plant expression. The genes were integrated at a single locus in the nuclear genome of GT73 and the transgenic DNA was stably inherited across generations.

Canada approved this transgenic line in 1994, followed by the European Union in 1997, and USA in January 1999 for commercialization as 'Westar Roundup Ready'. In 1997, Canadian Food Inspection Agency suspended the sale of GT73 because the seed of this transgenic line was contaminated with seed of the unapproved canola event GT200 [The Center for Food Safety 2002]. The US Department of Agriculture conducted an environmental assessment under the National Environmental Policy Act and issued a 'finding of no significant impact', leading to its deregulation in January 1999.

An important issue to consider for transgenic canola is its ability to transfer genes via pollen to conventional canola and related weed species such as wild mustard (*Sinapis arvensis* L.), India mustard {*Brassica juncea* (L.) Czern}, Ethiopian mustard (*Brassica carinata* A. Braun), black mustard {*Brassica nigra* (L.) W.D.J. Kock}, common dogmustard {*Erucastrum gallicum* (Willd.) O.E. Schultz}, etc. The frequency of outcrossing to weed species is low and any hybrids are less fit, but at least one canola transgene has become established in a stable wild population [Warwick et al. 2008].

GT200. Like its antecedent GT73, Monsanto's canola GT200 event has been genetically engineered to express *cp4 epsps* and *gox* genes that impart resistance to glyphosate. GT200 was developed in the early 1990s along with GT73. This transgenic line was also created by *Agrobacterium*-mediated transformation of 'Westar' cultivar using transformation vector PV-BNGT03, which is different from the one used in GT73. GT200 was produced using a plasmid that contained the *cp4 epsps* gene and the *gox* gene.

The T-DNA region of the Ti plasmid was "disarmed" by removal of *vir*A genes, normally associated with *A. tumefaciens*, and replaced with the genes coding for glyphosate resistance. During transformation, the T-DNA portion of the plasmid was transferred into the plant cells and stably integrated into the plant's genome. This vector contained a plasmid backbone which was not intended to be transferred to the recipient canola line. T-DNA contained these herbicide-tolerant two genes whose expression was placed under the control of FMV 35S promoter (vide GT73). Both genes were controlled by the same constitutive promoter and fused to a chloroplast transit peptide sequence to allow targeting of the newly translated CP4 EPSPS and GOX enzymes into the chloroplast, the site of aromatic amino acid biosynthesis and endogenous EPSPS and GOX activity.

Although both lines express *cp4 epsps* and *gox* genes, they were derived from distinct transformation events. The GOX protein expressed in this line differed from the GOX protein expressed in the GT73 line by three amino acids. GT200 (Roundup Ready) was approved by USA in 2002, preceded by Canada in 1997.

MON88302. This second generation glyphosate-tolerant product Event MON88302 canola was developed to provide farmers with improved weed control through greater flexibility for glyphosate application. It was produced by *Agrobacterium*-mediated transformation of 'Ebony' cultivar utilizing plasmid vector PV-BNHT2672. The T-DNA contained the *cp4 epsps* coding sequence under the control of the *FMV/Tsf1* chimeric promoter, the *Tsf1* leader and intron sequences, and the *E9* 3' untranslated region. The chloroplast transit peptide,

CTP2, directed transport of the CP4 EPSPS protein to the chloroplast and it was derived from *CTP2* target sequence of the *A. thaliana shkG* gene [Monsanto 2011a].

The backbone sequence of PV-BNHT2672, located outside of the T-DNA, contained two origins of replication for maintenance of plasmid vector in bacteria (*ori V* and *ori-pBR322*), a bacterial selectable marker (kanamycin) gene, *aadA*, and a coding sequence for repressor of primer protein (rop) for maintenance of plasmid vector copy number in *E. coli* [Monsanto 2011a]. The *cp4 epsps* expression cassette encodes a 47.6 kDa CP4 EPSPS protein consisting of a single polypeptide of 455 amino acids. The *cp4 epsps* coding sequence is the codon optimized coding sequence of the *aroA* gene from *Agrobacterium* sp. strain CP4 encoding CP4 EPSPS.

During transformation, the excised hypocotyl segments were co-cultured with *Agrobacterium* carrying the vector and then placed on callus growth medium containing carbenicillin, ticarcillin disodium, and clavulanate potassium to inhibit the growth of excess bacterium. The hypocotyls were placed in selection media containing glyphosate to inhibit the growth of untransformed cells and plant growth regulators conducive to shoot regeneration. Rooted plants that exhibited normal phenotypic characteristics were selected and transferred to soil for growth and self-pollinated to produce seed. Plants that arose from these seeds were evaluated for glyphosate resistance and screened for the presence of T-DNA (*cp4 epsps* expression cassette) and absence of backbone sequence (*ori V*). Subsequently, the *cp4 epsps* homozygous R1 plant was self-pollinated to give rise to second generation plants [Monsanto 2011a]. This transgenic canola line MON88302 (TruFlex Roundup Ready) canola contained a single copy of the *cp4 epsps* expression cassette stably integrated at a single locus, with the exception of the introduced trait.

Event 73496. This transgenic glyphosate-tolerant canola line 73496 (Optimum Gly) of Pioneer Hi-Bred was also produced in 2011 by using the microspores from the donor canola cultivar '1822B'. Gold particles coated with the purified PHP28281A DNA fragment were used for biolistic transformation to express the GAT4621 protein. The *gat4621* gene was optimized by a gene shuffling process of *gat* genes from *Bacillus licheniformis*. Fragment PHP28181A contains the *gat4621* cassette consisting the constitutive *UBQ10* (*A. thaliana* polyubiquitin) promoter, *gat4621* gene, and *Pin*II terminator [Pioneer 2011; Canadian Food Inspection Agency 2012b].

The transformed embryogenic microspores were cultured in fresh medium in the dark for 10–12 days and then under dim light for 1–4 weeks. Green embryos were transferred to fresh medium and cultured for 2 weeks at 4°C and 4 weeks at 25°C in the presence of glyphosate (0.1 mM) to select for glyphosate-resistant transformants. Germinated shoots or plants were transferred to growth medium supplemented with glyphosate for selection. The regenerated plants from transformation and tissue culture (T$_0$ plants) were selected for further characterization analyses, herbicide efficacy, and agronomic evaluations, followed by subsequent selfing and breeding for several generations. Molecular characterization of Event 73496 canola confirmed that a single, intact PHP28181A DNA fragment has been inserted into the genome with no plasmid backbone DNA [Pioneer 2011]. Canada approved this product in December 2012 [Canadian Food Inspection Agency 2012b].

Glufosinate

HCN92. Canola line HCN92 (Topas19/2) was produced by AgrEvo via *Agrobacterium*-mediated transformation of microspores from cultivar 'Topas'. The T-DNA region of the Ti plasmid pOCA/Ac was disarmed by removing *vir*A genes, normally associated with the pathogenicity and disease-causing properties of *A. tumefaciens*, and replacing them with the *pat* gene encoding for PAT and aminoglycoside resistance marker *npt*II coding for neomycin phosphostransferase II. Expression of the *pat* gene was regulated by CaMV 35S promoter and terminator. The *npt*II gene was driven by the *nos* promoter and *ocs* terminator sequence. During transformation, the T-DNA portion of the plasmid was transferred into the plant cells and stably integrated into the plant's genome. The transformed line was initially crossed with non-transgenic 'ACSN3', followed by a second cross with another non-transgenic cultivar 'AC Excel'. HCN92 was derived from a bulk of single F_3 plants selected from the cross [Health Canada 2000a]. This transgenic glufosinate-resistant rapeseed (Libert Link Innovator) was approved by Canada and USA in 1995, followed by many countries by 2005 [CERA 2006d].

T45 (HCN28). The T45 (synonym HCN28) transformant of AgrEvo was developed by *Agrobacterium*-mediated transformation of canola cultivar 'AC Excel' with additional genetic material. As in the case of HCN92, the T-DNA region of the Ti plasmid was disarmed by the removal of *vir*A genes and replacing them with the *pat* gene, a synthetic version of the gene isolated from *Streptomyces viridochromogenes*, strain Tu 494. The nucleotide sequence was modified to provide codons preferred by plants without changing the amino acid sequence of the enzyme. Expression of the *pat* gene was regulated by including CaMV 35S as promoter and terminator. The *pat* gene, the only gene transferred into the T45 transformant was stably integrated into the plant's genome. Later, T45 was backcrossed twice with 'AC Excel', yielding HCN28 from the resultant BC_2 using the single seed descent method [CERA 2009g]. The expression of *pat* gene in HCN28 (InVigor) remained stable even four generations after being developed from the original transformant T45. This glufosinate-resistant event was approved by Canada and Japan in 1997, USA in 1998 [CERA 2009g].

MS8 and RF3. The MS8 and RF3 canola lines were both produced by AgrEvo using *Agrobacterium*-mediated transformation of cultivar 'Drakkar'. MS8 was engineered for male sterility and RF3 for restoration of male fertility. Additionally, the *bar* gene which confers resistance to glufosinate has also been inserted in both events.

MS8 has been engineered to express a ribonuclease encoded by the *barnase* gene derived from the bacterium *Bacillus amyloliquefaciens*. The ribonuclease blocks pollen development and causes male sterility in MS8 canola or progeny containing the gene. RF3 has been engineered to express a specific inhibitor of this ribonuclease encoded by the *barstar* gene, also derived from *B. amyloliquefaciens*. The ribonuclease inhibitor restores male fertility in plants containing the *barnase* gene [APHIS-USDA 1999].

The T-DNA region of the Ti plasmid was disarmed by removal of *vir*A genes, and replaced with the genes of interest for each transgenic line. Line MS8 was produced using plasmid pTHW107, which contained a copy of the *barnase* gene whose transcription was regulated with another specific promoter pTa29 from *Nicotiana tabacum*, terminated by part of the 3′non-coding region of the *nos* gene of *A. tumefaciens* [CERA 2006e]. Line RF3 was produced using plasmid pTHW118, which contained the *barstar* gene under the control of the pTa29 anther-specific promoter from *N. tabacum* and the *nos* termination signal.

Besides, each T-DNA contained a copy of the *bar* gene from *S. hygroscopicus* which encodes PAT enzyme. Expression of the *bar* gene was regulated by the PSsuAra promoter from *A. thaliana* CTP and post-translational targeting of the gene product to the chloroplast organelles was accomplished by fusion of the 5′-terminal coding sequence with the chloroplast transit peptide DNA sequence from *A. thaliana*. Sequences outside the

T-DNA region contained: colE1 replication region from *E. coli*; pVS1 replication region isolated from *Pseudomonas*; and a fragment of plasmid R751 from *Klebsiella aerogenes* comprising the streptomycin/spectinomycin resistance gene *Sm/Sp* with its own promoter [APHIS-USDA 1999].

The male sterile MS8 line contains the PSsuAra-*bar*-3'g7-PTA29-*barnase*-3'nos gene construct, while the fertility restorer RF3 line contains the PSsuAra-*bar*-3'g7-PTA29-*barstar*-3'nos gene construct. During transformation, the T-DNA portion of each plasmid was transferred into the plant cells and stably integrated into the plant genome of MS8 and RF3 respectively. Southern blot analyses also showed that the incorporation has been limited to DNA sequences contained with the T-DNA borders. As a result, the *bar* and *barnase* genes were stably integrated into MS8 similar to the *bar* and *barstar* genes in RF3. There were no marker genes for antibiotic resistance present in the transformed plants. The events MS8 and RF3 were approved by the U.S. in 1996 and Canada and Japan in 1997 [CERA 2006e].

MS1 x *RF1* (PGS1). MS1 and RF1 events were developed to provide a pollination control system for production of hybrid PGS1. The novel hybridization system involves the use of two parental lines, a male sterile MS1 and a fertility restorer RF1. The transgenic MS1 plants do not produce viable pollen grains and cannot self-pollinate. In order to completely restore fertility in the hybrid progeny, the glufosinate-resistant MS1 must be pollinated by a modified plant containing a fertility restorer gene, such as RF1. The resultant F_1 hybrid seed derived from MS1 x RF1 cross (InVigor Canola), produces plants that produce pollen and are completely fertile [CERA 2006b].

Both events have been engineered by using *Agrobacterium*-mediated transformation of the cultivar 'Drakkar'. The T-DNA region of the Ti plasmid was "disarmed" by removing *vir*A genes, normally associated with the pathogenicity and disease-causing properties of *A. tumefaciens*, and replaced with the genes of interest for each transgenic line MS1 and RF1. During transformation, the T-DNA portion was transferred into the plant cells and stably integrated into the plant's genome.

MS1 was produced from the insertion of T-DNA containing the *barnase* gene together with an anther-specific promoter, pTa29, from *N. tabacum*, terminated by part of the *nos* gene from *A. tumefaciens*. Similarly, RF1 produced from T-DNA containing the *barstar* gene under the control of the pTa29 anther-specific promoter from *N. tabacum* and 3'*nos* terminator.

In addition, each T-DNA contained a copy of the *bar* gene from *S. hygroscopicus*, which encodes the PAT enzyme, and the NPTII encoding *neo* gene from the Tn5 transposon of *E. coli*, strain K12. Expression of the *bar* gene, which encodes PAT protein to confer resistance to glufosinate, was regulated by the PSsuAra promoter from *A. thaliana* and post-translational targeting of the gene product to the chloroplast organelles was accomplished by fusion of the 5'-terminal coding sequence with the chloroplast transit peptide DNA sequence from *A. thaliana*. The expression of NPTII activity, under control of the nopaline synthase promoter from *A. tumefaciens*, was used as a selectable trait for screening transformed plants for the presence of the *barnase* and *barstar* genes, respectively [CERA 2006b].

Southern blot and segregation analyses demonstrated the stable integration of the *barnase*, *bar*, and *neo* genes in MS1 and the *barstar*, *bar* and *neo* genes in RF1. This glufosinate resistant male-sterile cum fertility-restore hybrid PGS1 was approved by Canada in 1995, USA in 1996, South Africa 2001, Australia in 2002, China in 2004, and Korea and the European Union in 2005 [CERA2006b).

Bromoxynil

Oxy-235. This bromoxynil-resistant Oxy-235 was produced by Rhône-Poulenc Canada by using *Agrobacterium*-mediated transformation of cultivar 'Westar' with plasmid pRPA-BL-150a using the EHA 101 strain of *A. tumefaciens*. This double-border binary Ti vector contained the T-DNA region of an *A. tumefaciens* plasmid from which virulence and disease causing genes were removed, and replaced with the oxynil tolerance gene. The gene construct consisted of one copy of the *bxn* gene (1150 bp) under the control of CaMV 35S promoter and an enhancer region from the non-translated portion of the RuBisCo small subunit gene from maize. The construct also contained the *nos* 3' noncoding regions from *A. tumefaciens* (plasmid pTi37). The *bxn* gene was isolated from the bacterium *Klebsiella anthropic* subsp. *ozaenae,* which encodes a nitrilase enzyme that hydrolyzes oxynil herbicides to non-phytotoxic compounds. During transformation, one complete copy of the *bxn* gene was stably integrated into the host genome at a single insertion site [Health Canada 1999g; CERA 2006c].

Transformed plants were regenerated in tissue culture using bromoxynil as the sole selective agent. The original transformation event—Westar-Oxy-235—was used as a parental line in the production of commercial canola varieties marketed under the name of 'Navigator'. Event Oxy-235 was approved by Canada in 1997, followed by USA and Japan in 1999, and Australia in 2002 [CERA 2006c].

2,4-D

Indian mustard (*Brassica juncea*) is a major spice and oilseed crop in India grown over six million ha. Bhist et al. [2004] developed 2,4-D resistant transgenic mustard lines by introducing *tfdA* gene in the hypocotyl explants of cultivar 'Varuna' via *Agrobacterium*-mediated transformation. A binary vector containing the *tfdA* gene (864 bp; isolated from plasmid pBRL7) and CaMV 35S promoter was developed. The start codon of *tfdA* gene was changed from GTG to ATG by PCR-based site-directed mutagenesis. Besides, a leader sequence from RNA4 of alfalfa mosaic virus (AMV) was also incorporated for enhanced expression of the transgene in plants. The 35S-AMV-tfdA-35SpA expression cassette was finally cloned in binary vector pPZP200 which also contained a *nos*Pr-*npt*II-*ocs*pA {(*npt*II) driven by *nos* promoter (*nos*Pr) and carrying *ocs* gene polyadenylation (A) signal at mRNA 3' end} cassette as an additional marker gene cloned from the *lox*P sites. This vector was then transformed into disarmed *A. tumefaciens* strain GV3101 by electroporation and used for genetic transformation.

Shoots regenerated from the transformed hypocotyl explants were grown in kanamycin-containing medium before transferring the growing transgenic plants. The T_0 transgenics which had single copy inserts were selfed and backcrossed followed by testing of T_1 plants in the field for tolerance to 2,4-D at 10 mg L^{-1} to 1000 mg L^{-1} concentrations. Four transgenic lines tested showed resistance to the herbicide at 500 mg L^{-1}, while nontransformed wild type plants were affected even at 10 mg L^{-1} [Bhist et al. 2004]. This held promise for further development in developing 2,4-D resistant transgenic Indian mustard for commercial use.

Mono-Trait Events: Herbicides (Non-Transgenic)

Imidazolinones

Modifications of ALS genes in various plant species result in herbicide-tolerant phenotypes and these typically consist of one amino acid substitution, leading to alteration of the binding site for IMIs. Development of non-transgenic IMI-resistant canola lines began in the early 1980s in Canada.

Microspores of the rapeseed variety 'Topas' were isolated, mutagenized with ethyl nitrosourea, and developed into embryos and eventually double haploid plantlets, which were then doubled with colchicine to induce chromosome duplication [Tan et al. 2005]. Of the five double-haploids that survived the subsequent soil application of imazethapyr, two, P1 and P2 (PM1 and PM2), showed superior tolerance to the herbicide. All IMI-resistant rapeseed varieties developed subsequently originated from these two mutant lines before being marketed first as 'Smart Canola' in 1995 and later as 'Clearfield Canola' [Tan et al. 2005].

Of the five AHAS (ALS) loci (genes) reported in rapeseed [Rutledge et al. 1991; Tan et al. 2005], AHAS2, AHAS3, and AHAS4 originate from the A-genome while AHAS1 and AHAS5 from C-genome. AHAS1 and AHAS3 are the only genes that constitutively express and encode the primary AHAS activities essential to growth and development in *B. napus*. The alleles of PM1 and PM2 mutants correspond to AHAS1 and AHAS3 [Rutledge et al. 1991]. PM1 has a single nucleotide substitution at codon 653, where Ser was substituted by Asn (Ser653→Asn653) of the ALS protein as in the case of XI12 mutation in maize. In comparison, PM2 has a single nucleotide substitution at codon 574, replacing Trp by Leu (Trp574→Leu574) similar to XA17 mutation in maize [Tan et al. 2005].

PM1 was resistant to IMIs only, while PM2 was cross-resistant to both IMIs and sulfonylureas. Although both mutant lines confer resistance to IMIs, the tolerance level of PM2 was much higher than that of PM1. However, highest level of tolerance was obtained when both PM1 and PM2 were stacked and homozygous [Tan et al. 2005]. The IMI-resistant canola varieties NS738, NS1471, and NS1473, have been developed using 'Topas' cultivar similar to the process employed for PM1 and PM2. These imidazolinone-resistant non-transgenic lines were approved by Canada in 1995 [Health Canada 1999c].

Cotton

Mono-Trait Events: Herbicides

Glyphosate

MON1445. Commercialized in 1997, the MON1445 (Roundup Ready) trait was inserted into 'Coker 312' cultivar via *Agrobacterium*-mediated transformation. The T-DNA region of the Ti plasmid PV-GHGT07 contained sequences encoding CP4 EPSPS, GOX, NPTII, and AAD enzymes. This plasmid is a single-border vector that has only a right border containing the 0.4 kb *oriV* fragment from the RK2 plasmid fused to the 3.0 kb segment of pBR322 allowing maintenance in *E. coli* and in *A. tumefaciens*. This was fused to the 90 bp DNA fragment from pTiT37 plasmid, which contained the 25 bp nopaline-type T-DNA right border.

The nucleotide sequence of the *cp4 epsps* gene was modified to maximize protein expression in plant cells. Expression of this gene was under the control of the P-CMoVb, modified FMV promoter and the 3′ non-translated region of the *rubisco* (rbcS, small subunit ribulose bisphosphate carboxylase oxygenase) *E9* gene terminator sequence from pea. Additionally, the synthetic CP4 EPSPS gene was fused at the 5′ end to the region that codes for the CPT2 from *A. thaliana*. The CTP2 peptide sequence targets CP4 EPSPS to the chloroplasts where the aromatic amino acid biosynthetic pathway and endogenous EPSPS activity are located [CERA 2009i].

The *aad* gene, derived from *E. coli* bacterial transposon Tn7, encodes aminoglycoside adenyltransferase (AAD), which confers resistance to the antibiotics spectinomycin and streptomycin. The *aad* gene was under the control of its own bacterial promoter and terminator and was not expressed in plant tissues. The NPTII, encoding the selectable marker *npt*II gene, was located downstream of the *aad* gene. It was under the control of CaMV 35S promoter and the 3′ region of the *nos* gene from the pTiT37 plasmid of *A. timefaciens* strain T37. The *gox* gene was fused at the 5′ end to the sequence coding for the CTP from *A. thaliana* EPSPS and the P-CmoVb promoter. The 3′ region was derived from the 3′ non-translated region of the *nos* gene. The *gox* gene was not integrated into the genome of the transgenic cotton line [CERA 2009i].

Over-the-top applications of glyphosate after the four-leaf stage occasionally caused fruit abortion and yield loss because of insufficient expression of *cp4 epsps* gene [Dill et al. 2008]. However, cotton farmers have valued the early weed control and rapidly adopted the technology [Green and Castle 2010]. USA approved this glyphosate resistant line MON1445 in 1995 followed by Canada in 1996, Japan in 1997, and many countries later [CERA 2009i].

MON 88913. In order to improve crop tolerance and allow farmers to apply glyphosate during the critical reproductive phases of cotton growth compared to event MON1445, Monsanto developed a second generation glyphosate-resistant cotton variety MON88913. It was commercialized as 'Roundup Ready Flex' cotton in 2006. The increased level of glyphosate tolerance in MON88913 was achieved through the use of improved promoter sequences that regulate the expression of the *cp4 epsps* gene conferring glyphosate tolerance.

Unlike MON1445 which contained one *cp4 epsps* gene, MON88913 contains two *cp4 epsps* genes with chimeric promoters that more strongly express the trait in the 4- to 12-leaf vegetative stages and in the sensitive reproductive tissue [CaJacob et al. 2007]. The genes from the transformation binary plasmid vector pV-GHGT35 with two tandem *cp4 epsps* gene cassettes were inserted into the commercial variety 'Cocker 312' using the *Agrobacterium*-mediated transformation method. In the first cassette, the FMV e35S/TSF1 chimeric promoter regulates gene expression. In the second cassette, CaMV 35S/ACT8, a chimeric promoter from the actin gene *A CT8* of *A. thaliana*, and the CaMV 35S promoter, regulate the expression of the *cp4 epsps* gene. The transcriptional termination sequence was derived from the TE-9 DNA sequence of pea, containing the 3′ nontranslated region of the Rubisco small subunit E9 gene.

The CP4 EPSPS protein produced in MON88913 was targeted to the chloroplasts via an N-terminal fusion with the CTP2 to form a CTP2-CP4 ESPSP precursor protein. The precursor protein produced in the cytoplasm is processed to remove the transit peptide upon translocation into the plant chloroplast, resulting in the mature CP4 EPSPS protein. Finding the transgenic line MON88913 easier to manage weeds, cotton farmers rapidly transitioned to it [Dill et al. 2008]. USA, Canada, Philippines and Japan approved this transgenic line in 2005 [CERA 2009k].

GHB614. In 2006, another glyphosate resistant 'GlyTol' cotton event GHB614, developed by Bayer CropScience, used a modified *epsps* gene, *zm-2mepsps* [Bayer 2006b; Trolinder et al. 2008]. The Zm-2mEPSPS enzyme differs from the naturally occurring maize EPSPS by two amino acids and is the same EPSPS as in GA21 maize. The gene was inserted into the cultivar 'Coker 312' using *Agrobacterium*-mediated transformation.

The transformation plasmid vector pTEM2 used for gene transfer was derived from pGSC1700 [Cornelissen and Vandewiele 1989]. The vector backbone contained: a) a plasmid core comprising the plasmid core comprising the origin of replication from the plasmid pBR322 for replication in *E. coli* (ORI ColE1) and a restriction fragment comprising the origin of replication from the *Pseudomonas* plasmid pVS1 for replication in *A. tumefaciens* (ORI pVS1); b) a selectable marker gene, *aad*A, and c) a DNA region consisting of a fragment of the *npt*II gene from transposon Tn903.

The event has a single copy of the *zm-2mepsps* gene and a constitutive promoter of the *histone H4* gene from *A. thaliana* (Ph4a748At) [Chabouté et al. 1987], which controls expression of the *2mepsps* gene. The Ph4a748At promoter, combined with the first intron of gene II of the histone H3.III variant of *A. thaliana* [Chaubet et al. 1992] directs high level constitutive expression, especially in the rapidly growing plant tissues. The optimized transit peptide (TPotp C), which contained sequences from the RuBisCO small subunit genes of maize and sunflower, targets the mature protein to the plastids, where the wild-type protein is located [Lebrun et al. 1996].

The Event GHB614 has a stably integrated gene, with no marker gene and vector backbone sequences. USA, Canada and Mexico it in 2008 [CERA 2009b].

Glufosinate

LLCotton25. LLCotton25 was developed to allow the use of glufosinate as a weed control option in cotton production. Plant cells of cotton variety 'Coker 312' was transformed by Bayer CropScience (AgrEvo) with pGSV71 vector using *Agrobacterium*-mediated method. The transformation was achieved by culturing cotton tissue, excised between the hypocotyl and the radicle of 3-day old seedlings, with a culture of *A. tumefaciens* harboring the Ti plasmid pGV3000 and the plasmid vector pGSV71 [LLCotton25 USDA Petition 2002]. The transforming plasmid vector carried a T-DNA gene cassette consisting of the *bar* gene isolated from *S. hygroscopicus* (strain ATCC21705), CaMV 35S promoter, and the 3'*nos* terminator derived from the T-DNA of pTiT37 isolated from *A. tumefaciens*. After transformation, the explants were regenerated using tissue culture techniques. The transformed plants expressing the *bar* gene were selected from glufosinate. Genomic DNA blot analysis confirmed that the *bar* gene was stably integrated [Health Canada 2005]. The transgenic line LLCotton25 (Liberty Link) was approved by USA in 2003, Canada and Japan in 2004, Korea in 2005, Australia, New Zealand, China and Mexico in 2006, and the European Union and Brazil in 2008 [CERA 2009j].

Bromoxynil

BXN. BXN cotton was the first herbicide-resistant transgenic crop variety developed by Calgene and Rhône-Poulenc. This Davis, California-based biotechnology company Calgene (acquired by Monsanto in 1996), known for engineering the 'Flavr Savr' tomato in 1994 (vide Chapter 4), used the *Agrobacterium*-mediated transformation method in which the T-DNA contained the *bxn* gene isolated from *Klebsiella ozaenae* [CERA 2002c]. An antibiotic resistance marker gene, *neo*, encoding the enzyme NPTII, which inactivates aminoglycoside antibiotics such as kanamycin and neomycin, was also introduced into the genome of these transgenic lines. This gene was derived from a bacterial Tn5 transposon *E. coli*, strain K12 [CERA 2002c].

In the transformed plants, there was no incorporation of translatable plasmid DNA sequences outside of the T-DNA region. The nitrilase enzyme expressed at very low levels in unprocessed cotton seed while the refined oils were generally free of DNA and protein. This transgenic BXN cotton was resistant to bromoxynil at rates as high as 11.2 kg ha^{-1}, i.e., 20 times the normal field rate. This first herbicide-resistant transgenic variety was approved by USA in 1994, followed by Canada and Mexico in 1996, Japan in 1997, and Australia and New Zealand in 2002.

Calgene marketed BXN cotton in May 1995 for use on 250,000 acres (101,000 ha), about three percent of the total cotton area in the U.S., of transgenic cotton for three years before raising it to 10 percent of the U.S. crop in May 1998. At this time, Calgene has already been acquired by Monsanto which had its 'Roundup Ready' cotton and *Bt*-based insect resistant 'Bollgard' cotton (developed by its acquired company Delta & Pine Land Co.) approved by EPA for commercial use. Consequently, BXN cotton has been less widely grown than Roundup Ready cotton. Eventually, Calgene's BXN cotton was discontinued because of its limited weed control spectrum.

2,4-D

Cotton is one of the most sensitive crops to 2,4-D and the damage is quite extensive. One approach to reduce this damage is to incorporate resistance to 2,4-D into cotton germplasm. There have been some efforts in USA and Australia to develop 2,4-D-resistant transgenic cotton lines. Although plant biotechnologists were successful in engineering cotton for 2,4-D tolerance, their progress was bogged down amid consumer concerns about drift and volatilization problems associated with the herbicide.

In an effort to circumvent the volatilization problem that exists more with ester formulation, Dow CropSciences developed in 2011 technology to commercially produce the choline salt formulation of 2,4-D. With its ultra-low volatility, the potential for drift and decreased odor may be minimized. If the regulatory approvals are obtained, this formulation is expected to find acceptance in commercialization of 2,4-D resistant transgenic crops.

Bayley et al. [1992] engineered the 2,4-D resistance trait into precultured hypocotyl sections of cultivar 'Coker 312'. The 2,4-D monooxygenase gene *tfdA* (involved in the degradation of 2,4-D) from *Alcaligenes eutrophus* plasmid pJPS was isolated, modified, and expressed in transgenic cotton plants. In order to transfer the *tfdA* gene into an *Agrobacterium* vector, pRO17 was linearized by *Hind*III and ligated to HindIII-cleaved pBIN19 to form a co-integrate plasmid, with transcription of the *tfdA* gene towards the left T-DNA border. The cointegrate pBIN19::pRO17 was transferred into the disarmed *A. tumefaciens* strain GV3111 harboring the helper Ti plasmid pTiB6S3S3. The gene cassette was driven by CaMV 35S promoter and the *nos*3'. The transformants were unaffected up to three times the recommended 2,4-D doses, while the non-transformed plants were severely affected even at rates 300 times lower than the recommended rate. Besides, the activity of 2,4-D monooxygenase in the protoplasts of transformed plants was 17- to 38-fold above that of the non-transformed plants. The transformed plants survived 2,4-D even at 1.5 kg ha^{-1}.

Similar transgenic transformation of cotton was done in Australia by Lyon et al. [1993] in the case of 'Coker 315' variety. The transformed line contained a single copy of the chimeric *tfd*A gene and it was resistant to over 600 ppm of 2,4-D. T2 and T3 plants that were either hemizygous or homozygous. The hemizygous plants were slightly damaged initially, but eventually grew out of it and became normal in appearance and yield while homozygous were completely unaffected by this level of 2,4-D. The best transgenic plants were 100-fold more tolerant than non-transformed cotton plants.

Double-Trait/Event Stacks: Herbicides

Dicamba and Glufosinate

MON88701. In 2012, Monsanto developed MON88701 which is expected to enable growers to utilize both dicamba and glufosinate for effective control of weeds [European Union 2013]. This dicamba-cum-glufosinate resistant cotton was produced via *Agrobacterium*-mediated transformation of the hypocotyl segments using plasmid vector PV-GHHT6997 of 9.4 kb. It contains a demethylase gene, *dmo* (from *Stenotrophomonas maltophilia*), that expresses a dicamba mono-oxygenase (DMO) protein to confer tolerance to dicamba and a bialaphos resistance gene, *bar* (from *S. hygroscopicus*) that expresses the PAT protein conferring tolerance to glufosinate [European Union 2013].

The plasmid vector PV-GHHT6997 of 9.4 kb contains one T-DNA that is delineated by Left Border and Right Border regions. The T-DNA contains the *dmo* and *bar* expression cassettes. The *dmo* expression cassette is regulated by the *PC1SV* promoter, the *TEV* (tobacco etch virus) 5' leader sequence, and the 3' non-translated sequence of the *E6* gene. The chloroplast transit peptide (CTP2), derived from the CTP2 target sequence of the *A. thaliana shkG* gene directs transport of the DMO protein to the chloroplast. The *bar* expression cassette is regulated by the e35S promoter from the 35S RNA of CaMV, the heat shock protein 70 (*Hsp70*) leader, and the nopaline synthase (*nos*) 3'non-translated region [European Union 2013].

The backbone region of PV-GHHT6997, located outside of the T-DNA, contains two origins of replication for maintenance of plasmid vector in bacteria (*oriV* and *ori-pBR322*), a bacterial selectable marker gene (*aadA*), and a coding sequence for repressor of primer (*rop*) protein for maintenance of plasmid vector copy number in *E. coli* [European Union 2013].

Southern blot analyses confirmed that the T-DNA was inserted into the cotton genome at a locus containing one copy of the *dmo* and *bar* expression cassettes [European Union 2013]. No additional elements were detected other than those associated with the insert.

Glyphosate and Glufosinate

GHB614 x LLCotton 25. This double event cotton hybrid (GlyTol-Liberty Link cotton) resistant to both non-selective herbicides glyphosate and glufosinate was developed by Bayer CropScience by cross breeding of two transgenic parental donors GHB614 and LLCotton25 (vide individual events discussed earlier). This stacked variety inherited the glyphosate tolerance trait (conferred by *2mepsps* gene) from GHB614 and glufosinate tolerance trait (conferred by *bar* gene) from LLCotton25.

Double-Trait/Event Stacks: Herbicide-cum-Insect Resistance

Glyphosate and Lepidopteran Insects

MON15985 x *MON88913*. The stacked cotton hybrid MON15985 x MON88913 (Roundup Ready Flex Bollgard II) expresses three novel proteins: the δ-endotoxins of Cry1Ac and Cry2Ab, both of which confer resistance to the lepidopteran pests of cotton (the cotton bollworm, pink bollworm, and tobacco budworm), and the CP4 EPSPS protein which confers resistance to glyphosate. The insecticidal proteins Cry1Ac and Cry2Ab are produced by *cry1Ac* and *cry2Ab* genes respectively, both of which are from the lepidopteran insect-resistant MON15985. The CP4 EPSPS protein is produced by two *cp4 epsps* genes from MON88913 to confer resistance to glyphosate later in the growing season, specifically after the fifth true leaf stage. The novel traits of each parental line have been combined, through traditional plant breeding, to produce this new cotton line. For a full description of each parental line please refer to the individual product descriptions in the crop database for 15985 and MON88913. This insect- cum glyphosate-resistant stacked cotton variety was approved by Japan in 2005, Australia, Korea, Mexico, New Zealand and Philippines in 2006, South Africa and USA in 2007 [CERA 2008b].

Glufosinate-cum-Lepidopteran Insects

T304-40 x *GHB119*. The lepidopteran insect-cum-glufosinate resistant T304-40 and GHB119 cotton lines of AgrEvo (Bayer CropSciences), derived from cultivars 'Coker 312' and 'Coker 315' respectively, were both transformed by *Agrobacterium*-mediated gene transfer of the T-DNAs from pTDL008 and pTEM12 vectors respectively. Cotyledon explants from *in vitro* germinated seedlings were co-cultivated with *A. tumefaciens* strain EHA 101 containing the standard binary vectors [Bayer 2008].

For event T304-40, the vector pTDL008 is a derivative of the vector pGSV20 in which the *bar* gene cassette coding for the PAT protein was inserted together with the *cry1Ab* gene cassette encoding a fragment of the insecticidal *cry1Ab* crystal protein. Besides, the transgene cassette contained 5' e1-Ps7s7 promoter, 3'me1 terminator, P35S3 promoter, and 3'nos terminator. The resulting DNA fragments were separated by agarose gel electrophoresis, transferred to a membrane, and sequentially hybridized with six different probes, each representing a fragment of the transforming gene cassette, or the complete T-DNA probe. Genomic analysis demonstrated that the inserted transgenic sequence in the cotton line T304-40 consists of one nearly complete copy of the T-DNA flanked by an inverted incomplete copy of the *cry1Ab* gene cassette and one additional 3'me1 terminator [Bayer 2008].

In the case of GHB119 line, the vector pTEM12, derived from pGSC1700, contains the *bar* gene and *cry2Ae* gene encoding a fragment of the insecticidal *cry2Ae* crystal protein of *B. thuringiensis* (*B.t.*) subsp. *dakota*. Besides, the gene cassette contained ORI ColE1 (the origin of replication from the plasmid pBR322 for replication in *E. coli*) and ORIpVS1 (a restriction fragment comprising the origin of replication from the *Pseudomonas* plasmid pVS1 for replication in *A. tumefaciens*; the selectable marker gene *aadA* conferring resistance to streptomycin and spectinomycin for propagation and selection of the plasmid in *E. coli* and *A. tumefaciens*; and *npt*II gene (a DNA region consisting of a fragment of the neomycin phosphotransferase coding sequence from transposon Tn903).

As these elements were outside the T-DNA borders, they were not transferred into the cotton genome [Bayer 2008].

Following transformation, T_0 plants of Events T304-40 and GHB119 were independently back-crossed into a conventional BCS breeding line to obtain homozygous and stable lines (BC3F$_3$). At the BC2F$_1$ generation, the two events were combined and the progeny were tested for seven generations. The resulting lines were crossed to obtain 'TwinLink' which carried the insect-resistant *cry1Ab* and *cry2Ae* genes along with glufosinate-resistant *bar* gene [Bayer 2008]. US Environmental Protection Agency approved LibertyLink technology in 2012. This lepidopteran insect- and glufosinate-resistant transgenic cotton hybrid(GEM1) has also been approved by Canada, Australia, New Zealand, and Brazil.

DAS-21023-5 x *DAS-24236-5*. Another glufosinate-resistant and lepidopteran insect-protected stacked hybrid, 'WideStrike', was produced by Dow AgroSciences by cross-breeding two transgenic cotton lines: 'DAS-21023-5' and 'DAS-24236-5'. This maize variety expresses two novel proteins: Cry1F and Cry1Ac δ-endotoxins which confer resistance to lepidopteran insects. The Cry1F was produced by the *cry*1F gene inherited from cotton line 24236-5 and Cry1Ac was produced by the *cry*1Ac gene inherited from event 21023-5. The *pat* gene, inherited from cotton line 21023-5, produces the PAT protein, and this confers resistance to the herbicide glufosinate. This gene was inserted solely to be used as a selectable marker during the transformation that led to the production of Events '24236-5' and '21023-5'. This bi-event maize hybrid was approved by USA and Mexico in 2004 followed by Australia, Canada, Japan and Korea in 2005, and Brazil in 2009 [CERA 2009f].

Triple-Event/Trait Stack Hybrids: Herbicide-cum-Insect Resistance

Glufosinate, Glyphosate, and Lepidopteran Insects

DAS-21023-5 x *DAS-24236-5* x *MON88913*. The multigene-stacked cotton hybrid was derived from traditional breeding crosses between the lepidopteran-insect protecting and glufosinate-resistant parental line 'WideStrike' (DAS-21023-5 x DAS-24236-5) and the glyphosate-resistant parent line 'MON88913'. This maize hybrid expresses four novel proteins: Cry1F, Cry1Ac, CP4 EPSPS, PAT. The insecticidal Cry1F and Cry1Ac (from DAS24236-5 and DAS-21023-5 respectively) proteins are produced by *cry*1F gene and *cry*1Ac genes respectively. The CP4 EPSPS protein is expressed by two copies of the *cp4 epsps* gene, while the PAT protein is from DAS-21023-5 which contains the *pat* gene. This multi-transgenic cotton hybrid of Dow and Pioneer Hi-Bred was approved by Mexico, Japan, and Korea in 2006 for commercial use (Widerstrike Roundup Ready Flex) [CERA 2006a].

DAS-21023-5 x *DAS-24236-5* x *MON-01445-2*. The three-event cotton hybrid developed by Dow and Monsanto through conventional breeding between 'WideStrike' (DAS-21023-5 x DAS-24236-5) and MON1445 expresses four novel proteins: the δ-endotoxins Cry1F and Cry1Ac, the CP4 EPSPS protein, and the PAT protein which confer resistance to the lepidopteran insects, glyphosate, and glufosinate respectively. The Cry1F and Cry1Ac proteins, produced by the *cry1F* and *cry1Ac* genes respectively, are derived from 'WideStrike' developed by crossing DAS-24236-5 with DAS-21023-5. The *pat* gene to produce PAT from the transgenic 21023-5, but this expressed novel protein was intended solely for use as a selectable marker during plant transformation [CERA 2005b]. The CP4 EPSPS protein was produced by the *cp4 epsps* gene from MON1445. This transgenic herbicide-insect resistant cotton hybrid (WideStrike Roundup Ready) was approved by Mexico in 2005 and Japan and Korea in 2006.

Rice

Mono-Trait Events: Herbicides

Glufosinate

LLRICE06; LLRICE62. These two transgenic rice lines were produced by Bayer CropScience using biolistic method of DNA transfer system in which DNA sequences encoding PAT enzyme were introduced into the callus tissue of parental medium-grain rice varieties 'M202' and 'Bengal' in the case of LLRICE06 and LLRICE62 respectively. The plasmid pB5/35S-*bar* was digested with restriction enzymes *Pvu*I and *Hind*III and a 1501 bp fragment containing the *bar* gene cassette (P35S-*bar*-T35S) was purified and used in the transformation. The genetic elements of this cassette include the CaMV 35S promoter, the *bar* gene (derived from *S. hygroscopicus* and modified to improve expression in plant cells), and the CaMV T35S terminator [CERA 2009m]. The *bar* gene, including these regulatory sequences, was the only novel gene introduced into the parental rice varieties.

Genomic DNA analysis indicated that single copy of the *bar* gene cassette was stably integrated into LLRICE06 and LLRICE62 transgenic lines. The herbicide-resistant trait from these lines was introduced into commercial rice varieties via traditional backcrossing. Analyses of genomic DNA of T_2, T_3, and T_4 generations of both lines yielded identical hybridization patterns, confirming the genetic stability of the original transformation events [CERA 2009m]. These two 'LibertyLink' rice events were approved by USA in 2000, Canada in 2006, Mexico in 2007, and Australia and Colombia in 2008.

LLRICE601. This 'Liberty Link' rice event is similar to LLRICE62 and LLRICE06 as it also contains a single *bar* gene driven by the CaMV 35S promoter [Bayer 2006a]. However, the DNA construct into LLRICE601 was introduced by *Agrobacterium*-mediated transformation, with the double-stranded CaMV 35S promoter being slightly longer. The *bar* genes was isolated from *S. hygroscopicus*, strain HP632. Transformation was done using the long grain variety 'Cocodrie' unlike the other two lines. The plasmid used was pGSV71, derived from pGSC1700. It contains an artificial T-region consisting of the left and right border sequences of the TL-DNA from pTiB6S3 and multilinker cloning sites allowing the insertion of chimeric genes between the T-DNA border repeats. In pGSV71, the gene of interest, inserted between the T-DNA border repeats, is P35S-*bar*-3'nos. That is, the chimeric *bar* gene construct containing the 35S promoter CaMV is followed by the 3' non-translated region of *nos* [Bayer 2006a].

Southern blot analysis demonstrated that LLRICE601 contained only one copy of the *bar* gene. The PAT protein level was below the level in LLRICE06 and LLRICE62, while the same in the leaf tissue was much lower than the level in LLRICE62 and was slightly higher than the level in LLRICE06. This transgenic LLRICE601 was approved by USA in 2006 and Colombia in 2008 [CERA 2008c].

Mono-Trait Events: Herbicides (Non-Transgenic)

Imidazolinones

CL121, CL141, CFX51. The non-transgenic IMI-resistant 'Clearfield' rice events, CL121, CL141, and CFX51, were developed by BASF through chemically-induced accelerated mutagenesis of rice cultivar 'AS3510'. Chemical mutagenesis was done with 0.5 percent ethyl methanesulfonate for 16 hours prior to planting. The M_2 plants were sprayed with imazethapyr. One surviving IMI-herbicide resistant rice mutant, '93AS3510', was crossed with commercial varieties 'Cocodrie', 'Maybelle', and 'Cypress' to produce CL121, CL141, and CFX51 respectively [CERA 2002a]. Since recombinant techniques were not used, no foreign DNA was introduced to derive herbicide resistance. The mutation affects the ALS enzyme of rice at a specific location, resulting in an alteration to the binding site for IMI herbicides. The tolerance to IMI-containing herbicides was due to single-point mutation within the ALS encoding gene. The amino acid sequence of the mutated enzyme differs by one amino acid from that of the wild-type enzyme as in the case of maize and rapeseed [CERA 2002a]. These three IMI-tolerant Clearfield rice varieties developed by BASF were approved by USA and Canada in 2002.

PWC16. Another non-transgenic rice line PWC16 was developed by BASF to allow the use of imazethapyr in rice. This trait was developed using chemically induced seed mutagenesis and whole plant selection procedures. This rice line expresses a mutated form of the acetohydroxyacid synthase (AHAS/ALS) enzyme, which renders the plant tolerant to levels of imazethapyr used in weed control. Seeds of the rice cultivar 'Cypress' were mutagenized by BASF with 0.175 percent ethyl methane-sulfonate aqueous solution for 23 hours to induce point mutations within the ALS encoding gene, similar to that described in other IMI-tolerant lines (CL121, CL141, CFX51). This was followed by foliage application of plants with imazethapyr. Then, herbicide tolerant plants were made from whole plants treated with this IMI herbicide. The designation of PWC16 was given to the imazethapyr-tolerant plant selected from the population [CERA 2005k]. Seed increase was achieved through self-pollination or natural (wind) pollination occurring among plants of this line.

The modified AHAS gene, conferring tolerance to imazethapyr, was under control of the native AHAS promoter and it was believed to be constitutively expressed. Whole plant tolerance to this herbicide was expressed in PWC16 rice. The amino acid composition of this 'Clearfield' rice (CL rice) was compared to commercial cultivars, confirming that the AHAS activity of this IMI-tolerant rice was not affected by the mutation [Health Canada 2006b].

DNA sequencing reveals that PWC16 mutant has single codon changes in its AHAS gene which was responsible for herbicide tolerance [Croughan 2002]. The position of the mutation for 93AS3510 was codon 654 where glycine was substituted by glutamic acid (Gly654→Glu654) in the encoded AHAS protein [Croughan 2002]. The position of the target site mutation for the PWC16 was at codon 653 where serine was substituted by asparagine in the encoded AHAS protein [Croughan 2002].

IMINTA 1 and *IMINTA 4.* Like other IMI-resistant rice varieties, these two were developed by chemical mutagenesis of seeds of 'IRGA 417' rice variety treated with the mutagen sodium azide. These seeds were then grown under greenhouse conditions to produce the M_1 generation. Imazapic and imazapyr were applied postemergence. The herbicide-tolerant lines IMINTA 1 and IMINTA 4 were selected from the M_3 generation based on parental type and tolerance to IM herbicides. The IMINTA rice varieties have been generated using self-pollination for seven generations. These lines, developed by BASF, have been approved by Canada in 2006.

Wheat

Mono-Trait Events: Herbicides

Glyphosate

MON71800. The wheat line MON71800 was produced by *Agrobacterium*-mediated transformation of plant cells from spring wheat cultivar 'Bobwhite'. The binary vector PV-TXGT10 used for the transformation contained two *cp4 epsps* gene cassettes coding for glyphosate tolerance. Each gene cassette consisted of chloroplast transit peptide coding sequences from the *Arabidopsis thaliana epsps* gene (*Arab*TP) associated with the sequences of the *cp4 epsps* gene [CERA 2005i]. Two different promoters were used to regulate the expression of each *cp4 epsps* gene: a) the enhanced CaMV 35S promoter and b) the promoter, transcription start site, and first intron of the 5′ region of the rice *actin*1 gene. Terminator sequences in each gene cassette consisted of the 3′ non-translated region of the nopaline synthase gene (*nos*). The PV-TXGT10 vector backbone contained the origin of replication sequences *ori-V* and *ori-322/rop*. The vector backbone also contained the *aad* gene, which codes for streptomycin adenyltransferase, to allow the selection of bacteria containing the PV-TXGT10 vector [CERA 2005i].

Southern blot analysis confirmed the stable insertion of one intact copy of each *cp4 epsps* gene cassette, including the promoter, terminator, and chlorophyll transit peptide sequences. None of the vector backbone sequences were integrated into the genome of MON71800 wheat. The introduced glyphosate-tolerance trait was found even after 18 generations of selfing of the original homozygous glyphosate-resistant plants. There was no decrease in tolerance to glyphosate. This 'Roundup Ready' event of Monsanto was approved for commercialization in USA and Colombia in 2004 [CERA 2005i].

Mono-Trait Events: Herbicides (Non-Transgenic)

Imidazolinones

Several imidazolinone-resistant winter wheat varieties have been developed over the past 20 years. In all cases, a single, partially dominant nuclear gene conferred resistance. Four IMI-tolerant wheat plants were selected after having French cultivar 'Fidel' mutagenized with sodium azide [Newhouse et al. 1992]. These tolerant plants were named FS1 (Fidel Selection 1), FS2, FS3, and FS4. Subsequently, these four IMI-resistant Fidel selections have been used as trait donors for breeding IMI-resistant wheat varieties which were marketed in 2001.

Similar approach was also followed in developing IMI-resistant spring wheat varieties. Seeds of cultivar 'Teal' were treated chemical mutagen ethyl methanesulfonate, and M_2 plants were sprayed with imazamox. Six M3 lines with moderate to high levels of imazamox resistance were selected and designated 'TealIMI' lines, 1A, 9A, 10A, 11A, 15A, and 16A. Two distinctive mutations, different from FS4, were discovered from lines 11A and

15A. Line 15A had the FS4 mutation and another novel mutation, while 11A possessed a non-allelic mutation to FS4 [Pozniak and Hucl 2004].

The hexaploid wheat has three genomes: A, B, and C [Poehlman and Sleper 1995] and the three loci are located on the long arm of chromosomes 6D, 6B, and 6A [Pozniak and Hucl 2004; Anderson et al. 2004; Pozniak et al. 2004]. The mutation of the ALS gene on 6DL has been named 'Imi1', while two other mutations have been named 'Imi2' and 'Imi3' [Pozniak and Hucl 2004]. Inheritance studies on IMI-tolerance and allelism of traits indicated that Imi1 of FS4 or TealIMI 15A, Imi2 of TealIMI 11A, and Imi3 of TealIMI 15A were all semi-dominant and are unlinked [Tan et al. 2005]. Higher levels of IMI tolerance in wheat can be achieved by stacking two or more tolerant genes into a single genotype [Newhouse et al. 1992; Pozniak and Hucl 2004]. Imi2 and Imi3 are believed to be on genomes B and A, respectively [Pozniak et al. 2004]. DNA sequencing data showed that both Imi1 and Imi2 have an amino-acid substitution of Ser653→Asn653 in the AHAS enzyme that was analogous to the XI12 mutation in maize, the PM1 mutation in rapeseed, and the PWC16 mutation in rice [Pozniak et al. 2004]. Winter wheat FS4 with a single homozygous gene has an acceptable IMI tolerance. For spring wheat, two homozygous IMI-tolerant AHAS genes are to be stacked to achieve acceptable tolerance to IMI herbicides [Tan et al. 2005].

SWP965001. Of the four IMI-tolerant selection lines (FS1, FS2, FS3, FS4) of winter wheat developed following chemical mutagenesis with sodium azide, described earlier, the FS2 mutation affects the ALS enzyme of wheat at a specific location, while other properties of the enzyme are unaffected. The SWP965001 line was derived from an initial cross of FS2 to the spring wheat cultivar 'Grandlin', followed by two backcrosses to Grandlin. As a result, 87.5 percent of the genetic information was from the Grandlin parent, with the remaining 12.5 percent being from Fidel, the parent of FS2 [CERA 2001g]. The resulting SWP965001 became more resistant to IMI herbicide imazamox than F2. This Cynamid-developed wheat line was approved by Canada in 1999 for commercial availability in 2000.

AP205CL. AP205CL is another IMI-resistant spring wheat variety developed by BASF to allow the use of imazamox by using mutagenesis and conventional seed increasing techniques. There was introduction or incorporation of heterologous DNA into the plant genome. Seeds of wheat variety 'Gunner' were mutagenized by ethyl methancesulfonate and diethyl sulfate, known to induce point mutations within the genome of organisms. The selection of IMI-tolerant plants was made from whole plants treated with imazamox. The designation AP205CL was given to the imazamox-tolerant plant selected from the population. The tolerance to imazamox was due to a point mutation of a single nucleotide in one (*als2*) of the three AHAS genes [CERA 2005j]. The altered gene contains a mutation which results in a codon change from ACG to AAC, producing an amino acid change from serine to asparagine. As with the previously approved imidazolinone tolerant wheat, rice, maize, and rapeseed, the single amino acid change alters the binding site for the herbicide on the AHAS enzyme while having no effect on the normal functioning of the enzyme. This 'Clearfield Wheat' line was approved by Canada in 2003.

BW7. BW7 is a bread variety of wheat with an imidazolinone resistance trait *Als1b* gene. It was developed by BASF using seed mutagenesis with sodium azide from the parental winter wheat variety. Treated seeds were grown as the M_1 generation to produce M_2 generation seeds. The M_2 generation and each subsequent generation until M_6 where screened using imazamox for herbicide tolerance. Thereafter, BW7 has been bred to the M_9 generation by self-pollination. In this imidazolinone-tolerant variety, there was a single base substitution in the *Als1* gene coding region, with no alterations in the coding regions of *Als2* and *Als3* genes. The mutation in *Als1b* of the *Als1* gene was stably inherited in BW7 [Health Canada 2007]. Canada approved BW7 in 2007 for commercial purpose.

BW255-2 and BW238-3. The IMI-tolerant wheat lines BW255-2 and BW238-3 (Clearfield varieties), developed by BASF Canada, are products of mutagenesis and conventional seed increase techniques [Health Canada 2013]. Seed from spring wheat cultivars BW255-2 and BW238-3 were modified using mutagenesis with sodium azide, producing the M_1 generation. These seeds were grown under greenhouse conditions to produce M_2 and the subsequent M_3 generations. Plants were selected from the M_3 generation based on parental type and imidazolinone tolerance. Seeds from these plants were then used for generational increase; both varieties are currently in the M_9 generation. Tolerance of these non-transgenic lines to IMI herbicides is due to a mutation in the AHAS (ALS) produced by *Als3* (agglutinin-like sequence) gene that encodes a family of proteins which, with their adhesion function, in mediate attachment to epithelial cells, endothelial cells, and extracellular matrix proteins. The selection of herbicide resistance was made on whole plants. The designation BW255-2 and BW238-3 was given to one herbicide resistant mutant selected from each population. The selection of herbicide resistance was made on whole plants. These IMI-resistant lines were approved by Canada in 2006.

Sugar Beet

Mono-Trait Events: Herbicides

Glyphosate

GTSB77. The sugar beet event GTSB77, developed by Novartis Seeds and Monsanto, contains two sequences encoding the glyphosate-resistance conferred by *cp4 epsps* and the *gox* genes. Transformation was done via *Agrobacterium*-mediated method by using the proprietary cytoplasmic male sterile sugar beet line 'A1012' with plasmid PV-BVGT03 which contains well characterized DNA segments required for selection and replication of the plasmid in the bacteria as well as a right border for initiating the region of T-DNA into the plant genomic DNA.

The T-DNA region of the *Agrobacterium* Ti plasmid used for this transgenic variety contained *cp4 epsps*, *gox*, *npt*β-carotene is then converted into retinal via a cleavage reaction, which is catalyzed by an oxygenase, and *gus* genes. Associated with the *cp4 epsps* gene, encoding EPSPS enzyme, were promoter sequences derived from the FMV 35S, sequences encoding the chloroplast transit peptide (CTP2) from *A. thaliana*, and transcription termination and polyadenylation signal sequences from the RuBisCo (rbcS) encoding gene from pea. Expression of the GUS enzyme encoding *uidA* gene was regulated by the enhanced CaMV 35S promoter and the transcription termination sequences from the 3' region of the pea *rbcS* encoding gene. Associated with the *goxv247* gene were the FMV 35S promoter, encoding the CTP from *A. thaliana* and the 3'*nos* gene, which directs polyadenylation.

Following transformation, the genomic DNA of GTSB77 was found to contain the existence of a single copy of the T-DNA region containing complete functional copies of the CP4 EPSPS and GUS encoding genes, and a partial non-functional copy of the GOX encoding gene. This partial copy of the GOX encoding sequences was due to a 3' truncation of the T-DNA insert, which also resulted in no integration of NPTII encoding sequences into the host genome [CERA 2001a]. This glyphosate-resistant GTSB77 sugar beet line (InVigor) was approved by USA in 1998, and other countries later.

Event H7-1. In 2003, Monsanto and the German KWS SAAT AG Company developed glyphosate-resistant sugarbeet event H7-1. It was produced by transformation of '3S0057', the proprietary sugar beet cultivar of the German company, by using the *Agrobacterium*-mediated method.

The cotyledons were infected with *A. tumefaciens* strain CP4 containing plasmid PV-BVGT08. The plasmid contains a single gene expression cassette flanked by the LB and RB sequences from the Ti plasmid of this bacterium. The Event H7-1 contains: a) the FMV 35S gene promoter; b) the ctp2 terminal chloroplast peptide (CTP) from the *A. thaliana cp4 epsps* coding region that targets the protein to the chloroplast; c) the *cp4 epsps* gene; and d) 3´non-translated region (transcriptional terminator and polyadenylation site) of the *rbc*S E9 gene from pea. The inserted *cp4 epsps* gene has been stably integrated at the genome level [CERA 2008a].

In 2005, USA approved Event H7-1 (Roundup Ready Sugarbeet), while sugar from this glyphosate-resistant sugar beet has been approved for human and animal consumption in the U.S. and the European Union. The U.S. farmers quickly adopted this transgenic line by bringing 95 percent of the acreage under it by 2011. This transgenic sugarbeet Event H7-1 was approved by Australia, Canada, Colombia, New Zealand and Philippines in 2005, Korea and Mexico in 2006, and the European Union and Japan in 2007 [CERA 2008a].

Glufosinate

T120-7. This glufosinate-resistant event of sugar beet, developed in 1997 by Bayer CropScience {Aventis Crop Science (AgrEvo)}, was produced by *Agrobacterium*-mediated transformation of calli from the parent line 'R01' with plasmid vector pOCA18/Ac (CERA 2002b,c). The T-DNA portion of the tumor inducing (Ti) bacterial plasmid was engineered to contain a modified form the PAT encoding gene as well as the *npt*II gene, which inactivates the selectable marker gene aminoglycoside {(3') phosphotransferase type II derived from transposon Tn5 of *E. coli*} antibiotics such as kanamycin and neomycin. In order to enhance plant expression of the protein, the nucleotide sequence of the *pat* gene, derived from *S. viridochromogenes*, strain Tü494 was modified using site-directed mutagenesis to contain plant-preferred codons. These modifications did not result in changes to the predicted amino acid sequence of the PAT enzyme. Expression of NPTII was used as a selectable marker to screen for transformed plants during tissue culture regeneration and multiplication. Associated with the *pat* gene were promoter sequences, as well as transcription termination and polyadenylation signal sequences, derived from the CaMV 35S. Expression of the NPTII encoding gene was regulated using promoter sequences from the *A. tumefaciens* nopaline synthase encoding gene (*nos*).

Southern blot and PCR analyses of genomic DNA from event T120-7 indicated the presence of a single insertion of the T-DNA region, and no evidence of insertion of sequences outside the T-DNA borders. The glufosinate-tolerant trait was maintained over multiple generations. This sugar beet event T120-7 (Liberty Link) was approved by USA in 1998, Japan in 1999, and Canada in 2000 [Health Canada 2000b; CERA 2002b,c].

Lucerne (Alfalfa)

Mono-Trait Events: Herbicides

Glyphosate

J101, J163. In 2003, Monsanto and Forage Genetics developed two glyphosate-resistant alfalfa events 'J101' and 'J163'. Alfalfa is the first perennial crop to be genetically engineered. Both events (Roundup Ready Alfalfa) were generated by *Agrobacterium*-mediated transformation of the calli of the high-yielding fall-dormant FGI alfalfaclone 'R2336'. The T-DNA segment of the double-border, binary vector PV-MSHT4 contains a single *cp4 epsps* expression cassette flanked by left and right border sequences. The sequence, of approximately 3.8 Kb, was transferred into the alfalfa genome by *A. tumefaciens* during the transformation process. The *cp4 epsps* cassette contains the following elements: P-eFMV 35S promoter of the FMV with duplicated enhancer region; HSP70-Leader, the *Petunia* heat shock protein 70 5' untranslated leader sequence; CTP2, the chloroplast transit peptide coding sequence derived from the *A. thaliana epsps* gene; *cp4 epsps*, derived *Agrobacterium* sp. strain CP4; and F9 3', the terminator sequence from the pea ribulose-1,5-biphosphate carboxylase (Rubisco), small subunit E9 gene [CERA 2009l]. The last element terminates transcription and directs polyadenylation of the *cp4 epsps* mRNA.

The PV-MSHT4 vector contains the *ori-322* origin of DNA replication, which allows the replication of the vector in the intermediate host *E. coli*, and the *aad* gene (codes for streptomycin adenyltransferase and allows selection of bacteria on culture media containing streptomycin or spectinomycin). The *ori-322* and *aad* genes are located on the vector backbone, outside of the T-DNA border sequences. Therefore, the genes are not intended for transfer to the recipient plant line during transformation. Transformant cells were identified through selection for glyphosate tolerance.

Post-transformation assay suggested that for each event the intact copy of the *cp4 epsps* gene cassette was incorporated into the alfalfa genome at a different locus and they did not contain any detectable backbone sequence from the vector. These two glyphosate-resistant RRA varieties J101 and J163 were approved by USA, Canada, Japan, and Mexico in 2005 [CERA 2009l].

Double-Event Stack Hybrids: Herbicides

Glyphosate

J101 x J163. Monsanto and Forage Genetics also used the independent events, J101 and J163, to combine (J101 x J163) through a traditional breeding process [Samac and Temple 2004]. The breeding process involved one copy of the *cp4 epsps* transgene at each of two different, independently segregating loci.

Alfalfa is an outcrossing autotetraploid plant with four sets of eight chromosomes (n=8; 4x8=32), which means that this forage crop contains four copies of all gene loci, a key feature to the success of transgenic breeding program [Wright 2012]. Alfalfa also exhibits genetic self-incompatibility or self-sterility, and is affected very negatively by inbreeding. Because of this trait and the inability to achieve 'high trait purity' through self-crossing, a strategy was needed to minimize inbreeding depression. In 1997, Forage

Genetics in partnership with Monsanto developed a complex breeding system to prevent 'inbreeding depression' and after over five years of field trials they produced a viable seed [Wright 2012]. Their research found individual plants to have one to eight copies of the *cp4 epsps* gene inserted, and the Roundup Ready (RR) phenotype was exhibited no matter how many copies of the gene it had. The alfalfa stacked hybrid J101 x J163 has inherited, from the parent lines, two copies of the *cp4 epsps* transgene which expresses the CP4 EPSPS enzyme conferring tolerance to glyphosate.

Sunflower

Mono-Trait Events: Herbicides (Non-Transgenic)

Imidazolinones

IMISUN. An imazethapyr-resistant trait was discovered in wild sunflower population in a field near Rossville, Kansas, USA, where soya bean treated with imazethapyr for seven consecutive years was cultivated [Al-Khatib et al. 1998]. Seeds of sunflower population that expressed a mutation in ALS gene were collected and grown in IMI-tolerant gene donors to introduce the tolerant trait into cultivated sunflower varieties. IMISUN-1 is composed of BC_2F_2 (backcrossed twice second filial) seeds derived from IMI-resistant BC_2F_1 plants from the cross HA 89*3/*H. annuus*. It is an oilseed maintainer genetic stock with the single-headed characteristic. IMISUN-2 is composed of BC_2F_2 seeds derived from the IMI-resistant BC_2F_1 plants from the cross RHA 409//RHA 376*2/*H. annuus*. RHA 409 is an oilseed restorer line released by the USDA and the North Dakota Agricultural Experiment Station in 1995 [Al-Khatib and Miller 2000]. IMISUN-2 is an oilseed restorer genetic stock that will segregate for the recessive branching characteristic. Both genetic stocks lack anthocyanin pigmentation in seeds and plants [Al-Khatib and Miller 2000].

IMISUN-3 is composed of BC_2F_2 seeds derived from IMI-resistant BC_2F_1 plants from the cross HA 292*3/*H. annuus* [Al-Khatib and Miller 2000]. It is a confection maintainer genetic stock with the single-headed characteristic. IMISUN-4 is composed of BC_2F_2 seeds derived from the IMI-resistant BC_2F_1 plants from the cross RHA 324// RHA 280*2/*H. annuus*. It is a confection restorer genetic stock that will segregate for the recessive branching characteristic [Al-Khatib and Miller 2000]. Further pedigree breeding led to the development of maintainer HA425 (BC2F6) from IMISUN-1, and restorers RHA426 and RHA427 from IMISUN-2 [Al-Khatib and Miller 2000]. Several commercial seed companies have introduced the IMI-resistance trait into their own sunflower lines, and IMI-resistant 'Clearfield' sunflower varieties were first in USA, Argentina and Turkey in 2003. IMISUN lines soon became an essential component of hybrid sunflower oil and confectionary production.

As in other crops, tolerance of sunflower to IMIs is due to mutation in the ALS gene at codon 205, with valine substituting for alanine (Ala205→Val205). Total tolerance may be controlled by at least two genes, with one a semi-dominant gene (*Imr1*) and the other a modifier or enhancer gene (*Imr2*) [Miller and Al-Khatib 2000; Bruniyard 2001].

X81359. Sunflower variety X81359 of BASF Canada was the result of a cross between two IMI-tolerant parents and extensive backcrossing with commercial NuSun sunflower varieties. The oil from these sunflower lines have an altered fatty acid profile, and are termed 'mild-oleic'. The natural mutation conferring the IMI-resistance was detected in its parents in a field where soya bean has been cultivated and treated with imazethapyr for several seasons [CERA 2005a]. The mutant sunflower population expressed a mutation in AHAS gene, thus conferring tolerance to this IMI herbicide. Conventional plant breeding techniques were then used to introduce the IMI-resistance trait into sunflower germplasm, including extensive backcrossing into commercial sunflower cultivars including 'NuSun' hybrids. Canada approved this Event X81359 (commercial name 'Clearfield') in 2003 [CERA 2005a]. Only the meal and oil of X81359 hybrid sunflower will be imported into Canada since this hybrid is intended solely for use as human food and livestock feed. X81359 hybrid was not intended for cultivation in Canada. No environmental effects are therefore expected from the importation Canada [CERA 2005a].

CLHA-PLUS. This IMI-resistant sunflower event was developed by subjecting the seeds of cultivar 'BTK47' with 0.25 percent ethyl methanesulfonate, the mutagen, for 15 hr, followed by soaking them in 1.5 percent sodium thiosulphate solution for 30 min. The M_2 plants were sprayed with imazapyr and eight resistant plants were selected subsequently. After several generations of screening with IMI herbicides, one of the resistant lines closely resembling BTK 47 attributes was selected in 2006 and designated CLHA-Plus, which had an *ALS* gene mutation with the nucleotide substitution of GCG to ACG at codon 122, resulting in A122T amino acid substitution of the ALS enzyme [Tan and Bowe 2012]. The mutated *Ahasl* gene allele in CLHA-Plus is partially dominant and it confers a higher level of IMI-tolerance than the allele of IMISUN. The CLHA-Plus IMI resistance trait is under the control of the native AHAS promoter and is believed to be constitutively expressed.

The CL sunflower hybrid H4 was developed by conventional breeding of the previously approved IMI-resistant X81359 (CL IMISUN) event, with CLHA-Plus. This hybrid contains two distinct mutations derived from two parents (X81359 and CLHA Plus), each in different AHAS alleles that confer IMI tolerance. The single amino acid substitution in the AHAS gene of CLHA-PLUS is sufficient to alter the binding site such that IMI herbicides no longer bind to the AHAS enzyme, resulting in the herbicide tolerant phenotype for both CLHA-PLUS and the CL sunflower hybrid H4 [Tan and Bowe 2012].

Sugarcane

Mono-Trait Events: Herbicides

Glufosinate

Gallo-Meagher and Irvine [1996] developed a bialaphos-resistant transgenic sugarcane line by transforming commercial cultivar 'Nco 310'. Embryogenic calli derived from immature influorescences were used for transformation by biolistic method. Plasmid pAHC20 contained a coding region of *bar* gene fused with *ubi-1* promoter, first exon, and first intron followed by the *nos* terminator cloned into pUC8. The precipitated plasmid DNA was mixed with gold or tungsten particles. Two days after bombardment using a modified particle flow gun, calli were transferred to a culture medium containing bialaphos (1 mg L^{-1}). Four weeks later, bialaphos-resistant calli were grown in a medium to promote shoot regeneration. After subculturing on this medium every 2 weeks for 12 weeeks, shoots were placed in a rooting medium containing 3 mg L^{-1} bialaphos. When roots were well developed, shoots were placed in a growth chamber. After 6 weeks, plants were placed in the greenhouse. Single bud cuttings from a primary transformant were grown into plants. This process was repeated two cycles to produce third and fourth generation transformed clones.

Southern blot analysis in bialaphos-resistant sugarcane plants had the *bar* gene incorporated into the genomic DNA, while Northern blot analysis indicated that the proper size *bar* transcript was produced in these plants. A well-characterized transformant was submitted to three rounds of vegetative propagation, as well as through meristem culture. All of the resulting propagated clones displayed herbicide resistance comparable to the original transformant.

Integration of the *bar* gene following *Agrobacterium*-mediated gene transfer resulted in high-level herbicide resistance in most of the lines [Enriquez-Obregon et al. 1998; Manickavasagam et al. 2004]. However, some lines did not show resistance to the herbicide despite integration of one or two copies consequent to this transfer. Leibbrandta and Snyman [2003] found transgenic sugarcane line '22.2' containing nine copies of the *pat* gene expressing stable and high-level resistance to glufosinate during three rounds of vegetative propagation without any adverse effect on yield despite complex transgene integration following biolistic gene transfer. There were no significant differences in sugarcane yield between transgenic and non-transgenic lines to conventional preemergence and postemergence applications of glufosinate.

Mono-Trait Events: Herbicides (Non-Transgenic)

ALS-Inhibitors

Vyver et al. [2013] were the first to describe a mutated *als* gene as a selectable marker to produce non-transgenic sugarcane lines resistant to ALS-inhibiting sulfonylurea herbicides. In the first step, the sensitivity of sugarcane embryonic calli (produced by excising the basal part of the leaf roll) of cultivar 'Nco310' towards a range of ALS-targeting herbicides (sulfonylureas: chlorsulfuron, nicosulfuron, and rimsulfuron; and the imidazolinones imazapyr, imazethapyr, and imazaquin) was determined. Callus growth was most affected by chlorsulfuron (3.6 µg L^{-1}). Herbicide-resistant transgenic sugarcane plants containing mutant forms of a tobacco *als* gene were obtained following biolistic transformation. Post-bombardment, putative transgenic callus was selectively proliferated on MS medium containing 3 mg L^{-1} 2,4-D, 20 g L^{-1} sucrose, 0.5 g L^{-1} casein, and 3.6 µg L^{-1} chlorsulfuron. Plant regeneration and rooting was done on MS medium lacking 2,4-D under similar selection conditions. Thirty vigorously growing putative transgenic plants were successfully *ex vitro*-acclimatized and established under glasshouse conditions. Glasshouse spraying of putative transgenic plants with 100 mg L^{-1} chlorsulfuron dramatically decreased the amount of non-transgenic plants that had escaped the *in vitro* selection regime. PCR analysis showed that six surviving plants were *als*-positive, with five of these expressing the mutant *als* gene.

Tobacco

Mono-Trait Events: Herbicides

2,4-D

Tobacco provides a useful model system with which to test the function and specificity of introduced transgenes as it is the most commonly used plant for expression of transgenes from a variety of organisms. Besides, as it is easily grown and transformed, it provides abundant amounts of fresh tissue and has a well-established cell culture system. Many bacterial proteins involved in the synthesis of commercial products are currently engineered for production in tobacco [Jube and Borthakur 2007]. Bacterial enzymes synthesized in tobacco can incorporate resistance, *inter alia*, to herbicides.

One of the first attempts to engineer tobacco for 2,4-D resistance was made by Lyon et al. [1989]. The *tfd*A gene was cloned into three different gene expression vectors based on the CaMV 35S promoter *nos* 3′ polyadenylation signal, and introduced into leaves of tobacco cultivar 'Wisconsin 38' by *Agrobacterium*-mediated transformation. Oligomutagenesis of three bases in the 5′ region of the *tfd*A gene converted the GTG translational codon to ATG and produced a plant consensus sequence (TAAACA) immediately upstream of this codon. The 35D-*tfd*A-*nos* constructs were cleaved with *Hind*III and cloned into the *Hind*III site of the binary vector plasmid pGA470. Triparental mating with selection was employed to transfer the binary vector constructs from *E. coli* to the avirulent A. *tumefaciens* strain LBA 4404 Rif. The shoots were regenerated on kanamycin culture medium and the transgenic kanamycin-resistant plants transformed by these gene constructs were designated as TFD1, TFD5, or TFD8 (corresponding to *tfd*A in the coding orientation in pJ35SN, p3x35SN, or p35SAMVN respectively) or TFD3 (corresponding to *tfd*A in the inverse orientation of pJ35SN). Transgenic plants were maintained in antibiotic culture media until selected transgenic plants were grown in soil and propagated for seed [Lyon et al. 1989].

Many of the transgenic plants with modified *tfd*A gene demonstrated increased tolerance to 2,4-D, while plants transformed with an unmodified *tfd*A gene exhibited lower tolerance to 2,4-D, indicating that this gene was not expressed efficiently in the latter. Resistant plants were produced from all three promoter constructs. Transgenic tobacco plants expressing the highest levels of the monooxygenase enzyme exhibited increased tolerance to 2,4-D, with young plants surviving herbicide up to eight times the usual field application rate. The level of resistance in the transgenic plants could be further improved by selecting progeny carrying multiple copies of the *tfd*A gene and by judicious crossing [Lyon et al. 1989]. With further modifications and improvement in transformation process, the *fad* gene certainly offers a genuine possibility of developing 2, 4-D-resistant tobacco lines.

Phenmedipham

Phenmedipham competes with phenylureas for binding to the Q_B site on the D1 peptide of the photosystem II (vide Chapter 2). Sugar beet, spinach (*Spinacea oleracea*), and strawberry (*Fragaria vesca*) plants are resistant to this selective postemergence phenylcarbamate herbicide. Tolerance of sugar beet is due to hydroxylation of phenmedipham to the herbicidally active *N*-hydroxy-phenmedipham which is then glycosylyzed into the inactive glucoside [Davies et al. 1990]. However, soil bacterium *Arthrobactor oxidans* strain P52 degrades phenmedipham by ring cleavage catalyzed by phenylcarbamate hydrolase (PMPH) enzyme encoded by the *pcd* gene which can be inserted to develop phenmedipham-resistant transgenic lines.

Streber et al. [1994] genetically engineered to confer resistance to phenmedipham by inserting the *pcd* gene in tobacco via *Agrobacterium*-mediated transformation of cultivar 'W-38'. The plasmid constructs containing hybrid PMPH were integrated into the binary vector pBIN19. The *pcd* gene coding for PMPH on A. *oxidans* plasmid pHP52 was characterized by a 1479 nucleotide open reading frame. A *Hinc*II site downstream of the stop codon was used to link the gene to the polyadenylation signal of the octopine synthase gene (*OCS*-polyA). The GTG translational start codon of the native gene was converted into an ATG by removing the N-terminal coding region by cleavage with Ppu MI and replacing it with a synthetic double-stranded oligonucleotide.

For expression of PMPH in plants, the modified PMPH coding sequence was placed under the control of the CaMV 35S promoter. A series of constructs were prepared to achieve best conditions for high level expression. The first series of plasmids contained short 5'-untranslated sequences (pBCPO1027, pBCPD1033, pB-CPE1034, pBCPF1035, and pBCPK1037). Further constructs included viral untranslated leader sequence which is known to enhance translation of mRNA in plants from recombinant genes (pBOPO1039) or a synthetic *pin*II signal peptide sequence fused to the N-terminus of the coding region (pBAPO1028). The latter was designed to target the PMPH protein to the extracellular space. As the enzyme is active without the requirement of cofactors, extracellular localization of PMPH should allow the detoxification of the herbicide before it enters the cell [Streber et al. 1994].

Transformed shoots were selected on kanamycin and about 20 regenerated plants from each gene construct were propagated in soil. Transformants with good expression levels were identified by Northern blot hybridization of total RNA in leaves, and/or by spraying phenmedipham. Plants transformed with the PMPH gene constructs synthesized a transcript of ca. 1.7 kb that specifically hybridized with the carbamate hydrolase-coding sequence. There was a great variation among DNA constructs, with pBAPO1028, pBAP11025 and pBOPO1039 achieving the best expression [Streber et al. 1994].

Transformed plants with high expression of carbamate hydrolysis showed no injury even at 3 kg ha^{-1} while completely surviving even at 10 kg ha^{-1} despite a slight initial growth inhibition, suggesting that the transgenic plants can tolerate up to 10 times the usual field application rate. The transgenic plants also tolerated desmedipham, the other phenylcarbamate herbicide [Streber et al. 1994].

Paraquat

Gene construction to insert *pqrA* gene involved isolation of *Ochrobactrum anthropic* JW2 and its partial digestion with Sau3AI. DNA fragments (2 to 6 kb) were isolated and ligated into the Bam HI site of the pBluescript II SK+ vector. The resulting plasmids were introduced into *E. coli* DH5α. The transformants were plated out on LB (Luria broth) agar plates containing 1 mM paraquat plus 50 μg ml^{-1} ampicillin. Plasmids contained in resistant colonies were isolated, reintroduced into *E. coli* DH5α, and spread on the same type of plates again. The re-transformants were used for further analysis [Won et al. 2001].

The PqrA protein was expressed in *E. coli* using a glutathione *S*-transferase (GST) gene fusion system [Won et al. 2001]. A DNA fragment encoding the C-terminal 189 amino acid residues of the PqrA protein were amplified by PCR using an N-terminal primer, 5'-CCGAATTCCT-GGCACTTATC-3' containing an *Eco*RI site and a C-terminal primer, 5'-GACTCGAGTTACGGAGATTG-3' containing an *Xho*I site, and plasmid pBpq2.5 as the template [Won et al. 2001]. The amplified DNA fragment was digested with *Eco*RI and *Xho*I, and then subcloned into the expression vector pGEX4T-1 to produce a fusion protein, GST-PqrA. This fusion protein was then purified to homogeneity using glutathione Sepharose column chromatography, preparative SDS–PAGE, and electroelution (extraction of a nucleic acid or a protein sample from an electrophoresis gel). This process was followed by protein extraction and immunoblot analysis [Won et al. 2001].

Paraquat-resistant *E. coli* transformants were isolated through the transferring of *O. anthropic* JW2 chromosomal DNA fragments to *E. coli* DH5α [Won et al. 2001]. Three putative paraquat-resistant colonies were obtained, among which one contained pBluescript II SK+ with an insert DNA fragment of approximately 2.5 kb long which was designated as pBpq2.5. This transformant, which harbored the DNA fragment encoding a protein endowed with *E. coli* paraquat resistance, tolerated 0.5 mM paraquat even if its growth was slightly retarded. Resistance of the gene product(s) in the insert DNA fragment of pBpq2.5 was not related to detoxification of active oxygen species inside cells [Won et al. 2001]. The cloned DNA fragment was originated from the chromosomal DNA of *O. anthropic* JW2, which had a single gene corresponding to the cloned DNA fragment.

Jo et al. [2004] introduced the *pqr*A gene into the genome of tobacco cultivar 'Samsun' using the *Agrobacterium*-mediated transformation method to explore its capability to protect transgenic plants from paraquat damage. For this, they inserted A 2.1 kb *Hind*III fragment of pBpq2.5 [Won et al. 2001] containing the open reading frame of the *pqr*A gene into the binary vector pGA748, which contains the CaMV 35S promoter and the *npt*II gene for kanamycin resistance. The resulting expression vector, pGApqr2.1, was transferred into the *A. tumefaciens* strain, LBA4404 [Jo et al. 2004].

Following transformation, the *pqr*A gene was present in all T3 homozygous five transgenic lines, with the 2.2 kb pqrA transcripts found in all of them while no transcript was found in wild type plants. However, expression levels varied between the transgenic lines, with P5 line displaying the highest expression levels. Plants of P5 transgenic line remained green and grew normally even at 20 μM paraquat, whereas wild-type plants started to bleach even at 1 μM and died at 20 μM. Paraquat, typical of its toxicity, causes chlorophyll loss and cellular ion leakage. Paraquat induced considerable cellular ion leakage from leaf discs of untransformed plants and leakage increased with an increase in concentration. However, cellular leakage from leaf discs was lower in transgenic plants even at 50 μM paraquat, indicating that transgenic plants expressing the *Ochrobactrum anthropic pqr*A gene were resistant to this herbicide [Jo et al. 2004].

The PqrA protein, which exhibited properties of an integral membrane protein, functioned as a transporter and enhanced paraquat resistance of transgenic plants by either decreasing uptake or increasing efflux of paraquat, and not by detoxifying active oxygen species. The transformed tobacco plants accumulated lower levels of paraquat in their tissues compared to untransformed tobacco, establishing that the bacterial protein PqrA is effective in the detoxification of paraquat [Jo et al. 2004].

Linseed (Flax)

Mono-Trait Event: Herbicides

Sulfonylureas

CDC Triffid (FP967). Linseed is moderately tolerant to soil application of the residual sulfonylurea herbicides which bind to the ALS enzyme, thereby inhibiting the biosynthesis of the branched chain amino acids (Chapters 2 and 5). However, the roots are sensitive to sulfonylurea residues in the soil.

CDC Triffid was engineered by Crop Development Center, University of Saskatchewan, Canada via *Agrobacterium*-mediated transformation. The flax cultivar 'Norlin' was transformed using the disarmed *A. tumefaciens* strain C58 Ti-plasmid vector pGV3850. The T-DNA region of the vector contained modified *als* gene from a chlorsulfuron-tolerant line of *A. thaliana*. This variant *als* gene differs from the wild type *A. thaliana* gene by one nucleotide. Two selectable marker genes, *npt*II from the Tn5 transposon of *E. coli*, strain K12, and *nos* from *A. tumefaciens*, were also introduced in the plant. The *als* and *nos* genes were under the control of their native regulatory sequences. The *npt*II gene was under the control of the *nos* regulatory sequences from *A. tumefaciens*. Ampicillin and spectinomycin antibiotic resistance selectable marker genes from *E. coli* were introduced into the plant but they were not expressed [CERA 1999; Health Canada 1999e].

Successful transformants were selected *in vitro*, on the basis of expression of kanamycin resistance and presence of nopaline. These were later grown on selection medium containing chlorsulfuron to confirm the expression of the inserted chlorsulfuron-tolerant *als* gene. The inserted genes were integrated in at least two unlinked loci, with a possible third locus exhibiting partial linkage to one of the other two. The expression of the genes remained stable after eight generations of selfing CDC Triffid (FP967) plants. This sulfonylurea herbicide-resistant linseed line was approved as food crop by Canada and USA in 1998 [CERA 1999].

Potato

Mono-Trait Events: Herbicides

Bromoxynil

An attempt has been made by Eberlein et al. [1998] to insert the *bxn* gene, which encodes the bromoxynil-specific nitrilase and produce bromoxynil-resistant transgenic potato by using the cultivar 'Lemhi Russet' via *Agrobacterium*-mediated method of transformation.

Transcription of the chimeric *bxn* gene into RNA was directed by CaMV 35S promoter. The 3' end of the *bxn* gene preceded a polyadenylation signal from the terminator gene, *nos* from *A. tumefaciens*. The *bxn* gene with the 35S promoter and NOS polyadenylation signal was cloned into the *Hind*III site of the binary transformation vector pCGN1547. This vector contained the *npt*II gene that confers resistance to kanamycin which provides the selectable marker gene for transformation. The binary vector containing both the *npt*II and *bxn* genes was transformed into the *A. tumefaciens* strain PC2760 harboring Ti plasmid pAL4404. Microtuber slices of potato were coated with this soil bacterium followed by incubation for two days before transferring to a selective medium containing kanamycin which selects for transformed plant cells, and cefataxim which kills the *Agrobacterium*. The selective medium also contained hormones for plant generation zeatin riboside and NAA. As shoots developed, they were transferred to a rooting media containing kanamycin, but lacking hormones. Plantlets that developed roots transplanted to the greenhouse for further evaluation.

The transgenic clones were at least 70 times more resistant to bromoxynil than the untransformed control. Resistance was due to rapid metabolism of bromoxynil to 3,5-dibromo-4-hydroxybenzoic acid, followed by conjugation to polar compounds. The best performing transgenic clones has total tuber yields equal to the untreated, untransformed control. Lower yields in the transgenic lines were due to somaclonal variation rather than due to the direct effect of *bxn*-transgene expression [Eberlein et al. 1998].

Sweet Potato

Mono-Trait Events: Herbicides

Glufosinate

The *bar* gene has been successfully used to confer glufosinate-resistance to sweet potato in the recent past. However, no glufosinate-resistant transgenic line is available commercially at the present time.

Choi et al. [2007] infected the embryonic calli derived from apical meristems of genotype 'Yulmi' with *A. tumefaciens* strain EHA105 harboring the pCAMBIA3301 binary vector containing the *bar* gene encoding PAT and the *gusA* gene encoding GUS. These genes were under the control of CaMV 35S promoter. The plasmid vector was subsequently transferred to *Agrobacterium* cells by the freeze-thaw method. Following *Agrobacterium*-mediated transformation, the PPT-resistant calli and plants were selected with 5.0 and 2.5 mg L^{-1} phosphinothricin, respectively and the plants that survived were grown for 28–36 weeks. The transformed crop plants had one to three copies of the transgene, *bar*, stably integrated into the plant genome of each transgenic plant which showed resistance to glufosinate even at 900 mg L^{-1}.

A similar attempt has been made by Yi et al. [2007] and Zan et al. [2009] by using the *bar* gene to confer glufosinate resistance to sweet potato cultivar 'Lizixiang' by using *Agrobacterium*-mediated transformation. Cell aggregates from embryonic suspension cultures were co-cultivated with *A. tumefaciens* strain EHA105 harboring binary vector pCAMBIA3300 with *bar* gene and the fragment of *uidA* gene—which encodes the β-glucuronidase enzyme in *E. coli*—driven by a CaMV 35S promoter. The gene cassette was in the order of 35S::*bar*::*uidA*. Selection culture was prepared using 0.5 mg L^{-1} phosphinothricin. A total of 1,431 plants were produced from the inoculated 870 cell aggregates via somatic embryogenesis. The transgenic plants, grown in the greenhouse, were evaluated for glufosinate resistance 1.0 mg L^{-1} (field dosage) and 2.0 mg L^{-1}. The *bar* gene was found stably integrated into the genome of transgenic plants.

References

Al-Khatib, K. and J.F. Miller. 2000. Registration of four genetic stocks of sunflower resistant to imidazolinone herbicides. Crop Sci. 40: 869–870.

Al-Khatib, K., J.R. Baumgartner, D.E. Peterson and R.S. Currie. 1998. Imazethapyr resistance in common sunflower (*Helianthus annuus*). Weed Sci. 46: 403–407.

Anderson, J.A., L. Matthiesen and J. Hegstad. 2004. Resistance to an imidazolinone herbicide is conferred by a gene on chromosome 6DL in the wheat line cv. 9804. Weed Sci. 52: 83–90.

Anderson, P.C. and M. Georgeson. 1989. Herbicide-tolerant mutants of corn. Genome 31: 994–999.

ANZFA. 2000. Application A316: Food produced from insect-protected corn line MON 810. Draft Risk Analysis Report. 30 August 2000. 56 pp.

APHIS-USDA. 2013. Stine Seed Farm, Inc.; Extension of a Determination of Nonregulated Status of Corn Genetically Engineered for Herbicide Resistance. Federal Register. The Daily J. of the U.S. Government. 3 May.

APHIS-USDA Petition Number #11-244-01p. 2012. Plant Pest Risk Assessment for Pioneer 4114 Maize. www. aphis.usda. gov/brs/aphisdocs/ 11_24401p_dpra.pdf.

APHIS-USDA. 2011. Draft Plant Pest Risk Assessment for Double Herbicide-Tolerant Soybean (*Glycine max*) event FG72. Petition 09-328-01p. September 2011.

APHIS-USDA. 1999. Response to AgrEvo Petition 98-278-01p for Determination of Nonregulated Status for Canola Transformation Events MS8 and RF3 Genetically Engineered for Pollination Control and Tolerance to Glufosinate Herbicide. March 1999.

APHIS-USDA. 1995. USDA/APHIS Petition 94-035-01 for the determination of Nonregulated Status of Glufosinate resistant Corn Transformation Events T14 and T25. June 22, 1995. Federal Register 60 (134). 13 July.

Bayer CropScience. 2008. Petition for Determination of Nonregulated Status for Insect-Resistant and Glufosinate Ammonium-Tolerant cotton: TwinLink™ cotton (events T304-40 x GHB119) OECD Unique Identifier BCS-GH04-7 x BCS-GH005-8. 04 December.

Bayer CropScience. 2006a. LLRICE601 USDA Extension Petition Letter regarding corrections. APHIS-USDA. 28 August.

Bayer CropScience. 2006b. Petition for Determination of Nonregulated Status for Glyphosate-Tolerant Cotton: GlyTolTM Cotton Event GHB614. 20 November.

Bayley, C., N. Trolinder, C. Ray, M. Morgan, J.E. Quisenberry and D.W. Ow. 1992. Engineering 2,4-D resistance into cotton. Theor. Appl. Genet. 85: 645–649.

Biosafety Clearing House. 2006. DKB-89790-5 (DLL25/B16)-Herbicide-tolerant maize. Record No. 14771. Convention on Biological Diversity.

Biosafety Scanner. 2013. Event BT11 x 59122 x MIR604 x TC1507 x GA21 and Detection Methods. 16 December.

Bisht, N.C., P.K. Burma and D. Pental. 2004. Development of 2,4-D-resistant transgenics in Indian oilseed mustard (*Brassica juncea*). Current Sci. 87(3): 367–370.

Bruniard, J.M. 2001. Inheritance of imidazolinone resistance, characterization of cross-resistance pattern, and identification of molecular markers in sunflower (*Helianthus annuus* L.). Ph.D. Dissertation, North Dakota State Univ., USA.

CaJacob, C.A., P.C.C. Feng, S. Reiser and S.R. Padgette. 2007. Genetically modified herbicide-resistant crops. pp. 283–302. *In*: W. Kramer and W. Schirmer (eds.). Modern Crop Protection Chemicals, Vol. 1. Wiley-VCH Verlag GmbH and Co. KGaA. Weinheim, 1394 pp.

Canadian Food Inspection Agency. 2012a. Decision Document DD 2012-89: Determination of the Safety of Monsanto Canada Inc. Corn (*Zea mays* L.) Event MON 87427. 18 December.

Canadian Food Inspection Agency. 2012b. Decision Document DD 2012-88: Determination of the Safety of Pioneer Hi-Bred Production Ltd.'s Canola (*Brassica napus* L.) Events 73496 and 61061. 14 December.

Center for Environmental Risk Assessment. 2013. GM Crop Database. Event Name: SYN 5307 x MIR604 x Bt11 x TC1507 x GA21 x MIR162.

Center for Environmental Risk Assessment. 2012a. GM Crop Database: BT11 x MIR162 x MIR604 x GA21. 04 August.

Center for Environmental Risk Assessment. 2012b. GM Crop Database: SYN-BT011-1, SYN-IR162-4, MON-00021-9 (BT11 x GA21 x MIR162). 04 August.

Center for Environmental Risk Assessment. 2012c. GM Crop Database: DP-356043-5 (DP356043). 04 August.

Center for Environmental Risk Assessment. 2012d. GM Crop Data Base: DP-098140-6 (Event 98140). 04 August.

Center for Environmental Risk Assessment. 2012e. GM Crop Data Base: SYN-BT011-1, MON-00021-9 (BT11 x GA21). 04 August.

Center for Environmental Risk Assessment. 2010a. GM Crop Database: MON89034 x TC1507 x MON88017 x DAS-59122-7. 25 January.

Center for Environmental Risk Assessment. 2010b. GM Crop Database: BPS-CV127-9 (BPS-CV127-9). 23 January.

Center for Environmental Risk Assessment. 2009a. GM Crop Database: MON-88017-3 (MON88017) 08 November.

Center for Environmental Risk Assessment. 2009b. GM Crp Database: BCS-GH002-5 (GHB614). 16 September.

Center for Environmental Risk Assessment. 2009c. MON-00603-6 (NK603). GM Crop Database. 6 August.

Center for Environmental Risk Assessment. 2009d. GM Crop Database: ACS-GM005-3 (A2704-12, A2704-21, A5547-35). 04 April.

Center for Environmental Risk Assessment. 2009e. GM Crop Database: MON-04032-6 (GTS 40-3-2). 03 April.

Center for Environmental Risk Assessment. 2009f. GM Crop Database: DAS-21023-5 x DAS-24236-5. 20 March.

Center for Environmental Risk Assessment. 2009g. GM Crop Database: ACS-BN008-2 (T45 (HCN28). 13 March.

Center for Environmental Risk Assessment. 2009h. GM Crop Database: MON89788. 01 March.

Center for Environmental Risk Assessment. 2009i. GM Crop Database: MON-01445(2) (MON1445/1698). 01 March.

Center for Environmental Risk Assessment. 2009j. GM Crop Database: ACS-GH001-3 (LLCotton25). 01 March.

Center for Environmental Risk Assessment. 2009k. GM Crop Database: MON-88913-8 (MON88913). 23 February.

Center for Environmental Risk Assessment. 2009l. GM Crop Database: MON-00101-8, MON-00163-7 (J101, J163). 28 January.

Center for Environmental Risk Assessment. 2009m. GM Group Database: ACS-OS001-4, ACS-OS002-5 (LLRICE06, LLRICE62). 28 January.

Center for Environmental Risk Assessment. 2008a. GM Crop Database: KM-00071-4 (H7-1). 30 December.

Center for Environmental Risk Assessment. 2008b. GM Crop Database: MON15985 x MON88913. 30 December.

Center for Environmental Risk Assessment. 2008c. GM Crop Database: BCS-OS003-7 (LLRICE601). 30 December.

Center for Environmental Risk Assessment. 2007a. GM Crop Database: SYN-BT011-1, SYN-IR604-5, MON-00021-9 (BT11 x MIR604 x GA21). 25 November.

Center for Environmental Risk Assessment. 2007b. GM Crop Database: SYN-BT011-1, SYN-IR604-5 (BT11 x MIR604). 25 November.

Center for Environmental Risk Assessment. 2006a. GM Crop Database: DAS-21023-5 x DAS-24236-5 x MON 88913. 13 February.

Center for Environmental Risk Assessment. 2006b. GM Crop Database. ACS-BN004-7 x ACS-BN001-4 (MS1, RF1 =>PGS1). 30 January.

Center for Environmental Risk Assessment. 2006c. GM Crop Database: ACS-BN011-5 (OXY-235). 30 January.

Center for Environmental Risk Assessment. 2006d. GM Crop Database: ACS-BN007-1 (HCN92). 30 January.

Center for Environmental Risk Assessment. 2006e. GM Crop Database: ACS-BN005-8 x ACS-BN003-6 (MS8xRF3). 30 January.

Center for Environmental Risk Assessment. 2006f. GM Crop Database: ACS-ZM002-1/ACS-ZM003-2 (T14, T25) 30 January.

Center for Environmental Risk Assessment. 2005a. GM Database: X81359. 03 August.

Center for Environmental Risk Assessment. 2005b. GM Crop Database. DAS-21023-5 x DAS-24236-5 x MON-01445-2. 03 August.

Center for Environmental Risk Assessment. 2005c. GM Crop Database: ACS-GM001-8, ACS-GM002-9 (W62, W98). 21 July.

Center for Environmental Risk Assessment. 2005d. GM Crop Database: SYN-BT011-1 {BT11 (X4334CBR, X4734CBR)}. 21 July.

Center for Environmental Risk Assessment. 2005e. GM Crop Database: MON802. 21 July.

Center for Environmental Risk Assessment. 2005f. GM Crop Database: ACS-GM003-1 (GU262). 21 July .

Center for Environmental Risk Assessment. 2005g. GM Crop Database: DKB-89614-9 (DBT418). 21 July.

Center for Environmental Risk Assessment. 2005h. GM Crop Database: MON802. 21 July.

Center for Environmental Risk Assessment. 2005i. GM Crop Database: MON71800. 29 June.

Center for Environmental Risk Assessment. 2005j. GM Crop Database: AP205CL. 28 June.

Center for Environment Risk Assessment. 2005k. GM Crop Database. PWC16. 28 June.

Center for Environment Risk Assessment. 2002a. GM Crop Database. CL121, CL141, CFX51. 14 May.

Center for Environmental Risk Assessment. 2002b. GM Group Database: ACS-BV001-3 (T120-7). 12 May.

Center for Environmental Risk Assessment. 2002c. GM Crop Database. ACS-BV001-3 (T120-7). 12 May.

Center for Environmental Risk Assessment. 2002d. GM Group Database: BXN. 02 May.

Center for Environmental Risk Assessment. 2001a. GTSB77. 19 August.

Center for Environmental Risk Assessment. 2001b. GM Crop Database: MON832. 22 July.

Center for Environmental Risk Assessment. 2001c. GM Crop Database: DK404SR. 22 July.

Center for Environemntal Risk Assessment. 2001d. GM Crop Database. 676, 678, 680. 21 July.

Center for Environmental Risk Assessment. 2001e. GM Crop Database: ACS-ZM004-3 (CBH-351). 02 July.

Center for Environmental Risk Assessment. 2001f. GM Crop Database: DKB-89790-5 {B16 (DLL25)} 02 July.

Center for Environmental Risk Assessment. 2001g. GM Crop Database: SWP965001. 10 June.

Center for Environmental Risk Assessment. 1999. GM Crop Database: CDC-FL001-2 (FP967). 2 November.

Chaubet, N., B. Clement and C. Gigot. 1992. Genes encoding a histone H3.3-like variant in *Arabidopsis* contain intervening sequences. J. Mol. Biol. 225: 569–574.

Chabouté, M.E., N. Chaubet, G. Philipps, M. Ehling and C. Gigot. 1987. Genomic organization and nucleotide sequences of two histone H3 and two histone H4 genes of *Arabidopsis thaliana*. Plant Mol. Biol. 8: 179–191.

Choi, H.J., T. Chandrasekhar, H.-Y.n. Lee and K.-M. Kim. 2007. Production of herbicide-resistant transgenic sweet potato plants through *Agrobacterium tumefaciens* method. Plant Cell, Tissue and Organ Culture 91(3): 235–242.

Cogem. 2013. Advisory report concerning import of herbicide tolerant soybean SYHT0H2. Cogem advisory report CGM/130325-01. 25 March.

Cornelissen, M. and M. Vandewiele. 1989. Nuclear transcriptional activity of the tobacco plasmid psbA promoter. Nucleic Acids Res. 17(1): 19–29.

Crouch, M.L. 2012. The Center for Food Safety, Science Comments—FG72 Soybean: Addressed to APHIS (11 September 2012). Washington, DC, 84 pp.

Croughan, T.P. 2002. Herbicide resistant rice, US Patent Application 2002 0019 313.

Currie, R.S., C.S. Kwon and D. Penner. 1995. Magnitude of imazethapyr resistance of corn (*Zea mays*) hybrids with altered acetolactate synthase. Weed Sci. 43: 578–582.

Dahl, W.J., E.A. Lockert, A.L. Cammer and S.J. Whiting. 2005. Effects of flax on laxation and glycemic response in healthy volunteers. J. Med. Food 8(4): 508–511.

Della-Cioppa, G., S.C. Bauer, B.K. Klein, D.M. Shah, R.T. Fraley and G.M. Kishore. 1986. Translocation of the precursor of 5-enolpyruvylshikimate-3-phosphate synthase into chloroplast of higher plants *in vitro*. Proc. Natl. Acad. Sci. USA 83: 6873–6877.

Davies, H.M., A. Merydith, L. Mende-Mueller and A. Aapola. 1990. Metabolic detoxification of Phenmedipham in leaf tissue of tolerant and susceptible species. Weed Sci. 38: 206–214.

Dietrich, G.E. 1998. Imidazolinone resistant AHAS mutants, US Patent 5 767 361.

Dill, G.M., C.A. CaJacob and S.R. Padgette. 2008. Glyphosate-resistant crops: adoption, use, and future considerations. Pest Management Sci. 64: 326–331.

Dow AgroSciences. 2012. Application for authorization of DAS-40278-9 maize grain for all uses as for any other maize, excluding cultivation, according to Articles 5 and 17 of Regulation (EC) No. 1829/2003 on genetically modified food and feed. EFSA-GMO-NL-2010-89.

Dow AgroSciences. 2011. USDA Petition for Determination of Nonregulated Status for Herbicide Tolerant DAS-4406-6 Soybean. OECD Unique Identifier: DAS-44406-6. Revised 17 October.

Dow AgroSciences. 2010. Petition for Determination of Nonregulated Status for Herbicide Tolerant DAS-68416-4 Soybean OECD Unique Identifier: DAS-68416-4. 15 November.

Dow AgroSciences/Pioneer Hi-Bred International. 2004. Application for the determination of Nonregulated Status for B.t. Cry34/35Ab1 Insect-Resistant, Glufosinate-Tolerant Corn: Corn Line 59122. December 18, 2003; Revised 07 September 2004.

Eberlein, C.V., M.J. Guttieri and J.F. Steffen-Campbell. 1998. Bromoxynil resistance in transgenic potato clones expressing the *bxn* gene. Weed Sci. 46(2): 150–157.

Europa. 2005. Event NK603 x MON810. Monsanto International Services S.A./N.V. http://www.defra. gov.uk/ environment/gm/regulation/pdf/c-gb-02-m3-3.pdf.

European Union. 2013. Application for authorization to place on the market MON 88701 cotton in the European Union, according to Regulation (EC) No. 1829/2003 on genetically modified food and feed. EFSA-GMO-NL-2013-114; EFSA-Q-2013-00219. Part VII, Summary. http://www.miljodirektoratet. no/Global/dokumenter/horinger/ horing 2013-3791_sammendrag.pdf.

Fernandez-Cornejo, J. and M. Caswell. 2006. The First Decade of Genetically Engineered Crops in the United States (PDF). United States Department of Agriculture (1 April 2006).

Gallo-Meagher, M. and J.E. Irvine. 1996. Herbicide resistant transgenic sugarcane plants containing the *bar* gene. Crop Sci. 36(5): 1367–1374.

Genuity. 2013. Corn Traits. GENUITY® SMARTSTAX® RIB COMPLETE® CORN BLEND. 06 June 2013. (http: //www.genuity.com/Home.aspx#/home).

Green, J.M. and L.A. Castle. 2010. Transitioning from single to multiple herbicide-resistant crops. pp. 67–91. *In*: V.K. Nandula (ed.). Glyphosate Resistance in Crops and Weeds: History, Development, and Management. John Wiley & Sons, New York.

Halpin, C. 2005. Gene stacking in transgenic plants—the challenge for 21st century plant biotechnology. Plant Biotech. J. 3: 141–155.

Health Canada. 2013. Approved Products. Novel Food Decisions: Clearfield Bread Wheat Varieties BW255-2 and BW238-3. 02 October.

Health Canada. 2007. Clearfield Bread Wheat Variety BW7. 28 September 2007. www.hc-sc.gc.ca.

Health Canada. 2006a. *Bacillus thuringiensis* (*B.t.*) Cry34/35/Ab1 insect resistant, glufosinate-tolerant transformation corn event DAS-59122-7. 19 June.

Health Canada. 2006b. Imidazolinone Tolerant Rice (PWC16). www.hc-sc.gc.ca. 03 March.

Health Canada. 2005. Novel Food Information: Glufosinate Tolerant Cotton Event LLCotton25. 16 September.

Health Canada. 2002. Cry1F Insect-resistant/Glufosinate-tolerant Maize Line 1507: October.

Health Canada. 2001. Roundup Ready Corn Line 603. 07 June.

Health Canada. 2000a. Glufosinate Ammonium Tolerant Canola (HCN92). 01 January.

Health Canada. 2000b. Glufosinate Ammonium Tolerant Sugar Beet (Event T120-7). 20 November.

Health Canada. 1999a. Imidazolinone tolerant Corn, 3417R. 01 October.

Health Canada. 1999b. Imazethapyr-tolerant EXP1910IT. FD/OFB-097-189-A. 01 October.

Health Canada. 1999c. Imidazolinone Herbicide Tolerant Canola Lines NS738, NS1471, NS1473. FD/OFB-095-115-A. 01 October.

Health Canada. 1999d. Glyphosate Tolerant Corn, GA21. 01 October.

Health Canada 1999e. Sulfonylurea Tolerant Flax, Cdc Triffid-FP967. 01 October.

Health Canada. 1999f. Roundup ReadyTM Corn Line Mon 802. 01 September.

Health Canada. 1999g. Bromoxynil Tolerant Canola (Westar-Oxy-235). 01 September.

Health Canada. 1997. Sethoxydim Tolerant Corn (DK412SR and DK404SR). FD/OFB-97-04. 01 April.

Held, B.M., H.M. Wilson, P.E. Dykema, C.J. Lewnau and J.C. Eby. 2006. Glyphosate resistant plants. US Patent No. 7,045,684.

Held, B.M., H.M. Wilson, L. Hou, C.J. Lewnau and J.C. Eby. 2004. Methods and compositions for the introduction of molecules into cells. US Patent No. 6,809,232.

Herman, R.A., A.M. Phillips, M.D. Lepping, J. Fast Brandon and J. Sabbatini. 2010. Compositional safety of event DAS-40278-9 (AAD-1) herbicide-tolerant maize. GM Crops 1(5): 294–311.

Hirschberg, J. 2001. Carotenoid biosynthesis in flowering plants. Current Opinion in Plant Biol. 4(3): 210–218.

International Service for the Acquisition of Agri-biotech Applications. 2013a. APHIS Approves Non-regulated Status of HT Corn, Male Sterile and HT Canola. Crop Biotech Update. 2 October.

International Service for the Acquisition of Agri-biotech Applications. 2013b. Pocket K No. 42: Stacked Traits in Biotech Crops. 2013.

International Service for the Acquisition of Agri-Biotech Applications. 2013c. Event Name: BT11 x DAS59122 x MIR604 x TC1507 x GA21. Crop: *Zea mays* L. www.isaaa.org.GM Approval Database. GM Crops.

International Service for the Acquisition of Agri-Biotech Applications. 2013d. Event Name: SYN 5307 x MIR604 x Bt11 x TC1507 x GA21. Crop: *Zea mays* L. www.isaaa.org.GM Approval Database. GM Crops.

International Service for the Acquisition of Agri-Biotech Applications. 2013e. Event Name: VCO-01981-5. Crop: *Zea mays* L. www.isaaa.org.GM Approval Database. GM Crops.

International Service for the Acquisition of Agro-Biotech Applications. 2012. Event Name: SYN5307 x MIR604 x Bt11 x TC1507 x GA21 x MIR162.

Jo, J., S.-H. Won, D. Son and B.-H. Lee. 2004. Paraquat resistance of transgenic tobacco plants over-expressing the *Ochrobactrum anthropi pqr A* gene. Biotech. Letters 26(18): 1391–1396.

Jube, S. and D. Borthakur. 2007. Expression of bacterial genes in transgenic tobacco: methods, applications and future prospects. Electronic J. Biotechnol. 10(3): 451–467.

Lebrun, M., B. Leroux and A. Sailland. 1996. Chimeric gene for the transformation of plants. US Patent US5510471 (23 APRIL 1996). Rhone Poulenc Agrochimie (FR).

Lebrun, M., A. Sailland, G. Freyssinet and E. Degryse. 2003. Mutated 5-enolpyruvylshikimate-3-phosphate synthase gene coding for said protein and transformed plants containing said gene. U.S. Patent 6566587.

Lee, H.-S. and J. Jinki. 2001. An *Ochrobactrum anthropi* gene conferring paraquat resistance to the heterologous host *Escherichia coli*. Biochem. Biophys. Res. Commu. 285(4): 885–890.

Leibbrandt, N.B. and S.J. Snyman. 2003. Stability of gene expression and agronomic performance of a transgenic herbicide-resistant sugarcane line in South Africa. Crop Sci. 43(2): 671–677.

LLCotton25 USDA Petition. 2002. Molecular Charcetrization of Transformation Event LLCotton25. Federal Register. Vol. 67(241): 16 December.

Lyon, B.R., Y.L. Cousins, D.J. Llewellyn and E.S. Dennis. 1993. Cotton plants transformed with a bacterial degradation gene are protected from accidental spray drift damage by the herbicide 2,4-dichlorophenoxyacetic acid. Transgen. Res. 2: 162–169.

Lyon, B.R., D.J. Llewellyn, J.L. Huppatz, E.S. Dennis and W.J. Peacock. 1989. Expression of a bacterial gene in transgenic tobacco plants confers resistance to the herbicide 2,4-dichlorophenoxyacetic acid. Plant Mol. Biol. 13: 533–540.

Manickavasagam, M., A. Ganapathi, V.R. Anbazhagan, B. Sudhakar, N. Selvaraj, A. Vasudevan et al. 2004. *Agrobacterium*-mediated genetic transformation and development of herbicide-resistant sugarcane (*Sacchraum* species hybrids) using axillary buds. Plant Cell Rep. 23: 134–143.

Miller, J.F. and K. Al-Khatib. 2000. Development of herbicide resistant germplasm in sunflower. Proc. 15th Int. Sunflower Assoc. Conf. France 37–41.

Monsanto Company. 2011a. Petition for the Determination of Nonregulated Status for Glyphosate-Tolerant Canola MON 88302. Monsanto Petition Number: 11-CA-233U. 27 June.

Monsanto Company. 2011b. Application for authorization to place on the market MON 87708 soybean in the European Union, according to Regulation (EC) No. 1829/2003 on genetically modified food and feed. Part II-Summary-MON87708, 17 January 2011. The Netherlands, Notification No. EFCA-GMO-NL-2011-93.

Monsanto Company. 2002. Safety Assessment of Roundup Ready Corn Event NK603. September 2002. 37 pp.

Monsanto Company. 2000. Request for Extension on Determination of Nonregulated Status for Glyphosate Tolerant Line NK603.

Monsanto Company. 1997. Revised Version of Petition 97-099-01p for Determination of Nonregulated Status of Roundup Ready Corn Line GA21 (30 July 1997): To APHIS, USDA. Riverdale, MD, USA.

Newhouse, K., W.A. Smith, M.A. Starrett, T.J. Schaefer and B.K. Singh. 1992. Tolerance to imidazolinone herbicides in wheat. Plant Physiol. 100: 882–886.

Owen, M. 1998. North American developments in herbicide tolerant crops. The British Crop Protection Conf. Brighton, England. Available online at http://www.weeds.iastate.edu/weednews/Brighton.htm.

Parker, W.B., L.C. Marshall, J.D. Burton, D.A. Somers, D.L. Wyse, J.W. Gronwald and B.G. Gengenbach. 1990. Dominant mutations causing alterations in acetyl-coenzyme A carboxylase confer tolerance to cyclohexanedione and aryloxyphenoxypropionate herbicides in maize. Proc. Natl. Acad. Sci. USA 87: 7175–7179.

Pighin, J. 2003. Transgenic Crops: How Genetics is Providing New Ways to Envision Agriculture? scq. ubc.ca, accessed March 3.

Pioneer Hi-Bred International. 2011. Petition for the determination of Nonregulated Status for Herbicide-Tolerant 73496 Canola. Revised submission. 03 August.

Pioneer Hi-Bred International. 2007. Petition to APHIS for the determination of Nonregulated Status for Herbicide Tolerant 9814 Corn. 30 May.

Pozniak, C.J. and P.J. Hucl. 2004. Genetic analysis of imidazolinone resistance in mutation-derived lines of common wheat. Crop Sci. 44: 23–30.

Pozniak, C.J., I.T. Birk, L.S. O'Donoughue, C. Menard, P.J. Hucl and B.K. Singh. 2004. Physiological and molecular characterization of mutation-derived imidazolinone resistance in spring wheat. Crop Sci. 44: 434–1443.

Przybylski, R., T. Mag, N.A.M. Eskin and B.E. McDonald. 2005. Canola Oil. pp. 61–121. *In*: F. Shahidi (ed.). Bailey's Industrial Oil and Fat Products. Sixth Edition (Six Volume Set). John Wiley & Sons, Inc., New York.

Rutledge, R.G., T. Quellet, J. Hattori and B.L. Miki. 1991. Molecular characterization and genetic origin of the *Brassica napus* acetohydroxyacid synthase multigene family. Mol. Gen. Genet. 229: 31–40.

Samac, D.A. and S.J. Temple. 2004. Development and utilization of transformation in *Medicago* species. *In*: G.H. Liang and D.Z. Skinner (eds.). Genetic Transformation in Crops. Haworth Press Inc., Binghamton, NY.

Schaub, P., S. Al-Babili, R. Drake and P. Beyer. 2005. Why is Golden Rice golden (yellow) instead of red?. Plant Physiol. 138(1): 441–450.

Science Daily. 2007. Flaxseed Stunts The Growth Of Prostate Tumors. Science Daily. June 4.

SENASA. 2012. Food and feed safety assessment of maize event MON 89034 x TC1507 x NK603 OECD: MON-89034-3 x DAS-01507-1 x MON- 00603-6 (Includes all possible intermediate combinations). Directorate of Agrifood Quality. Office of Biotechnology and Industrialized Agrifood Products. 31 May.

Shaner, D.L. and B.K. Singh. 1997. Acetohydroxyacid synthase inhibitors. pp. 69–110. *In*: R.M. Roe et al. (eds.). Herbicide Activity: Toxicology, Biochemistry, and Molecular Biology. IOS Press, Amsterdam, Netherlands.

Shaner, D.L., N.F. Bascomb and W. Smith. 1996. Imidazolinone-resistant crops: selection, characterization and management. pp. 143–157. *In*: S.O. Duke (ed.). Herbicide-Resistant Crops: Agricultural, Environmental, Economic, Regulatory, and Technical Aspects. CRC Press (Lewis Publishers), Boca Raton, FL., USA.

Smartstax. 2013. Wikipedia. http://www.genuity.com/Traits/Corn/Genuity-SmartStax.aspx. 27 September.

Spencer, M., R. Mumm and J. Gwyn. 2000. Glyphosate resistant maize lines. U.S. Patent 6040497.

Stein, A.J. and E. Rodriguez-Cerezo. 2009. The global pipeline of new GM crops: implications of asynchronous approval for international trade. European Commission Joint Research Centre.

Stine. 2012. Request for Extension of Determination of Nonregulated Status to the Additional Article: Maize Line HCEM485. Stine Seed Farm. Stine Petition #SSF-09-061. 02 March.

Streber, W.R., U. Kutschka, F. Thomas and H.-D. Pohlenz. 1994. Expression of a bacterial gene in transgenic plants confers resistance to the herbicide phenmedipham. Plant Mol. Biol. 25: 977–987.

Tan, S. and S.J. Bowe. 2012. Herbicide-tolerant crops developed from mutations. pp. 423–436. *In*: Q.Y. Shu, B.P. Forster and H. Nakagawa (eds.). Plant Mutation Breeding and Biotechnology. CABI, Cambridge, MA, USA.

Tan, S., R.R. Evans, M.L Dahmer, B.K Singh and D.L. Shaner. 2005. Imidazolinone-tolerant crops: history, current status and future. Pest Manag Sci. 61: 246–257. DOI: 10.1002/ps.993.

The Center for Food Safety. 2002. Commenton USDA/APHIS environmental assessment accompanying APHIS decision on Monsanto Company request (01-324-01P) seeking extension of determination of nonregulated status for glyphosate tolerant canola event GT200 & 60 day notice of intent to sue under the endangered species Act. 27 March.

Trolinder, L., G. Jefferson and M. Debeuckeleer. 2008. Herbicide-tolerant cotton plants and methods of producing the same. *In*: Bayer Bioscience. eds. (US7442504): USA. http:// www. patents. com/ Herbicide-tolerant-cotton-plants-methods-producing-identifying the same/US7442504/en-US/.

United States Department of Agriculture. September 2011. Draft Plant Pest Risk Assessment for Double Herbicide- Tolerant Soybean (*Glycine max*) event FG72. Petition 09-328-01p.

United States Federal Register. 1997. Monsanto Co. and Dekalb Genetics Corp.; Availability of Determination of Non-regulated Status for Genetically Engineered Corn Line GA21. Vol. 62(234): 05 December.

United States Food and Drug Administration. 2013. Event VCO-01981-5 glyphosate-tolerant corn. Biotechnology Consultation—Note to the File Biotechnology Notification File BNF No. 000137. 30 April.

United States Food and Drug Administration. 2012. MON 87427, tissue-selective herbicide tolerant corn Biotechnology Consultation Note to the File BNF No. 000126. 23 March.

United States Food and Drug Administration. 2007. CFSAN/Office of Food Safety. Biotechnol. Consultation. Note to the File, BNF No. 000108. 21 September.

United States Food and Drug Administration. 2001. Biotechnology Consultation Note to the File BNF No. 000073. 08 June.

United States Food and Drug Administration. 1995. Biotechnology Consultation Note to the File No. 000020: Monsanto's Glyphosate Tolerant canola Line GT73 (26 September 2009).

USDA-APHIS. 2013. Petitions for Determination of Nonregulated Status: Glyphosate and Isoxaflutole Tolerant/FG72Glyphosate. 21 August.

USDA-APHIS. 1997. Petition 96-317-01p for Determination of Nonregulated Status for Insect-Resistant/ Glyphosate-Tolerant Corn Line MON 802. Environmental Assessment and Finding of No Significant Impact. 23 October.

USDA-NASS. 2005a. Acreage 2005 (June report). United States Department of Agriculture National Agricultural Statistics Service, Washington, D.C.

Venkatesh, T.V., M.L. Breeze, K. Liu, G.G. Harrigan and A.H. Culler. 2014. Compositional analysis of grain and forage from MON 87427, an inducible male sterile and tissue selective glyphosate-tolerant maize product for hybrid seed production. J. Agric Food Chem. 62(8): 1964–1973.

Visser, R.G.F., C.W.B. Bachem, J.M. Boer, G.J. Bryan, S.K. Chakrabati, S. Feingold et al. 2009. Sequencing the potato genome: outline and first results to come from the elucidation of the sequence of the world's third most important food crop. Amer. J. Potato Res. 86(6): 417–429.

Vyver, C. van der, T. Conradie, J. Kossmann and J.L. 2013. *In vitro* selection of transgenic sugarcane callus utilizing a plant gene encoding a mutant form of acetolactate synthase. *In vitro* Cell Dev. Biol. Plant. 49(2): 198–206.

Walters, F.S., C.M. deFontes, H. Hart, G.W. Warren and J.S. Chen. 2010. Lepidopteran-active variable-region sequence imparts coleopteran activity in eCry3.1Ab, an engineered *Bacillus thuringiensis* hybrid insecticidal protein. Appl. Environ. Microbiol. 76(10): 3082–3088.

Warwick, A.I., A. Légère, M.-J. Simard and T. James. 2008. Do escaped transgenes persist in nature? The case of herbicide resistance transgene in a weed population of *Brassica rapa*. Mol. Ecol. 17: 1387–1395.

Wright, A.M. 2012. Transgenic (GE) Alfalfa in the United States: a blessing or a crutch? A Growing Culture 2 May.

Yi, G., Y.-M. Shin, G. Choe, B. Shin, Y.S. Kim and K.-M. Kim. 2007. Production of herbicide-resistant sweet potato plants transformed with the *bar* gene. Biotechnol. Lett. 29: 669–675.

Zan, N., H. Zhai, S. Gao, W. Chen, S. He and Q. Liu. 2009. Efficient production of transgenic plants using the *bar* gene for herbicide resistance in sweet potato. Science Horticulture 122: 649–653.

CHAPTER 7

Transgenic Phytoremediation of Herbicides and Explosives in Soil and Environment

Phytoremediation is not just a growing science but a fast-expanding industry. It is the process by which green plants detoxify soils, sediments, and aquatic sites contaminated with organic and inorganic pollutants. Most of the organic pollutants are xenobiotic and manmade. These include pesticides such as herbicides, insecticides and fungicides, oil spills, explosives and military weapons, industrial chemicals, etc. Inorganic pollutants include natural elements (cadmium, cobalt, iron, lead, mercury, selenium, tungsten, etc.) released into the environment by human activities in such areas as mining, industry, traffic, agriculture (plant nutrients by way of fertilizers), military, etc. [Pilon-Smits 2005]. The contaminants vary in toxicity, but after long-term exposure they can be detrimental to human and animal health. Some of them cause damage to DNA and their carcinogenic effects in humans and animals are probably caused by mutagenic ability [Knasmuller et al. 1998; Baudouin et al. 2002; Hooda 2007]. Commercial phytoremediation involves 80 percent organic and 20 percent inorganic contaminants [Pilon-Smits 2005].

Plants remove harmful chemicals from the ground when their roots take up water and nutrients from polluted soil, streams, and groundwater. Once inside the plant, chemicals are stored in the roots, stems, and/or leaves. The absorbed pollutants are biochemically converted by the tolerant plants into non-phytotoxic metabolites that accumulate in plant tissue or changed into gasses that are released into the air as the plant transpires. The cleanup of the contaminated sites (soils and water systems) is essential in order to eliminate or minimize the entry of toxic elements into the food chain. Phytoremediation works best at sites with low to medium amounts of pollution.

Some metals can be recovered from plant tissues in a process called 'phyto-mining' [Meagher 2000]. In addition to accumulating toxic minerals in their tissues, plants also take up a range of harmful organic compounds, including some of the most abundant environmental pollutants such as polychlorinated biphenyl (PCB), the halogenated hydrocarbon like trichloroethylene (TCE), and nitroaromatic explosives such as TNT (2,4,6-trinitrotoluene), tetryl (2,4,6-trinitrophenyl-N-methyl nitramine), pentryl (2,4,6-trinitrophenyl-N-nitroaminoethylnitrate), and glyceryl trinitrate (GTN), etc.

Another nitroaromatic environmental contaminant of serious proportions in the world today is RDX (Royal Demolition Explosive: hexahydro-1,3,5-trinitro-1,3,5-triazine), a powerful, highly energetic chemical whose wide use in various military and civilian applications has resulted in severe soil and groundwater contamination. Currently, many countries are involved in wars, bombing, and terrorism. As a result, it ultimately ends up as a chemical waste contaminating the soil and water systems.

RDX and TNT are world's major explosives and environmental pollutants contaminating millions of hectares of land across the world. Within the soil, the carcinogenic RDX is highly mobile, readily leaching into groundwater with the potential to pollute subsequent waterways. Although plants take up RDX readily at high rates, they lack the ability to degrade it [Winfield et al. 2004].

The concept of phytoremediation was born in the 1980s when some plant species exhibited an extraordinary ability in accumulating higher quantities of toxic metals in their tissues or organs [Maestri and Marmiroli 2012]. This new, elegant technology, which has become increasingly popular over the past 20 years, has been employed at sites with soils contaminated with pesticides, explosives, lead, selenium, uranium, and arsenic. While phytoremediation has the advantage of treating environmental concerns *in situ*, its major impediment is that it requires a long-term commitment, as the process is dependent on a plant's ability to grow and thrive in an environment that is not ideal for normal plant growth.

The major advantages of phytoremediation are: a) less expensive than that of traditional processes both *in situ* and *ex situ*, b) easy monitoring of plants, c) possibility of the recovery and re-use (in the case of valuable metals by phyto-mining companies), d) preservation and stability of the soil and environment by virtue of using naturally occurring plants and organisms, and e) possibility of biofuel and fiber production.

Phytoremediation does have certain limitations. These include, a) confining the process to the surface area and depth the roots reach, b) lower biomass and slower growth of plants, c) long-term commitment, d) possibility of the contaminants leaching into the groundwater before the process is completed, and e) bioaccumulation of contaminants in the plants and passing them into the food chain.

In the case of pesticides, including herbicides, phytoremediation is best suited to sites contaminated to a depth of less than 5 m and those containing moderately hydrophobic pollutants (log K_{ow} = 0.5–3.0), short-chain aliphatic chemicals, or excess nutrients [Schnoor et al. 1995]. The process can be used to remove herbicides from water and soil because most herbicides are moderately hydrophobic.

Selection of the appropriate plants is crucial for successful phytoremediation. This depends on the ability of plants to be resistant to the chemicals or pollutants to be removed along with their ability to metabolize and immobilize the pollutants.

Phytoremediation Process

Depending on the underlying processes, applicability, and type of contaminant, phytoremediation can be broadly grouped into six, which include phytodegradation, rhizodegradation, phytovolatilization, phytoextraction, rhizofiltration, and phytostabilization (Fig. 7.1).

Phytodegradation

Phytodegradation, also called phytotransformation, is the breakdown of contaminants taken up by plants through metabolic processes within the plant, or the breakdown of those contaminants surrounding the plant through the effect of compounds, such as enzymes, produced by the plants. Complex organic pollutants are degraded into simpler molecules,

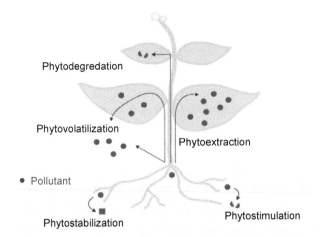

Fig. 7.1. Different processes of phytoremediation. The pollutant or contaminant (*red circles*) can be sequestered or degraded inside the plant tissue, extracted by the foliage, degraded or stabilized in the rhizosphere, or volatilized [Pilon Smits 2005].

Color image of this figure appears in the color plate section at the end of the book.

which are incorporated into the plant tissues to help the plant grow faster. Plants contain enzymes that catalyze and accelerate chemical reactions. Some enzymes break down and convert ammunition wastes while others degrade herbicides, chlorinated solvents (chloroform, carbontetrachloride, trichloroethylene, etc.), explosives (RDX, trinitrotoluene, etc.), nitrobenzene, nitroethane, etc. Certain aquatic and terrestrial plants take up, store, and degrade or transform organic compounds.

Phytodegradation is not dependent on microorganisms associated with the rhizosphere. For the type of phytodegradation that occurs within a plant, the plant must be able to take up the contaminant. Therefore, only moderately hydrophobic compounds, with an octanol-water partition coefficient log K_{ow} between 1 and 3.5, are subjected to phytodegradation [Paspaliraius et al. 2010].

The process of uptake, transport, degradation, or accumulation of organic pollutants is the key factor in phytoremediation. Therefore, it is important to optimize methods of phytodegradation to the prevailing environmental conditions, type of contamination, and other factors. The contaminants have a very strong impact on fundamental physiological processes like photosynthesis, respiration, and growth.

Rhizodegradation

Rhizodegradation, also called phytostimulation, rhizoremediation, or plant-assisted bioremediation/biodegradation, is the breakdown of contaminants in the rhizosphere through microbial activity that is enhanced by the presence of plant roots. Plant roots exude in the rhizosphere natural substances such as sugars, amino acids, organic acids, fatty acids, sterols, growth factors, flavanones, nucleotides, enzymes, etc. [Paspaliraius et al. 2010]. These exudates provide carbon to support the activity of microorganisms (yeast, fungi, or bacteria). The increased microbial population and activity in the rhizosphere can increase biodegradation of the contaminant in the soil, while degradation of the exudates can stimulate co-metabolism of contaminants in the rhizosphere [Paspaliraius et al. 2010].

Rhizodegradation is considered to be a symbiotic relationship existent between plants and microbes, in that the plants provide nutrients needed for the microbes to thrive and the microbes provide a healthier soil environment in which the plant roots can proliferate [Paspaliraius et al. 2010]. The choice of plants is likely to impact on the success of the rhizoremediation technology. Some plants such as *Cucurbita pepo* (zucchini) [Campanella et al. 2002] accumulate high level of hydrophobic chemicals while alfalfa (lucerne), with its extensive root systems, has high affinity towards them [Schwab et al. 1998].

Rhizodegradation is a much slower process than phytodegradation. Certain microbes digest organic substances like fuels or solvents that are hazardous to humans and break them down into harmless products. The microbial enzymes, mainly esterases and oxidoreductases, catalyze the degradation of organic compounds. Many of them have low substrate specificity, which ensures them a broader spectrum activity. Besides, plants also have significant metabolic activity in both roots and shoots, and some of the plant enzymes (laccases, nitroreductases, peroxidases, dehalogenases, etc.) involved in these metabolic processes are useful in the remediation process [Schnoor et al. 1995; Wolfe and Hoehamer 2003].

Aerobic microbial biodegradation of organic compounds with aromatic structure can be divided into three stages: ring hydroxylation, cleavage of the aromatic ring and oxidation of the aliphatic moiety to Krebs cycle intermediates, and their further breaking down to CO_2 and H_2O [Greń et al. 2008]. Natural substances (sugars, alcohols, and acids) released as exudates by the plant roots contain organic carbon that provides food for soil microorganisms and the additional nutrients enhance their activity. Rhizodegradation is also aided by the way plants loosen the soil and transport water to the area. The major advantage of this is that the plants do not need to take up the pollutants in order to detoxify them. Instead, they secrete enzymes which then transform the chemicals in the rhizosphere.

Phytovolatilization

Phytovolatilization is the uptake and transpiration of the contaminant by growing trees and plants, releasing the contaminant or its modified form from the tree/plant to the atmosphere. In this, plants take up both volatile and nonvolatile contaminants from the soil and transform them into volatile forms, thus facilitating their transpiration into the atmosphere. However, metabolic processes within the plant might alter the initial form of the contaminant, and in some cases, transforms them into less toxic forms.

Phytovolatilization may be useful in the case of both organic (chlorinated solvents, methyl tertiary butyl ether: MTBE) and inorganic (selenium, silver, arsenic, etc.) contaminants. Some of these contaminants can pass through the plants to the leaves and evaporate, or volatilize, into the atmosphere. In fact, plants serve as effective pump-and-treat systems for mobile contaminants. One major disadvantage of this process is that the transfer of contaminants to the atmosphere may have an impact on the ecosystem and human health.

Phytoextraction

Phytoextraction (phyto-accumulation) uses plants or algae to remove contaminants from soils, sediments, or water into harvestable plant biomass. The organisms that take larger-than-normal amounts of contaminants from the soil are called hyper-accumulators. Phytoextraction, also called phyto-mining, is used more for extracting heavy metals than for organics. Plants absorb contaminants through the root system and store them in the root biomass and/or transport them up into the stems and/or leaves. A living plant may continue to absorb contaminants until it is harvested. After harvest, a lower level of the contaminant will remain in the soil, so the growth/harvest cycle must usually be repeated through several crops to achieve a significant cleanup. After the process, the cleaned soil can support other vegetation.

Phytoextraction, applied to remove metals (silver, cadmium, cobalt, copper, mercury, manganese, nickel, nobelium, lead, and zinc), metalloids (arsenic and selenium), and radionuclides, is not considered useful for organic and nutrient contaminants as these can be metabolized, changed, or volatilized by the plant, thus preventing accumulation [Paspaliraius et al. 2010]. This process is usually carried out by certain plants, called hyper-accumulators, which absorb large amounts of metals compared to other plants. A hyper-accumulating plant species is capable of accumulating 100 times more metal than a normal plant. As phytoextraction occurs in the root zone of plants, the root zone may typically be shallow, with the bulk of the root system remaining at shallower depths. Plant species such as Indian mustard (*Brassica juncea*), sunflower, crucifers, etc. are considered to be good phytoextractants [Paspaliraius et al. 2010].

The major advantage of phytoextraction is its low cost than other clean-up processes besides its environmental friendliness. In some cases, the contaminant accumulated in the plant biomass can be recycled. However, the use of hyperaccumulating plants is limited by their slow growth and smaller biomass production, thus precluding absorption of larger concentrations of heavy metals by the roots and their translocation into the surface biomass.

Phytostabilization

Phytostabilization focuses on long-term stabilization and containment of the pollutant. The process involves using certain plant species to immobilize the contaminant in the soil through absorption and accumulation by roots, adsorption onto roots, or by way of precipitation, complexation and metal valence reduction within the root zone [Paspaliraius et al. 2010].

There are three mechanisms in the process of phytostabilization [Paspaliraius et al. 2010]. The first one involves the root exudates released into the rhizosphere, and these may precipitate or immobilize the contaminant in the root zone. The second one, facilitated by proteins and enzymes associated with root cell walls, can stabilize the contaminant on the exterior surfaces of the root membranes, thus preventing it from entering the plant. In the third mechanism, proteins and enzymes present on the cell walls may facilitate the transport of the contaminant across the root membrane, and upon uptake it is sequestered into the vacuole of the root cells, preventing further entry into the shoots.

Phytostabilization has greater utility in the case of metal contaminants (lead, chromium, and mercury), it can be used to remediate the soil of organic pollutants which are very hydrophobic in nature (> 3.5 log K_{ow}) and which can be attached to or incorporated into plant components [Paspaliraius et al. 2010]. This technique is very effective when rapid immobilization is needed in order to preserve ground and surface water. Besides, as the pollutant becomes less bioavailable, exposure of livestock, wildlife, and humans to it is reduced. Its main disadvantage is that the contaminants remain in the soil for a long time, thus preventing a faster remediation.

Indian mustard plants and poplar trees have the potential for effective phytostabilization. Of the two, poplar trees, with their wider and deeper root penetration, offer a better choice.

Rhizofiltration

Rhizofiltration is the adsorption or precipitation onto plant roots (or absorption into the roots) of contaminants that are in solution surrounding the root zone. Rhizofiltration is similar to phytoextraction, but the plants are used to clean up contaminated groundwater rather than soil. Rhizofiltration is typically exploited in groundwater (either *in situ* or extracted), surface water, or wastewater for removal of metals or other inorganic compounds [EPA 2000]. Rhizofiltration can be used for lead, cadmium, copper, nickel, zinc, and chromium, which are primarily retained within the roots [USEPA 2000].

The plants to be used for cleanup are raised in greenhouses with their roots in water rather than in soil. In order to acclimatize the plants, once a large root system has been developed, contaminated water is collected from a waste site and brought to the plants where it is substituted for their water source. The plants are then transplanted in the contaminated area where the roots take up the water and the contaminants along with it. As the roots become saturated with contaminants, they are harvested. Sunflower, Indian mustard, tobacco, rye, spinach, and maize have the ability to remove lead from water, with sunflower having the greatest ability. In one study, after only one hour of treatment, sunflower reduced lead concentrations significantly [Raskin and Ensley 2000]. Sunflower plants have been successfully used to remove radioactive contaminants from pond water in a test at Chernobyl, Ukraine.

Of the six phytoremediation processes described, phytodegradation, rhizodegradation, and phytovolatilization are suitable for organic contaminants including herbicides, while the other three, phytoextraction, rhizofiltration, and phytostabilization, are becoming more useful to remove metal contaminants. This chapter discusses the more pertinent phytodegradation in remediating the soil and environment of the ever increasing accumulation of herbicide residues and explosives.

Herbicides and Their Residues

Herbicides reach the soil through pre-planting and pre-emergence applications, as foliage run-off from post-emergence applications, and through the return of crop residues to the soil. It is desirable for the herbicides to control weeds during the season of application, but they should not be persistent in the soil long enough to affect the growth of subsequent sensitive follow-up crops. Herbicides vary in their potential to persist in the soil. Those belonging to triazine, uracil, phenylurea, sulfonylurea, imidazolinone, dinitroaniline, the plant-growth inhibiting, and pigment-inhibiting herbicide families have longer persistence. The half-life of most of the herbicides varies from one week to several months, depending on application rate, soil type, and environmental conditions.

In addition to herbicides and their degradation products being persistent in soil, they also reach surface water and groundwater through leaching. Most of the herbicide transport occurs during first rain and irrigation soon after application. Once herbicides reach the surface water system, they end up in lakes, reservoirs, canals, and rivers, where they show minimal loss by volatilization, sorption, or transformation. It is difficult to determine and quantify the long-term effects of exposure to low levels of herbicides in the soil and environment. However, herbicides and their metabolites reduce primary productivity by killing phytoplankton and thus altering the composition of the surrounding aquatic and terrestrial ecosystems [Relyea 2005].

In higher plants, herbicides can be detoxified through several biochemical processes which are grouped into four main phases: conversion, conjugation, transportation, and compartmentalization [Yuan et al. 2007]. Transgenic plants for high-efficiency phytoremediation can be produced when the genes involved in all these phases are properly transferred into the host plants [Kawahigashi 2009].

The first phase of detoxification (conversion and herbicide metabolism) transforms the initial properties of the toxic herbicide through chemical oxidation, reduction, or hydrolysis resulting in a substance more water soluble, and most likely, less phytotoxic than the herbicide itself. It also results in activation of inactive herbicide molecules. The primary class of enzymes functioning in herbicide conversion metabolism is cytochrome P450-dependent monooxygenases which are involved in substrate oxidation. They also convert hydrophobic herbicides into less hydrophobic metabolites. In plants, most of the P450-catalyzed reactions are conducted by hydroxylation, although oxidative dealkylation and epoxidation do also occur. This enzyme system, found in microsomes, comprises several cytochrome P450 isoforms and a non-specific NADPH-cytochrome P450 oxidoreductase [Kawahigashi 2009].

In the second phase of herbicide detoxification, plants make use of a battery of enzymes that could make additional detoxification steps through conjugation with a sugar (glucose, galactose, mannose, arabinose, or xylosyl-glucose), amino acid (glutamic acid, glycine, leucine, or cysteine), carboxylic acids (malonic acid), or peptide (glutathione). These steps usually involve adding a large polar group (conjugation reaction), such as glucuronide, to further increase the compound's solubility. Thus, the conjugated form of the herbicide results in increased water solubility which enhances successful subsequent compartmentalization. The resulting conjugates are often more hydrophilic than the initial molecules. This provides them with mobility in the cytoplasm. Transferase enzymes are responsible for most of the second phase reactions. These include uridine diphosphoglucuronosyl transferase (UGT), N-acetyl transferase (NAT), malonyltransfearse, glutathione S-transferase (GST), and sulfotransferase (ST). Glucosylglucosides or other derivatives are the predominant reaction products. Glutathione (GSH) conjugation via GSTs is one of the most studied mechanisms of herbicide conjugation.

The conjugation of xenobiotics and amino acids is catalyzed by N-acetyl transferases by transferring the acetyl group from acetyl-CoA to arylamines. As their substrates are essentially carboxylic acids, the resulting conjugates may still be biologically active, although they become less mobile. These substrates are then secreted into the cell wall [Leszczyński 2001; Różański 1998; Zakrzewski 2000].

In the third phase, the conjugated metabolites are recognized by the ATP-dependent membrane pumps and actively transported across the membranes through the tonoplast by Mg- and ATP-dependent transporters called ABC (ATP-binding cassette) transporters. The increase in transport activity in *Arabidopsis* treated with herbicides and herbicide safeners indicates that several glutathione-conjugate transporters exist in plants [Tommasini et al. 1997].

In the fourth and final phase of the detoxification process, the metabolites, which may also undergo secondary conjugation, are transported to and compartmentalized in the vacuoles before being transferred to the apoplast, or incorporated with the components of the cell wall such as pectin, lignin, polysaccharides, hemicellulose, and cellulose fractions, forming insoluble residues [Marrs 1996; Cole 1994; Edwards et al. 2005]. The reactions such as solubility, phytotoxicity, and mobility of metabolites occurring in plants during the herbicide detoxification process over the four phases are listed in Table 7.1. Secondary conjugations may occur during this phase.

Table 7.1. Herbicide detoxification phases in plants. Adapted with minorchanges based on publications by Eerd et al. [2003] and Shimabukuro [1985]; [Review by de Carvalho et al. 2009].

Charateristics	Initial Properties	Phase I	Phase II	Phase III	Phase IV
Reactions	Herbicidal Action	Oxidation, Reduction Hydrolysis, Oxygenation, Hydroxylation	Conjugation with amino acids, sugar, or glutathione	Transportation acrosscell membrane (sometimes secondary conjugation)	Compartmentalization in vacuoles and cell wall components
Solubility	Lipophilic	Intermediary	Hydrophilic	Hydrophilic	Insoluble
Phytotoxicity	High herbicidal toxicity	Modified or less toxic	Reduced or nontoxic	Non-phytotoxic	Non-phytotoxic
Mobility in Plants	Variable as function of herbicide	Modified or reduced	Limited or immobile	Immobile	Immobile

Enzymes in Phytoremediation

Plants produce a variety of enzymes that detoxify complex organic contaminants. They have evolved capabilities to break the contaminants down to such essential elements as nitrogen, phosphorous, and potassium. As with all living things, detoxification of complex contaminants is catalyzed by enzymes, leading to, as in many cases, breaking them down to carbon dioxide, water, inert gases, and other simple non-toxic molecules.

Phytoremediation of organics requires enzymes specific to contaminants. Each receptor site has a reporter site that identifies contaminants it can break down. When the enzyme comes in contact with an appropriate contaminant molecule, the molecule binds to the enzyme's receptor site. The enzyme then destroys the contaminant using the stored plant energy or energy derived from molecular bonds broken down in the process. Enzyme-mediated degradation may occur inside plant cells as a part of plant's metabolism or in the surrounding soil by enzymes exuded by plant exudates. Externally active enzymes are essential for phytoremediating contaminants that cannot be taken by up plants due to the size of molecules or chemical properties. The metabolites released due to the activity of external enzymes are exuded by plants under electrostatic pressure.

The genes identified so far for implantation into the plant genome include those that encode cytochrome P450 and glutathione S-transferase, nitroreductase, peroxidase, and laccase enzymes (vide Appendix: Table 5). These genes are derived from plants and other organisms.

Cytochrome P450s

The cytochrome P450 (CYP) represents a class of enzymes present in most plants, mammals, and bacteria. The letter 'P' in P450 represents the word 'pigment' as these enzymes are red because of their heme group, while the number '450' reflects wavelength of the absorption maximum of the enzyme when it is in the reduced state and complexed with CO (carbon monoxide).

CYP450s are involved in different types of mono-oxygenation/hydroxylation reactions in the primary and secondary metabolism of many natural and xenobiotic compounds. The number of P450 genes in plant genomes is estimated to be up to one percent of total gene annotations of each plant species [Mizutani and Ohta 2010]. This type of diversification within P450 gene superfamilies has led to the emergence of new metabolic pathways throughout land-plant evolution. P450 enzymes, which have a molecular mass ranging of 45–62 kDa, are a super family of membrane-bound ubiquitous hemoproteins involved in the first phase of metabolism of numerous xenobiotics including herbicides, therapeutic drugs, and industrial contaminants. Also referred to as 'heme-thiolate proteins', the P450s participate in the anabolism or catabolism of membrane sterols, oxylipins, structural polymers, hormones, and many secondary metabolites functioning as pigments, antioxidants, and defense compounds. These enzymes can also detoxify exogenous molecules such as pesticides and pollutants, thereby aiding in their remediation from soil and environment. Their action is similar to the way the human liver increases the polarity of drugs and foreign compounds. The human liver enzymes including cytochrome P450s are responsible for the initial reactions while in plants it is the enzymes like nitroreductases that carry out the same function.

P450s oxidize carbon and nitrogen, usually resulting in the formation of hydroxyl group and occasionally the subsequent removal of alkyl and amino groups. The reactions catalyzed by CYPs are extremely diverse, accounting for about 75 percent of different metabolic reactions. The most common reaction catalyzed by cytochrome P450 is a monooxygenase reaction, i.e., insertion of one atom of molecular oxygen into an organic substrate (RH) while the other oxygen atom is reduced to water at the expense of NADPH (nicotinamide adenine dinucleotide phosphate) or NADH (nicotinamide adenine dinucleotide) as shown below:

$$RH + O_2 + NADPH + H^+ \rightarrow ROH + H_2O + NADP^+$$

In addition to this hydroxylation reaction, CYPs can also catalyze oxidation, dealkylation, deamination, dehalogenation, and sulfoxide formation. CYPs use a variety of small and large molecules as substrates in enzymatic reactions. Often, they form part of multi-component electron transfer chains, called P450-containing systems.

The cytochrome P450s generally fall into two broad classes, depending on the nature of the auxiliary protein(s) [Nebert and Gonzalez 1987]. Class I P450s, found on the membranes of mitochondria or in bacteria, are three-component systems comprising a flavin adenine dinucleotide (FAD)-containing reductase, an iron-sulfur protein (ferredoxin), and the P450. Class II cytochrome P450s are two-component systems made up of an FAD-containing, flavin mononucleotide (FMN)-containing NADPH-dependent cytochrome P450 reductase and a P450 (Fig. 7.2: Class I and Class II) [Nebert and Gonzalez 1987]. These are typified by the liver microsomal enzymes in mammalian cells which are involved in both steroid metabolism and detoxification pathways. Roberts et al. [2002] proposed the class system made up of an FMN-containing reductase with a ferredoxin-like center linked to a P450 in a single polypeptide. The N and C terminals of the fused class III and IV systems (Fig. 7.2).

Over the last decade, there has been an increasing realization of the power of CYP catalysts for producing herbicide-resistant plants and for detoxification of soil and water contaminants via phytoremediation method [Kumar 2010; Abhilash et al. 2009; Van Aken 2009; Van Aken and Doty 2010; Kumar et al. 2012]. In plants, oxidative demethylation (and dealkylation) is followed by elimination of the hydroxylated methyl (alkyl) group as the aldehyde. CYP450 is reduced with cytosolic NADPH by NADPH: Cyt P-450 reductase, a membrane-bound flavoprotein. Thus, the electron flow is from cytosolic reductant to O_2. The P450s involved in transformation reactions may metabolize a number of substrates.

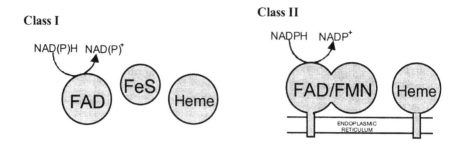

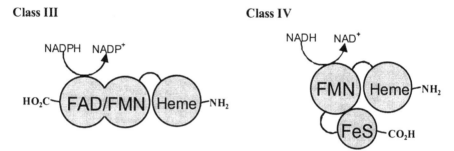

Fig. 7.2. Schematic representation of the different classes of cytochrome P450 systems. Class I systems comprise a FAD-containing flavodoxin reductase, an iron-sulfur protein (ferredoxin), and the P450. These eukaryotic class I enzymes are associated with the mitochondrial membrane. In a class II system, the P450 is partnered with a diflavin reductase, whereas in the class III system, the diflavin reductase is fused to the P450. The class IV system is made up of an FMN-containing reductase with a ferredoxin-like center linked to a P450 in a single polypeptide. The N and C terminals of the fused class III and IV systems are shown [Nebert and Gonzalez 1987; Roberts et al. 2002].

CYP enzymes have been identified in mammalian, bacterial (excluding *E. coli*), fungal, and plant systems (Table 7.2). Over 11,500 distinct CYP proteins are known to exist [Hanukoglu 1996].

Table 7.2. Number of CYP families and species found in selected groups [Abhilash et al. 2009].

Taxonomical groups	No. of CYP families	No. of sequences
Animals	99	2,279
Plants	97	1,932
Fungi	282	1,001
Bacteria	177	621
Protists	51	210
Arachaea	5	8

Mammalian CYPs

The mammalian P450s are localized predominantly in the smooth endoplasmic reticulum of the cell. They are entirely distinct from the cytochrome proteins that comprise the mitochondrial electron-transport function. The mammalian liver in which highest concentrations of P450s exist is the mammalian detoxification center for both exogenous and endogenous chemicals. P450s can oxidize both xenobiotics and endogenous compounds, and are important for detoxification of foreign substances, as well as for controlling the level of endogenous compounds, such as hormone synthesis and breakdown, cholesterol synthesis, and vitamin D metabolism.

Many mammalian enzymes also have the capability of converting herbicides and their metabolites into relatively safer products. Over the past decade, CYP1A1, CYP2B6, CYP2B22, CYP2C9, CYP2C19, CYP2C49, CYP2E1, CYP71A1, CYP76B1, and CYP105A1 have been transgenically introduced in various plants, such as rice and tobacco, for phytoremediation of numerous herbicides, petrochemicals, and industrial contaminants [James et al. 2008; Kawahigashi et al. 2007; Kawahigashi et al. 2008; Inui and Ohkawa 2005]. Inui et al. [2001] reported that each of the CYP1A1 and CYP2C19 enzymes can metabolize 16 herbicides including triazines, ureas, and carbamates while CYP2B6 can do so more than 10 herbicides, three insecticides, and two industrial chemicals. This trait has been used to engineer herbicide-resistant transgenic crop plants.

Although mammalian CYP enzymes have been used to create herbicide-resistant transgenic plants, there are some major limitations. One of them is that they require the redox partner chromosome P450 reductase (CPR), which transfers electrons from the NADPH to the heme of CYP, to oxidize the substrate [Anzenbacher and Anzenbacherová 2001]. Besides, mammalian CYP enzymes have relatively low turn-over (1–25), enzyme stability, and expression in heterologous systems, including plants [Anzenbacher and Anzenbacherová 2001; Kumar 2010; Kumar et al. 2012]. Therefore, there is a need to design mammalian CYPs to improve their catalytic efficiency, stability, and expression, besides the stability of P450-CPR fusion enzymes [Bernhardt 2006; Kumar 2010; Sakaki 2012; Kumar et al. 2012].

Although mammalian CYPs have lower activity and stability than bacterial CYP enzymes, they have the ability to metabolize a wide range of substrates including herbicides and environmental contaminants. Engineering of mammalian CYP enzymes with increased activity can produce CYP biocatalysts that are amenable to phytoremediation [Anzenbacher and Anzenbacherová 2001; Kumar 2010]. These enzymes can also be engineered using artificial fusion proteins that would couple between CYP and CPR for efficient transfer of electrons.

Bacterial CYPs

Many bacterial CYP enzymes, unlike mammalian CYPs, do not require external redox partner for the transfer of electrons from the NADPH to CYP, because they contain reductase domain within them (self-sufficient CYP). In addition, these enzymes have much higher turnover (> 100) and enzyme stability than the mammalian CYPs [Whitehouse et al. 2012; Girvan et al. 2006]. However, unlike mammalian CYP enzymes, bacterial CYP enzymes do not show substrate diversity and metabolize a limited number of compounds [Anzenbacher and Anzenbacherová 2001; Whitehouse et al. 2012]. Despite their being self-sufficient with high activity, they have a relatively rigid and small active site(s) which enable them to metabolize only smaller compounds. This makes them less suitable for detoxification of many large compounds including herbicides and industrial contaminants.

Bacterial CYPs need to be designed for enhanced stability, expression, and substrate diversity [Bernhardt 2006; Kumar 2010; Sakaki 2012; Kumar et al. 2012]. Transgenic plants expressing bacterial CYP genes have been used for enhanced degradation and remediation of herbicides and industrial contaminants. In one example, *Arabidopsis thaliana* expressing *Rhodococcus rhodochorus* xplA (fused to a flavodoxin redox partner) degraded RDX grown in RDX-contaminated soil and was found to be resistant to RDX phytotoxicity [Rylott et al. 2006; Kumar et al. 2012]. Thus, transgenic plants expressing the *xplA* gene may provide a phytoremediation strategy for sites contaminated by this type of explosive [Rylott et al. 2011; Kumar et al. 2012]. Another bacterial CYP gene *CYP105A1* isolated from *Streptomyces griseolus* was cloned in tobacco plants for the degradation of sulfonylurea herbicide [O'keefe et al. 1994]. In this, the expressed CYP105A1 appeared to interact with the host redox partner present in the chloroplast for the metabolism of sulfonylureas [Kumar et al. 2012].

The more prominent microbial CYP enzymes are P450 cam, P450 eryF, P450 BM3, and P450 119. Cytochrome P450cam originally from the Gram-negative rod-shaped saprotrophicsoil bacterium *Pseudomonas putida* catalyzes the regio- and stereo-specific hydroxylation of (1*R*)-(+)-camphor to 5-*exo*-hydroxy camphor [Poulos et al. 1987]. Its (*CYP101* gene) monooxygenase system consists of three soluble proteins: putidaredoxin reductase; putidaredoxin, an intermediary 2Fe-2S cluster-containing protein cofactor; and the cytochrome P450cam. P450cam has been established as the archetypal bacterial cytochrome P450.

Cytochrome P450 eryF (*CYP107A1* gene), originally from the actinomycete bacterium *Saccharopolyspora erythraea*, is responsible for the biosynthesis of the antibioticerythromycin by C6-hydroxylation of the macrolide 6-deoxyerythronolide. It lacks the otherwise highly conserved threonine that is thought to promote O-O bond scission.

Cytochrome P450 BM3 (*CYP102A1* gene) from the soil bacterium *Bacillus megaterium* catalyzes the NADPH-dependent hydroxylation of several long-chain fatty acids at the ω-1 through ω-3 positions. Unlike other CYPs (except CYP505A1, cytochrome P450 foxy), it constitutes a natural fusion protein between the CYP domain and an electron donating cofactor. Thus, P450 BM3 is potentially very useful in biotechnological applications [Narhi and Fulco 1986; Girvan et al. 2006]. Transgenic plants generated using CYPBM3 are capable of metabolizing a variety of gasoline, and they also have the potential in the clean up of petroleum pollutants [Kumar 2010; Whitehouse et al. 2012]. CYP BM3 can be engineered to metabolize novel petroleum substrates that are of similar sizes and shapes, which could further be used for phytoremediation of a variety of petroleum products [Whitehouse et al. 2012; Kawakami et al. 2011]. In the degradation of explosives, the introduced bacterial genes improve the phytoremediation efficiency by three different ways: increasing degradation, plant tolerance, and uptake of toxic compounds. The toxicity of explosives and chlorinated solvents is related to a reduced phytotoxicity and a steeper gradient of pollutant concentration inside plant tissues [Doty et al. 2000].

Cytochrome P450119 (*CYP119* gene), isolated from the thermophilic archea *Sulfolobus acidocaldarius* [Wright et al. 1996], has been used in a variety of mechanistic studies [Rittle and Green 2010]. As thermophilic enzymes are evolved to function at high temperatures, they tend to function more slowly at room temperature. This makes them good mechanistic models.

Fungal CYP450s

There is a greater and marked molecular diversity in fungal cytochrome P450 enzymes even though the fungal organisms are taxonomically similar. This diversity of fungal P450s appears greater than of animal, plant, or bacterial P450s. For example, *Aspergillus oryzae* contains 155 P450 genes (plus 13 pseudo-genes), and this number is significantly greater than the 74 and 11 genes in *A. fumigatus* and *A. nidulans* respectively [Ichinose 2012]. These differences reflect the evolution of *Aspergillus* species, in which *A. oryzae* expanded its genome by horizontal transfer and duplication [Michida et al. 2005]. However, the vast majority of fungal P450s are found only in certain species.

Fungal P450s appear to have continuously diversified after genetic separation, and perhaps after speciation. In addition, basidiomycetous P450s have been reported from comparative phylogenetic analyses of P450s in the white-rot basidiomycete *Phanerochaete chrysosporium* and the brown rot basidiomycete *Postia placenta* [Idi et al. 2012]. *P. chrysosporium* contains CYP5027, CYP5350, and CYP5348 genes while *P. placenta* has CYP53 [Idi et al. 2012].

Plant CYPs

Cytochrome P450s, one of the largest families of enzymatic proteins in higher plants, are found in subcellular locations like endoplasmic reticulum, plasma membranes, glyoxysomes and mitochondria. They also catalyze herbicide metabolism and contribute to detoxification or activation of other agrochemicals in crops plants [Morant et al. 2003; Komives and Gullner 2005]. These multifunctional enzymes catalyze more than 60 reactions. They, however, exhibit tissue and substrate specificity and there are many sites regulating the expression of Cyt P450 gene in plants. More than 1000 P450 gene sequences are obtained from a small subset of plant species, with most from rice, *Arabidopsis thaliana*, tomato, *Physcomitrella patens* (moss), etc. The *Arabidopsis* genome contains 272 P450 genes, with 246 of them being predicted full-length genes and 26 pseudo-genes [Nelson et al. 2004]. Rice has been identified with 356 P450 genes and 99 related pseudo-genes [Nelson et al. 2004] as against 57 in humans.

Plants adapt to various environmental conditions, ward off predators, and protect themselves from bacterial and fungal pathogens. They circumvent these challenges by synthesizing a myriad array of small compounds using P450s to catalyze oxidation steps in secondary metabolism that are simple hydroxylations or epoxidations as well as more complex dealkylations, isomerizations, aryl migrations, decarboxylations, dehydrations, ring formations, ring extensions, and carbon-carbon bond cleavages [Schuler 1996; Danielson 2002; Werck-Reichhart et al. 2002; Nielsen and Moller 2005]. Plant P450s also participate in several biochemical pathways to produce primary and secondary metabolites such as lipids, alkaloids, phenylpropanoids, terpenoids, cyanogenic glycosides, glucosinates, and hormones [Mizutani and Ohta 2010]. Some metabolites are important sources of antioxidants, phytoestrogens, and many plant defense compounds. The usefulness of these secondary metabolites and the ability of P450s to catalyze such diverse chemistries has generated much interest in plant metabolic engineering in the fields of crop improvement and phytoremediation in addition to pharmacology [Harvey et al. 2002; Morant et al. 2003].

Plant CYPs are also capable of metabolizing herbicides and environmental pollutants and this makes them attractive targets for engineering herbicide and insect tolerance in crops [Werck-Reichhart et al. 2002]. Most herbicide-detoxifying CYPs are expressed only at low levels in plants while molecular information regarding the metabolism of herbicides by plant CYPs is limited [Kumar et al. 2012]. Therefore, characterizations of plant CYP enzymes may help increase their expression levels followed by creation of herbicide-resistant plants useful for phytoremediation.

The first two plant CYPs identified are CYP76B1 and CYP71A10 from *Helianthus tuberosus* and soya bean respectively. These CYP enzymes actively metabolize herbicides, particularly phenylureas [Batard et al. 1998; Siminszky et al. 1999]. Furthermore, CYP81B2 and CYP71A11 isolated from tobacco metabolize chlortoluron [Yamada et al. 2000]. Although the CYP enzymes play some role in herbicide metabolism, these plants do not contain a complete and efficient catabolic pathway to detoxify herbicides and other toxic compounds [Morant et al. 2003; Komives and Gullner 2005; Batard et al. 1998; Siminszky et al. 1999; Yamada et al. 2000; Kumar et al. 2012]. Therefore, bacterial and mammalian CYP enzymes have been introduced in various plants for effective phytoremediation of these chemicals [Abhilash et al. 2009; Van Aken 2009; Van Aken and Doty 2010].

CYP450-Mediated Herbicide Metabolism

The cytochrome P450s mediate hydroxylation and demethylation reactions in plants to detoxify a range of herbicides [Durst and Benvenist 1993] such as phenylureas and sulfonylureas [Fonne-Pfister and Kruez 1990], chloroacetanilides [Moreland et al. 1993], imidazolinones, and triazolopyrimidine sulfonanilides [Frear and Still 1968] in addition to bentazon, a buthiadiazole herbicide [McFadden et al. 1990]. Many of the herbicide-metabolizing activities are enhanced by the treatment of plant tissue with herbicide safeners. Differential herbicide-metabolizing P450 activities are believed to represent one of the mechanisms that enable certain crop species to be more tolerant of a particular herbicide than other crop or weedy species.

Glutathione S-Transferases

Glutathione S-transferases (GSTs) are dimeric, multifunctional enzymes found in plants involved in the cellular detoxification and excretion of many physiological and endogenous substances [Wilce and Parker 1994]. Their major function is detoxification of a wide variety of hydrophobic, electrophilic, and usually cytotoxic compounds by catalyzing their conjugation with the tripeptide glutathione (γ-glutamylcysteinyl-β-glycine: γ-Glu-Cys-Gly: GSH), found in the reduced form. Glutathione protects plant cells from free radicals produced through oxidation. It can only do this by remaining in its naturally reduced state so that it is readily available to neutralize free radicals by bonding with them. Only a small proportion of glutathione is present in its fully oxidized state (glutathione disulfide: GSSG) [Dixon et al. 1998a, 1998b].

GSH is a tripeptide with a gamma (γ) peptide linkage between the amino group of cysteine (which is attached by normal peptide linkage to a glycine) and the carboxyl group of the glutamateside-chain. Concentrations of GSH vary during plant development, but it is found in relatively high concentrations in most plant tissues. The phloem-mobile glutathione is degraded by carboxypeptidases and transpeptidases in the cytoplasm and vacuoles. Generally, GSH synthesis is limited by availability of cysteine and hence by the concentration of sulfate ions.

GSH conjugation (R-X + GSH $\rightarrow$ R-SG + HX) is an important pathway in the metabolism of several herbicides including those belonging to triazine, chloroacetanilide, imidazolinone, diphenyl ether, aryloxyphenoxypropionate, and sulfonylurea families. GSTs catalyze the nucleophilic attack by the sulfur atom of glutathione on the electrophilic group of a variety of hydrophobic substrates, including herbicides. The glutathione anion GS⁻serves as a nucleophile, while chlorine, *p*-nitrophenol, or an alkyl-sulfoxide is the group involved in conjugation. GSTs are known to be induced by herbicides and safeners

[Schroder et al. 2008; Brazier-Hicks et al. 2008]. The resulting *S*-glutathionylated products are non-toxic.

As with cytochrome P450s, a number of GSTs seem to be involved in herbicide metabolism in plants. A rate-limiting step in the biosynthesis of glutathione is catalyzed by γ-glutamylcysteine synthetase (γ-ECS) or glutathione synthetase [Noctor and Foyer 1998]. Overexpression of GST genes enhances the potential for phytoremediation of herbicides.

In plants, GSTs are present at every stage of development from early embryogenesis to senescence and in every tissue. They presumably function to protect the cell from oxidative damage by quenching reactive molecules with the addition of GSH [McGonigle et al. 2000]. Molecules conjugated with GSH are efficiently imported into vacuoles via ATP-binding cassette transporters [Martinoia et al. 1993; Rea 1999]. The import of GSH-conjugated compounds into the vacuole acts to limit the effects of GST end-product inhibition and to further protect plant cells from danger by sequestration of those compounds whose conjugation with GSH does not cause detoxification [Rea et al. 1998].

In higher plants, reduced GSH and GSTs play crucial roles in the degradation of several pollutants. The level of GSH in plants is important in defense reactions against biotic and abiotic stresses [May et al. 1998]. The efficiency at which cells can raise the cytoplasmic GSH pool after depletion may influence their degree of stress tolerance or susceptibility to herbicides [May et al. 1998; DeRidder et al. 2002]. GSTs are abundant in plants, often exceeding 1 mM concentration in the leaf cell cytoplasm, where it functions as a scavenger of free radicals, protecting photosynthetic cells from oxidative damage.

During the conjugation process with TNT, sugars, amino acids, and organic acids available for substitution at the amino group reactive sites of reduced TNT metabolites. Plants attach glucose (by glucosyl transferases) or malonate (by malonyltransferases) to amino groups, thus forming the TNT conjugates [Coleman et al. 1997]. The enzymes responsible for conjugation with glucose are the glycosyl-transferases (GTs) which, like GSTs, can conjugate herbicides directly, by glycosylation of specific functional groups of lipophilic molecules, such as -OH, -COOH, -NH$_2$ and -SH [Yuan et al. 2007].

Plant GSTs

Plant GSTs are encoded by a large gene family with around 50 members in *Arabidopsis* and rice [Soranzo et al. 2004; Edwards and Dixon 2005], highlighting the importance of GS conjugate formation for the metabolism of endogenous compounds and the detoxification of noxious compounds such as herbicides. GS conjugates are predominantly generated in the cytosol, with minor activities in the nucleus, chloroplast, and mitochondria [Dixon et al. 2002; Dixon and Edwards 2009].

The soluble GSTs form a diverse super family of enzymes characterized by a shared overall structure and a well-defined glutathione binding domain. They can be divided into six families based on sequence similarity, catalytic function, and expression. These include *phi* (F), *tau* (U), *zeta* (Z), *theta* (T), *lambda* (L), and dehydroascorbate reductases [Dixon et al. 2002; Moons 2005; Jiang et al. 2010].

The plant-specific *phi* and *tau* GSTs are the large gene families primarily responsible for herbicide detoxification, showing class specificity in substrate preference. The much smaller *lambda* class has members that are strongly induced by xenobiotic treatment [Dixon et al. 2002], but their likely redox-related role does not extend to the detoxification of synthetic compounds. Other classes including the *theta* and dehydroascorbate reductases have no known ability to conjugate or detoxify synthetic compounds. However, GSTs belonging to *zeta* class have been characterized as glutathione-dependent *cis-trans* isomerases involved in tyrosine detoxification; they can also catalyze the glutathione-dependent dechlorination of dichloroacetate.

The *tau* and *phi* class GSTs catalyze reactions that typically result in the formation of a glutathione-conjugated substrate. *GST* genes of *tau* class are plant-specific and have been well characterized for their role in detoxification of herbicides in plants [McGonigle et al. 2000; Dixon et al. 2002; Wagner et al. 2002; Killili et al. 2004]. The *tau*-GSTs generally have narrower substrate specificity than other classes of GSTs. *Phi* enzymes (GSTFs) are highly active towards chloroacetanilide and thiocarbamate herbicides, whereas the *tau* enzymes (GSTUs) are efficient in detoxifying diphenylethers and aryloxyphenoxy-propionates [Jepson et al. 1994; Thom et al. 2002]. In maize, GSTFs are the major class of expressed GST [Jepson et al. 1994], while in soya bean GSTUs predominate, with the difference being the differential detoxification of different classes of herbicides in the two crops [McGonigle et al. 2000]. Of the nine GST genes present in tobacco, the more prominent ones include NbGSTU1, NbGSTU2, NbGSTU3 (*tau* class), and NbGSTF1 (*phi* class).

Of the *tau*-GSTs present in maize, *Zm*GSTU1 and *Zm*GSTU3 enzymes constitutively expressed, although considerably less abundant than the *Zm*GSTFs [Dixon et al. 1998a,b; Dixon et al. 1999]. Two other GSTs, *Zm*GSTU2 and *Zm*GSTU4 accumulate following treatment of herbicide safeners.

GST-mediated detoxification is dependent on the availability of the co-substrate glutathione, with higher levels resulting in both increased enzymatic and spontaneous conjugation. Glutathione depletion can become a problem when detoxifying large quantities of xenobiotics, particularly when such xenobiotics are highly relative and/or efficiently conjugated [Brazier-Hicks et al. 2008]. When considering using GSTs for phytoremediation applications, it would be therefore very desirable to ensure these enzymes were supplied with an adequate supply of glutathione. In addition to providing enough glutathione for these enzymes, it is also necessary to provide the optimal thiol substrate. Many major crops such as soya bean, wheat, and rice partially or completely replace glutathione with closely related thiols, and that can give rise to surprising alterations in pesticide detoxification. Soya bean uses homoglutathione (hGSH: γ-glutamyl-cysteinyl-β-alanine) in place of glutathione, with the respective GSTs showing a marked preference for this alternative thiol in detoxifying diphenylether herbicides. Their detoxification in transgenic tobacco could be engineered when the plants were transformed with the soya bean GSTs along with a homoglutathione synthetase to provide the preferred thiol substrate.

In leguminous plants, GSH is partially or completely substituted by homoglutathione. Because of structural similarities with GHS, hGSH is also conjugated with a variety of substrates in the presence of GSTs, and, therefore, it is one of the components of detoxification [Sugiyama and Sekiya 2005].

Nitroreductases

Nitroreductases have raised enormous interest due to their role in mediating in nitroaromatic toxicity and their potential use in phytoremediation. They belong to a family of evolutionarily related proteins involved in the reduction of nitrogen-containing compounds, including those with the nitro functional group. Members of this family utilize flavin mononucleotide (FMN) as a co-factor. These enzymes catalyze the reduction of nitro-substituted compounds using FMN or FAD as prosthetic groups and NADH or NADPH as reducing agents [Bryant et al. 1981; Bryant and DeLuca 1991]. They break down complex nitrogen-containing molecules to provide plants with a useable form of nitrogen.

In the nitro group, the bond between the oxygen and nitrogen atoms is polar. This is because oxygen is more electronegative than nitrogen, attracting nitrogen's electrons to form partially negative and positive poles. As the positive pole tends to attract electrons, it has a great tendency to undergo reduction [Spain 1995]. The reduction of nitro groups can be catalyzed by nitroreductase enzymes that can perform by transferring one electron or two [Bryant et al. 1981; Bryant and DeLuca 1991]. Thus, the nitroreductases have been divided into two categories based on their ability to reduce nitro groups in the presence of oxygen by one-electron or two-electron transfers.

In the presence or absence of oxygen, the type I (oxygen-insensitive) nitroreductases catalyze the sequential transfer of two electrons from NAD(P)H to the nitro groups of nitro-substituted compounds, resulting in nitroso and hydroxylamine intermediates and finally primary amines. The type II (oxygen-sensitive) nitroreductases catalyze one-electron reductions of the nitro group in the presence of oxygen, producing a nitro anion radical that subsequently reacts with molecular oxygen, forming a superoxide radical and regenerating the original nitroaromatic compound. This 'futile redox cycle' can cause oxidative stress by producing large amounts of superoxides [de Oliveira et al. 2010].

Type I nitroreductases are involved in the reduction of a variety of nitrocompounds, including nitrofurans, nitrobenzene, nitrophenols, nitrobenzoate, nitrotoluenes (TNT), and nitroimidazoles [Race et al. 2005; Olekhnovich et al. 2009]. Based on their similarity with *E. coli* nitroreductases, type I nitroreductases are grouped into two: NfsA (group A) and NfsB (group B) [Bryant et al. 1981].

NfsA, encoded by *nfsA* gene, is the major 'oxygen-insensitive' nitroreductase which uses NADPH as an electron source, while the *nfsB* gene encoded NfsB, the minor 'oxygen-insensitive' nitroreductase, can use either NADH or NADPH as a source of reducing equivalents. Most of the nitroreductases share similar biochemical properties. They are usually homodimeric proteins of approximately 30 kDa and have broad substrate specificity. They catalyze the reduction of nitrocompounds using a 'two-electron' transfer mechanism [Race et al. 2005; Bryant et al. 1981; Race et al. 2007]. Of the two nitroreductases present in *Pseudomonas putida*, PnrA belongs to group A and PnrB to group B [Caballero et al. 2005].

Enterobacter cloacae strain 96-3 nitroreductase (NfsI) produces retro-nitroreductase which catalyzes the pyridine nucleotide-dependent four-electron reduction of a variety of nitroaromatic compounds, including the explosives TNT, RDX, tetryl, and pentryl [Koder et al. 2001]. This enzyme displays a catalytic efficiency for nitroreduction at least 10-fold higher than that of several highly homologous bacterial nitroreductases. Retro-nitroreductase is 96.7 percent identical to *E. cloacae* strain 96-3 nitroreductase [Koder et al. 2001].

Together with flavonitroreductases, a large variety of nitroreductases, including NfsA and NfsB of *E. coli*, PnrA of *P. putida*, and NfsI from *E. cloacae*, catalyzes the sequential reduction of TNT nitro groups to hydroxylamino and amino derivatives, potentially followed by the release of ammonium through a Bamberger-like rearrangement.

Phytoremediation by Transgenic Engineering

The efficiency of transformation and degradation of pollutants by plants can be enhanced by the introduction of bacterial genes encoding target proteins that biodegrade pollutants [Perumbakkam et al. 2006]. Considerable progress has been made over the past two decades in inserting certain new traits in plants and developing newer crop varieties by way of transgenic engineering which is now being extended to phytoremediation by inserting genes encoding factors of phase I and II of detoxification.

To date, only a few plants have been genetically engineered to degrade herbicides for the purpose of cleaning up the environment. In fact, transgenic plants for phytoremediation were first developed for cleaning up soil sites contaminated with heavy metals. Tobacco plants were found to be expressing a yeast *metallothionen* gene, *CUP1*, for greater tolerance of cadmium [Misra and Gedama 1989], while *Arabidopsis thaliana* expressed a mercuric ion reductase gene, *merA*, which produced MerApe9 enzyme for tolerance to toxic levels of mercury [Rugh et al. 1996]. The first attempts to develop transgenic plants for phytoremediation of organic pollutants targeted explosives and halogenated organic contaminants in tobacco [French et al. 1999; Doty et al. 2000]. Since then, numerous efforts have been made to identify transgenes that cause increased tolerance to xenobiotics including herbicides and develop plants that contain transgenes responsible for their detoxification and ultimate remediation from soil and environment. Transgenic plants were also found successful in detoxifying and removal of mercury [Rugh et al. 1998], lead [Song et al. 2003], and selenium [LeDuc et al. 2006] in addition to other metals [Kramer and Chardonnens 2001].

The success of using transgenic engineering in phytoremediation depends primarily on the type of plants used. Candidate plants for genetic engineering for phytoremediation should have large biomass and highly branched root system with large surface area for efficient uptake of large amounts of contaminants in the field. Besides, they should be easily cultivated and maintained [Kawahigashi 2009]. The plants should also have an inherent capability for phytoremediation. Some of the high biomass accumulators identified and for which regeneration protocols have been developed include tobacco, rice, Indian mustard, potato, sunflower, tomato, poplar, etc.

Genetic engineering offers a major means of improving the potential ability of these and other plants to remediate environmental pollutants. For example, the nitroaromatic explosives like TNT and RDX are phytotoxic and cannot be effectively treated by using conventional phytoremediation. By introducing bacterial genes involved in the metabolism of TNT and RDX, the tolerance to and uptake of these pollutants by transgenic plants were considerably improved [Hannink et al. 2001; Rylott et al. 2006].

CYP P450-Mediated Phytoremediation

Although the association of P450s with herbicide metabolism has been investigated for several decades, reports on the successful cloning and expression of P450 cDNAs encoding herbicide-metabolizing isozymes are relatively recent. Many attempts have been made to successfully insert genes encoding mammalian cytochromes P450 into the plant genomes. These herbicide-tolerant transgenic plants can be commercially utilized for phytoremediation of herbicide residues in the environment.

The process of inserting the genes into plants is relatively easy to perform using *Agrobacterium tumefaciens* as a vector, or the biolistic method [Eapen et al. 2007; Doty 2008]. The first attempts were connected with obtaining herbicide-resistant plants, exhibiting increased tolerance mainly to atrazine and simazine and remove them. Similar attempts have also been made to obtain plants which would remove some volatile chlorinated derivatives of hydrocarbons from the soil and groundwater.

Engineering of various combinations of vertebrate CYP1A1, CYP1A2, CYP2B6, CYP2C9, CYP2C19 enzymes into transgenic plants has been shown to confer tolerance to phenylureas (chlortoluron, methabenzthiazuron), triazines (atrazine), chloroacetanilides (acetochlor, metolachlor, etc.), oxyacetamides (mefenacet), thiocarbamates (pyributicarb), phosphoamidates (amiprofos-methyl), 2,6-dinotroanilines (trifluralin, pendimethalin), and pyridazinones (norflurazon) in various transgenic crops including tobacco, rice, and potato [Shiota et al. 2000; Yamada et al. 2002; Kawahigashi et al. 2005b]. Two other plant CYP450s, one from *Helianthus tuberoses* (Jerusalem artichoke), CYP76B1, and the other from soya bean, CYP71A10, have been shown to confer resistance to phenyl urea herbicides, linuron and chlortoluron [Siminszky et al. 1999; Didierjean et al. 2002].

However, despite this and other advantages, the progress and application of this technology to tackle these widespread problems is being hampered by regulatory issues surrounding the use of bacterial and mammalian P450s in transgenic plants. It then becomes essential to identify additional, naturally-occurring plant P450s that can potentially be used for bioremediation and/or herbicide detoxifications.

Tobacco

Plants expressing CYP76B1 is a potential tool for phytoremediation of contaminated sites [Didierjean et al. 2002] as well as for engineering herbicide resistance or increasing the bioremediation potential of some plants. This enzyme, derived from yeast (*Saccharomyces cerevisiae*), metabolizes a broad range of exogenous molecules with a catalytic efficiency often comparable with that observed for P450-dependent metabolism of endogenous compounds [Robineau et al. 1998]. CYP76B1 was particularly effective at one or more *N*-demethylation reactions with phenylurea and sulfonylurea herbicides.

Transformation with CYP76B1 conferred on tobacco (and also *Arabidopsis*), a 20-fold increase in tolerance to linuron, a herbicide detoxified by a single dealkylation step, and a 10-fold increase in tolerance to isoproturon and chlortoluron, which needed successive catalytic steps for detoxification [Didierjean et al. 2002]. Plants with two constructs for expression of translational fusions of CYP76B1 with 450 reductase had lower herbicide tolerance than CYP76B1 alone, which was apparently a consequence of reduced stability of the fusion proteins [Didierjean et al. 2002]. In all cases, increased herbicide tolerance resulted from more extensive metabolism. Besides increased herbicide tolerance, expression of CYP76B1 has no other visible phenotype in the transgenic plants [Didierjean et al. 2002]. These results suggested that CYP76B1 can function as a selectable marker for plant transformation, allowing efficient selection *in vitro* and in soil-grown plants.

Ohkawa et al. [1997] examined the expression of the fused gene between rat CYP1A1 cDNA and yeast reductase gene under the control of CaMV 35S promoter and *nos* (nopaline synthase) terminator in transgenic tobacco plants. The fused enzyme was mainly located on the microsomes, which showed about 10-times higher monooxygenase activities towards 7-ethoxycoumarin and benzo-pyrene than those of control plants. The transgenic tobacco expressing the fused enzyme exhibited tolerance to chlortoluron. The transgenic plants metabolized the herbicide more rapidly than control plants through ring-methyl hydroxylation and *N*-demethylation, although they metabolized the herbicide through *N*-demethylation [Shiota et al. 1996]. Since the *N*-demethylated chlortoluron was still phytotoxic, the activity in ring methyl hydroxylation of the transgenic plants appeared to be attributable to tolerance of the herbicide.

Bode et al. [2004] introduced two species of human P450, CYP1A1 and CYP1A2, in tobacco cells via *Agrobacterium*-mediated transformation method. The transgenic plant cell cultures were selected by a combination of kanamycin-resistance, 7-ethoxycoumarin *O*-de-ethylase activity, and PCR and Western blot analyses. Using ^{14}C-labelled atrazine for metabolism studies, they found de-ethylatrazine and de-isopropylatrazine in the control culture as well as in the transgenic culture, while the non-phytotoxic metabolite de-ethyl-de-isopropylatrazine was found only in the transgenic cell cultures. Both foreign enzymes CYP1A1 and CYP1A2 catalyzed *N*-dealkylation of atrazine, with CYP1A2 exhibiting a higher conversion rate than CYP1A1. CYP1A2 catalyzed predominantly *N*-de-ethylation followed by de-isopropylation. The extent of metabolism was considerably higher than in non-transformed cell cultures. The transgenic cell cultures can therefore be suitable tools for the production of large quantities of primary oxidized pesticide metabolites.

Overexpression of human CYP2E1 in modified tobacco seedlings significantly increased the capacity of the plants in the metabolism of volatile hydrocarbons such as trichloroethylene, benzene, carbon tetrachloride, etc. [Doty et al. 2000]. The modified plants were able to metabolize xenobiotics over 640-times faster than seedlings with an unaltered genome.

The human P4502B6 and yeast reductase fused enzyme was also expressed in transgenic tobacco plants [Gorinova et al. 2005]. Molecular analysis of transgenic plants revealed that the fused enzyme cDNA seemed to be integrated into the tobacco genome and transferred into its mRNA [Gorinova et al. 2005]. The microsomal fraction of the transgenic plants where the fused enzyme protein was localized, exhibited 4 to 6 times higher monooxygenase activity to towards 7-ethoxycoumarin than that of the control plants. These results suggested that the P4502B6/reductase fused enzyme was functionally expressed in the transgenic tobacco plants [Shiota 1996].

The transgenic plants, expressing P4501A1 and P4502B6 were expected to metabolize a number of foreign chemicals. There seem to be apparent differences in toxicity and the metabolism of xenobiotics between human and experimental animals. The present strategy to express human P450 monooxygenases in transgenic plants may be useful for production of human gene-inserted transgenic crops having the ability to detoxify agrochemicals and environmental contaminants.

Potato

Modification of the potato genome is characterized by overexpression of human CYP1A1. Inui et al. [1999] generated transgenic potato plants expressing human CYP1A1 and human CYP1A1/yeast NADPH-cytochrome P450 reductase (YR) fused enzyme by using microtuber discs of cultivar 'MayQueen' via the *Agrobacterium*-mediated transformation. The P450-dependent monooxygenase activity of the transgenic plants S1384 (S = single expression of human CYP1A1), S1386, and F1515 (F = expression of CYP1A1/YR fused enzyme) was 3.5, 4.2, and 3.8 times greater in 7-ethoxycoumarin *O*-deethylation *in vitro* and 6.4, 5.8, and 5.3 times higher in chlortoluron metabolism *in vivo* than non-transgenic plants, respectively. Although four metabolites were found during degradation of atrazine, the proportion of the non-phytotoxic de-isopropylated de-ethylated metabolite (6-chloro-2,4-diamino-1, 3, 5-triazine) was five time higher in the transgenic P1384 plants than in non-transgenic potato plants. The amount of atrazine in the transgenic plants was remarkably lower than in non-transgenic plants. These atrazine-resistant transgenic potato plants expressing human CYP1A1 showed remarkable cross-resistance towards chlortoluron and pyriminobac methyl.

Later, Inui and Ohkawa [2005] accomplished co-expression of three human P450 species, CYP1A1, CYP2B6, and CYP2C9 in transgenic potato plants. These transgenic potato plants exhibited remarkable cross-resistance towards photosynthesis-inhibiting atrazine, chlortoluron, metabenathiazuron, the lipid biosynthesis-inhibiting acetochlor and metolachlor and the carotenoid biosynthesis-inhibiting norflurazon, probably by co-operative herbicide metabolism of the three P450 species. These results suggested that transgenic potato plants could serve as a useful phytoremediation system to clean up these herbicide contaminants.

Rice

Rice is a plant candidate for metabolizing herbicides and reducing herbicide contaminants in paddy fields and streams which can remove contaminants from both soil and stream-water in significant quantities. In many cases, overexpression of endogenous plant genes or transgenic expression of bacterial or animal genes is required to enhance the phytoremediation properties of plants [Lincare et al. 2003]. Insertion of rat's *CYP1A1* gene as well as yeast's gene *NCP1* encoding NADPH-dependent cytochrome P450 oxidoreductase into the genome of rice allowed the transgenic plants to metabolize chlortoluron. Plants thus transformed became resistant to this herbicide [Shiota et al. 1994; Shiota 1996].

Mammalian cytochrome P450 genes have been incorporated in transgenic rice plants to detoxify herbicides [Inui and Ohkawa 2005]. Rice transformed with genes encoding human CYP1A1, CYP2B6, and CYP2C19 are more tolerant of various herbicides than non-transgenic rice plants, due to increased metabolism by the introduced P450 enzymes [Kawahigashi et al. 2005a, 2007, 2008; James et al. 2008]. Transgenic rice plants carrying CYP1A1 show tolerance towards atrazine, chlortoluron, diuron, quizalofop-ethyl, and other herbicides, and they reduce the levels of atrazine and simazine in hydroponic solutions [Kawahigashi et al. 2007]. Transgenic rice carrying CYP2B6 germinates well in a medium containing chloroacetanilide herbicides like alachlor, metolachlor, and thenylchlor, but not the non-transgenic rice plants [Kawahigashi et al. 2005a,b]. Large-scale greenhouse experiments confirmed the ability of CYP2B6 rice plants to remove metolachlor and these plants appeared to reduce the residual quantities of metolachlor in water and soil [Kawahigashi et al. 2005b].

Kawahigashi et al. [2006] found that transgenic rice plants (cultivar 'Nipponbare' introduced with the plasmid pIKBACH) expressing all the three human CYP P450s (CYP1A1, CYP2B6, and CYP2C19) metabolized the substrate herbicides more rapidly than the non-transgenic rice plants and they have shown phytoremediation activity against atrazine and metolachlor. The rice plants were grown in soil contaminated with 4.2 mM atrazine and 2.9 mM metolachlor. Both the transgenic and control plants showed healthy growth over a period of one month. The residual atrazine in soil with pIKBACH rice plants was 70.1 percent of that in soil without any plants, while that with non-transgenic plants was 93.2 percent. Similarly, residual metolachlor concentration in soil with pIKBACH plants also came down to 73.5 percent in soil from 100 percent in soil with non-transgenic rice plants. These results showed that pIKBACH rice plants can get rid of substantial quantities of atrazine and metolachlor from contaminated paddy soil. The transgenic rice plants show cross-tolerance to eight herbicides belonging to five different groups [Kawahigashi et al.

2005a]. These results suggested that transgenic rice plants expressing P450 might be a useful and valuable method to reduce concentrations of various organic chemicals in the soil and environment.

Poplar

Poplar and aspen (*Populus* sp.) are widely distributed, fast-growing, high biomass plants ideal for phytoremediation applications [Schnoor 2000]. Hybrid poplar (*Populus tremula* x *Populus alba*) trees are characterized by extensive roots, and they are able to cover much larger space and penetrate deeper layers of the soil than small shrubs. Doty et al. [2007] found transgenic poplar plants inserted with rat-derived CYP2E1 enhancing the phytodegradation of several toxic organic compounds. Two of the CYP2E1 transgenic lines were able to metabolize the industrial solvent trichloroethylene (TCE), also called chloroethene, more than 100-fold faster than non-transgenic plants. The transgenic cuttings grew normally and did not display any adverse reaction to the TCE or its metabolites. The transgenic hybrid poplar plants exhibited increased removal rates of TCE, vinyl chloride, carbon tetrachloride, benzene, and chloroform from hydroponic solution. When these transgenic plants were exposed to gaseous trichloroethylene, chloroform, and benzene, they also demonstrated superior removal of the pollutants from air. Transgenic poplars trees removed 79 percent TCE, 49 percent vinyl chloride, and around 40 percent benzene as against 0, 29 percent, and 13 percent by non-transgenic plants respectively.

Arabidopsis

The strain 11Y of *Rhodococcus rhodochrous* strain 11Y has the ability to degrade RDX [Seth-Smith et al. 2002]. The RDX-degrading ability of this bacterium is encoded by *XplA* gene, and it has been shown to be the result of a fused flavodoxin-cytochrome P450-like enzyme [Rylott et al. 2006; Seth-Smith et al. 2002], which catalyzes the aerobic degradation of RDX by de-nitration, leading to ring cleavage and the release of small aliphatic metabolites including 4-nitro-2,4-diazabutanol (NDAB), nitrite, and formaldehyde [Rylott et al. 2006]. Its anaerobic degradation produces methylenedinitramine instead of NDAB [Jackson et al. 2007].

The plants of *A. thaliana* were engineered to express *XplA* gene. The donor strain 11Y, *Rhodococcus rhodochrous* strain, isolated from RDX-contaminated soil, was capable of achieving a 30 percent-mineralization of RDX in pure culture [Seth-Smith et al. 2002]. Liquid cultures of the transgenic *A. thaliana* plants expressing XplA removed 32–100 percent of RDX at 180 mM concentration, while wild-type plants removed less than 10 percent. Transgenic lines, grown on RDX-contaminated soil, showed no phytotoxic effect even up to 2000 mg kg^{-1}, while growth of wild-type plants was significantly inhibited at even at 250 mg kg^{-1}. The transgenic *Arabidopsis* plants growing in RDX-contaminated soil increased their biomass suggesting that these XplA-expressing plants utilize the nitrite released from the degradation of the explosive as a nitrogen source for growth [Rylott et al. 2006]. Conversely, RDX analysis in exposed plants showed significantly higher concentrations in wild-type plants, suggesting that transgenic lines were capable of more efficient transformation of RDX.

Plants expressing *XplA* gene along with *XplB*, a partnering reductase gene for *XplA* from *R. rhodochrous* strain 11Y, exhibited an additional 30-fold increase in the rate of RDX removal rate [Jackson et al. 2007]. Rylott et al. [2011a,b] transformed *A. thaliana* with the

RDX-degrading *xplA*, and the associated reductase gene *xplB* from *R. rhodochrous* strain 11Y, in combination with the TNT-detoxifying nitroreductase (NR) gene, *nfsI*, derived from *Enterobacter cloacae*. The transgenic XplA-XplB-NR plants removed significantly more RDX from soil leachate than either wild-type or NR-only plants. In addition, XplA-NR plants grown in soil contaminated with both RDX and TNT (two diverse organic compounds) had higher biomasses than plants expressing NR alone, indicating that RDX can be utilized as a nitrogen source by XplA-expressing plants. They also found that the transgenic plants remediated RDX while tolerating the toxicity of TNT. Besides, the levels of RDX in the shoots of transgenic plants expressing XplA were 34- to 94-fold lower than in wild type, non-transgenic plants.

One limitation in using the annual *Arabidopsis* for phytoremediation is its relatively smaller root system penetrating only the top few centimeters of the soil. Instead, other plant species, including weeds, which have extensive root systems and deeper penetration in the soil, would be more suitable for remediating the soil and environment of various contaminants including RDX, TNT, and xenobiotics.

GST-Mediated Phytoremediation

Besides cytochromes, glutathione s-transferase gene factors of phase II cellular detoxification are used in the transgenic modification. As GSTs have a clear role in detoxifying halogenated xenobiotics, they are excellent means for engineering enhanced tolerance to pesticides and pollutants. Several studies have over-expressed GSTs in a range of plants and shown both enhanced tolerance to, and increased detoxification of, various xenobiotics. Some of the recent notable examples include the demonstration that over-expression of maize GSTs has conferred tolerance to chloroacetanilide and thiocarbamate herbicides in tobacco [Karavangeli et al. 2005] and wheat [Milligan et al. 2001]. The detoxifying activities of native GSTs can be enhanced by forced evolution approaches, with maize *tau* class GSTs engineered with an enhanced ability to detoxify diphenylethers, conferring resistance to these herbicides in *Arabidopsis* plants [Dixon et al. 2003].

High hopes are also attached to the use of plants transformed with GSTs to remove residues of explosives from soil contaminated by intense military action [Richman 1996]. Such substances as nitroglycerin, TNT, aminodinitrotoluene (DNT) or RDX are highly toxic and mutagenic compounds that, due to violent reactions during explosion, often undergo incomplete combustion and are shed along with the shock wave [Bruns-Nogel et al. 1996]. Given that the substances are phytotoxic, phytoremediation of contaminated sites using conventional techniques is very difficult to achieve.

Tobacco

The transgenic tobacco plants expressing soya bean hGSHS (homoglutathione synthase: γ-glutamyl-cysteinyl-β-alanine synthase) in the cytosol have higher hGSHS activity than glutathione synthetase activity and contain hGSH at a detectable level [Sugiyama et al. 2004]. These transgenic plants expressing hGSHS in the cytosol have enhanced tolerance to acifluorfen but not to fomesafen or metolachlor, and those expressing hGSHS in the chloroplasts do not have enhanced tolerance to any of the herbicides [Sugiyama and Sekiya 2005]. The transgenic plants which expressed soya bean hGSH in the cytosol have greater tolerance to acifluorfen than the hGSH-absent wild-type plants. However, this enhanced tolerance to fomesafen was not observed in these transgenic plants and this is possibly because this herbicide can utilize GSH in a GST-catalyzed reaction [Sugiyama and Sekiya 2005].

Karavangeli et al. [2005] established that the expression of a *gstI* gene from maize, encoding for an enzyme lacking peroxidase activity, provides significant protection against the chloroacetanilide herbicide alachlor. They used the *gstI-6His* gene, which encodes for 6His-tagged GST I, for transgenic transformation of leaf discs of young tobacco (cultivar 'Basmas Xanthi') plants. This gene was excised from the plasmid pGSTI and sub-cloned in the EcoRI restriction site of the pART27 binary vector, which carried the CaMV 35S promoter and the 3′ UTR (untranslated region) of *ocs* (octopine synthase) gene as a terminator. The entire CaMV35S-*gstI-ocs* cassette was digested by NotI restriction enzyme, followed by its insertion in the T-DNA region of the vector. The *npt*II gene was under the control of the *nos* gene promoter. The transgenic plants thus developed showed substantially higher tolerance to alachlor compared to non-transgenic plants in terms of root, leaves, and vigorous development. These transgenic plants are potentially useful as biotechnological tools in the development of phytoremediation systems for the degradation of herbicide pollutants in agricultural fields.

In the U.S., Wang et al. [2005] transferred into tobacco (cultivar 'Samsun') plants the modified *atzA* gene, *p-atzA*, by directionally cloning into the *Xba*I and *Bam*HI restriction sites of the binary vector pILTAB 357 behind the constitutive CsMV (cassava vein mosaic virus) promoter to yield pPW1. The vector also contained the *npt*II gene controlled by the *nos* promoter for the selection of transgenic plants. Plasmid pPW1 was transformed into *E. coli* DH5α by conjugating it into *A. tumefaciens* strains LBA4404 and C58C1 (pMP90) by triparental mating. In this *Agrobacterium*-mediated transformation, leaf pieces of tobacco (cv. 'Samsun') were co-cultivated with *A. tumefaciens* LBA4404 (pPW1). The transgenic tobacco plants were able to grow in the presence of 38 times more atrazine than the wild-type parent plants.

In China, Wang et al. [2010] studied the potential use of transgenic tobacco plants for phyto-remediation of atrazine residues in the soil by transferring *atzA* genes from *Pseudomonas* sp. strain ADP and *Arthrobacter* strain AD1 into tobacco. Transformation was done using pET21b-NK plasmid for *atzA*-AD1 from *Arthrobacter* strain AD1 and pMD4 plasmid for *atzA*-ADP from *Pseudomonas* sp. strain ADP. These genes were amplified using pET21b-NK and pMD4 respectively and then directionally inserted into the *Bam*HI and *Sal*I restriction sites of pBin438 binary vector. The resulting plasmid, pBin438-*atzA*, contained a double 35S promoter-driven *atzA* expression cassette and a *nos*-driven hygromycin gene expression cassette. Tobacco seeds were grown on a culture medium for 30 days, and the young leaves were cut into pieces before co-cultivating them with *A. tumefaciens* LBA4404 (pBin438-atzA) in the dark for 3 days. The explants were grown in a selection medium consisting of kanamycin for regeneration and the antibiotic-resistant shoots were transferred for rooting.

Following transformation, *atzA*-ADP and *atzA*-AD1 genes were found to have been expressed by three lines (401, 402, 403) and four lines (701, 702, 703, 704) respectively [Wang et al. 2010]. The *atzA* gene was stably integrated into the tobacco genome and even expressed in the progenies. Barring line 401, no atrazine residue remained in the soil in which the T_2 transgenic lines were grown, while, the untransformed plants had 0.91 mg (81.3 percent) and 1.66 mg (74.1 percent) of the atrazine still remaining in the soil containing 1 and 2 mg kg^{-1} of atrazine, respectively, indicating that the transgenic lines could degrade atrazine effectively. The transgenic tobacco lines inserted with *atzA* gene could be useful for phytoremediation of atrazine-contaminated soil and water [Wang et al. 2010].

This suggests that transgenic tobacco holds a great promise in the phytoremediation of atrazine from soils as it is widely cultivated around the world, produces a large biomass and well-developed root system, and poses lower risk of transferring *atzA* gene to the wild relatives [Wang et al. 2010].

Simultaneous insertion of the human cytochrome *P450 2E1* gene and the *gstI* gene encoding the GST derived from the *Trichoderma virens* fungus into tobacco genome resulted in more efficient metabolism of chlorpyrifos (insecticide) and anthracene (solid polycyclic aromatic hydrocarbon, a component of coal tar) in the modified plants [Dixit et al. 2008]. Dixit et al. [2011] cloned a glutathione transferase gene, *TvGST*, from *Trichoderma virens*, a biocontrol fungus and introducing it into tobacco plants by *Agrobacterium*-mediated gene transfer. When transgenic plants expressing *TvGST* gene were exposed to different concentrations of cadmium, they were found to be more tolerant than wild type plants, with transgenic plants showing lower levels of lipid peroxidation. Levels of different antioxidant enzymes such as GST, superoxide dismutase, ascorbate peroxidase, guiacol peroxidase, and catalase were found enhanced in *TvGST* gene-expressed transgenic plants compared to control plants when exposed to cadmium [Dixit et al. 2011]. Cadmium accumulation in the plant biomass in transgenic plants was similar or lower than wild-type plants.

Indian Mustard

The rapidly growing high-biomass plant species mustard (*Brassica juncea*) cultivated in India is a good candidate for transgenic phytoremediation. Zhu et al. [1999a,b] found transgenic lines of Indian mustard overexpressing glutathione synthetase in the cytosol (cytGS) and γ-glutamylcysteine synthetase in the chloroplast (cpECS). Both transgenic cpECS and cytGS plants contained two-fold higher concentration of GSH and total non-protein thiols than wild-type plants.

Flocco et al. [2004] reported that the transgenic mustard plants overexpressing genes related to γ-glutamyl-cysteine synthetase and GSH biosynthesis showed increased tolerance to atrazine and a moderate increase in tolerance to metolachlor, phenanthrene (a polycyclic hydrocarbon), and 1-chloro-2,4-dinitrobenzene (CDNB). The mustard seedlings grew in relatively high concentrations of these organic compounds (50–200 mg L^{-1}). Thus, overexpression of enzymes involved in GSH biosynthesis may be a promising approach to create plants with enhanced capacity to tolerate—and possibly remediate—not only heavy metals, but also certain organics including herbicides.

Rice

In rice, GSTs conjugate the herbicide pertilachlor with glutathione, a reaction induced by the safener fenclorim [Scarponi et al. 2003]. Using RNAi, GST activity towards pertilachlor can be reduced by as much as 77 percent in transgenic calli [Deng et al. 2003].

Soya bean

Transgenic soya bean plants synthesizing hGSH in the cytosol were more tolerant to acifluorfen than wild-type plants [Sugiyama and Sekiya 2005]. However, transgenic plants synthesizing hGSH in the chloroplasts showed no enhanced tolerance to acifluorfen or fomesafen, possibly because the hGSH synthesized in the chloroplasts was not transported to the cytosol at a level to be effective for detoxifying acifluorfen [Sugiyama and Sekiya 2005].

Arabidopsis

Wang et al. [2005] modified the *atzA* gene, *p-atzA*, and transformed *A. thaliana* (ecotype 'Columbia') with *A. tumefaciens* strain C58C1 (pMP90) using the floral-dip vacuum infiltration method. The gene was cloned into the *Xba*I and *Bam*HI restriction sites of the binary vector pILTAB 357 behind the constitutive CsMV promoter to yield pPW1. The vector also contained the *npt*II gene controlled by the *nos* promoter for the selection of transgenic plants. Plasmid pPW1 was transformed into *E. coli* DH5α by conjugating it into *A. tumefaciens* strains LBA4404 and C58C1 (pMP90) by triparental mating. The resultant transgenic *Arabidopsis* tolerated 50 times more atrazine than the wild-type parent plants.

Poplar

Poplar (*Populus tremula* x *P. alba*) plants with the bacterial synthetase gene, *γ-ECS* (from *E. coli*), encoding γ-glutamyl-cysteine synthetase, an enzyme involved in the synthesis of glutathione (GSH), caused a significant increase in the production of GSH in plant cells which, in turn, increases the protection against oxidative stress caused by various harmful external influences [Noctor et al. 1996; Noctor 1998]. Glutathione contents are unaffected by high chloroplastic glutathione synthetase activity, with the effects being similar to those observed in poplars overexpressing these enzymes in the cytosol [Noctor et al. 1998]. Gulner et al. [2001] found enhanced tolerance to acetochlor and alachlor, by elevating GSH content through overexpression of γ-glutamyl-cysteine synthetase in poplar plants. They indicated that the elevated GSH content enabled transgenic poplar plants to conjugate with herbicides in the presence of GST at a higher rate. This increased GSH level was possibly due to the increased scavenging of reactive oxygen species in transgenic poplar plants [Sugiyama and Sekiya 2005].

Brenter et al. [2008] showed that expression of glutathione S-transferases in poplar (*Populus trichocarpa*) trees resulted in a significant increase of gene expression to two GST, peaking at levels of 25- and 10-fold, the expression level of non-exposed plants after 24 hours of each of the GST genes, respectively.

Nitroreductase-Mediated Phytoremediation

There are no known reports of nitroreductase-catalyzed degradation of herbicides in plants. However, nitroreductases, present aplenty in microorganisms, are able to degrade the explosives by reducing their nitro side groups.

TNT

In the case of TNT, the reduced product, aminonitro-toluene, can bind irreversibly to clay and organic matter in the soil. The transgenic introduction of bacterial catabolic genes into plants significantly improves their capability to take up and metabolize the toxic explosive TNT.

Enterobactor cloacae PB2 utilizes TNT as its primary source of nitrogen source for growth and metabolism because of the presence of two unique enzymes, pentaerythritol tetranitrate (PETN) reductase and nitroreductase [French et al. 1998]. Both these enzymes utilize NADPH as an electron donor source, and thereby can easily reduce TNT into less toxic compounds. PETN reductases are bacterial flavonitroreductases. They catalyze degradation of TNT by either sequential reduction of nitro groups or aromatic hydride addition, resulting in de-nitration and, eventually, its detoxification.

French et al. [1999] introduced the PCR-modified gene encoding PETN reductase, *onr*, into the tobacco genome with a plant consensus start sequence for better expression in the plant system. They found seeds from transgenic lines germinating and growing successfully in the presence of 1 mM GTN or 0.05 mM TNT, compared to their non-transformed wild types. The resultant transgenic seedlings grown in liquid medium containing 1 mM GTN exhibited faster and complete degradation (de-nitration) of GTN than non-transformed lines.

In another related study, the PCR-modified bacterial (*E. cloacae* NCIMB101011) gene, *nfs1*, encoded for nitroreductase enzyme with a consensus start sequence to facilitate translation of tobacco plants. This enzyme catalyzed the reduction of TNT to hydroxyaminodinitrotoluene, following the sequential reduction of aminodinitrotoluene derivatives. Transgenic plants expressing nitroreductase exhibited a significant increase in TNT uptake, tolerance, and subsequent detoxification compared to wild-type plants [Review by Basu et al. 2010].

Hanninck et al. [2001] found transgenic tobacco plants that constitutively expressed the nitroreductase NfsI from *E. cloacae* removing high amounts of TNT from the solution (100 percent of 0.25 mM after 72 h), as compared to negligible amounts by wild-type plants. Transgenic plants were also able to tolerate higher TNT concentrations, up to 0.5 mM, which was lethal for wild-type seedlings. The same transgenic system showed a reduction of TNT into 4-hydroxylamino-2,6-dinitrotoluene and conjugation to plant macromolecules to a greater extent in genetically modified plants than in wild-type plants [Hannink et al. 2007].

When the plants of *Arabidopsis thaliana* plants were transformed by the introduction of *nfsA* gene which encodes the *E. coli* nitroreductase, NfsA, they exhibited a 20-time higher nitroreductase activity and 7–8 times higher TNT uptake compared with wild-type plants [Kurumata et al. 2005]. Besides, the modified plants grew at 0.1 mM concentration TNT and showed in-tissue reduction of TNT into 4-amino-2,6-dinitrotoluene, which was not observed with wild-type organisms. These examples demonstrate that the introduction of bacterial catabolic genes into plants significantly improved their capability to take up and metabolize the toxic TNT.

The fast growing poplar and aspen (*Populus* sp.) plants, with high biomass production capability, are ideal for phytoremediation of several organic contaminants including explosives. Van Dillewijn et al. [2008] found the transgenic hybrid aspen (*P. tremula* x *P. tremuloides* var. Etropole) that expressed the bacterial nitroreductase gene *pnrA* from *P. putida* showing tolerance to, and absorbing greater amounts of, the toxic and recalcitrant TNT from contaminated waters and soil than wild-type plants. Transgenic aspen tolerated TNT up to 57 mg L^{-1} in hydroponic media and more than 1000 mg kg^{-1} soil, whereas the parental aspen could not endure it in hydroponic culture with more than 11 mg L^{-1} or soil with more than 500 mg kg^{-1}. Using U-ring ^{13}C-TNT, they showed a rapid adsorption of greater amounts of TNT on the root surface from contaminated water and soil, followed by a slow entrance into the plant, with most ^{13}C-carbon from the labeled TNT taken up by the plant (> 95 percent) remaining in the roots.

Introduction of exogenous genes involved in xenobiotic transformation can induce unexpected side effects, such as an enhancement of the rhizosphere microfloral activity. Travis et al. [2007] showed transgenic tobacco plants overexpressing type I nitroreductase gene from *E. cloacae* were able to detoxify TNT in soil, resulting in increased microbial biomass and metabolic activity in the rhizosphere of transgenic plants as compared with wild-type plants.

RDX

Besides TNT, transgenic plants are also capable of enhancing the detoxification of RDX, the most widely used military explosive. Plants of *Arabidopsis thaliana* were engineered to express a bacterial gene, *xplA*, encoding a RDX-degrading fused flavodoxin-cytochrome P450-like enzyme [Rylott et al. 2006; Seth-Smith et al. 2002]. The donor strain, *Rhodococcus rhodochrous* strain 11Y, originally isolated from RDX-contaminated soil, was capable of achieving a 30 percent mineralization of U-^{14}CRDX in pure culture [Seth-Smith et al. 2002]. Rhodococcal bacteria degrade RDX by de-nitration, leading to ring cleavage and the release of small aliphatic metabolites [Rylott et al. 2006]. Liquid cultures of *A. thaliana* expressing XplA removed 32–100 percent of RDX from the initial concentration of 180 mM, while less than 10 percent was removed by wild-type plants. When grown on RDX-contaminated soil, the transgenic lines showed no phytotoxic effect up to 2000 mg kg^{-1}, while significant inhibition was observed in wild type even at 250 mg kg^{-1}. On the other hand, wild-type plants had significantly higher non-detoxified RDX than transformed plants, suggesting that transgenic lines were capable of more efficient transformation of RDX.

Although *A. thaliana* and tobacco are well characterized laboratory model plants, they are not well adapted for phytoremediation applications, given their small stature and shallow root system.

References

Abhilash, P.C., S. Jamil and N. Singh. 2009. Transgenic plants for enhanced biodegradation and phytoremediation of organic xenobiotics. Biotechnol. Adv. 27: 474–488.

Alderete, L.G.S., M.A. Talano, S.G. Ibannez, S. Purro and E. Agostini. 2009. Establishment of transgenic tobacco hairy roots expressing basic peroxidases and its application for phenol removal. J. Biotechnol. 139: 273–279.

Anzenbacher, P. and E. Anzenbacherová. 2001. Cytochromes P450 and metabolism of xenobiotics. Cell Mol. Life Sci. 58: 737–747.

Arshad, M., M. Saleem and S. Hussain. 2007. Perspectives of bacterial ACC deaminase in phytoremediation. Trends in Biotechnol. 25: 356–362.

Basu, S.K., F. Eudes and I. Kovalchuk. 2010. Better crops for ecology and environment. pp. 301–340. *In*: C. Kole, C. Michler, A.G. Abbott and T.C. Hall (eds.). Transgenic Crop Plants. Vol. 2: Utilization and Safety. Springer-Verlag, New York, 516 pp.

Batard, Y., M. LeRet, M. Schalk, T. Robineau, F. Durst et al. 1998. Molecular cloning and functional expression in yeast of CYP76B1, a xenobiotic-inducible 7-ethoxycoumarin O-de-ethylase from *Helianthus tuberosus*. Plant J. 14: 111–120.

Baudouin, C., M. Charveron, R. Tarrouse and Y. Gall. 2002. Environmental pollutants and skin cancer. Cell Biol. Toxicol. 18: 341–348.

Bode, M., P. Stöbe, B. Thiede, I. Schuphan and B. Schmidt. 2004. Biotransformation of atrazine in transgenic tobacco cell culture expressing human P450. Pest Manag. Sci. 60: 49–58.

Brazier-Hicks, M., E.M. Evans, O.D. Cunningham, D.R. Hodgson, P.G. Steel and R. Edwards. 2008. Catabolism of glutathione conjugates in *Arabidopsis thaliana*. Role in metabolic reactivation of the herbicide safener fenclorim. J. Biol. Chem. 283: 21102–21112.

Brenter, L.B., S.T. Mukherji, K.M. Merchie, J.M. Yoon and J.L. Schnoor. 2008. Expression of glutathione S-transferase in poplar trees (*Populus trichocarpa*) exposed to 2,4,6-trinitrotoluene (TNT). Chemosphere 73: 657–662.

Bruns-Nagel, D., J. Breitung, E. von Low, K. Steinbach, T. Gorontzy, M. Kahl, K. Blotevogel and D. Gemsa. 1996. Microbial transformation of 2,4,6-trinitrotoluene in aerobic soil columns. Applied and Environ. Microbiol. 62: 2651–2656.

Bryant, C. and M. DeLuca. 1991. Nitroreductases. pp. 291–304. *In*: F. Muller (ed.). Chemistry and Biochemistry of Flavoenzymes. Vol II. CRC Press, Boca Raton, FL.

Bryant, D., D. McCalla, M. Leeksma and P. Laneuville. 1981. Type I nitroreductases of *Escherichia coli*. Can. J. Microbiol. 27: 81–86.

Caballero, A., J.J. Lázaro, J.L. Ramos and A. Esteve-Nuñez. 2005. PnrA, a new nitroreductase-family enzyme in the TNT-degrading strain *Pseudomonas putida* JLR11. Environ. Microbiol. 8: 1211–1219.

Campanella, B., C. Bok and P. Schroeder. 2002. Phytoremediation to increase degradation of PCBs and PCDD/Fs potential and limitations. Environ. Sci. Pollut. Res. 9: 73–85.

Chrastilova, Z., M. Mackova, M. Novakova, T. Macek and M. Szekeres. 2007. Transgenic plants for effective phytoremediation of persistent toxic organic pollutants present in the environment. Abstracts: J. Biotechnol. 131S: S38.

Cole, D.J. 1994. Detoxification and activation of agrochemicals in plants. Pesticide Sci. 42: 209–222.

Dai, M. and S.D. Copley. 2004. Genome shuffling improves degradation of the anthropogenic pesticide pentachlorophenol by *Sphingobium chlorophenolicum* ATCC 39723. Appl. Environ. Microbiol. 70: 2391–2397.

Danielson, P.B. 2002. The cytochrome P450 superfamily: biochemistry, evolution and drug metabolism in humans. Curr. Drug Metab. 3: 561–597.

de Carvalho, S.J.P., M. Nicolai, R.R. Ferreira, A.V.O. Figueira and P.J. Christoffoleti. 2009. Herbicide selectivity by differential metabolism: considerations for reducing crop damages. Scientia Agricola 66(1): 136–142.

Deng, F., J. Jelesko, C.L. Cramer, J.R. Wu and K.K. Hatzios. 2003. Use of an antisense gene to characterize glutathione S-transferase functions in transformed suspension-cultured rice cells and calli. Pesti. Biochem. Physiol. 75: 27–37.

de Oliveira, I.M., A. Zanotto-Filho, J.C. Moreira, D. Bonatto and J.A. Henriques. 2010. The role of two putative nitroreductases, Frm2p and Hbn1p, in the oxidative stress response in *Saccharomyces cerevisiae*. Yeast 27: 89–102.

DeRidder, B., D.P. Dixon, D.J. Beussman, R. Edwards and G.B. Goldsbrough. 2002. Induction of glutathione S-transferases in *Arabidopsis* by herbicide safeners. Plant Physiol. 130: 1497–1505.

Didierjean, L., L. Gondet, R. Perkins, S.-M.C. Lau, S.M. Schaller, J. O'Keefe and D.P. Werck-Reichhart. 2002. Engineering herbicide metabolism in tobacco and *Arabidopsis* with CYP76B1, a cytochrome P450 enzyme from Jerusalem artichoke. Plant Physiol. 130: 179–189.

Dixit, P., S. Singh, P. Mukherjee and S. Eapen. 2008. Development of transgenic plants with cytochrome P450E1 gene and glutathione-S-transferase gene for degradation of organic pollutants. J. Biotechnol. 136S: S692–693.

Dixit, P., P.K. Mukherjee, V. Ramachandran and S. Eapen. 2011. Glutathione transferase from *Trichoderma virens* enhances cadmium tolerance without enhancing its accumulation in transgenic *Nicotiana tabacum*. PLoS One 6(1): e16360.

Dixon, D.P. and R. Edwards. 2009. Selective binding of glutathione conjugates of fatty acid derivatives by plant glutathione transferases. J. Biol. Chem. 284: 21249–21256.

Dixon, D.P., A. Lapthorn and R. Edwards. 2002. Plant glutathione transferases. Genome Biol. 3004.1– 3004.10.

Dixon, D.P., D.J. Cole and R. Edwards. 1999. Dimerization of maize glutathione transferases in recombinant bacteria. Plant Mol. Biol. 40: 997–1008.

Dixon, D.P., I. Cumminus, D.J. Cole and R. Edwards. 1998a. Glutathione-mediated detoxification system in plants. Curr. Opin. Plant Biol. 1: 258–266.

Dixon, D.P., D.J. Cole and R. Edwards. 1998b. Purification, regulation and cloning of a glutathione transferase (GST) from maize resembling the auxin-inducible type-Ill GSTs. Plant Mol. Biol. 36: 75–87.

Doty, S.L. 2008. Enhancing phytoremediation through the use of transgenic plants and entophytes. New Phytologist 179: 318–33.

Doty, S.L., C.A. James, A.L. Moore, A. Vajzovic, G.L. Singleton, C. Ma, et al. 2007. Enhanced phytoremediation of volatile environmental pollutants with transgenic trees. Proc. Natl. Acad. Sci. USA 23: 104(43): 16816–16821.

Doty, S.L., T.Q. Shang, A.M. Wilson, J. Tangen, A.D. Westergreen, L.A. Newman, S.E. Strand and M.P. Gordon. 2000. Enhanced metabolism of halogenated hydrocarbons in transgenic plants containing mammalian cytochrome P450 2E1. Proc. Natl. Acad. Sci. USA 97: 6287–6291.

Durst, F. and I. Benvenist. 1993. Cytochrome P450 in Plants. pp. 293–310. *In*: J.B. Schenkman and H. Grein (eds.). HandBook of Experimental Pharmacology: Cytochrome P450. Springer-Verlag, New York.

Eapen, S., S. Singh and S. D'Souza. 2007. Advances in development of transgenic plants for remediation of xenobiotic pollutants. Biotechnol. Adv. 25: 442–451.

Edwards, R. and D.P. Dixon. 2005. Plant glutathione transferases. Methods Enzymol. 401: 169–186.

Edwards, R., M. Brazier-Hicks, D.P. Dixon and I. Cummins. 2005. Chemical manipulation of antioxidant defenses in plants. Adv. Bot. Res. 42: 1–32.

Environmental Protectin Agency. 2000. A Citizen's Guide to Phytoremediation. EPA 542-F-98-011. United States Environmental Protection Agency, p. 6. Available at: http//www.bugsatwork. com/ XYCLONYX /EPA_ GUIDES/ PHYTO.PDF.

Eerd, L.L. van, R.E. Hoagland, R.M. Zablotowicz and J.C. Hall. 2003. Pesticide metabolism in plants and microorganisms. Weed Sci. 51: 472–495.

Fonne-Pfister, R. and K. Kruez. 1990. Ring-methyl hydroxylation of chlortoluron by an inducible cytochrome P450-dependent enzyme from maize. Phytochemistry 29: 2793–2796.

Francova, K., M. Sura, T. Macek, M. Szekeres, S. Bancos, K. Demnerova et al. 2003. Preparation of plants containing bacterial enzyme for degradation of poly chlorinated biphenyls. Fresenius Environ. Bull. 12: 309–313.

Frear, D.S. and G.G. Still. 1968. The metabolism of 3,4-dichloropropionanilide in plants. Partial purification and properties of an aryl acylamidase from rice. Phytochemistry 7: 913–920.

French, C.E., S. Nicklin and N.C. Bruce. 1998. Aerobic degradation of 2,4,6-trinitrotoluene by Enterobacter cloacae PB2 and by pentaerythritol tetranitrate reductase. Appl. Environ. Microbiol. 64(8): 2864–2868.

Gandia-Herrero, F., A. Lorenz, T. Larson, I.A. Graham, J. Bowles, E.L. Rylott et al. 2008. Detoxification of the explosive 2,4,6-nitrotoluene in *Arabidopsis*: discovery of bifunctional *O*- and *C*-glucosyltransferases. Plant J. 56: 963–974.

Girvan, H., T. Waltham, R. Neeli, H. Collins, K. McLean, N. Scrutton et al. 2006. Flavocytochrome P450 BM3 and the origin of CYP102 fusion species. Biochem. Soc. Trans. 34(Pt 6): 1173–1177.

Gorinova, N., A. Nedkovska and A. Atanassov. 2005. Cytochrome P450 monooxygenases as a tool for metabolizing of herbicides in plants. Biotechnol. & Biotechnol. Eq. 19(Spl. Issue): 105–115.

Greń, I., U. Guzik, D. Wojcieszyńska and S. Łabużek. 2008. Molekularne podstawy rozkładu ksenobiotycznych związków aromatycznych. Biotechnologia 2: 58–67.

Gullner, G., T. Komives and H. Rennenberg. 2001. Enhanced tolerance of transgenic poplar plants overexpressing gamma-glutamylcysteine synthetase towards chloroacetanilide herbicides. J. Exp. Bot. 52: 971–979.

Hannink, N.K., M. Subramanian, S.J. Rosser, A. Basran, J.A.H. Murray, J.V. Shanks and N.C. Bruce. 2007. Enhanced transformation of TNT by tobacco plants expressing a bacterial nitroreductase. Int. J. Phytorem. 9: 385–401.

Hannink, N.K., S.J. Rosser, C.E. French, A. Basran, J.A.H. Murray, S. Nicklin and N.C. Bruce. 2001. Phytodetoxification of TNT by transgenic plants expressing a bacterial nitroreductase. Nat. Biotechnol. 19: 1168–1172.

Hanukoglu, I. 1996. Electron transfer proteins of cytochrome P450 systems. Adv. Mol. Cell Biol. 14: 29–55.

Harvey, P.J., B.F. Campanella, P.M. Castro, H. Harms, E. Lichtfouse, A.R. Schaffner et al. 2002. Phytoremediation of polyaromatic hydrocarbons, anilines and phenols. Environ. Sci. Pollut. Res. Int. 9: 29–47.

Hirose, S., H. Kawahigashi, T. Inoue, H. Inui, H. Ohkawa and Y. Ohkawa. 2005. Enhanced expression of CYP2C9 and tolerance to sulfonylurea herbicides in transgenic rice plants. Plant Biotechnol. 22(2): 89–96.

Hooda, V. 2007. Phytoremediation of toxic metals from soil and waste water. J. Environ. Biol. 28(2): 367–376.

Ichinose, H. 2012. Molecular and functional diversity of fungal cytochrome P450s. Biol. Pharm. Bull. 35(6): 833–837.

Idi, M., H. Ichinose and H. Wariishi. 2012. Molecular identification and functional characterization of cytochrome P450 monooxygenases from the brown-rot basidiomycete *Postia placenta*. Arch. Microbiol. 194: 243–253.

Iimura, Y., S. Ikeda, T. Sonoki, T. Hayakawa, S. Kajita, K. Kimbara et al. 2002. Expression of a gene for Mn-peroxidase from Coriolus versicolor in transgenic tobacco generates potential tool for phytoremediation. Appl. Microbiol. Biotechnol. 59: 246–251.

Inui, H. and H. Ohkawa. 2005. Herbicide resistance in transgenic plants with mammalian P450 monooxygenase genes. Pest Manag. Sci. 61: 286–291.

Inui, H., N. Shiota, Y. Motoi, Y. Ido, T. Inoue, T. Kodama et al. 2001. Metabolism of herbicides and other chemicals in human cytochrome P450 species and in transgenic potato plants co-expressing human CYP1A1, CYP2B6 and CYP2C19. J. Pestic. Sci. 26: 28–40.

Inui, H., Y. Ueyama, N. Shiota, Y. Ohkawa and H. Ohkawa. 1999. Herbicide metabolism and cross-tolerance in transgenic potato plants expressing human CYP1A1. Pestic. Biochem. Physiol. 64: 33–46.

Jackson, E.G., E.L. Rylott, D. Fournier, J. Hawari and N.C. Bruce. 2007. Exploring the biochemical properties and remediation applications of the unusual explosive-degrading P450 system XplA/B. Proc. Natl. Acad. Sci. USA 104: 16822–16827.

James, C.A., G. Xin, S.L. Doty and S.E. Strand. 2008. Degradation of low molecular weight volatile organic compounds by plants genetically modified with mammalian cytochrome P450 2E1. Environ. Sci. Technol. 42: 289–293.

Jepson, I., V.J. Lay, D.C. Holt, S.W.J. Bright and A.J. Greenland. 1994. Cloning and characterization of maize herbicide safener-induced cDNAs encoding subunits of glutathione S-transferase isoforms I, II and IV. Plant Mol. Biol. 26: 1855–1866.

Jung, S., H.-J. Lee, Y. Lee, K. Kang, Y.S. Kim, B. Grimm and K. Back. 2008. Toxic tetrapyrrole accumulation in protoporphyrinogen IX oxidase-overexpressing transgenic rice plants. Plant Mol. Biol. 67(5): 535–546.

Karavangeli, M., N.E. Labrou, Y.D. Clonis and A. Tsaftaris. 2005. Development of transgenic tobacco plants overexpressing maize glutathione S-transferase I for chloroacetanilide herbicides phytoremediation. Biomol. Eng. 22: 121–128.

Kawahigashi, H. 2009. Transgenic plants for phytoremediation of herbicides. Current Opinion in Biotechnology 20(2): 225–230.

Kawahigashi, H., S. Hirose, H. Ohkawa and Y. Ohkawa. 2008. Transgenic rice plants expressing human p450 genes involved in xenobiotic metabolism for phytoremediation. J. Mol. Microbiol. Biotechnol. 15: 210–219.

Kawahigashi, H., S. Hirose, H. Ohkawa and Y. Ohkawa. 2007. Herbicide resistance of transgenic rice plants expressing human CYP1A1. Biotechnol. Adv. 25: 75–84.

Kawahigashi, H., S. Hirose, H. Ohkawa and Y. Ohkawa. 2006. Phytoremediation of the herbicides atrazine and metolachlor by transgenic rice plants expressing human CYP1A1, CYP2B6, and CYP2C19. J. Agric. Food Chem. 54: 2985–2991.

Kawahigashi, H., S. Hirose, H. Inui, H. Ohkawa and Y. Ohkawa. 2005a. Enhanced herbicide cross-tolerance in transgenic rice plants coexpressing human CYP1A1, CYP2B6, and CYP2C19. Plant Sci. 168: 773–781.

Kawahigashi, H., S. Hirose, H. Ohkawa and Y. Ohkawa. 2005b. Phytoremediation of metolachlor by transgenic rice plants expressing human CYP2B6. J. Agr. Food Chem. 53: 9155–9160.

Kawahigashi, H., S. Hirose, K. Ozawa, Y. Ido, M. Kojima and H. Ohkawa. 2005c. Analysis of substrate specificity of pig CYP2B22 and CYP2C49 towards herbicides by transgenic rice plants. Transg. Res. 14: 907–917.

Kawakami, N., O. Shoji and Y. Watanabi. 2011. Use of perfulorocarboxylic acids to trick cytochrome P450BM3 into initiating the hydroxylation of gaseous alkanes Angew Chem. Int. Ed. Engl. 50: 5315–5318.

Killili, K.G., N. Atanassova, A. Vardanyan, N. Clatot, K. Al-Sabarna, P.N. Kanellopoulos et al. 2004. Differential roles of tau class glutathione S-transferases in oxidative stress. J. Biol. Chem. 279: 24540–24551.

Knasmuller, S., E. Gottmann, H. Steinkellner, A. Fomin, C. Pickl, A. Paschke et al. 1998. Detection of genotoxic effects of heavy metal contaminated soils with plant bioassays. Mutat. Res. 420: 37–48.

Koder, R.L., O. Oyedele and A.F. Miller. 2001. Retro-nitroreductase, a putative evolutionary precursor to *Enterobacter cloacae* strain 96-3 nitroreductase. Antioxid. Redox Signal 3(5): 747–755.

Komives, T. and G. Gullner. 2005. Phase I xenobiotic metabolic systems in plants. Z. Naturforsch. C 60: 179–185.

Kramer, U. and A.N. Chardonnens. 2001. Appl. Microbiol. Biotechnol. 55: 661–672.

Kumar, S., M. Jin and J.L. Weemhoff. 2012. Cytochrome P450-mediated phytoremediation using Transgenic plants: A Need for Engineered Cytochrome P450 Enzymes. J. Pet. Environ. Biotechnol. 3(5): 127–131.

Kumar, S. 2010. Engineering cytochrome P450 biocatalysts for biotechnology, medicine, and bioremediation. Expert. Opin. Drug Metabol. Toxicol. 6: 115–131.

Kurumata, M., M. Takahashi, A. Sakamoto, J.L. Ramos, A. Nepovim, T. Vanek et al. 2005. Tolerance to, and uptake and degradation of, 2,4,6-trinitrotoluene (TNT) are enhanced by the expression of a bacterial nitroreductase gene in *Arabidopsis thaliana*. Z. Naturforsch. [C] 60: 272–278.

LeDuc, D.L., M. AbdelSamie, M. Montes-Bayon, C.P. Wu, S.J. Reisinger and N. Terry. 2006. Overexpressing both ATP sulfurylase and selenocysteine methyltransferase enhances selenium phytoremediation traits in Indian mustard. Environ. Pollut. 144: 70–76.

Leszczyński, B. 2001. Wybrane zagadnienia z biochemii i toksykologii środowiska. Wydawnictwo Akademii Podlaskiej, Siedlce.

Lutz, K.A., J.E. Knapp and P. Maliga. 2001. Expression of *bar* in the plastid genome confers herbicide resistance. Plant Physiol. 125: 1585–1590.

Machida, M., K. Asai, M. Sano, T. Tanaka, T. Kumagai et al. Genome sequencing and analysis of *Aspergillus oryzae*. Nature 438: 1157–1161.

Maestri, E. and M. Marmiroli. 2011. Genetic and molecular aspects of metal tolerance and hyperaccumulation. pp. 41–63. *In*: D.K. Gupta and L.M. Sandalio (eds.). Metal Toxicity in Plants: Perception, Signaling and Remediation. Springer-Verlag Berlin Heidelberg.

Marrs, K.A. 1996. The functions and regulation of glutathione S-transferases in plants. Ann. Rev. Plant Physiol. Plant Mol. Biol. 47: 127–158.

Martinez, B., J. Tomkins, L.P. Wackett, R. Wing and M.J. Sadowsky. 2001. Complete nucleotide sequence and organization of the atrazine catabolic plasmid pADP-1 from *Pseudomonas* sp. strain ADP. J. Bacteriol. 183: 5684–5697.

Martinoia, E., E. Grill, R. Tommasini, K. Kreuz and N. Amrhein. 1993. ATP-dependent glutathione S-conjugate 'export' pump in the vacuolar membrane of plants. Nature 364: 247–249.

May, M.J., T. Vernoux, C. Leaver, M. Van Montagu and D. Inze. 1998. Glutathione homeostasis in plants: implications for environmental sensing and plant development. J. Exp. Bot. 49: 649–667.

McFadden, J.J., J.W. Gronwald and C.V. Eberlein. 1990. *In vitro* hydroxylation of bentazon by microsomes from napththalic anhydride-treated corn shoots. Biochem. Biophys. Res. Commun. 168: 206.

McGonigle, B., S.J. Keeler, S.-M.C. Lau, M.K. Koeppe and D.P. O'Keefe. 2000. A genomics approach to the comprehensive analysis of the glutathione *S*-transferase gene family in soybean and maize. Plant Physiol. 124(3): 1105–1120.

Meagher, R.B. 2000. Phytoremediation of toxic elemental and organic pollutants. Curr. Opin. Plant Biotechnol. 3: 153–162.

Milligan, A.S., A. Daly, M.A.J. Parry, P.A. Lazzeri and I. Jepson. 2001. The expression of maize glutathione S-transferase gene in transgenic wheat confers herbicide tolerance, both in plants and *in vitro*. Mol. Breed. 7: 301–315.

Misra, S. and L. Gedama. 1989. Heavy metal tolerant transgenic *Brassica napus* L. and *Nicotiana tabacum* L. plants. Theor. Appl. Genet. 78: 161–168.

Mizutani, M. and D. Ohta. 2010. Diversification of P450 genes during land plant evolution. Ann. Rev. Plant Biol. 61: 291–315.

Mohammadi, M., V. Chalavi, M. Novakova-Sura, J.F. Laliberté and M. Sylvestre. 2007. Expression of bacterial biphenyl-chlorobiphenyl dioxygenase genes in tobacco plants. Biotechnol. Bioeng. 97: 496–505.

Moons, A. 2005. Regulatory and functional interactions of plant growth regulators and plant glutathione S-tansferases (GSTs). Plant Hormones 72: 155–202.

Morant, M., S. Bak, B.L. Moller and D. Werck-Reichhart. 2003. Plant cytochromes P450: tools for pharmacology, plant protection and phytoremediation. Curr. Opin. Biotechnol. 14: 151–162.

Moreland, D.E., F.T. Corbin and J.E. McFarland. 1993. Oxidation of multiple substrates by corn shoot microsomes. Pestic. Biochem. Physiol. 47: 206–210.

Narhi, L. and A. Fulco. 1986. Characterization of a catalytically self-sufficient 119,000-dalton cytochrome P-450 monooxygenase induced by barbiturates in *Bacillus megaterium*. J. Biol. Chem. 261(16): 7160–7169.

Nebert, D.W. and F.J. Gonzalez. 1987. P450 genes: structure, evolution and regulation. Annu. Rev. Biochem. 56: 945–993.

Nelson, D.R., M.A. Schuler, S.M. Paquette, D. Werck-Reichhart and S. Bak. 2004. Comparative genomics of rice and *Arabidopsis*. Analysis of 727 cytochrome P450 genes and pseudogenes from a monocot and a dicot. Plant Physiol. 135: 756–772.

Nielsen, K.A. and B.L. Moller. 2005. Cytochrome P450s in plants. pp. 553–583. *In*: P.R. Ortiz de Montellano (ed.). Cytochrome P450: Structure, Mechanism, and Biochemistry, 3e. Kluwer Academic/Plenum Publishers, New York.

Noctor, G., A.-C.M. Arisi, L. Jouanin and C.H. Foyer. 1998. Manipulation of glutathione and amino acid biosynthesis in the chloroplast. Plant Physiol. 118(2): 471–482.

Noctor, G. and C.H. Foyer. 1998. Ascorbate and glutathione: keeping active oxygen under control. Ann. Rev. Plant Physiol. Plant Mol. Biol. 49: 249–279.

Noctor, G., M. Strohm, L. Jouanin, K.J. Kunert, C.H. Foyer and H. Rennenberg. 1996. Synthesis of glutathione in leaves of transgenic poplar (*Populus tremula* × *P. alba*) overexpressing γ-glutamylcysteine synthetase. Plant Physiol. 112: 1071–1078.

Novakova, M., M. Mackova, Z. Chrastilova, J. Viktorova, M. Szekeres, K. Demnerova et al. 2009. Cloning of the bacterial *bphC* gene into *Nicotiana tabacum* to improve the efficiency of PCB phytoremediation. Biotechnol. Bioeng. 102: 29–37.

Ohkawa, H., N. Shiota., H. Inui, M. Sigiura, Y. Yabusaki, Y. Ohkawa and T. Ishige. 1997. Transgenic plant analysis as a tool for the study of maize glutathione s-transferases. pp. 307–312. *In*: K.K. Hatzios (ed.). Regulation of Enzymatic Systems Detoxifying Xenobiotics in Plants. Kluwer Academic Publishers, Proc. Of the NATO Advanced Research Workshop, Kriopigi, Halkidiki, Greece, 22–28 September 1996.

O'Keefe, D.P., J.M. Tepperman, C. Dean, K.J. Leto, D.L. Erbes and J.T. Odell. 1994. Plant Expression of a Bacterial Cytochrome P450 That Catalyzes Activation of a Sulfonylurea Pro-Herbicide. Plant Physiol. 105: 473–482.

Olekhnovich, I., A. Goodwin and P. Hoffman. 2009. Characterization of the NAD(P)H oxidase and metronidazole reductase activities of the RdxA nitroreductase of *Helicobacter pylori*. FEBS J. 276: 3354–3364.

Oller, A.L.W., E. Agostini, M.A. Talano, C. Capozucca, S.R. Milrad, H.A. Tigier et al. 2005. Overexpression of a basic peroxidase in transgenic tomato (*Lycopersicon esculentum Mill.* cv. Pera) hairy roots increases phytoremediation of phenol. Plant Sci. 169: 1102–1111.

Paspaliraius, I., N. Papassiopi, A. Xenidis and Y.-T. Hung. 2010. Soil remediation. pp. 519–570. *In*: L.K. Wang, Y.-T. Hung and N.K. Shammas (eds.). Advanced Industrial and Hazardous Wastes Treatment. CRC Press, Taylor & Francis Group, Boca Raton, FL.

Pilon-Smits, E.A.H., S.B. Hwang, C.M. Lytle, Y.L. Zhu, J.C. Tai et al. 1999. Overexpression of ATP sulfurylase in *Brassica juncea* leads to increased selenate uptake, reduction and tolerance. Plant Physiol. 119: 123–132.

Poulos, T.L., B.C. Finzel and A.J. Howard. 1987. High-resolution crystal structure of cytochrome P450cam. J. Mol. Biol. 195: 687–700.

Race, P.R., A.L. Lovering, S.A. White, J. Grove, P.F. Searle, C.J. Wrighton and E.I. Hyde. 2007. Kinetic and structural characterization of *Escherichia coli* nitroreductase mutants showing improved efficacy for the prodrug substrate CB1954. J. Mol. Biol. 368: 481–492.

Race, P.R., A.L. Lovering, R.M. Green, A. Ossor, S.A. White, P.F. Searle et al. 2005. Structural and mechanistic studies of *Escherichia coli* nitroreductase with the antibiotic nitrofurazone. J. Biol. Chem. 280: 13256–13264.

Rao, V.S. 2000. Principles of Weed Science. Second Edition. Science Publishers, Inc., Enfield, NH, USA, 555 pp.

Raskin, I. and B.D. Ensley. 2000. Phytoremediation of Toxic Metals: Using Plants to Clean Up the Environment. John Wiley & Sons Inc., New York, 304 pp.

Rea, P.A. 1999. MRP subfamily ABC transporters from plants and yeast. J. Exp. Bot. 50: 895–913.

Rea, P.A., Z.-S. Li, Y.-P. Lu, Y.M. Drozdowicz and E. Martinoia. 1998. From vacuolar GS-X pumps to multispecific ABC transporters. Ann. Rev. Plant Physiol. Plant Mol. Biol. 49: 727–760.

Relyea, R.A. 2005. The lethal impact of Roundup on aquatic and terrestrial amphibians. Ecolog. Appl. 15(4): 1118–1124.

Richman, M. 1996. Terrestrial plants tested for cleanup of radionuclides, explosives residue. Water Environ. Technol. 8: 17–18.

Rittle, J. and M.T. Green. 2010. Cytochrome P450 compound I: capture, characterization, and C-H bond activation kinetics. Science 330(6006): 933–937.

Roberts, G.A., G. Grogan, A. Greter, S.L. Flitsch and N.J. Turner. 2002. Identification of a new class of cytochrome P450 from a *Rhodococcus* sp. J. Bacteriol. 184(14): 3898–3908.

Robineau, T., Y. Batard, S. Nedelkina, F. Cabello-Hurtado, M. LeRet, O. Sorokine et al. 1998. The chemically inducible plant cytochrome P450 CYP76B1 actively metabolizes phenylureas and other xenobiotics. Plant Physiol. 118: 1049–1056.

Różański, L. 1998. Przemiany pestycydów w organizmach żywych i środowisku. Wyd. AGRA-ENVIROLAB, Poznań.

Rugh, C.L., J.F. Senecoff, R.B. Meagher and S.A. Merkle. 1998. Development of transgenic yellow poplar for mercury phytoremediation. Nat. Biotechnol. 16: 925–928.

Rugh, C.L., D. Wilde, N.M. Stack, D.M. Thompson, A.O. Summer and R.B. Meagher. 1996. Mercuric ion reduction and resistance in transgenic *Arabidopsis thaliana* plants expressing a modified bacterial *merA* gene. Proc. Natl. Acad. Sci. USA 93: 3182–3187.

Rylott, E.L., R.G. Jackson, F. Sabbadin, H.M. Seth-Smith, J. Edwards et al. 2011a. The explosive-degrading cytochrome P450 XplA: biochemistry, structural features and prospects for bioremediation. Biochim. Biophys. Acta 1814: 230–236.

Rylott, E.L., M.V. Budarina, A. Barker, A. Lorenz, S.E. Strand and N.C. Bruce. 2011b. Engineering plants for the phytoremediation of RDX in the presence of the co-contaminating explosive TNT. New Phytologist 192: 405–413.

Rylott, E.L., R.G. Jackson, J. Edwards, G.L. Womack, H.M.B. Seth-Smith, D.A. Rathbone et al. 2006. Strand SE, Bruce NC: An explosive-degrading cytochrome P450 activity and its targeted application for the phytoremediation of RDX. Nat. Biotechnol. 24: 216–219.

Scarponi, L., D. Del Buono and C. Vischetti. 2003. Persistence and detoxification of Pretilachlor and Fenclorim in rice (*Oryza sativa*). Agronomie 23: 147–151.

Schnoor, J. 2000. Degradation by plants—Phytoremediation. pp. 371–384. *In*: H.J. Rehm and R.G. Reed (eds.). Biotechnology. Vol. 11b, edn. 2. Wiley-VCH, Germany.

Schnoor, J.L., L.A. Licht, S.C. McCutcheon, N.L. Wolf and L.H. Carreira. 1995. Phytoremediation of organic and nutrient contaminants. Environ. Sci. Technol. 29: 318–323.

Schroder, P., D. Daubner, H. Maier, J. Neustifter and R. Debus. 2008. Phytoremediation of organic xenobiotics-Glutathione dependent detoxification in Phragmites plants from European treatment sites. Bioresour. Technol. 29: 7183–7191.

Schwab, A.P., A.A. Al-Assi and A.K. Banks. 1998. Adsorption of naphthalene onto plant roots. J. Environ. Qual. 27: 220–224.

Seth-Smith, H.M.B., S.J. Rosser, A. Basran, E.R. Travis, E.R. Dabbs, S. Nicklin and N.C. Bruce. 2002. Cloning, sequencing, and characterization of the hexahydro-1,3,5-trinitro-1,3,5-triazine degradation gene cluster from *Rhodococcus rhodochrous*. Appl. Environ. Microbiol. 68(10): 4764–4771.

Shimabukuro, R.H. 1985. Detoxification of herbicides. pp. 215–240. *In*: S.O. Duke (ed.). Weed Physiology. Vol. II: Herbicide Physiology. CRC Press, Boca Raton, FL.

Shiota, N.S., S. Kodama, H. Inui and H. Ohkawa. 2000. Expression of human cytochromes P450 1A1 and P450 1A2 as fused enzymes with yeast NADPH-cytochrome P450 oxidoreductase in transgenic tobacco plants. Biosci. Biotechnol. Biochem. 64: 2025–2033.

Shiota, N. 1996. Studies on Genetically Engineered Tobacco Plants Resistant to Herbicides. Ph.D. Dessertation. AgroBioInstitute 75 pp.

Shiota, N., H. Inui and H. Ohkawa. 1996. Metabolism of the herbicide chlortoluron in transgenic tobacco plants expressing the fused enzyme between rat cytochrome P4501A1 and yeast NADPH-cytochrome P450 oxidoreductase. Pestic. Biochem. Physiol. 54: 190–198.

Siminszky, B., F.T. Corbin, E.J. Ward, T.J. Fleischmann and R.E. Dewey. 1999. Expression of a soybean cytochrome P450 monooxygenase cDNA in yeast and tobacco enhances the metabolism of phenylurea herbicides. Proc. Natl. Acad. Sci. USA. 96: 1750–1755.

Song, W.Y., E.J. Son, E. Martinoia, Y.J. Lee, Y.Y. Yang, M. Jasinski et al. 2003. Engineering tolerance and accumulation of lead and cadmium in transgenic plants. Nat. Biotechnol. 21: 914–919.

Sonoki, T., S. Kajita, S. Ikeda, M. Uesugi, K. Tatsumi, Y. Katayama and Y. Iimura. 2005. Transgenic tobacco expressing fungal laccase promotes the detoxification of environmental pollutants. Appl. Microbiol. Biotechnol. 67: 138–142.

Soranzo, N., M. Sari Gorla, L. Mizzi, G. De Toma and C. Frova. 2004. Organisation and structural evolution of the rice glutathione S-transferase gene family. Mol. Genet. Genomics 271: 511–521.

Spain, J. 1995. Biodegradation of nitroaromatic compounds. Annu. Rev. Microbiol. 49: 523–555.

Sugiyama, A. and J. Sekiya. 2005. Homoglutathione confers tolerance to acifluorfen in transgenic tobacco plants expressing soybean homoglutathione synthetase. Plant Cell Physiol. 46(8): 1428–1432.

Sugiyama, A., J. Nishimura, Y. Mochizuki, K. Inagaki and J. Sekiya. 2004. Homoglutathione synthesis in transgenic tobacco plants expressing soybean homoglutathione synthetase. Plant Biotechnol. 21: 79–83.

Sylvestre, M., T. Macek and M. Mackova. 2009. Transgenic plants to improve rhizoremediation of polychlorinated biphenyls (PCBs). Current Opinion Biotechnol. 20(2): 242–247.

Thom, R., I. Cummins, D.P. Dixon, R. Edwards, D.J. Cole and A.J. Lapthorn. 2002. Structure of a tau class glutathione S-transferase from wheat active in herbicide detoxification. Biochemistry 41: 7008–7020.

Tommasini, R., E. Vogt, J. Schmid, M. Fromentau, N. Amrhein and E. Martinoia. 1997. Differential expression of genes coding for ABC transporters after treatment of *Arabidopsis thaliana* with xenobiotics. FEBS Lett. 411: 206–210.

Travis, E.R., N.K. Hannink, C.J. Van der Gast, I.P. Thompson, S.J. Rosser and N.C. Bruce. 2007. Impact of Transgenic tobacco on trinitrotoluene (TNT) contaminated soil community. Environ. Sci. Technol. 41: 5854–5861.

Uchida, E., T. Ouchi, Y. Suzuki, T. Yoshida, H. Habe, I. Yamaguchi, T. Omori and H. Nojiri. 2005. Secretion of bacterial xenobiotic-degrading enzymes from transgenic plants by anapoplastic expressional system: an applicability for phytoremediation. Environ. Sci. Technol. 39: 7671–7677.

United Nations Development Program. 2002. Phytoremediation: An Environmentally Sound Technology Prevention, Control and Remediation: An Introductory Guide to Decision Makers. Newsletters and Technical Publications: Freshwater Management Series No. 2. Division of Technology, Industry, and Economics.

United States Environmental Protection Agency. 2004. Environmental Protection Agency Toxic Release Inventory (TRI) program. Available from http:// www. epa.gov/tri/chemical/index.htm.

United States Environmental Protection Agency (USEPA). 2000. Introduction to Phytoremediation. EPA 600/R-99/107, U.S. Environmental Protection Agency, Office of Research and Development, Cincinnati, OH.

Van Aken, B. and S.L. Doty. 2010. Transgenic plants and associated bacteria for phytoremediation of chlorinated compounds. Biotechnol. Genet. Eng. Rev. 26: 43–64.

Van Aken, B. 2009. Transgenic plants for enhanced phytoremediation of toxic explosives. Curr. Opin. Biotechnol. 20: 231–236.

van Dillewijn, P., J.L. Couselo, E. Corredoira, A. Delgado, R.M. Wittich, A. Ballester and J.L. Ramos. 2008. Bioremediation of 2,4,6-trinitrotoluene by bacterial nitroreductase expressing transgenic aspen. Environ. Sci. Technol. 42(19): 7405–7410.

Wagner, U., R. Edwards, D.P. Dixon and F. Mauch. 2002. Probing the diversity of the Arabidopsis glutathione S-transferase gene family. Plant Mol. Biol. 49: 515–532.

Wang, G.D., Q.J. Li, B. Luo and X.Y. Chen. 2004. Ex planta phytoremediation of trichlorophenol and phenolic allelochemicals via an engineered secretory laccase. Nat. Biotechnol. 22: 893–897.

Wang, H., X. Chen, X. Xing, X. Hao and D. Chen. 2010. Transgenic tobacco plants expressing atzA exhibit resistance and strong ability to degrade atrazine. Plant Cell Rep. 29(12): 1391–1399.

Wang, L., D.A. Samac, N. Shapir, L.P. Wackett, C.P. Vance, N.E. Iszewski and M.J. Sadowsky. 2005. Biodegradation of atrazine in transgenic plants expressing a modified bacterial atrazine chlorohydrolase (*atzA*) gene. Plant Biotechnol. J. 3(5): 475–486.

Wang, X.X., N.F. Wu, J. Guo, X.Y. Chu, J. Tian, B. Yao et al. 2008. Phytodegradation of organophosphorus compounds by transgenic plants expressing a bacterial organophosphorus hydrolase. Biochem. Biophys. Res. Commun. 365: 453–458.

Werck-Reichhart, D., S. Bak and S. Paquette. 2002. Cytochrome P450. pp. 1–38. *In*: C.R. Somerville and E.M. Meyerowitz (eds.). The Arabidopsis Book, Amer. Soc. of Plant Biol., Rockville, MD, doi/10.1199/tab.0028, http://www. aspb. org/publications/Arabidopsis.

Whitehouse, C.J., S.G. Bell and L.L. Wong. 2012. P450(BM3) (CYP102A1) connecting the dots. Chem. Soc. Rev. 41: 1218–1260.

Wilce, M.C.J. and M.W. Parker. 1994. Structure and function of glutathione S-transferases. Biochem. Biophys. Acta 1205: 1–18.

Winfield, L.E., J.H. Rodgers, Jr. and S.J. D'Surney. 2004. The responses of selected terrestrial plants to short (< 12 days) and long term (2, 4 and 6 wk) hexahydro-1,3,5-trinitro-1,3,5-triazine (RDX) exposure. Part I: Growth and developmental effects. Ecotoxicology 13: 335–347.

Wolfe, N.L. and C.F. Hoehamer. 2003. Enzymes used by plants and microorganisms to detoxify organic compounds. pp. 158–197. *In*: S.C. Mc-Cutcheon and J.L. Schnoor (eds.). Phytoremediation: Transformation and control of contaminants. Wiley, New York.

Wright, R.L., K. Harris, B. Solow, R.H. White and P.J. Kennelly. 1996. Cloning of a potential cytochrome P450 from the archaeon *Sulfolobus solfataricus*. FEBS Lett. 384(3): 235–239.

Yamada, T., T. Ishige, N. Shiota, H. Inui, H. Ohkawa and Y. Ohkawa. 2002. Enhancement of metabolizing herbicides in young tubers of transgenic potato plants with the rat CYP1A1 gene. Theor. Appl. Genet. 105: 515–520.

Yamada, T., Y. Kambara, H. Imaishi and H. Ohkawa. 2000. Molecular cloning of novel cytochrome P450 species induced by chemical treatments in cultured tobacco cells. Pestic. Biochem. Physiol. 68: 11–25.

Yuan, J.S., P.J. Tranel and C.N. Stewart, Jr. 2007. Non-target-site herbicide resistance: a family business. Trends in Plant Sci. 12: 6–13.

Zakrzewski, S. 2000. Podstawy toksykologii środowiska. PWN, Warszawa.

Zhu, Y.L., E.A.H. Pilon-Smits, L. Jouanin and N. Terry. 1999a. Overexpression of glutathione synthetase in *Brassica juncea* enhances cadmium accumulation and tolerance. Plant Physiol. 119: 73–79.

Zhu, Y.L., E.A.H. Pilon-Smits, A.S. Tarun, S. Weber, L. Jouanin and N. Terry. 1999b. Cadmium tolerance and accumulation in Indian mustard is enhanced by overexpressing γ-glutamylcysteine synthetase. Plant Physiol. 121: 1169–1177.

CHAPTER 8

Adoption and Regulation of Transgenic Crops

The discovery of herbicide-resistant weeds in the early 1970s has triggered plant scientists into developing strategies, both chemical and non-chemical, to manage and minimize the spread of resistance genes. This fast-growing serious field problem has been mitigated, at least to a limited extent so far, by the simultaneous progress in molecular genetics that has enabled incorporation of genes from unrelated organisms into a susceptible crop. This means that plant scientists have taken a giant step by moving away from linking chemistry of a herbicide to biology of the crop to adapting biology to chemistry. The consequent development of transgenic herbicide-resistant (THR) crops in preference to traditional development of herbicide-resistant crops of the past has unintentionally led to a fierce debate in favor of, and against, the introduction and commercialization of the genetically modified crops.

Before the commercialization of glyphosate-resistant maize in 1996, transgenic bromoxynil-tolerant cotton line and a wide array of non-transgenic herbicide resistant crop lines were available, but farmers did not widely adopt them [Green and Castle 2010]. The key difference was the ability to use glyphosate, a relatively inexpensive and effective herbicide with an excellent environmental profile [Green 2012]. It goes to the credit of the glyphosate-resistant Event NK603 (Roundup Ready) maize for ready acceptance of not only the subsequently developed THR crop varieties but all biotech crops that carried other desired traits. Currently, glyphosate-resistant crops represent well over 80 percent of all THR crops grown worldwide [Duke and Cerdeira 2010].

Each country has a significant regulatory framework to assess and manage the risks and issues associated with the use of genetic engineering technology, and development and certification of genetically modified organisms (GMOs) including transgenic crops and foods derived from them. Regulation also depends on the intended use of the products of genetic engineering. For example, a crop intended for food use is generally reviewed and assessed by regulating authorities from a perspective different from the one used for non-food or feed purpose. Many a time, assessment, approval, and regulation in some nations are based on factors not entirely related to technology.

A key issue concerning regulators is whether a particular transgenic product should be labeled. Labeling can be mandatory up to a threshold level of transgenically modified (TM) content—this varies between countries—or voluntary. For example, 31 percent of the products labeled as GMO-free in South Africa have GM content above 1.0 percent [Botha and Viljoen 2009]. In USA and Canada, the labeling of TM food is voluntary, while in Europe all food (including processed food) or feed which contains greater than 0.9 percent of approved GMOs must be labeled [Davison 2010]. Although there is a broad consensus among proponents that foods derived from transgenic crops are safe to eat, many scientists and consumer groups have called for greater vigorous testing of transgenic foods over a much longer period than now.

Adoption at Global Level

Since the first commercialization of THR crops on 1.73 million ha in 1996 in the U.S., beginning with the glyphosate-resistant (GR) maize, farmers around the world have readily accepted and rapidly adopted transgenic crops such as soya bean, maize, cotton, rapeseed (canola), lucerne (alfalfa), and sugar beet. With area under biotech crops reaching 175.2 million ha in 2013, global agriculture witnessed more than 100-fold growth during the past 18 yr (Fig. 8.1; Table 8.1) [ISAAA 2013). This makes biotech crops the fastest adopted crop technology in recent history. The herbicide-resistant transgenic (including the stacked herbicide-cum-insect-resistant ones) lines accounted for 80 percent of the global biotech acreage.

The area under transgenic cultivation is doubling every 5 years and now accounts for some 12 percent of global arable land [GM Science Update 2014]. Biotech crops represented 35 percent of the global commercial seed market. Most of the commercially grown transgenic crops have one or both of two traits: herbicide (glyphosate) resistance and *Bt* insect resistance [GM Science Update 2014].

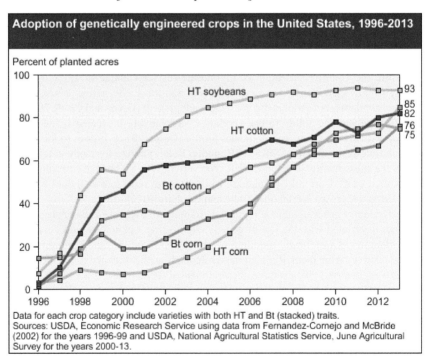

Fig. 8.1. Area under different herbicide-tolerant (HT) and *Bt* insect-tolerant transgenic crops and their adoption rates in USA during 1996–2013 [Fernandez-Cornejo 2013].

Color image of this figure appears in the color plate section at the end of the book.

Table 8.1. Area under transgenic crops in the top 13 biotech crop-growing nations [ISAAA 2013].

Country	Area (million ha %)		Transgenic Crops (all traits)
USA	70.1	40.8	Maize, soya bean, cotton, canola, sugar beet, alfalfa, papaya, squash
Brazil	40.3	21.5	Soya bean, maize, cotton
Argentina	24.4	14.0	Soya bean, maize, cotton
Canada	11.6	6.8	Canola, maize, soya bean, sugar beet
India	11.0	6.3	Cotton
China	4.2	2.3	Cotton, papaya, poplar, tomato, sweet pepper
Paraguay	3.6	2.0	Soya bean, maize, cotton
South Africa	2.9	1.7	Soya bean, maize, cotton
Pakistan	2.8	1.6	Cotton
Uruguay	1.5	0.8	Soya bean, maize
Bolivia	1.0	0.6	Soya bean
Philippines	0.8	0.5	Maize
Australia	0.6	0.5	Cotton, canola
Total	175.2*	-	

*Total does not include areas in 14 other countries.

In 2013, 18 million farmers, with over 91 percent of them (16.5 million) being risk-averse, and resource-limited small farmers in 27 countries planted transgenic crops. Nineteen of these were developing nations which have overtaken the eight developed nations by contributing more than 54 percent to the area under biotech crops. The nine major countries adopting transgenics include USA, Brazil, Argentina, Canada, India, China, Paraguay, South Africa, and Pakistan in that order which together account for 97 percent of biotech crops (Table 8.1) (ISAAA 2013). These nations excluding USA planted 100.8 million ha, accounting for over 57 percent of the global biotech crop area. Indonesia and Panama approved biotech crops in 2013 for commercialization in 2014.

The current growth engines of biotech crops, the developing nations China, India, and Pakistan in Asia, Brazil, Argentina, and Paraguay in South America, and South Africa in the African continent, which together grew biotechs on 89.2 million ha (51 percent), are home to 43 percent of the current world population of 7.2 billion. Although the European nations have largely shunned transgenic crops so far, Spain, Portugal, Czech Republic, Slovakia, and Romania planted 129,000 ha with *Bt* maize, with more than 90 percent of it being accounted for by Spain [GM Science Update 2014].

Two transgenic traits dominate the global biotech crops: herbicide resistance accounting for 65 percent, insect resistance 15 percent, and a combination of the two (stacked) for 15 percent. Stacked-trait transgenics, whose use has been on the rise since 2000, are currently an important feature of biotech crops. In 2013, 13 countries, including 10 developing ones, adopted 'stacks' over an area of 47 million ha, equivalent to 27 percent of the area under biotech crops [ISAAA 2013]. This growing trend of more stacked traits is expected to continue.

Among the four major transgenic crops, soya bean accounted for the largest share (49 percent) in 2011 followed by maize (32 percent), cotton (14 percent), and canola (rapeseed; 5 percent) [Mannion and Morse 2012; Brookes and Barfoot 2013]. In terms of the share of biotech crops as compared to gross global plantings in 2011, transgenic traits accounted for 72 percent of soya bean, 28 percent for maize, 56 percent for cotton, and 23 percent for canola (Table 8.2). Across all four crops, transgenics had a share of 44 percent of the global area under them. Over 81 percent of the global area under soya bean and cotton

Table 8.2. Share of transgenics in global gross plantings of key crops in 2011 [modified from Brookes and Barfoot 2013).

Biotech Crop	Gross cultivated area (million ha)	Area under Non-transgenics (million ha)	Area under Transgenics (million ha)	Share (%)
Soya bean	102.179	29.007	73.172	72
Maize	168.560	121.487	47.073	28
Cotton	35.760	15.620	20.140	56
Canola	33.110	25.410	7.700	23
Total	**339.609**	**191.525**	**148.085**	**44**

was accounted by transgenic varieties [GM Science Update 2014]. These were followed by 35 percent for maize and 30 percent for canola.

Nation-wise Adoption and Regulation

This part of the chapter deals with adoption and regulatory procedures followed by the top six transgenic-crop growing countries (USA, Brazil, Argentina, Canada, India, and China) besides the European Union which largely shunned biotech crops.

USA

Adoption

The U.S. is one of the first countries to adopt biotech crop technology in 1996 by planting transgenic soya bean, maize, and cotton, followed in 1999 by canola. In 2013, USA was the lead country with 70.1 million ha under transgenic crops, with an adoption rate of 90 percent across all traits (Table 8.1) [ISAAA 2013]. Herbicide resistant (HR) soya bean went up from 17 percent of total area under the crop in 1997 to 68 percent in 2001 and 93 percent in 2013 (Fig. 8.1) [Fernandez-Cornejo 2013]. The HR cotton expanded from 10 percent of the gross area under soya bean in 1997 to 56 percent in 2001 and 82 percent in 2013. The adoption of HR maize, which crept up slowly in the early years, has accelerated from 2005, reaching 85 percent of the gross maize acreage in 2013 [Fernandez-Cornejo 2013]. The U.S. also produces 95 percent of its sugar beet (465,000 ha) and 99 percent of canola (390,000 ha) from herbicide resistant transgenics.

Insect-resistant crops containing the *Bt* gene, targeting European maize borer, have been available for maize and cotton since 1996. Plantings of *Bt* maize rose from about 8 percent of the maize acreage in 1997 to 26 percent in 1999, then fell to 19 percent in 2000 and 2001, before climbing back to 29 percent in 2003 and 76 percent in 2013. Area under this *Bt* cotton also expanded rapidly, from 15 percent of U.S. cotton acreage in 1997 to 37 percent in 2001, and 75 percent in 2013. In fact, these *Bt* maize and cotton varieties also included herbicide tolerance trait as well. Stacked maize and cotton reached 71 percent and 67 percent of maize and cotton plantings in 2013 [Fernandez-Cornejo 2013]. Adoption of all transgenic varieties (HR, *Bt*-HR, and stacked) of both cotton and maize reached 90 percent of cotton area in 2013.

Herbicide resistant alfalfa was first commercialized in 2007 on about 100,000 ha. After resolving the legal issues surrounding transgenic alfalfa by 2010, planting of commercial planting of HR varieties was resumed in 2011 on approximately 200,000 ha. It is likely that many more GM crops will be cultivated in the USA, with more permissive regulatory systems combined with a supportive political system [GM Science Update 2014]. In USA, between 500–1,000 field trial applications are approved per annum and 96 applications for commercialization have been approved since 1990 [GM Science Update 2014]. Several North and South American countries have followed USA. In 2013, there were 601 field trials in USA, with herbicide-tolerant maize representing the largest category.

Currently, around 400 field trials have been approved for genetic engineering of forest trees in the U.S. [Häggman et al. 2013]. These transgenic trees include *Populus* accounting for 35 percent of all field trials, while *Eucalyptus*, *Liquidambar*, and *Malus* together account for another 37 percent. Other genera being tested include *Castanea* (American chestnut), *Pinus*, *Ulmus* (American elm), *Prunus*, *Musa*, *Citrus*, and *Juglans*. Forty field trials are being conducted by private companies, with ArborGen alone accounting for 36 of the 40 active field trials. Cornell University, North Carolina State University, Oregon State university, Purdue University, the United States Department of Agriculture and the University of California, Davis are among the public institutions currently conducting field trials. One of ArborGen's field trials with cold tolerant *Eucalyptus* covers 197.2 acres.

These include tests of a transgenic *Eucalyptus* variety with improved cold tolerance [Häggman et al. 2013] and designed for improved productivity and biofuel use [ArborGen 2013]. Also of interest is the U.S. trial of GM chestnut expressing a wheat oxalate oxidase (*oxo*) gene that provides resistance to chestnut blight (*Cryphonectria parasitica*) [Zhang et al. 2013].

Regulation

Biotechnology products in USA are regulated under the same laws that govern health, safety, and efficacy, and environmental impacts of similar products derived by traditional methods.

The federal policy that no new laws were needed to regulate the products of biotechnology was first adopted in 1986 by the federal regulatory agencies in the Coordinated Framework for Regulation of Biotechnology. The policy had three tenets: a) U.S. policy would focus on the product of genetic modification techniques, not the process itself, b) only regulation grounded in verifiable scientific risks would be tolerated, and c) GM products are on a continuum with existing products and, therefore, existing statutes are sufficient to review the products [Marden 2003]. This policy also states that a commercial product, regardless of its manner of production, should be regulated based on the product's composition and its intended use. It means that foods developed via biotechnology would be regulated in the same way as other foods developed through conventional processes. This may also mean that foods or products derived from transgenic crops are treated on par with those derived through conventional technology, regardless of the fact that the transgenic crops were the recipients of genes from non-plant sources.

However, these regulations were developed before the advent of biotechnology crops and products. As the technology has advanced, federal regulatory agencies have responded with additional regulations and guidance specific to particular biotechnology products. There is no single statute or single federal agency that governs the biotechnology products which include foods, drugs, and chemicals. Under the U.S. laws, three federal agencies have primary responsibility for the regulation of these products. These are: the United States Department of Agriculture (USDA), the Food and Drug Administration (FDA), and the Environmental Protection Agency (EPA). Of the three, the USDA assesses the plant's potential to become a weed, and the FDA reviews plants that could enter or alter the food supply while the EPA regulates genetically modified plants with pesticide properties, as well as agrochemical residues [Pew Charitable Trust 2012]. Most genetically modified plants are reviewed by at least two of the agencies, with many subjected to all three [McHughen and Smyth 2008; USFDA 1997].

Within the organization are departments that regulate different areas of GM food, including the Center for Food Safety and Applied Nutrition (CFSAN), and the Center for Biologics Evaluation and Research (CBER) [Pew Charitable Trust 2012]. Final approval can still be denied by individual counties within each state. For example, Mendocino County, California, was the first county to impose a ban on the "Propagation, Cultivation, Raising, and Growing of Genetically Modified Organisms" in 2004 [Walsh-Dilley 2009].

USDA. Within the USDA, a genetically modified organism must be assessed by the Animal and Plant Health Inspection Service (APHIS). Under the Federal Plant Pest Act, APHIS must determine whether a transgenic plant variety is likely to become a pest, i.e., to have negative agricultural or environmental effects. The agency regulates the import, transportation, and field testing of transgenic seeds and plants through notification and permitting procedures.

To commercialize a transgenic plant, the developer petitions APHIS for non-regulated status. This requires extensive data on the introduced gene construct, effects on plant biology, and effects on the ecosystem, including spread of the gene to other crops or wild relatives. For most common crops, developers notify APHIS of their intention to transport or field test a transgenic plant. Certain criteria must be met for the introduced gene. These include: a) stable integration in the plant's chromosomes, b) non-pathogenicity to animals or humans, c) unlikely to be toxic to other non-target organisms, and d) low risk of creating new plant viruses.

For less common crops and for genes or traits posing a greater risk, developers must file a formal application for permission to transport or plant the material. Field testing, typically conducted in many locations and several years, requires procedures to minimize spread of the transgene and keep it out of the food supply. Once the crop is on the market, APHIS has the authority to halt its sale if there is evidence that the plant is becoming a pest. In case some transgenic plants might not fall under APHIS's current regulatory oversight, it is recommended that the agency clarify its scope of coverage.

APHIS considers most bio-engineered plants as "regulated articles", while their progeny arising from a specific rDNA-transformed cell are called "events". For example, a maize plant with a gene construct carrying an inserted glyphosate-tolerant *epsps* gene would be a regulated article until such time the APHIS approves a petition for non-regulated status. Another maize plant with the same gene construct but regenerated from a transformed cell would be a different 'event' [McHughen and Smyth 2012]. This is justified on the basis that the locus of insertion, which varies from one transformation event to another, even when using identical DNA constructs and host plant genotypes, may give rise to different gene expression patterns and gene product levels, and perhaps affect other features (e.g., insertional knockout of endogenous genes) [McHughen and Smyth 2012]. At some stage of research and development of a regulated article that is intended for use as a food crop, a developer requests from APHIS a determination of the article's regulatory status. Once a 'regulated article' achieves 'non-regulated status', the transgenic plant can be released commercially with no further regulatory oversight. Two deregulated plants may be bred together to produce a hybrid combining the novel features of each parent. This hybrid, a 'stack', does not require USDA oversight.

FDA. Under the Federal Food, Drug, and Cosmetics Act, FDA has authority to determine the safety of foods or food ingredients besides feed. FDA authorities consult with the plant developer, review safety and nutritional data, and request additional information considered appropriate for each product. If the introduced gene is from a known allergenic source, then the transgenic food must be assessed for allergenicity. For example, if a gene from peanut (groundnut, which causes allergic reactions in some people) were introduced into soya bean, the FDA would require extensive allergenicity tests. Additional investigation may be required for transgenic crops if they involve: a) known toxicants, b) altered nutrient levels, c) new substances, and d) antibiotic resistance markers [Transgenic Crops 2004]. The developers submit a scientific and regulatory assessment of the bio-engineered food 120 d before it is marketed.

At the end of the consultation and review process, FDA sends a letter to the developer stating its satisfaction with the data regarding food safety of the product. Once a product is on the market, FDA has authority to order its removal if the food is deemed unsafe.

Critics of the FDA's regulation of transgenic foods believe that the voluntary nature of the consultation process between FDA and crop developers is inadequate. Although all companies to date have consulted extensively with FDA staff and presented large amounts of data on food safety satisfying FDA's concerns, the lack of a formal review and approval process prevents that data from being part of the public record, and this may reduce consumer confidence [Transgenic Crops 2004].

EPA. EPA has regulatory authority over substances with pesticidal (including herbicidal) characteristics with regard to their effects on human health and environment. This agency does not actually regulate GM plants, but rather regulates the pesticidal properties associated with them as in the case of *Bt*-insect resistant maize or virus-resistant squash, as well as agrochemical residues. It ensures the safe use of pesticidal substances. Thus, a bio-engineered food that is the subject of a consultation with FDA may contain an introduced pesticidal substance, also known as a 'plant-incorporated protectant' (PIP), which is subject to review by EPA. In EPA terminology, these transgenic plants contain PIPs (also known as 'plant-pesticides'). With regard to transgenics, EPA has the authority to regulate a) the distribution, sale, use, and testing of plants and microbes containing pesticidal substances, b) the pesticide content of foods, and c) commercial use of genetically engineered micro-organisms, which includes a bacterium that allows a plant to fix nitrogen or produce a novel chemical [Transgenic Crops 2004].

In order to implement its oversight of transgenic crops [Transgenic Crops 2004], EPA

- examines data characterizing the plant-incorporated protectant, e.g., the biochemical nature of the product, its mode of action, and the time and tissues in which the product is expressed.
- reviews environmental effects (both risks and benefits) of the proposed plant-incorporated protectant, including effects on non-target organisms and environmental fate.
- may require a "resistance management plan" containing measures to slow down development of resistance in the target pest.
- determines whether the introduced gene or its product is toxic, typically based on toxicity testing in animals.
- sets tolerance levels for pesticide residues, if there is evidence of toxicity. Because of lack of toxicity in plant-incorporated protectants evaluated to date, they have been exempt from this requirement.
- regulates new uses of existing pesticides, such as use of herbicides together with herbicide-resistant transgenics.

In August 2002, the federal government formulated several changes for better regulation of biotechnology-derived crops by USDA, FDA, and EPA. According to these changes, developers of biotech crops need to submit protein safety information, particularly in regard to allergenicity and toxicity at an early stage in the evaluation process than practiced now. Besides, the changes also call for field testing protocols to minimize the potential for gene flow to seed and fields of commercial production and for establishment of procedures to ensure low level presence of new proteins in food and feed products that have not completed regulatory review.

Brazil

Transgenic glyphosate-tolerant soya bean was first grown illegally in the southern states of Brazil in 1997, a year after its legal adoption in Argentina. Following approval of Event GTS 40-30-2-6 (MON40302-6: Roundup Ready) soya bean in 1998, National Technical Mission Biosafety (Portuguese: Comissão Técnica Nacional de Biossegurança; acronym: CTNBio), the regulatory agency, faced legal battles from national and international organizations trying to stop the plantings of GMO crops in Brazil [Griffin et al. 2005]. In December 1999, some HR soya bean illegally planted in the southern state of Rio Grande Do Sul was destroyed by federal authorities [Lehman and Pengue 2000]. The ensuing legal action, lasting over 3 years, led to a moratorium on planting of GMO crops. In March 2003, the Brazilian government issued Provincial Measure (No. 113/2003) amnesty to soya bean farmers who had illegally planted during the ban imposed earlier.

This measure, converted into Law No. 10668/2003 in July 2003 [USDA 2013], legalized planting of transgenic soya bean. By then, more than 10 percent of the country's soya bean acreage had been under this biotech crop. The issuance of Provisional Measure and its conversion into a Law immediately led to a spurt in transgenic soya bean plantings in the country, particularly in Rio Grande Do Sul in southern Brazil. Yet the dissent in different states remained, especially in Parana which banned biotech soya bean [Griffin et al. 2005]. Since then, transgenic technology has been extended to cotton in 2006 and maize in 2008.

Brazil has a cropped area of 67.7 million ha and 59 percent of it (40.3 million ha) was planted with transgenics in 2013 [ISAAA 2013; Biosafety Scanner 2013]. Herbicide resistant soya bean is the largest transgenic crop, with a planting of 24.37 million ha in 2013, accounting for 82 percent of the gross area planted under this crop. Brazil has more varieties of transgenic soya bean than any other crop. Transgenic HR maize was planted on 5.18 million ha, accounting for 34 percent of the total area in 2011. Biotech maize containing HR, IR, and stacked traits was grown on 12.2 million ha (summer: 5.2 million ha; winter: 7.0 million ha). The transgenic cotton, with HR-IR stacked traits, covered an area of 550,000 ha in 2013, accounting for 50.5 percent of the gross cultivated area. Of the 44.7 million ha area under these three crops, transgenic varieties accounted for 83 percent.

Regulation

Integration and regulation of biotech crops in Brazil has not been an orderly process. It approved a biosafety law (8974/95) in 1995 that provided for a horizontal type of regulation that interfaces with other existing regulatory frameworks in the areas of agriculture, health, and environment [Fontes 2003]. However, planting of transgenic crop varieties has not been legalized. Realizing the growing popularity of transgenic HR soya bean in Rio Grande do Sul where farmers planted with seed smuggled from cross Argentina from 1997 to 2003, the Brazilian Congress passed a resolution in 2005 establishing an orderly process for the authorization and introduction of new GMO products in the country. The legislation created, by Law No. 11.105 of 24 March 2005, CTNBio which is vested with responsibility to review and authorize all new GMOs. In fact, CTNBio was initially established in 1995.

The 27-member—12 scientists, 9 Ministerial representatives, and 6 specialists—CTNBio has a mandate to develop and implement biosafety policies, propose the Code of Ethics on genetic manipulation, and determine GMO risk levels. It provides technical and advisory assistance to the Federal Government in the formulation, updating, and implementation of the national biosafety concerning GMOs. It also establishes technical safety standards and prepares guidelines for research, cultivation, handling, transportation, marketing, consumption, storage, release, disposal, and import of GMOs and food products derived from them. CTNBio, based at the Ministry of Science and Technology, also oversees the risk assessment made by the Institutional Biosafety Commissions, and considers other technical and scientific issues on biosafety [Varella et al. 1998].

Every private and public institution working on genetic engineering must establish an Internal Biosafety Commission (Comissão Interna de Biossegurança: CIBio) and apply for a Certificate of Quality on Biosafety (CQB) [Fontes et al. 1998]. CIBio, an essential component for monitoring and surveillance of the work of genetic engineering, manipulation, production, and transport of GMOs and to enforce biosafety regulations, assigns one principal technical officer for each research project.

The approval of any new GMO product involves a five-step process. The first step is to submit a research proposal to the CTNBio commission which, in turn, conducts an on-site inspection of the research facilities to ensure that there are adequate security protocols in place at the facility and that there is no danger to the outside environment while the research is being conducted. Once the project is approved, development and testing of the new GMO product can proceed under controlled conditions. The Minister of Agriculture routinely monitors the experiments to ensure safety procedures are in place. After the experiments are concluded, the CTNBio evaluates the data to ensure that it meets the criteria set up for biosecurity. Before a product is approved for commercialization, a commission of eleven federal Ministers examines the data and determines if the new product would be beneficial for the country.

Glufosinate-resistant maize Event T25 (Liberty Link) was the first biotech product approved, in 2010, by CTNBio since the enactment of the Biosafety Law of 2005. This third transgenic crop approval was preceded by the glyphosate-resistant Event GTS 40-30-2 (Roundup Ready) soya bean. It was approved in 1998. This was followed by insect-resistant *Bt* cotton Event MON531 [GAIN 2007].

Currently, farmers are confined to growing only soya bean, maize, and cotton despite approval for 36 distinct transgenic introduction/commercialization events (including 18 for maize, nine for cotton, five for soya bean, and one for virus-resistant edible bean) since 2003 [Katovich 2012; Brazil 2013]. Brazil, the first Latin American country to develop its own unique genetically modified crop, Golden Bean Virus Resistant Soya bean, approved it for commercialization in September 2011 [James 2012]. It was developed by the Brazilian Agricultural Corporation (Empresa Brasileira de Pesquisa Agropecuária: EMBRAPA), a public research institution linked to federal government. Currently, EMBRAPA is on track to incorporate Monsanto's Roundup Ready technology into its Golden Bean soya bean.

Brazil has sequenced the sugarcane genome as a first step towards developing more efficient biotech sugarcane for sugar and ethanol production with insect resistance. CTN Bioapproved the tri-event maize hybrid MON89034 x TC1507 x NK603 (commercial name: 'Power Core') which is resistant to glyphosate and glufosinate as well as lepidoptern insects (vide Chapter 6) [CTNBio 2010].

Labeling

Executive order 4680/2003 established a one percent tolerance limit for "food and food ingredients destined for human or animal consumption containing or being produced through biotech events" [Silva, JF 2010]. A later executive order (2658/2003) established guidelines for the use of transgenic logo "T" (no smaller than 1 cm) to be used to label all foods/food products/feeds that contain more than one percent tolerability benchmark. It imposes fines of US$1 to 65 million on producers that flout labeling regulations. Three organizations are responsible to enforce the law: the Ministry of Agriculture and the National Health Surveillance Agency to take care of agricultural and industry matters respectively, while PROCONS (Programa de Defasa do Consumidor) controls commerce of GM products. Despite the rhetoric of a strong mandatory labeling law for all GM products, there has been very little evidence of enforcement of the labeling requirement in Brazil [Silva, GR 2010]. The labeling enforcement remains apparently nonexistent, and the 'transgenic' crop is never seen in supermarkets despite the obvious existence of GM products with less than one percent content in these markets [Silva, JF 2010].

Argentina

Argentina, along with the U.S. and Canada, was among the first three big adopters of transgenic crops upon their introduction in 1996. Its ready acceptance of biotech crops was largely due to greater technical expertise, fewer small-land holders, and more reliance on modern monocrop agriculture than most other Latin American countries [Silva, GR 2010; Paarlberg 2001; Katovich 2012]. Argentina approved 27 transgenic crop events from 1996 to 2012, with 4, 20, and 3 of these belonging to soya bean, maize, and cotton respectively.

Of the 24.4 million ha (14 percent of global biotech crop acreage) planted with biotech crops in 2013, soya bean accounted for 20.6 million, amounting to 84 percent. Many Argentine farmers cultivate two crops (wheat followed by soya bean) in one season. Transgenic HR maize was planted commercially in 2004. The total maize area was 4.0 million ha. Of this, varieties containing a HR (glyphosate and glufosinate trait) were planted on 3.4 million ha, with 80 percent of it in the traditional maize production belt and 20 percent in newer area. Stacked biotech maize became available in 2007 and this contributed to a significant increase in the usage HR maize subsequently. In 2012, stacked transgenic maize covered over 85 percent of the total maize biotech area [Brookes and Barfoot 2013]. The stacked hybrids include glyphosate-resistant cum insect-resistant (*Bt*), glufosinate-resistant cum insect-resistant, glyphosate and glufosinate-resistant cum insect-resistant events.

Regulation

The competent authority for the administration and regulation framework of transgenic crops is the Secretariat of Agriculture, Livestock, Fisheries and Food (Spanish: Secretaría de Agricultura, Ganadería, Pesca y Alimentos: SAGyPA) within the Ministry of Agriculture, Livestock, and Fisheries [Burachik 2012]. National Advisory Committee on Agricultural Biotechnolgy (Comisión Nacional de Biotecnología Agropecuaria: CONABIA) is a multidisciplinary organization vested with responsibility of regulation of GMOs. Created in 1991, CONABIA is under the purview of SAGyPA. Initially, it was responsible for the regulatory and assessment processes and it had the administrative support of a branch of SAGyPA called Technical Coordination of CONABIA, which later became the Biotechnology Office. As the assessment and evaluation activity increased, these processes became the responsibility of the Biotechnology Directorate, created in 2009, to conduct the follow-ups and pre-assessment of applications submitted to develop GMO-related activities.

CONABIA evaluates the impact of GMO in the agricultural ecosystem. Its main responsibility is to assess, from a technical and scientific perspective, the potential environmental impact of the introduction of biotech crops in Argentine agriculture [Yankelevich 2012]. It also reviews and advises the Secretariat of Agriculture on issues related to trials and/or the release into the environment of transgenic crops and other products that may be derived from, or may contain, biotech crops. This multi-sectorial organization is made up of representatives from public sector, academia, and private sector organizations related to agricultural biotechnology. CONABIA members perform their duties as individuals and not as representatives of the sector they represent, and they are active participants in the international debate on biosafety and the related regulatory processes [Yankelevich 2012].

Besides CONABIA, there are other governmental organizations that are responsible in the regulatory framework. These include National Service of Agriculture and Food Health Quality (Servicio Nacional de Sanidad y Calidad Agroalimentaria: SENASA), National Direction of Agricultural Markets (del Instituto Nacional de Alimentos: DNMA), and National Seed Institute (Instituto Nacional de Semillas: INASE). SENASA evaluates the bio-safety of food products derived from bio-tech crops for human and animal consumption [Yankelevich 2012]. DNMA assesses the commercial impact on export markets by preparing a technical report in order to avoid negative impact on Argentine exports. It also analyzes the status of the transgenic event under study in destination markets. Under the new framework, the DNMA will evaluate the commercial impact on export markets within 45 days [Yankelevich 2012]. There was no timeframe for this evaluation in the past. INASE establishes requirements for registration in the National Registry of Cultivars.

The evaluation of a GMO starts during its developmental stage and continues until its eventual commercial production. Assessment is done in two stages. At the initial stage, the Biotechnology Directorate and CONABIA analyze the genetic information inserted into the GMO with all its characteristics and behavior in the agro-ecosystem (genotype and phenotypic expression). Once the safety of the GMO is established, its potential behavior in the agricultural production and the potential effects on health are assessed in relation to GMO management. During this process, not only is the GMO analyzed and measures are taken to prevent risks but also the applicant's (institution or person who intends to conduct trials) capacity regarding GMO use and management is evaluated. After completing these steps, the Biotechnology Office of CONABIA compiles all scientific and technical information and quantitative data about the safety of the GMO before preparing a final report for submission to the Secretary of Agriculture, Livestock, Fisheries and Food for final decision on GMOs and GMO-related activities.

Approval of stacked events is based on a case by case evaluation under which the applicant insists on submitting a letter to the Ministry of Agriculture and SENASA simultaneously. The evaluation is based on possible metabolic interactions between the individual events contained in the stacked event. It is also evaluated regarding its possible effects on the ecosystem as well as the food biosafety.

Argentinian scientists have developed a drought tolerant biotech sugarcane and are exploring co-operation to further develop this product with Brazil, which is also working on drought tolerant sugarcane. The product from this joint program could be ready by 2013 and approved for production by 2017 [Argentina 2013]. Such a product would allow Argentina to increase sugarcane area from the current 350,000 ha to 5 million ha in the future.

CONABIA is currently evaluating two other sugarcane products–glyphosate-resistant (Roundup Ready) sugarcane and *Bt* sugarcane. The glyphosate-resistant sugarcane could be approved for commercialization in 2014, and this could be the first biotech sugarcane to be commercialized globally [Argentina 2013]. CONABIA is currently evaluating biotech potatoes resistant to potato virus Y (PVY) and potato leafroll virus (PLRV) which cause significant yield loss, as well as herbicide tolerance. This product could be approved for commercial production in the near future [Argentina 2013].

Argentinian scientists have also transferred a drought tolerant gene from sunflower to maize, soya bean and wheat. BioCeres, an Argentinian company, has been granted a license for this gene and has a joint venture named Verdeca, with Arcadia Biosciences from the U.S. Field trials with the new seeds have increased yield by 15 percent or more and Verdeca has indicated that the drought tolerant seeds could be in the market as early as 2015–16 [Argentina 2013].

Argentina approved the tri-event maize hybrid MON89034 x TC1507 x NK603 ('Power Core' developed by Monsanto and Dow AgroScience) which is resistant to glyphosate and glufosinate as well as lepidopteran insects (vide Chapter 6) [SENASA 2012].

Labeling

There is no specific regulation in Argentina regarding labeling of biotech products. The current regulatory system is based on the characteristics and identified risks of the product and not in the production process of that product [Yankelevich 2012]. This policy of the Ministry of Agriculture on labeling of transgenic products is that it should be based on the type of food product derived from a specific biotech seed by taking into account that any food product obtained through biotechnology and substantially equivalent to a conventional food product should not be subjected to any specific mandatory label [Yankelevich 2012]. Besides, if a food product obtained through biotechnology is substantially different from a conventional food product for any specific characteristic, it may be labeled according to the characteristics of the product, but not according to aspects concerning the environment or production process.

Canada

Canada was the first country to commercialize HR canola (rapeseed) in 1996. The other three transgenic crops currently grown are maize (first generation adopted in 1997 and the second in 2009), soya bean (1997), and sugar beet (2005), all resistant to herbicides. In 2013, these four biotech crops were planted on 10.8 million ha, a share of 6.2 percent in global biotech crop area. The adoption rates in HR canola, HR and IR maize, HR soya bean, and HR sugar beet stood at 95 percent, 71 percent, 63 percent, and 96 percent respectively [Evans and Pupescu 2012]. Most of canola's production is centered in the western provinces of Manitoba, Saskatchewan, and Alberta. Canadians consume only 15 percent of the canola in various forms, with the rest being exported to the U.S., Japan, Mexico, and China. Traditionally, Quebec and Ontario are the primary maize and soya bean growing regions, accounting for 86 percent and 92 percent of the area under these crops respectively [Evans and Lupescu 2012].

There have been concerted efforts to seek approval for glyphosate-resistant (MON71800: Roundup Ready) wheat from 2002. However, these have been blocked by fear of lack of consumer acceptance of biotech wheat and farmers' unwillingness to adopt it because it would cost them their international wheat markets. As Canadian wheat groups continue to oppose GM wheat, there is no biotech variety that is in the regulatory approval pipeline at the present time [Evans and Lupescu 2012]. Nevertheless, Canada has approved field trials on transgenic forest trees belonging to *Populus* spp. (poplar), *Picea mariana* (black spruce) and *Picea glauca* (white spruce) [GM Science Update 2014].

Regulation

Health Canada is responsible, under the Food and Drugs Act and its Regulations, for the approval and regulation of genetically modified foods and other novel foods proposed for sale in the country. This agency is also responsible for evaluating the safety and the nutritional value of genetically modified foods. The Canadian regulatory system is based on whether a product has novel features regardless of method of origin. In other words, a product is regulated as genetically modified if it carries some trait not previously found in the species whether it was generated using traditional breeding methods (e.g., selective breeding, chemical mutagenesis, cell fusion, etc.) or genetic engineering [Evans and Lupescu 2012; Health Canada 2005].

Three agencies are responsible for the regulation and approval of products derived from biotechnology. These are the Canadian Food Inspection Agency (CFIA), Health Canada (HC), and Environment Canada (EC). All three agencies work together to monitor development of plants with novel traits, novel foods, and all plants or products with new characteristics not previously used in agriculture and food production [Evans and Lupescu 2012].

The CFIA is responsible for regulating the import, environmental release, variety registration, and use of livestock feed of plants with novel traits. It provides all federal inspection services related to food and enforces the food safety and nutritional quality standards established by Health Canada. This agency also oversees the environmental release of GM crops (under the Seeds Act) and feeds derived from them (under the Feeds Act), as well as makes and enforces decisions on safety. Health Canada is responsible for assessing the human health safety of all GM foods and other novel foods, and approving their use in commerce. Environment Canada administers the 'New Substances Notification Regulations' and performs the environmental risk assessments under Canadian Environmental Protection Act (CEPA) of toxic substances including organisms and microorganisms that may have been derived through biotechnology [Evans and Lupescu 2012]. Provincial governments support the leadership role played by the federal government in regulating agricultural products of biotechnology.

Canadian law requires that manufacturers and importers submit detailed scientific data to Health Canada for safety assessments and approval. This data includes: information on how the GM plant was developed; nucleic acid data that characterizes the genetic change; composition and nutritional data of the novel food compared to the original non-modified foods regarding their potential for new toxins; and potential for being an allergen. A decision is then made whether to approve the product for release along with any restrictions or requirements. The timeline from development to the point at which the product has been approved for human consumption may take 7 to 10 years and even longer in some instances.

Transgenic crops and foods have been the subject of much controversy in Canada [Sharma and John 2012]. Proponents argue that transgenic foods will help provide food to the ever expanding global population because of higher yields through resistance to herbicides, drought, cold and diseases as well as through improvement in the foods' nutritional content. Opponents have many criticisms on several aspects such as long-term health effects, fate of foreign DNA upon digestion, labeling and consumer choice, intellectual property rights, ethics, and food security. Critics also have environmental concerns about the potential harm that GM foods can unintentionally cause to other non-target organisms, or the fact that transgenic plants could crossbreed with wild plants and the introduced gene be transferred to non-target plants [Sharma and John 2012]. For instance, a herbicide-resistant crop could outbreed with a weed and transfer the herbicide-resistant gene thus creating a weed resistant to that specific herbicide. Insects could also become insecticide-resistant, making them difficult to control in the near future.

Labeling

Canada has a voluntary labeling policy for GM foods. However, mandatory labeling is required if the introduced gene poses an allergy risk or if the food's nutritional content is changed. Health Canada and the CFIA are responsible for all federal food labeling policies under the Food and Drugs Act. Health Canada sets food labeling policies with regard to health and safety matters, while the CFIA develops non-health and safety food labeling regulations and policies. It is the CFIA's responsibility to protect consumers from misrepresentation and fraud with respect to food labeling, packaging, and processing as well as for prescribing basic food labeling and advertising requirements applicable to all foods [Evans and Lupescu 2012]. There have been unsuccessful efforts by certain consumer activists to push for mandatory labeling of all genetically engineered foods containing biotechnology components.

India

Adoption

TransgenicIR (*Bt*) cotton {Monsanto's lepidopteran-insect tolerant Event MON531 (Bollgard I) and its three modified hybrids (Mech-12 *Bt*, Mech-162 *Bt*, and Mech-184 *Bt*} developed by Monsanto and its Indian partner Maharashtra Hybrid Seed Company (Mahyco) was first planted commercially in 2002 on 50,000 ha. These hybrids were developed by crossing Monsanto's Event MON531 with local elite Indian varieties. In 2006, Monsanto commercialized Event MON15985 (Bollgard II) which carried two IR genes, *cry1Ac* and *cry2Ab2*.

In 2013, MON15985 covered over 60 percent of insect resistant (IR) cotton. In 12 years time, the area under Monsanto's IR cotton has spread to 11 million ha, covering 91 percent of the gross cotton area of 12.1 million ha, the largest in the world. During this period, Indian companies developed several *Bt* cotton hybrids under sub-licenses of Monsanto. Over 1,100 *Bt* cotton hybrids have been released for commercial adoption until

2012. Of these, 215 hybrids were based on MON531 and 528 hybrids on MON15985. In 2011, Mahyco-Monsanto developed a glyphosate-resistant transgenic cotton hybrid that contained *cp4 epsps* gene. This original event MON88913 has already been approved in the U.S. in 2005.

The initial success of *Bt* cotton encouraged other international biotech companies set up shop in India either directly or by stitching partnerships with domestic biotech companies. Currently, there are around 140 of them conducting genetic engineering research to develop transgenic crop hybrids/varieties carrying such traits as insect resistance, herbicide resistance, disease resistance (fungal, bacteria, virus), yield increase, drought tolerance, nutrition enhancement, etc. The crops include cotton, rice, maize, sorghum, groundnut (peanut), potato, sugarcane, chickpea, *brinjal* (eggplant: *Solanum melongena*), tomato, watermelon, papaya, cabbage, cauliflower, *bhendi* (okra: *Abelmoschus esculentus* Moench), chillies, capsicum, pomegranate, banana, papaya, etc. Over 50 universities and 45 research institutions funded by central and state governments are also engaged in transgenic research.

Regulation

Transgenic engineering is a highly debated and contested field in India like almost in every nation in the world. There are many governmental and non-governmental organizations and actors involved in the debate on the benefits and risk the genetically modified crops and foods. As a result, India's regulation policy on transgenic crops and foods has undergone various shifts.

India's regulation of transgenic crops stems from the Environmental Protection Act (EPA) 1986 which covers research, development, large-scale use as food or feed, export, and import of transgenic events and products besides storage of genetically engineered organisms or cells. As per the Rules framed in 1989 under this Act, there are six regulatory authorities at the national and local levels. The premier national level regulator is the Recombinant DNA Advisory Committee (RDAC) functioning within the Department of Biotechnology. It monitors and reviews developments in biotechnology at national and international levels and makes recommendations for implementation of appropriate safety regulations in transgenic research, use, applications, and products. This committee is supported by a three-tier system consisting of Institutional Biosafety Committee, Review Committee on Genetic Manipulation, and Genetic Engineering Appraisal Committee.

The Institutional Biosafety Committee (IBSC) is the nodal point established by each of the research institutions and universities involved in transgenic research for implementation of the recombinant DNA guidelines and safety procedures. The Review Committee on Genetic Manipulation (RCGM), functioning in the Department of Biotechnology at the national level establishes, monitors, and issues specific guidelines on the safety aspects of research projects and activities involving genetically engineered organisms/hazardous microorganisms. RCGM lays down procedures restricting or prohibiting production, sale, import, and use of genetically modified organisms. The Genetic Engineering Approval Committee (GEAC) in the Ministry of Environment and Forests of Government of India, is responsible for approval of activities involving large-scale use of GMOs in research and industrial production. The committee is also responsible for approval of research proposals relating to release of genetically engineered organisms and products into the environment.

Each state within the country has a State Biotechnology Coordination Committee (SBCC) vested with authority to inspect, investigate, and take punitive action in case of violations of statutory provisions through other related departments (pollution, health/

medical). It monitors and reviews safety measures followed by institutions and industries involved in genetically engineered organisms and hazardous microorganisms. District Level Biotechnology Committees assists SBCC in implementing state policies.

Approval of the insect-resistant *Bt* cotton hybrids developed by Mahyco-Monsanto Biotech (MMB) was denied by GEAC in early 2001. What forced the Indian regulator to grant approval for their commercialization in 2002 was the news that *Bt* cotton had already been planted illegally by farmers in the western state of Gujarat later in 2001, with seed supplied by a local biotech company. When the 3-year period ended in 2005, GEAC approved renewal of these and many more *Bt* cotton varieties, except in the southern state of Andhra Pradesh where there was a sudden spurt in suicides by farmers who are believed to have grown *Bt* cotton.

Before MMB received GEAC's approval for *Bt* cotton, Monsanto had approached the Department of Biotechnology of the Ministry of Science and Technology in 1990 for transfer to India, technology of two constructs containing *cry1Ac* and *cry1Ab* genes and import of cotton seeds containing *cry1Ac* [Ghosh 2001]. Monsanto's offer was refused because the government felt the technology transfer fee was too high. In fact, the American biotech giant tried to sell *Bt* cotton technology to India for $45 million which the government felt could be developed within the country at a cost 20 times less [Bhargava 2003]. This decision was revised in 1995 when the government granted permission to Mahyco to import 100 g of transgenic cotton as per its agreement with Monsanto. In 1996, Mahyco imported seeds to use them for backcrossing into Indian cotton cultivars.

The 1993 rejection of Monsanto's offer of technology transfer led the company to build alliances with the indigenous not-for-profit biotech institutions such as Tata Energy Research Institute, New Delhi in December 2000, to develop 'golden mustard' that would produce cooking oil high in β-carotene (pro-vitamin A) and Indian Institute of Science, Bangalore in 1998 to conduct research in agricultural biotechnology in the areas of genome sequencing, and functional genomics, raising the nutritive value of cereals. These attempts by Monsanto were seen by sceptics more as goodwill gestures to cash in on potential future monetary benefits.

One of the most contentious issues roiling currently in India is whether the country should permit commercial planting of genetically modified food crops [News & Analysis 2013]. The turning point came in February 2010 when the Ministry of Environment and Forests called for an indefinite moratorium on the cultivation of brinjal (eggplant: *Solanum melongena*) engineered to contain the *Bt* gene, *Cry1Ac*. The ban would remain effective until studies establish "the safety of the product from the point of its long-term impact on human health and the environment" [Mercola 2012]. The GEAC approved *Bt* brinjal, Event EE1, developed by MMB in October 2009 for commercial cultivation, only to be followed by serious concerns expressed by some scientists, farmers, and anti-GM activists. As brinjal is an important vegetable crop in India, commercial cultivation of this transgenic brinjal was also severely opposed by consumer activists. Added to this was Monsanto's alleged attempt, in collaboration with Mahyco, to resort to 'biopiracy'—which, in essence, is using native brinjal varieties for the purpose of genetic modification—thus violating the country's Biological Diversity Act, 2002 [Mercola 2012].

In response to a public interest petition filed in 2005 for banning GM crops in India because of approval of field trials by GEAC without proper scientific evaluation of biosafety issues, the Supreme Court appointed a five-member Technical Expert Committee (TEC) on 10 May, 2012. In its report submitted to the court on 7 October, 2012, TEC recommended a 10-year moratorium on commercial release of all GM crops till all the systems are in place for independent research and regulation. It also recommended a moratorium on field trials of herbicide-tolerant crops until an independent assessment evaluates its impact and suitability. The country now awaits the final decision of the Supreme Court.

Despite the government's earlier imposition of indefinite moratorium on approval of transgenic crops and the advice of TEC appointed by the Supreme Court for a 10-year moratorium, GEAC—in March 2014—approved field trials of 11 varieties of GM crops that had come to it for re-validation. This came despite its decision to allow trials for only non-food crops and severe opposition by a majority of states [Mukherjee 2014]. This contentious issue awaits a final resolution by the Supreme Court. What is, however, evident has been that over two-thirds of India's farming community is against transgenic crops.

Regardless of what is happening in India, the neighboring Bangladesh approved *Bt* brinjal in October 2013 when its National Committee on Biosafety cleared four transgenic brinjal varieties BARI Bt Brinjal-1 (Uttara), BARI Bt Brinjal-2 (Kajla), BARI Bt Brinjal-3 (Nayontara), and BARI Bt Brinjal-4 (ISD006). Farmers are free to save seeds from their crop produce without the need to buy them from companies every year. These *Bt* brinjal varieties were developed to provide resistance to the lepidopteran shoot/fruit borer (*Leucinodes orbonalis*), the most destructive insect pest in this highly popular vegetable crop in South Asia. The farmers began cultivating these new transgenic varieties in January 2014.

New Regulations. The existing regulation rules have been heavily criticized for incompetence and non-transparency in the decision-making process related to GM organisms. In response to demands from biotech and pharmaceutical industry for a simpler regulation procedure, the Indian government proposed in 2008 to replace the current regulatory regime with a bill known as the Biotechnology Regulatory Authority of India Bill, 2013. Under the BRAI Bill, awaiting approval by the parliament, the proposed National Biotechnology Regulatory Authority (NBRA) will act as a single window fast track clearance body. This autonomous and statutory agency regulates the research, transport, import, manufacture, and use of genetically engineered organisms and products derived thereof. The Bill also provides, unlike in the past, for a Regulatory Authority Appellate Tribunal to hear appeals against NBRA's decisions, orders, and directions.

Opponents of NBRA contend that the Bill lowers the bar for the approval of genetically modified/engineered crops because it bypasses the Right to Information Act, provides opportunity for conflict of interest, curtails participation of public in decision making, deviates from task force report, overrides state governments' role, lacks socio-economic assessments, and kills consumer choice [CLRA 2013]. This Bill is in conflict with the provisions of Cartagena Protocol, the binding international agreement on biosafety, which India ratified on 17 January, 2003. The Protocol stipulates that governments shall consult the public in decision-making processes regarding GMOs and make all relevant decisions available to the public. It also stipulates that information about a summary of the risk assessment cannot be made confidential. Much against the spirit of this protocol, the BRAI Bill does not make any reference to the risks associated with 'modern biotechnology' and its potentially adverse effects on biological diversity and human health [CLRA 2013]. India is required to put in place a safety protocol that mandates openness, transparency, and public participation.

Labeling

Currently, there are no restrictions in marketing either domestically produced transgenic cottonseed oil and meal or imported foods. Besides, India has approved in 2006, imports of soya bean oil derived from glyphosate-resistant transgenic soya bean for consumption after refining. In order to placate consumer concerns, the national government has made it mandatory, effective 01 January 2013, for packaged foods derived from genetically modified crops to carry GM labels. The letters 'GM' should be displayed at the top of the label. The decision was made because many products in India are either derived from, or processed in, countries where a large majority of cultivated crops are genetically modified.

While this decision was a step in the right direction, critics contend that in a country where 90 percent of the food consumed is unpackaged and unprocessed there is no way for people to know whether a product is genetically modified or not. Besides, this notification fails to take note of the negative impact of GM crops and foods on human health and the environment as also contamination of non-transgenic crops by open transgenic crop fields.

China

The insect-resistant *Bt* cotton is China's biotech program's success story [Huang and Yang 2011]. In 2013, this IR cotton, commercialized in 1996, was grown by an estimated 7.2 million small, resource-poor farmers on an area of 4.0 million ha out of the total cotton planting of 4.933 million ha at an adoption rate of 81 percent. Despite approving several biotech events for commercial production, *Bt* cotton is the only crop that has been widely adopted by farmers. By 2006, China had approved the commercialization of late-ripening tomato, virus-resistant sweet pepper and tomato, and Papaya Ring Spot Virus-resistant papaya events besides transgenic petunia and poplar varieties.

Among the hundreds of biotech products under development that have been approved for productive testing are insect resistant rice (Event GM Shanyou 63; commercial name: Bt63), bacterial blight resistant rice (Xa21), high phytase maize, and high oil content canola. Other major crops undergoing field trials include insect resistant maize, high lysine content maize, pre-harvest germination-resistant wheat, and insect resistant soya bean [Cummins 2010].

China approved two varieties of insect-resistant *Bt* poplar varieties which have been planted since 2002 and by 2011 they occupied a total of 490 ha [Häggman et al. 2013] to meet its future timber needs. One variety *Populus nigra* was transformed with the *cry1Ac* gene and the second variety is a hybrid white poplar which is transformed with a fusion of *cry1Ac* and *API* (a gene coding for a proteinase inhibitor from *Sagittaria sagittifolia*). The transgenic *P. nigra* has also been hybridized with non-transgenic *Populus deltoides* to generate insect-resistant germplasm for a breeding programme designed to generate new hybrid varieties, that could expand the planting area of *Bt* poplar [Häggman et al. 2013]. Besides, field trials are in progress on *Robinia pseudoacacia* (black locust), *Larix* spp. (cotton names: tamarack, hackmatack, eastern lurch, black lurch, or red lurch), and *Paulownia* (common names: empress tree, princess tree or foxglove tree) [GM Science Update 2014].

Chinese leaders and scientists view transgenic technology as an important tool to boost agricultural productivity and improve national food security. China's R&D of agricultural GMOs, which began in the mid-1980s, heavily relies on funding from the government. This accounts for one-third of global funding on biotechnology. China is developing the largest biotechnology capacity outside North America. There are over 200 institutions focusing on research and development of agricultural biotechnology. These are divided into four groups. The first category includes institutions or firms affiliated to China Academy of Agricultural Science (e.g., Biocentury Research Institute), while the second group encompasses institutions and firms affiliated to national and provincial governments. The joint ventures such as Andai Seed Co., Judai Seed Company, etc. belong to the third category. Private institutions and companies are grouped in the last. In 2008, China initiated a new GM program with a total budget of US$3.8 billion for 2009–2020, focusing on transgenic rice, wheat, maize, cotton, soybean, pig, cattle, and sheep [Huang and Yang 2011]. China is pushing forward with GM technology and has successfully tested more than 1,000 transgenic events.

Biosafety approval for *Bt* rice, phytase-containing maize, and other biotech crops in the pipeline in 2009 by the Ministry of Agriculture represents a major milestone in China's biotech development which is expected to have significant implications for biotech development on the rest of the world and for trade flows between China and its major trade partners. As *Bt* rice and phytase maize still need to go through regional varietal demonstration and registration processes, it is expected that they will be cultivated extensively in the near future. Additionally, biotech maize, wheat, and soya bean are also in pre-production stage, the last stage of biosafety regulation before they will be issued biosafety certifications for production [Huang and Yang 2011].

Regulation

Currently, China has a comprehensive biosafety regulation and monitoring system for commercialization of domestic transgenic crops and imported transgenic crop events. Although a significant number of domestic transgenic crop events have been in the regulatory pipeline, China has not sought their approval in any foreign country. Chinese commodity exporters are concerned about trade disputes arising from low level presence of China-developed transgenic crop events in the importing country [Huang and Yang 2011]. China has been importing transgenic soya bean meal and oil for over 12 years, particularly from Brazil, and it is considered to be the largest biotech soya bean importer.

In 1993, Ministry of Science and Technology (MOST) of China issued the first biosafety regulation 'Measures for Safety Administration of Genetic Engineering'. This regulation consisted of general principles, safety categories, risk evaluation, application and approval, safety control measures, and legal responsibilities. Following MOST's guidelines, the Ministry of Agriculture (MOA) issued the 'Implementation Measures for Safety Control of Agricultural Organism Biological Engineering' in 1996. It covered plants, animals, and microorganisms. The "Implementation Measures" provided detailed procedures of biosafety regulation at each stage of development of a GM organism (GMO). Safety regulation has followed a case-by-case procedure. However, labeling was not part of this regulation, nor was any restriction imposed on imports or exports of GMO products. The regulation also did not regulate processed GM food products. Under this regulation, the Biosafety Committee was established in 1997 to provide MOA with expert advice on biosafety assessments.

With the continued development of agricultural biotechnology, increasing GMO imports, and rise in consumer concerns, China has periodically amended its biosafety regulations. The amended regulations, effective 20 March 2002, encompassed trade, labeling of GM food products. These regulations also established a "zero tolerance policy for unapproved GM products" as well [Huang and Yang 2011]. However, this policy was the reason for exporters and biotech countries from the U.S. and other countries to raise concerns about potential disruptions to trade [Huang and Yang 2011].

Under these regulations, the Ministry of Agriculture (MOA) is the primary institution in charge of implementing the agricultural biosafety regulations. Leading Group on Agricultural GMO Biosafety Management, the governing body under MOA, oversees the Agricultural GMO Biosafety Management Office (BMO). The biosafety assessments are conducted by the National Agricultural GMO Biosafety Committee (BC). Currently, the BC meets three times each year to evaluate all biosafety assessment applications related to three phases of experimental research {field trials (small scale), environmental release (medium scale field trials), and pre-production testing (large scale field trials)}, commercialization of GMOs, and events of import. Based on the BC's technical assessments and other considerations (social, economic, and political factors), the BMO prepares the recommendations to the MOA's Leading Group which is tasked with taking approval or disapproval decisions. The application process could take two years for completion, but it depends on the product's intended use and potential risk to human or animal health and the environment [Huang and Yang 2011].

Although the Chinese approach on regulating transgenic food crops is slower and time-consuming, it is considered more reliable than regulatory approval in the U.S. where a transgenic event once approved can be bred into other varieties or land races without further approval [Petry and Bugang 2009]. Approval by variety may leave a dangerous loophole if the 'variety' actually includes more than one transgenic event (stacked), as it is in the unreliable, unpredictable nature of genetic modification that each event is unique and differs according as to where and in what form the transgene has landed, and what kind of collateral damage has been done to the host genome [Ho 1999].

Foreign investment on research and production of biotech plants, livestock, and aquatic products is prohibited; but it is allowed in conventional seed production. China has approved four transgenic crops/products for import as processing materials (soya bean, maize, rapeseed mustard/canola, and cotton). The first batch of safety certificates was granted to imported biotech crops in 2004. The 28 varieties/events approved for import processing include those with the following traits: 15 herbicide-tolerant, five insect-resistant, and five herbicide-cum-insect resistant, in addition to three canola events with reduced undesirable fatty acids. Production of seeds for crops that are not genetically modified can be undertaken in partnership with multinational seed producers; but multinational corporations are minor stock holders in joint ventures with Chinese companies.

In spite of the restrictions on production or import of seed of transgenic crops, international biotechnology companies have extensive presence in China because they also produce herbicides, insecticides, etc. as well as non-transgenic crops seed. These companies include Dow-Pioneer, Monsanto-Seminis, and Syngenta Corporation, Bayer CropScience, etc. [Petry and Bugang 2009].

Europe

The first transgenic crop to be adopted by Europe was the insect resistant *Bt* maize event MON810 in 1998. The only other GM crop allowed for cultivation in the European Union (EU) was in 2010 when it approved the amylopectin starch-rich event potato, event EH92-527-1, known as the 'Amflora' potato, which has been genetically modified by BASF to produce starch for use in papermaking. It was, however, grown only in small quantities in Germany and Sweden in 2011. Although BASF stopped its commercialization activities in Europe in January 2012 due to overwhelming opposition to the technology, it continues to seek regulatory approval for its products in Asia and the Americas [Kanter 2012].

Biosafety-issues and food-safety concerns that arose shortly after planting of the insect-resistant *Bt* maize event MON810 have delayed adoption of transgenic crops in the food chain of the EU [Jensen et al. 2009]. These were caused by a series of unrelated food crises that created consumer apprehension about food safety in general and the consequent erosion of public trust in government oversight of the food industry [Pew 2005]. This spurred the EU to restrict and even ban imports of GM food. In May 2003, the U.S. and 12 other countries filed a formal complaint with the World Trade Organization that the EU was violating international trade agreements in blocking imports of U.S. farm products through its long-standing ban on genetically modified food. In response, the EU Parliament ratified in June 2003, a U.N. biosafety protocol regulating international trade in GM food, and a month later agreed to new regulations requiring labeling and traceability, as well as an opt-out provision for individual countries. Following this, the approval of new GMOs began again in May 2004. While a number of other GMOs have been approved since then, approvals remain controversial and various countries have utilized the opt-out provisions. In 2006, the WTO ruled that the pre-2004 restrictions had been violations [Genetically Modified Food Controversies 2013], although the ruling had little immediate effect since the moratorium had already been lifted.

Regulation

Initially, regulation of transgenic crops and foods in Europe was not as strict as it was in the U.S. Later, the European Union (EU) has adopted the most stringent GMO regulations in the world [Davison 2010]. The turning point in Europe's decision to initiate stricter regulation guidelines came at the end of 1996 when Europe received shipments of American glyphosate-resistant soya bean event GTS 40-30-2 (MON40302-6: Roundup Ready). This attracted considerable media attention and significantly increased public awareness and concern throughout Europe. In these imports, transgenic soya bean made up about 2 percent of the total harvest at the time, and EuroCommerce and European food retailers required that to be separated [Maitland 1996]. When that was not possible, it paved the way for a marked change in risk assessments by regulatory authorities in a number of Member States.

In 1998, the U.K. press described the genetically modified foods as "Frankenstein's foods" [Urry and Parker 1998]. In Switzerland, a June 1998 referendum was to ask its citizens whether they wanted to "protect life and the environment against genetic manipulation" [Urry and Parker 1998]. In Germany, over 80 percent of the public expressed a negative opinion of GMOs [Jacob 1998]. In the Netherlands, participants in a demonstration organized by the Alternative Consumers' Union dressed up as a genetically-engineered strawberry, the "Grim Reaper", and the "Devil" to protest biotechnology [Declaration to the US Government 1997]. In 1998, MON810, a *Bt* maize expressing resistance to the European corn borer, was approved for commercial cultivation in Europe. However, shortly thereafter the EU enacted a *de facto* moratorium on new approvals of GMOs pending new regulatory laws passed in 2003. Currently, MON810 is the only transgenic food crop with approval for cultivation in Europe. As of August 2012, the European Union had authorized 48 GMOs, with most of these being for use in food, feed, and food additives.

In order to ensure that the development of modern biotechnology, and more specifically of GMOs, takes place in complete safety, the EU has established a legal framework in September 2003. The 'Regulation (EC) 1829/2003' on genetically modified food and feed provides general framework for regulating genetically modified (GM) food and feed in the EU. The framework of these laws, effective April 2004, pursues the global objective of ensuring a high level of protection of human life and health and welfare, environment and consumer interests, whilst ensuring that the internal market works effectively. This Regulation was supplemented by 'Regulation (EC) 1830/2003' which ensures labeling and traceability of GMOs placed on the market and the traceability of GM food and feed products.

According to laws that apply to all 27 EU Member States, a GM food can only be allowed onto the market if it can be documented using scientific data that it is just as safe and healthy as a comparable conventional product. When evaluating the safety of food from a genetically modified organism, two areas are looked at in particular: the safety of the novel GM trait and unforeseen changes in plant metabolism as a result of gene transfer [GMO Compass 2006].

The criteria for authorization, vested in European Food Safety Authority (EFSA), fall in four broad categories: a) safety, b) freedom of choice, c) labeling, and d) traceability [GMO Compass 2012]. EFSA was established in 2002 (based in Parma, Italy) as the central authority for the scientific evaluation of food and feed safety in the European Union. It reports to the European Commission (EC) which then drafts a proposal for granting or refusing the authorization. This proposal is submitted to the Section on GM Food and Feed of the Standing Committee on the Food Chain and Animal Health and if accepted it will be adopted by the EC or passed on to the Council of Agricultural Ministers. The Council has three months to reach a qualified majority for or against the proposal, and if no majority is reached the proposal is passed back to the EC which will then adopt the proposal [Davison 2010; Wesseler and Kalaitzandonakes 2011]. However, even after authorization, individual European Union member states can ban individual varieties under a 'safeguard clause' if there are "justifiable reasons" that the variety may cause harm to humans or the environment. The Member State must then supply sufficient evidence for the Commission to investigate the cases before either overturning the original registration or recommending the country to withdraw its temporary restriction.

The EU laws also stipulated that member nations establish co-existence regulations [Beckman et al. 2009]. In many cases, national co-existence regulations include minimum distances between fields of transgenic crops and non-transgenic crops. For example, the distances for transgenic maize from non-transgenic maize for the six largest biotechnology countries are: France: 50 m; Britain: 110 m for grain maize and 80 m for silage maize; the Netherlands: 25 m for general maize and 250 m for organic or GM-free maize; Sweden: 15–50 m; and Germany: 150 m and 300 m from organic fields [Cooper 2012]. Larger minimum distance requirements discriminate against adoption of GM crops by smaller farms [Beckman et al. 2006, 2010; Groeneveld et al. 2013]. The EC has issued guidelines to allow the co-existence of GM and non-GM crops through buffer zones (where no GM crops are grown) [GMO Safety 2010].

One consumer concern is regarding allergens that transgenic foods could possible trigger in human and animal systems. When a new gene from bacteria is introduced into a plant—as is the case with herbicide resistance and insect resistance—the general outcome is the formation of a new protein. These proteins are often times new for human consumption. The safety of these proteins regarding toxicity is assessed using animal feeding tests. These tests are routine for herbicide residues and food additives. When results from animal trials are applied to humans, considerable extra safety measures must be taken. Safety evaluations must include tests to find out if the new protein could trigger allergies.

Labeling and Traceability

Labeling provides information for consumers and farmers, and allows them to make an informed choice. Under the rules of EFSA, all transgenic foods and feeds must carry a label referring to the presence of GMOs. Any food (including processed food) or feed which contains greater than 0.9 percent of approved GMOs must be labelled. In the case of pre-packaged products consisting of, or containing, GMOs, the list of ingredients must indicate 'genetically modified' or 'produced from genetically modified (name of the organism)'. For products without packaging, these words must be clearly displayed in close proximity of the product. The margins of admixing of GMO and non-GMO products should be in line with market demands. Depending on the country and region as well as the existing markets, the measures can be taken to ensure adherence to the designated threshold value of 0.9 percent or lower. If no ecological harm is expected, no restrictions to cultivation should be applied. In reality, Europe does not cultivate significant quantities of GM crops. Hence, the labeling threshold particularly applies to imported food and feed from the GM producing countries [Davison 2010].

In the light of EU's zero-tolerance policy, shipments of unapproved GM foods, which arrived in the EU twice in the past, have been forced to return to their port of origin in the U.S. [Davison 2010]. The first shipment was in 2006, containing an experimental transgenic glufosinate-resistant rice variety 'LLRice601' (Liberty Link) not meant for commercialization, which arrived at Rotterdam (Netherlands). The second shipment was in 2009 when trace amounts of a transgenic maize variety, approved in the U.S. but unapproved in EU, were found in a GM soya bean (originally approved for import into the EU) cargo.

Traceability is the ability to trace all GM food to its origin at all stages of supply chain. Traceability makes it possible to label all GM food/feed, to closely monitor the potential effects on the environment and on health and, where necessary, to withdraw products if an unexpected risk to human health or to the environment is detected. Under the traceability rules, all operators involved, including those who introduce a product to the supply chain or receive such a product (farmers, food/feed producers, etc.), must be able to identify their supplier and the companies to which the products have been supplied. Traceability archives must be kept for five years. European research programs such as Co-Extra, Transcontainer, and SIGMEA are investigating appropriate tools and rules for traceability. The 34-nation OECD (Organization for Economic Cooperation and Development) has introduced a "unique identifier" which is given to any GMO when it is approved, which must be forwarded at every stage of processing. A 5-digit 'Price Look-Up' (PLU) code (printed on a small sticker) beginning with the digit '8' indicates genetically modified food. However the absence of this digit does not necessarily indicate the food is not genetically modified since no retailer to date has elected to use the digit voluntarily to label genetically modified foods. Such measures are generally not used in North America because they are considered expensive and the industry admits of no safety-related reasons to employ them.

EU regulations also call for measures to avoid mixing of foods and feed produced from GM crops and conventional or organic crops, which can be done via isolation distances or biological containment strategies [Regulation of the Release of GMOs 2013]. In 2012, the EFSA Panel on Genetically Modified Organisms released a "Scientific opinion addressing the safety assessment of plants developed through cisgenesis and intragenesis" in response to a request from the European Commission [European Food Safety Authority 2012]. The opinion was, that while "the frequency of unintended changes may differ between breeding techniques and their occurrence cannot be predicted and needs to be assessed case by case

...similar hazards can be associated with cisgenic and conventionally bred plants, while novel hazards can be associated with intragenic and transgenic plants." In other words, cisgenic genetic engineering approaches should be considered similar in risk to conventional breeding approaches, each of which is less risky than transgenic approaches.

References

ArborGen. 2013. Superior Hardwood Seedlings for Multiple Uses. Available at: http://www.arborgen.us/ index.php/ products/hardwoods.

Argentina. 2013. Biotech Facts & Trends. http://www.isaaa.org

Beckmann, V., C. Soregaroli and J. Wesseler. 2010. Ex-ante regulation and ex-post liability under uncertainty and irreversibility: governing the coexistence of GM crops. Economics: The Open-Access, Open-Assessment E-J. 4: 2010–2019.

Beckmann, V., C. Soregaroli and J. Wesseler. 2006. Co-existence rules and regulations in the European Union. Amer. J. Agri. Econ. 88(5): 1193–1199.

Bhargava, P. 2003. High stakes in agro research: resisting the Push, Economic, and Political. Weekly 38(34): 23 August 2003.

Biosafety Scanner. 2013. GM Crop report relating to Brazil. 19 August 2013.

Botha, G.M. and C.D. Viljoen. 2009. South Africa: a case study for voluntary GM labeling. Food Chemistry 112(4): 1060–1064.

Brazil. 2013. Biotech Crops & Trends. http://www.isaaa.org/.

Brookes, G. and P. Barfoot. 2013. GM crops: global socio-economic and environmental impacts 1996–2011. PG Economics Ltd. Dorchester, UK, 191 pp.

Burachik, M. 2012. Regulation of GM crops in Argentina. GM Crops and Food: Biotechnol. in Agri. and Food Chain 3(1): 48–51.

Center for Legislative Research and Advocacy. 2013. The Biotechnology Regulatory Authority of India Bill. A threat to our Food and Farming. Policy Brief Series: 19: June–August.

Cooper, A. 2012. Political Indigestion: Germany Confronts Genetically Modified Foods. German Politics. Academic Search Premier. Web. 6 Dec. 2012. 18.4(2009): 536–558.

CTNBio. 2010. Technical Opinion nº 2753/2010—Commercial release of insect resistant and herbicide tolerant maize containing events MON 89034 X TC1507 X NK603. Proceedings: 01200.001455/2010-39.

Cummins, J. 2010. Genetically Modified China. ISIS Report 08-09-2010.

Davison, J. 2010. GM plants: Science, politics and EC regulations. Plant Sci. 178(2): 94–98.

Declaration to the US Government. 15 October 1997. Alternative Consumers' Union and A SEED Europe.

Duke, S.O. and A.L. Cerdeira. 2010. Transgenic crops for herbicide resistance. pp. 133–166. *In*: C. Kole, C.H. Michler, A.G. Abbott and T.C. Hall (eds.). Transgenic Crop Plants, Vol. 2: Utilization and Biosafety. Springer-Verlag, Berlin/Heidelburg, Germany.

European Food Safety Authority. 2012. Scientific opinion addressing the safety assessment of plants developed through cisgenesis and intragenesis. EFSA J. 10(2): 2561ff.

Evans, B. and M. Lupescu. 2012. Canada: Agricultural Biotechnology Annual-2012. GAIN (Global Agricultural Information Network) report CA12029, U.S. Dept. Agri., Foreign Agricultural Service. 15 July 2012.

Fernandez-Cornejo, J. 2013. Adoption of Genetically Engineered Crops in the U.S.: Recent Trends in Adoption. U.S. Dept. Agri. Economic Research Service. 09 July 2013.

Fontes, E.M.G. 2003. Legal and regulatory concerns of transgenic plants in Brazil. J. Invertebrate Pathol. 30. doi:10.1016/S0022-2011(03)00060-0.

Fontes, E.M.G., M. Varella and A.L. Assad. 1998. A regulamentação de Biossegurança e sua Interface com Outras Legislações. OEA Report. 1998. www.bdt.org.br/bdt/oeaproj/biosseguranca.

GAIN. 2007. Brazil Biotechnology: Brazil Approves Commercialization of Biotech Corn 2007. Gain Report No. BR7623.

Genetically Modified Food Controversies. 2013. Wikipedia. 30 December 2013.

Ghosh, P.K. 2001. Bt. cotton: Government procedures. 80(3): 323.

GMO Compass. 2012. The European Regulatory System. 28 July 2012.

GMO Compass. 2006. Food Safety Evaluation. Evaluation Safety: A Major Undertaking. 15 February 2006.

GMO Safety. 2010. New coexistence—Guidelines in the EU: Cultivation bans are now permitted. EU Commission.

GM Science Update. 2014. A report to the Council for Science and Technology. March 2014. https://www. gov.uk/.../ cst-14-634a-gm-science-update.pdf.

Green, J.M. 2012. The benefits of herbicide-resistant crops. Pest Manage. Sci. 68(10): 1323–1331.

Green, J.M. and *L.A.* Castle. 2010. Transitioning from single to multiple herbicide resistant crops. pp. 67–91. *In*: V.K. Nandula (ed.). Glyphosate Resistance in Crops and Weeds: History, Development, and Management. Wiley, Hoboken, NJ, USA.

Griffin, J.J., M. McNulty and W. Schoeffler. 2005. Shaping Brazil's emerging GMO policy: opportunities for leadership. J. Public Affairs. 5: 287–298.

Groeneveld, R., J. Wesseler and P. Berentsen. 2013. Dominos in the dairy: an analysis of transgenic maize in Dutch dairy farming. Ecological Econ. 86(2): 107–116.

Häggman, H., A. Raybould, A. Borem, T. Fox, L. Handley, M. Hetzburg, M.-Z. Lu et al. 2013. Genetically engineered trees for plantation forests: key considerations for environmental risk assessment. Plant Biotechnol. J. 11(7): 785–798.

Health Canada. 2005. The Regulation of Genetically Modified Food Glossary: Definition of Genetically Modified: Retrieved. 2 November 2012.

Ho, M.W. 1999. FAQ on genetic engineering. ISIS Tutorial, 1999. http://www.i-sis.org.uk/FAQ.php.

Huang, J. and J. Yang. 2011. China's Agricultural Biotechnology Regulations—Export and Import Considerations. Interntl. Food & Agri. Trade Policy Council, 22 pp.

ISAAA. 2013. Global Status of Commercialized Biotech/GM Crops: 2013. ISAAA Brief 46-2013: Executive Summary.

ISAAA. 2011. Global Status of Commercialized Biotech/GM Crops: 2012. ISAAA Brief 43-2011: Executive Summary.

Jacob, R. 1998. Monsanto Research Finds Deep Hostility to GM Foods. Financial Times, 18 November 1998, p. 10.

James, C. 2012. Global Status of Commercialized Biotech Crops, 2003–2012. ISAAA Global Brief, and Country Briefs, International Service for the Acquisition of Agri-biotech Applications.

Jensen, H.T., H.G. Jensen and M. Glylling. 2009. Adoption of GM Crops in the European Union.

Kanter, J. 2012. BASF to Stop Selling Genetically Modified Products in Europe. Global Business. The New York Times. 16 January 2012.

Katovich, E. 2012. The Regulation of Genetically Modified Organisms in Latin America: Policy Implications for Trade, Biosafety, and Development. Undergraduate dissertation, Univ. Minn., 73 pp.

Lehman, V. and W.A. Pengue. 2000. Herbicide tolerant soybean: just another step in a technology treadmill? Biotechnol. Develop. Monitor 43: 11–14.

Maitland, A. 1996. European News Digest: Call for Ban on Biotech Beans. Financial Times, 8 October 1996.

Mannion, A.M. and S. Morse. 2012. GM Crops 1996–2012: A Review of Agronomic, Environmental and socio-economic impacts. Univ. of Surrey Center for Environmental Strategy Working Paper 04/13, 39 pp.

Marden, E. 2003. Risk and Regulation: U.S. Regulatory Policy on Genetically Modified Food and Agriculture, 44 B.C.L. Rev. 733.

McHughen, A. 2012. Impact of herbicide tolerant crops on weed management in the Asia Pacific Region. Pak. J. Weed Sci. Res. Special Issue: October 2012. 18: 341–355.

McHughen, A. and S. Smyth. 2008. US regulatory system for genetically modified [genetically modified organism (GMO), rDNA or transgenic] crop cultivars. Plant Biotechnol. J. 6(1): 2–12.

Mercola. 2012. Monsanto to face biopiracy charges in India. Organic Consumers Assoc. 31 January 2012.

Mukherjee, S. 2014. GMOs in India face trials despite initial approval. Business Standard. March 31.

News & Analysis. 2013. Scientists Clash Swords Over Future of GM Food Crops in India. Science 340: 3 May 2013.

Paarlberg, R.L. 2001. The real threat to GM crops in poorer countries: consumer and policy resistance to GM foods in rich countries. Food Policy 27: 247–250.

Petry, M. and W. Bugang. 2009. China Peoples Republic Agricultural Biotechnology Annual GAIN Report 8/3/2009. http://gain.fas.usda.gov/Recent%20GAIN%20Publications/AGRICULTURAL%20 BIOTECHNOLOGY %20 ANNUAL_Beijing_China%20-%20Peoples%20Republic%20of_8-3-2009.pdf.

Pew Charitable Trust, The. 2012. Guide to U.S. Regulation of Genetically Modified Food and Agricultural Biotechnology Products. The Pew Initiative on Food and Biotechnology. Washington, DC. 2001. Retrieved 2012-06-02.

Pew Initiative on Food and Biotechnology. 2005. U.S. vs. E.U.: An Examination of the Trade Issues Surrounding Genetically Modified Food. December, 57 pp.

Regulation of the Release of Genetically Modified Organisms. Wikipedia 12 December 2013.

SENASA. 2012. Food and feed safety assessment of maize event MON 89034 x TC1507 x NK603 OECD: MON-89034-3 x DAS-01507-1 x MON- 00603-6 (Includes all possible intermediate combinations). Directorate of Agrifood Quality. Office of Biotechnology and Industrialized Agrifood Products. 31 May 2012.

Sharma, R. and J.S. John. 2012. Genetically Modified Foods. The Canadian Encyclopedia.

Silva, J.F. 2010. Brazil-Agricultural Biotechnology Annual, Biotech Annual 2011, United States Department of Agriculture Global Agricultural Information Network.

Silva Gilli, R. 2010. Genetically Modified Organisms in MERCOSUR, from "The Regulation of Genetically Modified Organisms: Comparative Approaches," Oxford University Press.

Transgenic Crops: An Introduction and resource Guide. 2004. The Regulatory Process for Transgenic Crops in the U.S. 28 February 2004.

Urry, M. and G. Parker. 1998. Enforcer to Monitor GM Crops: Farming Cunningham Heads Committee to Monitor Tests on 'Frankenstein Foods'. Financial Times, 22 October 1998, 11 pp.

US Food and Drug Administration. 2013. Biotechnology: Genetically Engineered Plants for Food & Feed. 31 May 2013.

US Food and Drug Administration. 1997. Consultation Procedures under FDA's 1992 Statement of Policy —Foods Derived from New Plant Varieties. June 1996; Revised October 1997.

Varella, M., E.M.G. Fontes and F.G. Rocha. 1998. Biosseguranca & Biodiversidade—Contexto Científico e Regulamentar. Ed. Del Rey, Belo Horizonte.

Walsh-Dilley, M. 2009. Localizing control: Mendocino County and the ban on GMOs. Agriculture and Human Values 26(1-2): 95–105.

Wesseler, J. and N. Kalaitzandonakes. 2011. Present and Future EU GMO policy. pp. 23–323. *In*: A. Oskam, G. Meesters and H. Silvis (eds.). EU Policy for Agriculture, Food and Rural Areas. Second Edition. Wageningen Academic Publishers, Wageningen.

Yankelevich, A. 2012. Argentina: Biotechnical Annual Report. GAIN Report. U.S. Dept. Agri. Agricultural Service. 18 July 2012.

Zhang, B., A.D. Oakes, A.E. Newhouse, K.M. Baier, C.A. Maynard, W.A. Powell et al. 2013. A threshold level of oxalate oxidase transgene expression reduces *Cryphonectria parasitica*-induced necrosis in a transgenic American chestnut (*Castanea dentata*) leaf bioassay. Transgenic Res. 22(5): 973–982.

CHAPTER 9

Benefits, Risks, and Issues Associated with Transgenic Crops and Foods

It has been claimed that transgenic crops have made a positive contribution to global crop production and food security and improved the economic status of farmers who adopted them. However, they have also accelerated changes in farming styles, affecting genetic diversity in agro-ecosystems of many countries that have adopted the biotechnology. For example, the adoption of herbicide-resistant transgenic crops has changed traditional weed management practices and the biodiversity of crop and weed species. It has also raised concerns about the bio-safety to consumers, besides long-term profitability to farmers.

This chapter deals with benefits derived by a diverse group of people following cultivation of transgenic crops over the past two decades while assessing and discussing the risks and issues associated with these crops and foods derived from them.

Benefits at Global Level

Increase in area under transgenic crops does not necessarily guarantee their success at the farm level. Good indicators of their success are the pecuniary and non-pecuniary benefits derived by farmers over long run. This part of the chapter discusses the proponents' views on transgenic crops regarding the benefits accrued at the farmer and national level following their adoption.

Pecuniary Benefits

Pecuniary or direct benefit includes net farm income or profitability which is based on crop yields, market value of crop produce, production costs (seed and crop protection expenditure), and costs of fuel and labor. The most obvious pecuniary benefit is yield increase which is tangible and quantifiable.

Farm Income

Proponents believe that transgenic crops have certainly increased the incomes of farmers who adopted them and countries which commercialized them. The incomes rose when biotech crops first became available in 1996 and they continued to rise even after 18 years of their adoption. The cumulative benefit on global income is also on the rise.

A study conducted by Brookes and Barfoot [2013] on the global impact of transgenic crops during the 16-year period 1996–2011 revealed that genetically modified technology has had a significant positive impact on farm income derived from a combination of enhanced productivity and efficiency gains. The direct global farm income benefit from transgenic crops during this period was US$98.2 billion {it rose to $116.9 billion for the 1996–2013 period [ISAAA 2013]} which was equivalent to an additional 6.3 percent of the value of the combined global production of soya bean, maize, cotton, and rapeseed (canola). The major contributors to income gains in 2011 were insect-resistant (IR) *Bt* maize (US$7.1 billion) and cotton ($6.56 billion) which together accounted for 69 percent of the benefit and the herbicide-resistant (HR) soya bean ($3.88 billion) maize ($1.54 billion), canola ($433 million), and HR cotton ($167 million) which added 30.4 percent value. However, certain parts of the world—particularly USA—have been growing cotton and maize varieties stacked with both herbicide tolerance and insect resistance traits from 2000. In USA, the area under stacked cotton and maize with IR and HR traits reached 67 percent and 73 percent in 2013 from 6 percent in 2000. Hence, over one-half of the gains attributed earlier to IR cotton and maize certainly belonged to HR trait as well.

The six major beneficiaries of transgenic crops include USA, Argentina, China, India, Brazil, and Canada in that order. These nations contributed 95 percent of the global income benefits between 1996 and 2011. The U.S. which grew all six transgenic crops accounted for 44.6 percent ($43.38 billion) of global farm income, while China ($13.07 billion) and India ($12.58 billion) which cultivated only the insect-resistant cotton added 27.3 percent to gross income gains during this 16-year period. Over 92 percent of the income gain by transgenic HR canola was from Canada. Argentina and Brazil derived 90 percent and 65 percent of their net benefits from HR soya bean.

Aside from growing on a larger area, developing nations overtook the developed ones by contributing to 51.5 percent of the total global farm income in 2011. They also reaped 55 percent of the total global benefits; and the bulk of this has been derived by just two transgenic crops, herbicide-resistant soya bean and insect-resistant cotton. In spite of being slow in adopting transgenic crops initially, farmers of developing countries finally accounted for a greater cumulative income gain of 50 percent over the 16-year period. This suggests that having tasted the success of transgenic crops, the developing nations may dominate the future biotech market.

The total cost farmers had to pay for accessing transgenic technology across the four biotech crops in 2011 was equal to 21.5 percent of the total technology gain (inclusive of farm income gain plus the cost of technology payable to seed supplier network) [Brookes and Barfoot 2013]. The cost of technology to farmers of developing countries was just 14.5 percent of the total technology gain as against 28 percent for farmers in developed nations. Whilst circumstances between countries vary, the higher share of total technology gain derived by developing countries than developed countries reflects factors such as a) weaker provision and enforcement of intellectual property rights in developing countries—hence, lower technology cost—and b) the higher gross income gain per unit of land derived by developing country farmers as compared to developed country farmers [Brookes and Barfoot 2013].

Transgenic crops, across HR and IR traits, have raised the global production significantly [Brookes and Barfoot 2013]. Production of soya bean during 1996–2011 was to the tune of 110.2 million tons, followed by maize, cotton, and canola at 195.0 million tons,

15.85 million tons, and 6.55 million tons respectively. The primary impact of transgenic HR technology has been to provide more cost-effective (less expensive) and better weed control [Brookes and Barfoot 2013]. The main source of additional production was due to the facility of adopting no tillage production systems, shortening the production cycle, and the possibility of taking a second crop in a relatively weed-free situation. Growing another crop following an HR crop would certainly raise farm income.

Pressure of weeds and insects on a crop depends on the region and season/year. Brookes and Barfoot [2012] analyzed the impact of weeds and insects on the yields of these biotech crops under two assumed pressure levels: 'lower than average' and 'higher/above than average'. The results suggested that regardless of weed pressure, the cumulative income benefits derived by farmers from using transgenic HR soya bean and canola between 1996 and 2010 remained the same. However, the income gains due to the transgenic HR maize depended on intensity of weed pressure: the benefits being $12 billion higher at above-average than at below-average weed pressure. Similar results were noticed even with transgenic IR cotton. This suggests that transgenic crops generally perform better under higher intensive weed/insect infestation than normal and below-normal infestations, which means "the greater the weed/insect problem the better the performance of biotech crop."

Non-Pecuniary Benefits

Non-pecuniary or indirect benefits include the intangible impacts influencing the adoption of transgenic crops. These include greater management flexibility, reduced crop toxicity, increased savings in time and equipment usage, improved quality of the crop produce, lesser impact on the environment, lower potential damage of soil-incorporated residual herbicides to rotation crops, etc. [Brookes and Barfoot 2012]. Some of these benefits are discussed in two categories: farm level and environmental level.

Farm Level

The primary impact of transgenic HR technology is on providing more cost-effective (less expensive) and easier and better weed control as against just a better weed control—regardless of cost—obtained from conventional technology, even if crop yields remain the same in both technologies. The cost-effective avenues include convenience of application, flexibility of management, prevention or reduction of environmental contamination, etc. HR biotech crops engineered to tolerate certain broad spectrum herbicides do provide these avenues.

In conventional cultivation, broad spectrum, non-selective post-emergence herbicides such as glyphosate, glufosinate, etc. are applied after the crop is established. When these are applied, the crop is very likely to be sensitive so as to suffer a setback in growth. This problem is eliminated when transgenic HR crop variety is used because the crop has already been engineered to be resistant to the herbicide.

Herbicide tolerant crops also facilitate the adoption of conservation or no-tillage systems. This provides for additional monetary savings in the form of lower labor and fuel costs associated with plowing, besides aiding in additional soil moisture retention, and reduced soil erosion [Brookes and Barfoot 2012]. Improved weed control may contribute to reduced harvesting time and enhanced quality of the harvested crop. Improved quality of crop produce may fetch higher market prices. Adoption of transgenic HR crop avoids the potential damage caused by soil-incorporated residual herbicides to follow-on (rotational) crops while reducing the need to apply herbicides to them because of earlier improved levels of weed control.

Brookes and Barfoot [2012] summarized the values for non-pecuniary benefits derived from biotech crops in USA during the 15-yr period of 1996–2010. The estimated (nominal value) of benefits were to the tune of $1.02 billion in 2010, with a cumulative benefit of $7.622 billion during 1996–2010. These non-pecuniary benefits were equal to 18.5 percent of the total direct (pecuniary) income benefits in 2010 and 21.6 percent of the cumulative direct income over the 15-year period. Non-pecuniary benefits boost the overall farm income even if the direct benefits are relatively small.

Impact on Environment

In order to assess the broader impact from changes in herbicide and insecticide use on the environment, Brookes and Barfoot [2013] analyzed the quantities of herbicides and insecticides used in the form of active ingredients and calculated the Environmental Impact Quotient (EIQ). EIQ distils the various environmental and health impacts of individual pesticides in different transgenic and conventional production systems into a single 'field value per hectare' and brings out key toxicity and environmental exposure data related to individual products. Herbicides and insecticides are not equal in terms of their toxicity and potential for environmental impact, which means that the toxicity of a pesticide is not directly related to the amount of active ingredient applied. Therefore, EIQ provides a better measure to contrast and compare the impact of various herbicides and insecticides on the environment and human health than the quantity when used alone. EIQ just serves as an indicator of toxicity only; it does not, however, take into account all environmental issues and impacts.

The data suggested that transgenic traits have contributed to a significant reduction in the environmental impact associated with herbicide and insecticide use in the areas planted with biotech crops. The quantity of pesticides used in seven transgenic crops between 1996 and 2011 came down by 437.7 million kg, which amounted to 8.9 percent. The environmental impact associated with herbicide and insecticide use in these crops, as measured by EIQ, fell by 18.3 percent.

The largest environmental gain has been derived from the adoption of IR technology, with *Bt* cotton contributing to a 24.8 percent reduction (188.7 million kg) in the volume of insecticides used and 27.3 percent reduction in EIQ indicator. Cotton receives multiple applications of insecticides. Any reduction in their use is expected to have a significant beneficial effect on the environment.

Among the HR crops, transgenic maize decreased herbicide use by 193.1 million kg, a 10.1 percent reduction, and this led to a concomitant reduction (12.5 percent) in the impact on the environment. Regarding transgenic soya bean, with the largest area under it, herbicide use came down by 12.5 million kg, which translated into a 15.5 percent decrease in impact on the environment. The impact on the environment was even lower in the case of HR canola (27.1 percent), despite having been grown on a much larger area (7.3 million ha) than HR soya bean and maize.

Of the environmental benefits derived by using transgenic crops, developed countries have been the major beneficiaries (55 percent) than developing nations (45 percent) during the 16-year period. This situation may very soon turn in favor of developing countries as they bring in larger area under transgenic crops. Over three-quarters of the environmental gains in developing nations (India, China, and Pakistan) have been from the use of transgenic IR cotton.

Impact on Greenhouse Gas Emissions

Transgenic crops have the potential to lower greenhouse gas (GHG) emissions by saving on fuel, by reducing the number of herbicide or insecticide applications. In the case of HR crops, particularly those engineered for resistance to glyphosate and glufosinate, adopting conservation, reduced, or no-tillage farming systems would lead to savings in CO_2 emissions. Brookes and Barfoot [2013] estimated that the reduction of CO_2 emissions consequent to growing biotech crops in 2011 was to the tune of 1,886 million kg by lowering the fuel usage by 706 million L. The cumulative reduction in gas emission over the period of 1996–2011 was 14,610 million kg arising from a saving of 5,472 million L of fuel.

Adoption of transgenic HR technology eliminates or reduces, in certain crops, the need to rely on the pre-planting soil cultivation or seed-bed preparation to eliminate weed growth. This makes "no-till" or "reduced till" a useful practice. As a result, tractor fuel use for tillage is reduced, soil quality is possibly enhanced, and soil erosion lowered. In turn, more carbon remains in the soil and this leads to lower gas GHG emissions. Based on savings arising from the rapid adoption of no till/reduced tillage farming systems in North America and South America in 2011, an extra 5,751 million kg of soil carbon is estimated to have been sequestered. This is equivalent to 21,107 million tons of CO_2 that has not been released into the global atmosphere. The cumulative savings over a longer period of growing transgenic HR crops would certainly be much higher. The reduction in GHG emissions and its quantification are dependent on several variables like crop type, crop duration, cropping system, soil type, and environmental conditions, etc. Thus, transgenic herbicide-tolerant crops have the potential to reduce emissions of greenhouse gases in substantial quantities.

Brookes and Barfoot [2013] examined the carbon sequestration benefits within the context of the carbon emissions from cars and arrived at the following conclusions:

a) In 2011, the permanent CO_2 from reduced fuel use was the equivalent of removing 838,000 cars from the roads.
b) The additional probable soil carbon sequestration gains in 2011 were equivalent to taking 9.381 million cars off the roads.
c) The combined transgenic crop-related CO_2 emission savings from reduced fuel use and soil carbon sequestration were equal to the removal from roads of 10.2 (838,000 + 9,381,000) million cars, equivalent to 36 percent of all registered cars in the United Kingdom.
d) If the transgenic HT crop was planted under reduced/no tillage system for 16 years (1996–2011), this would have resulted in a cumulative saving of 170,961 million kg of CO_2. This is equivalent to taking 76 million cars off the road.

These conclusions may appear to be overly optimistic, but adoption of transgenic herbicide tolerant crops translates into saving on precious fuel, leading to lower carbon emissions. As mentioned earlier, the saving is dependent on transgenic crop, cropping system, type of pesticide (herbicide, insecticide, etc.) used, soil type, environmental conditions, etc.

Risks

Commercial production of transgenic crops with various desired, beneficial traits has aroused concerns about their biosafety, followed by debates worldwide. Biosafety issues have become a crucial factor in the further development of transgenic biotechnology and its utility and wider application of transgenic products in global agriculture. Genetically engineered crops, however, are a heterogeneous group. As such, it is not reasonable to lump all of them together. Therefore, it would be prudent to assess the biosafety of each of the transgenic crops separately.

Development of transgenic crops by the agricultural biotechnology industry is seen more as a profit driven rather than need driven process. Therefore, the thrust of the genetic engineering industry is not really to solve agricultural problems, but to create profitability [Altieri 1998] as is evident by the fact that over the last 30 years, scores of multinational corporations have initiated transgenic research on a variety of crops around the world. Although several universities and research institutions are also simultaneously involved in this field, their research agenda is being increasingly influenced by the private sector in ways never seen in the past [Altieri 1998]. The challenge these organizations now face is how to ensure that ecologically sound aspects of biotechnology are researched and developed while at the same time carefully monitoring and controlling the provision of applied non-proprietary knowledge to the private sector, farmers, and consumers and making such knowledge available in the public domain for the benefit of society [Altieri 1998].

Currently, there is a great deal of confusion on the risks, both real and perceived, attributed to transgenic crops now being grown in global agriculture. Protagonists of these crops highlight their virtues while ignoring risks and issues, while antagonists sometimes find it hard to separate facts from fiction and half-truths. It thus becomes incumbent upon scientists to examine the key issues scientifically, systematically and dispassionately and evaluate the merits of various positions before arriving at meaningful decisions in the interest of farmers and consumers.

Agro-Ecological Concerns

Transgene Flow

Transgene flow is a primary concern for regulatory oversight of transgenic plants around the world. The main factor behind the agro-ecological concerns of antagonists regarding the use of herbicide-resistant transgenic crops is the outflow of transgenes to non-transgenic conventional crop varieties, weedy/wild relatives of the crop, and to non-target weed species, leading to unintended consequences. In fact, gene flow takes place between crop cultivars and between crops and compatible relatives. It occurs whether a gene is a transgene or not, and does not automatically result in a negative outcome [Mallory-Smith and Olguin 2011].

As discussed in Chapter 2, gene flow occurs '**horizontally**' and '**vertically**'. In horizontal gene movement or transfer, the acquisition of genes is passed over, i.e., 'horizontally' from one organism to another by means other than inheritance. In this case, genes move between disparate, unrelated species as in the case of plants and microbes. Conversely, transgenes, or more likely pieces of transgenes, could theoretically be transferred from decaying plant parts in the field to soil or aquatic microbes, or from transgenic food to bacteria in the gut or gut cells themselves. However, entire transgenes with the regulatory portions of the DNA have never been found to be horizontally transferred [Stewart 2008].

On the other hand, vertical gene flow, central to the evolution of species, is the exchange of genes between closely related species. It occurs in only one generation between varieties or types of plants within the same species, and sometimes even between species. Vertical gene flow is restricted to organisms that can mate with one another and make offspring. In the case of crop plants, which are the domesticated forms of wild plants, a high degree of compatibility can therefore exist between the crop and wild and weedy relatives.

Vertical gene flow often results in **introgression** which is the stable integration of a (trans)gene into a related plant genome. As in the case of hybridization, introgression can also occur within or between species. However, introgression, also known as introgressive hybridization, requires several generations of crosses. The first step in introgression involves transgenic hybrids crossing with the non-transgenic host. Typically, for effective gene escape the non-transgenic host would need to be the female parent and the transgenic male parent would produce the pollen. So, biotechnologists often study how pollen moves from transgenic plants to non-transgenic plants and how many transgenic seeds are produced from the non-transgenic parent. Since hybrids are often sexually non-viable, introgression is relatively rare [Stewart 2008]. Genes are known to move among crop populations through seeds and pollen. Transgenes from a genetically modified crop can flow to a non-transgenic crop, to a landrace, to a wild relative(s), and to a weedy relative.

From Transgenic Crop to Landraces. There is a good possibility that genes from a transgenic crop could escape to a landrace growing in the vicinity. There was one publicized report in 2001 [Quist and Chapela 2001] about high level of transgenic flow (as was evident by the presence of CaMV 35S promoter used in glyphosate-resistant event NK603) in maize progenitor landraces grown by farmers in the Sierra Norte de Oaxaca region in Southern Mexico in October-November 2000. Mexico is the home to 59 distinct landraces [Wellhausen et al. 1952; Sanchez et al. 2000] and 209 varieties of maize [Mann 2004]. Contamination with transgenes was believed to be due to an inadvertent admixture of transgenic maize seeds imported from USA with landrace seeds. Although subsequent seed samples taken from this region in 2003 and 2004 could not confirm the existence of transgenes, thus casting doubts on the earlier results [Ortiz-Garcia et al. 2005], Piñeyro-Nelson et al. [2009a,b] later confirmed their presence in three of the 23 localities sampled in 2001 at low frequencies. Differences in sampling, analytical, and detection methods may explain the discrepancies between the earlier and later reports. Despite these contradictory reports, the combined evidence suggests that there is significant potential for the presence of transgene(s) within the plant tissues of structured populations of landraces of transgenic crops around the world.

From Transgenic Crop to Wild/Weedy Relatives. Most of the major crop plants are subject to pollen-mediated gene flow and introgression with their wild relatives. This phenomenon is of common occurrence worldwide, although the frequencies of the gene flow can be variable. This situation poses a serious challenge for maintaining genetic diversity, genetic integrity, and evolutionary potential of *in situ* conserved crop wild relative populations and species [Lu 2013]. At least 44 cultivated crops have demonstrated the capacity for hybridization with wild and weedy relatives, including 12 of the 13 most widely cultivated crops [Ellstrand et al. 1999]. These include rice, wheat, sorghum, soya bean, rapeseed, sunflower, squash, etc. Gene flow depends on the availability of such species near the area of cultivation [Messeguer 2003].

Migration of herbicide resistant transgenes could make weedy relatives of the herbicide resistant crop much more problematic to the farmer. This problem has not been reported with maize, cotton, and soya bean because there are few or no weedy species with which they are sexually compatible in the places they are grown [Duke and Cedeira 2010]. A good example of transgene flow between a crop and its wild relative is that of genetically engineered *Brassica napus* (canola) and its wild relative *Brassica rapa*. Rapeseed cross-pollinates with several weedy related *Brassica* species, including related *Brassica* crops (cabbage, cauliflower, broccoli, etc.), but their crosses are generally unfit. However, introgression of a glyphosate resistant transgene from rapeseed to its weedy relative (*Brassica rapa*), has occurred in field situations, and the introgressed gene appears to be stable in the population, even in the absence of glyphosate application [Warwick et al. 2008]. This herbicide resistance gene in weedy non-transgenic feral *B. rapa* has wide dispersal and it persists outside of cultivated fields for many years [Pessel et al. 2001; Simard et al. 2002]. This posed a number of distinct challenges for weed management in agricultural lands in the vicinity of transgenic rapeseed crop. Besides, transgene flow to wild populations would lead to production of fertile hybrids and further dissemination of transgenes. These hybrids have a drop in fitness relative to their parents, but that fitness is rapidly regained when these hybrids mate with one another or backcross to either parent [Ellstrand 2006].

Rice is another crop that readily cross-pollinates with wild relatives. Cross-pollination of cultivated rice occurs readily with feral red rice as well as weedy related species. Although imidazolinone-(IMI) resistant weeds were seen growing in the vicinity where non-transgenic IMI-resistant rice was grown [Valverde 2007], it was unclear if it was due to evolved resistance or gene flow because IMI resistance evolves quickly in some species when exposed to this herbicide class [Heap 2008]. What is clear, however, is that genes from rice do flow to feral rice species and to sexually compatible species. This could pose a serious challenge in future as and when herbicide-resistant transgenic rice varieties are grown in rice-producing countries.

Sugar beet (*Beta vulgaris*) is a highly variable group, in which it is often difficult to distinguish between cultivated and wild forms [Bartsch and Ellstrand 1999]. This is mainly due to the extensive use of sea beet (*B. vulgaris* ssp. *maritima* ARCANG.) gene resources in conventional breeding programs. As cultivated beet (*B. vulgaris* ssp. *vulgaris*), sea beet, and wild sea beet (*B. macrocarpa*) are allogamous, they can easily produce hybrids, and they are completely interfertile [Boudry et al. 1993]. Darmancy et al. [2007] found glyphosate and glufosinate resistant seeds from the progeny of the weed sugar beet 112 m away from the closest transgenic pollen donor sugar beet (*Beta vulgaris*) within a few years of its adoption in France. Direct pollen flow from sugar beet bolters to weed beet species growing within the same field, as well as in a neighboring field that was left fallow, accounted for only 0.4 percent of the resistant seeds. Transmission of glyphosate-resistant gene from transgenic sugar beet line H7-1 into wild beet in California, USA has resulted in a protracted legal battle between farmers and Monsanto and KWS SAAT AG (vide Chapter 6).

Transgenic Volunteer Weeds. Seed shatter before harvest and the seed lost during harvest may result in herbicide-resistant crop volunteer in succeeding crops. Many transgenic herbicide-resistant crops become 'volunteer' weeds that infest crop rotations. These transgenic volunteers have the ability to become more effective, aggressive weeds and complicate cultivation, contaminate crops, and transfer herbicide-resistant transgenes to weedy relatives. Recurrent or sequential transgene outflow from a transgenic crop may cause transgene stacking in volunteers. Volunteers of canola are considered among the 20 most common weeds in fields in Canada, occurring as a residual weed in wheat and barley fields [Hall et al. 2000].

The spread of transgenes via volunteer weeds can be mitigated by maintaining the fitness of the recipients below the fitness of the volunteer and wild type [Gressel and Al-Ahamd 2006]. A concept of 'transgenic mitigation' (TM) was proposed in which mitigator genes are tandemly linked to the desired primary transgene, which would reduce the fitness of hybrids and their rare progeny, considerably reducing risk [Gressel 2002]. This TM approach is based on the premises that, 1) tandem constructs act as tightly linked genes with exceedingly rare segregation from each other, 2) the TM traits chosen are neutral or favorable to crops, but deleterious to non-crop progeny, and 3) individuals bearing even mildly harmful TM traits will remain at very low frequencies in weed/wild populations because weeds typically have a very high seed output and strongly compete among themselves, eliminating even marginally unfit individuals [Gressel and Al-Ahmad 2006].

In the case of rice, when herbicide resistance genes are genetically linked with mitigation genes that are neutral or good for the crop, the resultant offspring becomes unfit to compete [Gressel and Valverde 2009]. Mitigation genes confer traits such as non-shattering, dwarfism, and no secondary dormancy and herbicide sensitivity [Gressel and Valverde 2009].

From Transgenic Crop to Non-Transgenic Crop. Transgenic 'pollution' as a result of gene flow from transgenic crop to a non-transgenic cultivated crop of the same species has become a serious global commercial and political problem in addition to a crop ecological threat. Besides, transgenic introgression of conventional and organic crop varieties has raised serious global, biological, socio-economic, ethical, regulatory, and intellectual property concerns. Inadvertent flow of transgenes to a non-transgenic crop may ultimately end up in food supplies and processed foods. Furthermore, another problem that could arise is the loss of non-transgenic varieties, many of which are 'heirloom varieties' (landraces) of important crop diversity [Quist 2010]. There have been several reports of 'gene spill' events of transgenics 'contaminating' non-transgenic crops, resulting from cross pollination [Friesen et al. 2003; Mellon and Rissler 2004; Quist 2010], and sometimes seed mixing [Mellon and Rissler 2004].

When transgenes move from transgenic crops to their non-transgenic counterparts through seed-, vegetative organ-, or pollen-mediated gene flow, a major concern is the 'adventious mixing' of varieties of both crops. This often occurs where both are planted in close proximity. The frequencies of transgene movement mediated by pollen depend essentially on the breeding (mating) systems and quantity of pollen of crops [Lu 2008]. A significant amount of gene flow to non-transgenic crops has the potential for subsequent movement to weedy and wild relative populations.

The pollen-mediated gene flow is dependent on crop. For soya bean, cross-pollination is not a problem, but considerable outcrossing can occur with maize, rice, sugar beet, and canola. Wheat and rice are predominantly self-pollinating, but cross-pollination does occur at a low range. In wheat, cross-pollination rate is nearly 7.0 percent [Hucl 1996] and movement of resistance gene to non-resistant wheat has been documented under field conditions [Gaines et al. 2007]. In rice, cross-pollination occurs at less than 1.0 percent between adjacent plants or fields. Pollen-mediated gene flow from a transgenic herbicide-resistant rice variety to adjacent plants of a non-transgenic counterpart was 0.05–0.53 percent in Italy [Messeguer et al. 2001] and 0.02–0.80 percent in China [Rong et al. 2005].

Gene flow frequency is dramatically reduced with the increase in spatial isolation distance from transgenic rice donors by only a few meters [Rong et al. 2007].

In 2007, the Arkansas Rice Board in USA and Animal Plant Health Inspection Service of U.S. Department of Agriculture detected trace levels of '*bar*' gene (driven by CaMV 35S promoter) from the glufosinate-resistant LLRice62 line of rice developed by Bayer CropScience in non-transgenic event CL131 (Clearfield 131) rice adopted in 1999 [APHIS 2007]. Contamination was confirmed in 20 percent of the 500 samples. A similar contamination incident had occurred with Bayer's LL601 transgenic rice in 2006, leading to financial damage amounting to millions of dollars in the southern states of USA, blocking off the export market, and filing of 13 lawsuits against the company in USA and Europe.

Another widely known example was Aventis Crop Sciences's (later Bayer) insect- and glufosinate-resistant double stack variety CNH-351 (StarLink) transgenic maize that was approved in USA in 1998. In September 2000, the *Bt* toxin (Cry9C) from this transformed maize was detected in taco shells being sold for human consumption even though it was only approved for use in animal feed, sparking a wholesale product recall, and the transgene was still present in detectable concentrations in the USA food supply [Heinemann 2007]. It was detected in seed even in 2003, possibly because contaminated seed has been used in hybrid seed production. Although this CNH-351 maize was banned in USA, it also found its way into Saudi Arabia [Meridian 2013].

Organic farmers cannot retain the organic status of their crops if transgene presence is above a set limit, nor can crops be sold to markets that require the product to be non-transgenic if transgenic occurrence is above the level set by the regulatory jurisdiction [Duke and Cerdeira 2010]. Two years after approval of Monsanto's transgenic glyphosate-resistant alfalfa varieties J101 and J163 in USA in 2004, many organic growers found they cross-pollinated with, and contaminated, organic alfalfa. This has led to lawsuits (vide Chapter 6). In order to arrest or minimize contamination of non-transgenic crops, each nation has set its own threshold levels to allow defined levels of adventious presence of transgenic products in non-transgenic products, provided that the transgenic products are legally permitted for commercialization in the exporting countries [Lu 2008].

From Transgenic Crop to Unrelated Organisms. A different concern of agro-ecological importance is the possibility of gene flow from herbicide resistant transgenic crops to totally unrelated organisms, especially such as microbes. In this case, genes move horizontally between disparate, unrelated species as in the case of plants and microbes (vide earlier discussion on 'Transgene Flow'). Almost all of the transgenes currently used for herbicide resistant crops are from soil microbes. These genes are much more likely to be transferred to unrelated microbes from the natural, microbial sources than from herbicide resistant crops [Kim et al. 2005]. Levy-Booth et al. [2008] found that degradation of the *cp4-epsps* transgene from glyphosate-resistant soya bean leaf material in the soil was rapid, but could still be detected in soil after 30 days. Degradation of the transgene and a natural soya bean gene in soil were similar. Dale et al. [2002] found no compelling scientific arguments to demonstrate that transgenic crops are innately different from non-transgenic crops in regard to passing of recombinant and novel combinations of DNA into the environment and the possibility of their being taken up by microorganisms or other live biological material. However, entire transgenes with the regulatory portions of the DNA have never been found to be horizontally transferred [Stewart 2008].

Evolution of Herbicide Resistant Weeds

In the recent past, there have been several reports that transgenic herbicide-resistant crops (both single and stacked lines) have significantly increased the development of 'superweeds' [Gilbert 2013], but with inadequate scientific evidence to support it. This word is considered rather a misnomer to present the one-sided view of some of the antagonists of bio-tech crops. The following evidences support this point.

Herbicides that contributed most to evolution of resistant weeds over the past 30 yr in the world belonged to ALS-inhibiting family with 145 weed species followed by PS II-inhibiting and ACCase-inhibiting groups with 72 and 46 species respectively (vide Fig. 2.3, Chapter 2) [Heap 2014]. This suggested that 62 percent of the global weed resistance was contributed by herbicides of these groups. However, only less than 15 percent of the global area under transgenic herbicide-resistant crops was covered by those conferred with resistance to them.

Another evidence was presented by Kniss [2013] who analyzed all unique cases of herbicide resistant weeds between 1986 and 2012 by fitting a linear regression of the data from 1986 to 1996 and (time period before transgenics became widespread) another regression to the time period 1997 to 2012. The slope of the linear regression gives an estimate of the number of new herbicide resistant weeds documented each year. In the 11-year pre-transgenic crop period, 13.1 new cases of herbicide resistance were documented annually, and this number decreased to 11.4 cases per year during the 16-year transgenic crop period. This indicated that adoption of transgenic crops has not caused an increase in development of *superweeds* compared to non-transgenic crops [Owen 2008; Kniss 2013].

This finding is further supported by the data on the number of herbicide-resistant species found in various crops [Heap 2014]. The transgenic herbicide resistant crops that have been widely grown over the past two decades include maize, soya bean, rapeseed (canola), and cotton (Fig. 9.1). These four crops have accounted for just 32 percent of the resistant weed species. All other crops (plus roadsides), in which adoption of the transgenics is either negligible or nil, have contributed to over two-thirds of the resistant weed species.

Another point of contention was that glyphosate-resistant transgenics contributed more to the sudden rise of the so-called 'super' weeds. The first glyphosate-resistant weed *Lolium rigidum* found in Australia was documented in 1996, at about the time glyphosate-resistant transgenics were being adopted. This monocot weed cannot be considered a *superweed* because it was found in non-transgenic canola, cereals, and wheat crops as well [Heap 2014]. Kniss [2013] analyzed and found that of the 24 glyphosate-resistant species documented worldwide until 2012, 11 have evolved in transgenic crops compared to 13 in non-transgenic crops/sites. Resistance to glyphosate has evolved in many species in both conventional and transgenic crop situations. This emphasizes that there is little reason to support the contention that glyphosate-tolerant biotech crops has contributed to evolution of resistant weed species.

Any agricultural practice and production system will impart selection pressure on weed communities that will inevitably result in weed population shifts. The weed shifts caused by any herbicide, including glyphosate, are attributable to both the natural tolerance of a particular weed species to it and the evolution of herbicide resistance within the weed population. Both situations encourage a change in agricultural practice. For example, glyphosate enabled farmers to adopt minimum-till and no-till technology which imposes selection pressure on weed communities that will lead, over a time period, to population shifts in both conventional and transgenic cropping systems.

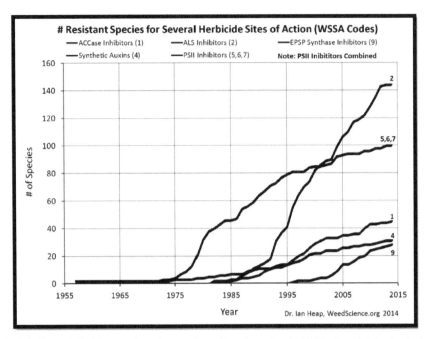

Fig. 9.1. Chronological increase in resistance of weed species to five herbicide sites of action beginning 1957 [Heap 2014].

Currently, 31 weed species have been found worldwide to be resistant to glyphosate [Heap 2014]. The rise of resistant species was particularly steep from year 2005 (Fig. 9.1, line 9) possibly because of two factors. One was the greater and more widespread use of this non-selective herbicide in conventional crops and cropping systems around the world over the past 10 years. The other factor was adoption of glyphosate-resistant transgenic crops, particularly soya bean, maize, and rapeseed (canola) in USA, Brazil, Argentina, and Canada. Of these, soya bean accounted for 72 percent of gross global plantings in 2011, followed by maize and canola with 28 percent, and 23 percent respectively, with most of the area being contributed by these four nations [Brookes and Barfoot 2013]. This further suggests that transgenic crops are not any more problematic than conventional crops in the evolution of herbicide-resistant weed species.

In reality, biotech crops do not select for herbicide resistant weeds, but herbicides do [Kniss 2013]. Thus, development of herbicide resistant weeds is not due to a transgenic crop, but it is a due to the herbicide used. When a herbicide is used continuously over a time period, evolution of resistance is a natural phenomenon, regardless of transgenic or non-transgenic crop culture. However, as more global acreage is treated continuously with herbicides, including glyphosate as well as ALS, ACCase and PS II inhibitors, in both cropping systems, faster and greater emergence of weed species resistant to them may become a serious problem in future. This will invariably cause weed shifts, thus requiring newer weed management strategies to combat the problem.

Food Safety Concerns

The widespread consumer concern about biotech crops is the potential risks they have on human and animal health. There is a wide variation among governments worldwide in the assessment and management of risks associated with consumption of the edible parts of crops and foods derived from them. This is particularly true between the USA and Europe. Generally in USA, unlike in the European Union, crops not intended for food use are generally not reviewed by regulatory authorities responsible for food safety. Food derived from transgenic crops is also not tested before it is marketed as it is not a single chemical, nor is it intended to be ingested in specific doses and times, which makes it difficult to design meaningful clinical studies [Winter and Gallegos 2006].

Bio-informatic tools are increasingly being used in the evaluation of transgenic crops. The guidelines proposed by WHO/FAO [2001] and the European Union [EFSA 2010] include the use of bio-informatic screening to assess the risk of potential allergenicity and toxicity.

The issues surrounding transgenic crops and foods and feeds are broadly grouped into a) changed nutrient levels, b) allergenicity, c) horizontal transfer and antibiotic resistance, d) consumption of foreign DNA, and e) CaMV promoter.

Nutrient Levels

One of the concerns about transgenic crops is whether the transgene will alter nutrient levels of foods and feeds derived from them. Transgene integration and/or transformation and tissue culture during transgenic process may induce unintended genomic alterations in transgenic plants such as deletions, insertions, and rearrangements, which may generate secondary or pleiotropic effects [Kuiper et al. 2001; Cellini et al. 2004; Garcia-Canas et al. 2011; Herman and Price 2013].

The current basis of assessing the nutritional status of transgenic crops is the concept of '**substantial equivalence**' used by regulatory agencies in Europe and the U.S. In this, if a transgenic food is substantially equivalent in composition and nutritional characteristics to an existing food, it is considered to be as safe as the conventional food and therefore does not require extensive testing [FDA 1992; OECD 1993]. This concept recognizes the fact that existing traditional foods (grains, nuts, seeds, etc.) often contain many chemicals that have toxic or carcinogenic effects and that our existing diets therefore have not yet been proven safe. For example, some of the crops contain inherent plant toxins and anti-nutrients. These include phytate in maize; protease-inhibitors, lectins, isoflavones, and phytase in soya bean; glucosinolates and phytate in canola; solanine, chaconine, and protease-inhibitors in potato; and tomatine, solanine, chaconine, lectins, and oxalate in tomato [Novak and Haslberger 2000]. Lack of adequate knowledge on these in traditional foods poses a problem because transgenically modified foods may have differences in anti-nutrients and natural toxins that have never been identified in the original plant, raising the possibility that harmful changes could be missed [Kuiper et al. 2002]. The possibility also exists that positive modifications may be missed.

Although the concept of substantial equivalence is an accepted standard in assessing the safety of transgenic foods [Bakshi 2003], the Royal Society of Canada [2001] recommended that it should only be considered if there is equivalence in the genome, proteome, and metabolome of the modified food when compared with the non-modified food. The potential unintended effects of the transgene and its insertion on food quality and safety can be evaluated by profiling techniques which allow simultaneous characterization and comparison of the genome, metabolome, and proteome of an organism [Kuiper et al. 2003; Ruebelt et al. 2006]. Of the three analytical techniques, proteomic analysis would provide more important information for understanding changes in biological processes after genetic modification [Gong and Wong 2013]. However, this principle left much scope for individual (and national) interpretation [Kok and Kuiper 2003].

Several transgenic crops, such as herbicide-resistant maize, canola, soya bean, and cotton, as well as insecticide-protected maize and cotton, have undergone this assessment and have been shown to be substantially equivalent to commercial crop varieties [Munro 2002]. In 2010, Organization for Economic Cooperation and Development (OECD) suggested that molecular characterization by itself is not the best way to predict the safety of transgenic crops, but can focus on the other safety assessment procedures [OECD 2010].

The U.S. Food and Drug Administration evaluated all of the 148 transgenic events developed and found them all substantially equivalent to their conventional counterparts, as have the Japanese regulators for 189 submissions including combined-trait products. This equivalence was confirmed by over 80 peer-reviewed publications [Herman and Price 2013].

While evaluating substantial equivalence of the transgenic food, protein level, fat and starch content, amino acid composition, and vitamin and mineral equivalency, along with levels of known allergens and other potentially toxic components, are considered [Bakshi 2003]. Of these, proteins, key players in gene function, are directly involved in metabolism and cellular development or have roles as toxins, anti-nutrients, or allergens, which are essential for human health. The insertion of a new gene can sometimes lead to an increase in the existing levels of anti-nutrients, some of which cannot be reduced by heat treatment [Bakke-McKellep et al. 2007]. Glyphosate-resistant transgenic soya bean, a source of widely available commercial modified products may cause an increase in anti-nutrients [Padgette et al. 1996]. An increase in the anti-nutrient level is not desirable as the transgenic food may be consumed as raw material.

In soya bean, genetic modification itself could be a stress factor, provoking changes in the activity of some enzymes as well alterations in seed proteome [Barbosa et al. 2012]. Production of malondialdehyde, an indicator of lipid peroxidation and oxidative stress, was higher in transgenic soya bean seeds than in non-transgenic seeds. Furthermore, higher enzyme activities for ascorbate peroxidase, glutathione reductase, and catalase were found in transgenic seeds [Barbosa et al. 2012].

In glufosinate-resistant transgenic canola line HCN92, Health Canada reported that the transgene-contained refined oil was as safe and nutritious as refined oil from the current commercial varieties and concluded that this does not raise concerns related to safety [Health Canada 2000]. The nutritional properties of glufosinate-resistant transgenic sugar beet and maize grains were found to be essentially equivalent to non-transgenic cultivars in feeding studies with swine and ruminants [Daenicke et al. 2000; Bohme et al. 2001]. Similar results have been reported with glufosinate-resistant rice in swine feeding studies [Cromwell et al. 2005].

Albo et al. [2007] found identical seed proteomes between insect-resistant MON810 *Bt* maize varieties and near-isogenic control varieties, while Zolla et al. [2008] reported about differently expressed proteins in seed proteomes in the presence and absence of the transgene *cry1Ab*. This differential response was considered to be due to differences in growing and cultural conditions. Balsamo et al. [2011] found leaf proteomes of four MON810 varieties similar to those in their two non-transgenic counterparts. These findings suggested that the expression of transgenes had no significant effect on the proteome of insect-resistant maize.

Isoflavones are thought to play a role in preventing heart disease, breast cancer, and osteoporosis. The estrogenic isoflavones of soya bean and their glycosides are products of the shikimate pathway, the target pathway of glyphosate [Duke et al. 2003]. The isoflavone content of glyphosate-resistant transgenic (MON40302-6) soya bean has been investigated by several researchers. Lappe et al. [1999] found reductions in isoflavone levels in glyphosate-resistant soya bean varieties in the absence of glyphosate. This study, however, was not done by comparing isogenic lines. Padgette et al. [1996] found no effects of the transgene on isoflavone content of soya bean. Additional evidence may clarify the arguments for and against glyphosate applications as a risk factor in soya bean cultivation.

Consumer groups contend that independent research in these areas is systematically blocked by biotech companies which develop transgenic crops and own modified seeds and reference materials. Diels et al. [2011] found a significant correlation between author affiliation to industry and study outcome in scientific work published on health risks or nutritional assessment studies of genetically modified products. Furthermore, consumer advocates suggest that the supposed ill effects of transgenes on human and animal health due to altered nutrient composition are more subtle and take time to show up. This requires long-term controlled studies on larger populations, and perhaps even taking longer time to determine the damaging effects, if any, just as it took decades before the adverse "slow poison" effects of trans-fats (another type of artificial food) came to be recognized. However, most of the tests on nutrient composition and food quality conducted so far have simply examined transgenic herbicide resistant crops for equivalence with conventional herbicide resistant crops.

Allergenicity

The possibility of an increase in allergic reactions to food as a result of genetic engineering is a powerful emotional issue. It is because exposure of individuals to biologically active transgenes can have major effects on their gastrointestinal tract. Even people who have never experienced an allergic reaction may worry that they are being exposed to new substances for which there is little track record of safety or harm. It is also likely that in addition to the effects on the gastrointestinal tract, the size, structure, and function of the internal organs will be affected, particularly in young and rapidly growing humans and animals.

Generally, food allergens are proteins or glycoproteins with isoelectric points and a molecular range of 10,000–80,000 Da [Lehrer 1999b]. As most proteins are immunogenic, their consumption may trigger immune/allergic effects both on mucosal immune system of the gut and the body. These food allergens are stable to digestion and processing, and many major allergens are generally proteins that are present in large amounts in allergenic foods [Lehrer 1999b]. People with food allergens, whose symptoms range from mild effects to sudden death, may likely be exposed to foreign proteins introduced into foods by genetic engineering [Bakshi 2003]. Transgenic foods can introduce novel proteins into the food supply from organisms that are never consumed as foods. Some of these foods could be allergenic. It is difficult to predict whether a particular protein will be a food allergen if consumed by humans. The only reliable method to determine this is through consumption of the engineered food [Bakshi 2003]. Therefore, incorporating genes that produce novel proteins into crops by genetic engineering, especially from non-food sources, might pose a health risk [Union of Concerned Scientists 2000; Lachman 1999].

Many proteins in bio-engineered foods have been derived from microbial sources, and producers of transgenic foods have shown that these proteins do not possess characteristics associated with food allergens—that is, these proteins do not share structural similarity to known allergens and are not resistant to digestive enzymes and acid [Bakshi 2003]. In addition, it is known which foods trigger the majority of allergic reactions [Metcalfe et al. 1996].

In order to determine and reduce the possibility that a newly introduced protein will be an allergen, its structure can be compared with structures of known allergic proteins [Bakshi 2003]. If similarity is found and if sera from sensitive individuals are available, possible cross-reaction can be determined. If the reaction is positive, that engineered crop is not fit for consumption, and further modification is necessary.

The World Health Organization (WHO) and Food and Agricultural Organization (FAO) of the United Nations [FAO/WHO 2001] and Lehrer [1999a] have described a hierarchical approach to identify allergens and evaluate the allergenicity of genetically modified crops and foods. These include a) characterization of amino acid, b) identification of the amino acid sequences that define allergenic epitopes to develop more precise sequence-screening criteria, and c) development of an animal model(s) that can recognize food allergens in a manner similar to that which occurs in human disease [Bakshi 2003]. Other factors in determining potential allergenicity of modified food products include molecular mass (most known allergens are between 10,000 and 40,000 Da), heat and processing stability, pH, and gastric juices (most allergens are resistant to gastric acidity and digestive proteases) [Bakshi 2003]. There is good correlation between the resistance of proteins to proteolytic digestion and their allergic potential [Astwood et al. 1996]. Genetically modified food should be labeled to make people aware of what they are buying, and individuals who have allergies would read the labels and not buy foods they think harmful to them [Miller 1999].

Testing for allergens is part of the regulatory requirements. The proponents of transgenic crops note that because of the safety testing requirements imposed on GM foods by regulatory agencies, the risk of introducing a plant variety with a new allergen or toxin using genetic modification is much smaller than using traditional breeding processes. Toxicologists note that conventional food is not risk-free; allergies do occur with many known and even new conventional foods. For example, the kiwi fruit was introduced into the U.S. and the European markets in the 1960s with no known human allergies. However, today there are people allergic to this fruit [Hollingworth et al. 2003]. Transgenic engineering can also be used to remove allergens from foods, potentially reducing the risk of food allergies [Herman 2003]. A hypo-allergenic strain of soya bean was tested in 2003 and shown to lack the major allergen that is found in the beans [Herman 2003]. A similar approach has been tried in ryegrass (*Loliumrigidum*) which produces pollen that is a major cause of hay fever: here, a fertile GM grass was produced that lacked the main pollen allergen, demonstrating that the production of hypoallergenic grass is also possible [Bhalla et al. 2003; Herman et al. 2003].

Soya bean is one of the main foods known to cause allergic reactions worldwide. Glyphosate potentially translocates to metabolic sinks such as seeds [Duke 1988]. Batista et al. [2007] compared seed proteome of glyphosate-resistant transgenic soya bean carrying *cp4 epsps* gene and non-transgenic soya bean and found that soya bean endogenous allergen expression did not seem to be altered after genetic modification.

However, there have been transgenic crop products found to cause allergic reactions before halting their marketing. For example, Pioneer Hi-Bred attempted to improve the protein content of soya bean intended for animal feed by inserting a gene from the Brazil-nut. In an *in vitro* test and a skin-prick test, the transgenic soya bean reacted with immunoglobulin E (IgE), a class of antibody molecules involved in allergic reactions, of individuals with Brazil-nut allergy in a way that indicated that these individuals would have an adverse, perhaps even fatal, reaction to transgenic soya beans [Nordlee et al.

1996]. However, this case was not an accurate representation of foods causing allergic reactions, because the donor Brazil-nut was known to be allergenic. The company therefore discontinued further development of the product [Streit et al. 2001].

The double-trait (insect- cum glufosinate-resistant) variety CNH351 (StarLink)— discussed earlier (vide "*From Transgenic Crop to Non-Transgenic Crop*")—which contained Cry9C protein has been approved only for animal feed. However, maize has been modified in a way that makes it harder to break down in the human gastrointestinal tract that regulatory agencies have refused to approve it for human use [Kaufman 2000]. Since the amino acid sequence of the protein was not similar to known allergens, its allergenicity was considered to be low. Furthermore, for people to become allergic to a protein they need to be exposed to it multiple times over an extended period of time. Since the slow-digesting Cry9C protein (slower than Cry 1Ab) is only a small fraction of maize protein, the probability that the protein would sensitize an individual is low [StarLink Corn 2004; StarLink Corn 2013]. According to Aventis CropScience, developer of CBH351, cooking would destroy most of the protein and people who ate the transgenic maize products would be exposed to extremely small quantities of the protein, not enough to cause alarm [StarLink 2004]. During recall, the U.S. Center for Disease Control evaluated reports of allergic reactions to StarLink corn, and determined that no allergic reactions to the maize had occurred [StarLink 2004]. These cases of products that failed safety testing have been viewed as evidence that genetic modification can produce unexpected and dangerous changes in foods.

Horizontal Gene Transfer and Antibiotic Gene Resistance

As mentioned earlier, horizontal gene transfer (HGT) refers to the transfer of genetic material between organisms as in the case of plants and microorganisms, unlike the parent-to-offspring channel in vertical transfer. The main concern of HGT is the possibility of transfer of transgenes to humans from plants used directly as food (also processed food) or indirectly as feed to animals that are used for food. Transfer occurs by the passage of donor genetic material across cellular boundaries, followed by heritable incorporation to the genome of the recipient organism. HGT plays an important role in the evolution of bacteria that can degrade novel compounds such as insecticides, herbicides, etc.

HGT from transgenic crops to gut (intestinal) microflora most likely occurs with transgenes of microbial origin [Kleter et al. 2005]. The mechanisms of horizontal transfer include phage transduction (from one bacterium to another by a virus), conjugation (cell-to-cell contact), and transformation by free DNA. The gene transfer between transgenic crops and microorganisms is, however, limited to transformation with free DNA [Kleter et al. 2005]. This transfer most likely occurs only if transgenes contain sufficient similarity with the corresponding genes in the recipient because homologous recombination is the most probable mechanism of transfer [Tepfer et al. 2003]. In most cases, the occurrence of HGT from transgenic plants to other organisms is expected to be lower than background rates [Keese 2008]. Therefore, horizontal gene transfer from transgenic plants poses negligible risks to human health and environment [Keese 2008].

Another consumer concern is whether selectable antibiotic resistance marker genes used in the development of transgenic crops and foods derived from them will play a part in human loss of ability to treat illnesses with antibiotic drugs. During several stages of development of transgenic crop varieties, biotechnologists use DNA that codes for resistance to certain antibiotics, and this DNA becomes a permanent feature of the final product. Antibiotic resistance genes are present in transgenic plants as a result of their use as marker genes to select transformed plant cells (vide Chapter 4). These selectable marker genes are under the control of plant promoters and are expressed in the transgenic plant. The main concern is that the presence of these genes in transgenic plants could provide a reservoir for the appearance of new drug-resistant bacteria through horizontal gene transfer from plants to pathogenic bacteria. Other concerns include whether they code for toxic products or allergen, compromise the use of therapeutic drugs, or create unwanted changes in the composition of the crop.

Two of the most common antibiotic resistance selectable marker genes (vide Chapter 4; Appendix: Table 2) present in transgenic plants developed so far are, a) *npt*II which encodes neomycin phosphotransferase type II (NPTII) and b) *hpt* (*hph*) which encodes the hygromycin phosphotransferase (HPT). HPT is the second most selectable antibiotic marker found in transgenic crops after NPT II. Both genes, which have no therapeutic relevance, are used as selectable markers in the early laboratory stages of development of the plants to enable selection of plant cells containing the desired genetic modification. Other antibiotic resistance genes commonly used in transgenic plants are *aad* (resistance to streptomycin and spectinomycin) and *bla* (resistance to ampicillin). These two belong to a different group which includes resistance to antibiotics restricted to defined areas of human and veterinary medicine.

Despite the concerns about the presence of selectable antibiotic resistance marker genes, *npt*II and *aadA*, their transfer from transgenic plants does not occur either in natural conditions or in the laboratory [EFSA 2009]. The key barrier to their stable uptake from transgenic plants to bacteria is the lack of DNA sequence identity between plants and bacteria. There is also no evidence that NPTII and HPT proteins are toxic or allergenic because they have failed to find homology to any known toxin and allergen [EFSA 2009; Lu et al. 2007]. Furthermore, neither protein is known to be involved in the production of a toxic or allergenic compound. Toxicity experiments with animals (mainly mice and rats), often involving the feeding of exaggerated doses of these proteins by gavage (use of a small tubing to insert food), have failed to establish any deleterious effects of either NPTII [Flavell et al. 1992; Fuchs et al. 1993] or HPT [Lu et al. 2007; Zhuo et al. 2009].

Another point of contention is that the dietary intake of the protein products of antibiotic selection genes could conceivably reduce the therapeutic efficacy of antibiotics taken orally. This is especially important in regard to the *npt*II gene, as kanamycin/neomycin has human and animal therapeutic importance [EFSA 2009]. Hygromycin is not used in humans, but may be used in animals such as pigs and poultry. Like most proteins, NPTII and HPT are rapidly inactivated in simulated mammalian gastric juice [FSANZ 2004; Fuchs et al. 1993]. Therefore, under normal digestion, it would be expected that any antibiotic resistance protein would be degraded before it could inactivate the corresponding antibiotic, negating any possible interference with oral administration of the antibiotic [EFSA 2009]. The expression level of HPT is extremely low in the transgenic rice line which contains below the lower limit of the detectable range of $0.03~\mu g~mL^{-1}$ [Yang et al. 2005; Wang et al. 2005]. As such, HPT protein, as a selectable marker of the transgenic rice is not considered to pose any safety problems to human health.

Consumption of Foreign DNA. When biotechnologists make a transgenic plant, they insert pieces of DNA that do not belong to a plant. When a food derived from a transgenic crop is consumed, we eat the DNA of bacteria and viruses without knowing that we do so. Some of this DNA is similar to human DNA, but much of it is foreign to us. While most of the ingested DNA is broken down into more basic molecules during digestion process, a small amount is not. This may either be absorbed into the blood stream or excreted in the feces.

In fact, DNA can persist in the gastrointestinal tract and it will become available for uptake by intestinal competent bacteria. Although the colon is the preferential site for transformation of these bacteria because it contains the largest population of bacteria within the gastrointestinal tract, the amount of DNA reaching it may only be a fraction of what is consumed. Besides, DNA is less rapidly degraded there. Wilcks et al. [2004] simulated human gut conditions in *ex vivo* and *in vivo* rat models and found that DNA was rapidly degraded in the upper part of the gastrointestinal tract, and to a lesser degree in the lower part.

The estimated intake of transgenic DNA from maize, soya bean, and potato amounted to approximately 0.38 µg day^{-1}, assuming that only GM crops are consumed [Jonas et al. 2001]. This is about 0.00006 percent of the total DNA intake of 0.6 g day^{-1}. Netherwood et al. [2004] reported evidence of low-frequency transfer of a small fragment (180 bp) of an introduced gene derived from transgenic soya bean to microorganisms within small intestine of human ileostomists (individuals in which terminal ileum is restricted and digested material is diverted from the body to a colostomy bag). However, only very low concentrations of (1–3 copies per 10^6 bacteria) of the small fragment were detected in samples of microorganism taken from the small bowel of three of seven ileostomists. Furthermore, the small fragment was only detected after two steps of amplification: a) extensive culturing of the samples and b) Polymerase Chain Reaction analysis. The introduced gene could not be detected in feces of human volunteers with intact digestive tracts following the consumption of the meal consisting of transgenic soya bean, indicating that the introduced gene was completely degraded in the large intestine.

CaMV Promoter. The cauliflower mosaic virus 35S is used as a preferred promoter in the development of transgenic crops. Of the 86 single transgenic plant events that have been approved in the U.S. till 2011, 54 contain one or more copies of the CaMV P35S. This promoter is used to "turn on" the gene inserted in the host genome. It causes cauliflower mosaic virus disease in cauliflower, broccoli, cabbage, and rapeseed. It can be horizontally transferred and cause disease, carcinogenesis, mutagenesis, reactivation of dormant viruses, and even generation of new genes [Hodgson 2000; Artemis and Arvanitoyannis 2009].

Ho et al. [2000] found CaMV in normal foods is not highly infectious and cannot be absorbed by mammals. Humans have been ingesting CaMV and its 35S promoter at high levels, but have never been reported to cause disease in humans or recombine with human viruses [Paparini and Romano-Spica 2004]. The transient expression in mammalian cells of transgenes transcribed from the CaMV35S promoter [Tepfer et al. 2004] raised the possibility that genes controlled by the 35S promoter have the potential for expression in animals. Paparini and Romano-Spica [2006] failed to detect DNA transfer in mice and CaMV35S transcriptional activity with real time polymerase chain reaction, although they emphasized the need for further studies.

Recently, some concerns were raised about using the CaMV 35S promoter for expression in transgenic plants because sequence overlap exists between this promoter and the coding sequences of P6 protein [Podevin and du Jardin 2012] encoded by viral gene, called "gene VI". The CaMV genome encodes seven genes (including gene VI) and has a large (~700 bp) and a small (~150 bp) inter-genic region that contain regulatory sequences and single-stranded interruptions. The coding sequences are either separated or overlapped by several nucleotides, except for gene VI, which lies between the two intergenic regions. The product of gene VI is a multifunctional viral protein (P6, 62 kDa) that harbors nuclear targeting and export signals [Haas et al. 2005] and ssRNA-, dsRNA- and protein-binding domains [Podevin and du Jardin 2012]. The gene VI-encoded protein of a viral pathogen is a plant toxin with three basic functions and possible risk implications: RNA silencing, transactivating the long RNA (35S RNA) produced by CaMV, and interfering with host defense mechanisms. These three functions, which cause unintended phenotypic changes and disturb the natural pest resistance, may have serious ramifications for crop biotechnology and its regulation, and even greater ones for consumers and farmers.

Soil Ecosystem

Soil ecosystem, 80 percent of which is accounted by soil-borne communities dominated by microbes, is one of the least understood areas in the risk assessment of transgenic crops. Rhizosphere microbes play a major role in nutrient mobilization, and cycling and decomposition of wastes. Any impact that genetically modified plants have on the dynamics of the rhizosphere and root-interior microbial community may result in either positive or negative effects on plant growth and health and, in turn, ecosystem sustainability. Soil microbial communities have several opportunities to interact with novel plant gene products during crop growth. After harvest, decomposition of plant litter and straw can release novel proteins into the soil environment.

A manifold increase in glyphosate application in glyphosate-resistant transgenic crops following their adoption since 1996 has been reported to have several adverse effects, including immobilization of nutrients, increase in plant diseases due to weakened plant defenses, and enhancement of pathogen virulence. These changes have apparently been caused by root exudates released by transgenic glyphosate-resistant crops following glyphosate application [Bromilow et al. 1993]. Thus, considerable concern exists regarding the potential detrimental effects of rhizosphere microorganisms on glyphosate-resistant transgenic crop productivity resulting from either direct effects of glyphosate or its indirect effects on plant physiological functions [Zobiole et al. 2011].

Zablotowicz and Reddy [2004] found that application of glyphosate had resulted in reduced root nodulation in glyphosate-resistant soya bean crop, while delaying nitrogen fixation and plant biomass accumulation. However, the severity of these effects was dependent upon formulation and number of applications of glyphosate aside from glyphosate-resistant cultivar. Powell et al. [2009] reported significant differences in nodulation among six glyphosate-resistant and three near-isoline glyphosate-resistant cultivars, but these were not related to glyphosate resistance.

Glyphosate application in glyphosate-resistant soya bean field has resulted in reduced nutrition uptake, leading to enhanced occurrence of many diseases which, in turn, detrimentally impacted many beneficial soil microorganisms [Kremer et al. 2005; Johal and Huber 2009; Duke et al. 2007]. The EPSPS enzyme present in glyphosate-resistant soya bean was considerably less efficient than wild-type enzyme, producing insufficient amounts of phytoalexins (the key defense components associated with shikimate pathway) to prevent fungal infection [Gressel 2002]. Besides, EPSPS also decreases the shikimate-dependent lignification of cell walls at or around the infection site. Decreased lignin content may also be due to the reduced photosynthesis in soya bean caused by glyphosate [Zobiole et al. 2010a].

Glyphosate is a strong broad-spectrum chelator forming complexes with nutrient metal ions such as Ca, Mg, Co, Cu, Fe, Mn, Ni, and Zn in the soil. Because of this ability, glyphosate has been postulated to affect plant uptake of trace nutrients such as Mn^{2+} or Zn^{2+} [Duke et al. 2012b]. This would make plants deficient of these nutrients. Manganese, like other essential nutrients, is involved in many physiological and biochemical processes that can affect the defenses of plants against pathogens causing root and foliar diseases. Glyphosate, a patented synergist for mycoherbicides to enhance the virulence and pathogenicity of organisms used for biological control, predisposes plants to infectious disease-causing organisms [Johal and Huber 2009]. The toxic microbial effects of glyphosate are cumulative, and its continuous application had led to Mn deficiency in areas that were previously Mn sufficient because of reduced populations of Mn-reducing soil organisms [Johal and Huber 2009]. Manganese is absorbed by plants in reduced state Mn^{2+}. High soil pH limits Mn availability due to its oxidation to Mn^{4+} state under alkaline conditions.

The virulence mechanism of some pathogens such as *Corynespora*, *Gaeumannomyces*, *Phymatotrichum*, *Magnaporthe*, and *Streptomyces* involves Mn oxidation at the infection site to compromise the plant's resistance mechanisms involving the shikimate pathway [Thompson and Huber 2007]. Isolates of these pathogens that cannot oxidize the physiologically available Mn^{2+} to the non-available Mn^{4+} are avirulent and not be able to cause significant tissue damage [Roseman et al. 1991] but they can become virulent once the plant defenses are weakened by the glyphosate-mediated Mn deficiency. As many plant defenses rely on the shikimate pathway that produces the anti-bacterial phytoalexins which are known to be blocked by glyphosate, it is not surprising that this herbicide would render plants more susceptible to pathogens.

Notwithstanding the aforementioned review on the reported adverse effect glyphosate on mineral nutrition in glyphosate-resistant crops, other studies found no such effect [Andrade and Rosolem 2011; Duke et al. 2012a; Bailey et al. 2002; Rosolem et al. 2010; Loecker et al. 2011; Cavalieri et al. 2012; Serra 2011]. These seemingly contradictory results could be entirely or in part due to differences in the soils (soil type, pH, nutrient status, etc.), climatic conditions, and/or glyphosate resistant cultivars [Duke et al. 2012b]. Further field studies by taking these variables into considerations are required before arriving at meaningful and un-biased conclusions.

Although glyphosate is generally known to be rapidly inactivated by soil adsorption, it may still serve as a substrate for some microorganisms [Araújo et al. 2003; Kuklinsky-Sobral et al. 2005]. Kremer and Means [2009] found higher colonization of roots by *Fusarium* spp. when field-grown glyphosate-resistant transgenic soya bean cultivars were applied with glyphosate over a 10-year period (1997–2007), while plants receiving no or conventional postemergence herbicides exhibited low *Fusarium* colonization. The non-transgenic cultivars had the lowest root colonization by *Fusarium*. This colonization increased as soya bean growth progressed and glyphosate rate increased [Zobiole et al. 2011]. Reduced production of both lignin and phytoalexin allows increased root colonization by *Fusarium* in plants injured by glyphosate [Johal and Rahe 1988].

After reviewing published reports evaluating interactions between glyphosate and *Fusarium*-caused diseases, Powell and Swanton [2008] concluded that field research has not documented a causative link between the two although it is impossible to rule it out. Further research is needed to establish if glyphosate-mediated root exudates stimulate the growth *Fusarium* spp. in transgenic glyphosate-resistant soya bean and other crops grown under various environmental conditions and soil types.

Insect-Resistant Transgenic Crops

Like the herbicide-resistant transgenic crops, the insect-resistant *Bt* crops also have the potential to affect the soil microbial dynamics because they produce insecticidal Cry proteins (Cry δ-endotoxins) in all plant parts. The insect-resistant transgenic plants manufacture, within their cells, one or more forms of the natural bio-insecticide. The rate of synthesis of *Bt* Cry protein δ-endotoxins is nearly proportional to the rate of plant growth, but the expression falls as plants mature and enter senescence [Benbrook 2012].

Some scientists have quantified the Cry δ-endotoxins manufactured by crops. For example, Nguyen and Jehle [2009] projected that a hectare of *Bt*-maize expressing *cry3Bb1* gene in MON88017 maize produces 905 g ha^{-1}. Benbrook [2012] reported that in USA many *Bt* maize events resistant to the lepidopteran European Corn Borer (ECB) synthesize as much or more insecticidal protein per ha (79,040 plants) than the weighted average rate of conventional insecticide application (0.15 kg ha^{-1}). Maize event MON810 expresses 0.2 kg ha^{-1} Cry1Ab δ-endotoxin, while BT11 synthesizes 0.28 kg ha^{-1}. The newer event MON89034 for ECB control produces a total of 0.62 kg ha^{-1} Cry 1A.105 and Cry 2Ab2 δ-endotoxins. However, TC1507 event expresses the least δ-endotoxin (Cry1F), 0.1 kg ha^{-1}, just below the rate of insecticides applied [Benbrook 2012].

In the case of the coleopteran Corn Rootworm (CRW), greater volumes of δ-endotoxins are expressed than the average insecticide application of ~0.2 kg ha^{-1} [Benbrook 2012]. MON88017 expresses 0.62 kg ha^{-1} of Cry 3Bb1 (in addition to CP4 EPSPS protein conferring tolerance to glyphosate), while DAS59122-7 expresses two Cry proteins (Cry34Ab1 and Cr35Ab1 besides the PAT protein to confer tolerance to glufosinate) totaling 2.8 kg ha^{-1}, which is 14-fold more than the insecticides they displaced. On the other hand, 'SmartStax' maize (MON89034 x TC1507 x MON88017 x DAS-59122-7), which carries eight genes (three each for ECB and CRW resistance and two for tolerance to glyphosate and glufosinate), expresses 4.2 kg ha^{-1}, 19 times more than the average conventional rate of insecticide application in 2010 [Benbrook 2012]. The systemic delivery of *Bt* Cry proteinaceous endotoxins poses more significant risks to animals and humans than application of *Bt* insecticides via liquid sprays [Benbrook 2012]. Systemic delivery also enhances the range of environmental and ecological risks [Stotzky 2000] than *Bt* application to the foliage which exposes Cry proteins for rapid breakdown by sunlight and rain.

The soil microbial communities are part of complex food webs. Along with numerous and varied soil-dwelling invertebrates (earthworms, collembolans, mites, woodlice, nematodes, etc.) they carry out several plant-related processes in soil ecosystems [Moore et al. 1988]. The transgenic Cry1Ab protein is strongly adsorbed by soil particles and tightly bound to clay minerals [Stotzky 2002; Pagel-Wieder et al. 2007]. This reduces its availability in quantities required for inhibition of microbial degradation [Koskella and Stotzky 1997]. The concentration of Cry1Ab protein found in bulk soil ranged between 3 ng g^{-1} and 4.4 ng g^{-1} of soil [Baumgarte and Tebbe 2005; Hopkins and Gregorich 2003], while the same in transgenic *Bt*-maize rhizosphere soil was positive for the presence of *Bt* toxin [Saxena and Stotzky 2000].

Saxena et al. [2004] showed that *Bt* maize, potato, and rice contributed to the presence and persistence of Cry proteins in soil via root exudation, whereas *Bt* cotton, canola, and tobacco did not. Transgenic *Bt* maize line NK4640, expressing the *Bt* toxin gene *cry1Ab* exudes some of the toxin protein from roots into the surrounding rhizosphere and soil, along with other exudates normally present in roots [Saxena et al. 1999]. Proteolyting microbes present in the rhizosphere degrade the novel proteins and assimilate the components

[Dunfield and Germida 2004]. Cry proteins are resistant to biodegradation in soil, thus facilitating a potential longer exposure of non-target organisms to the toxin [Icoz and Stotzky 2008; Koskella and Stotzky 1997; Stotzky 2004]. However, Cry root exudates have been shown to have little effect on non-target organisms (earthworms, nematodes, and protozoa) although fungal mycorrhizae did colonize *Bt* maize roots more efficiently than non-*Bt* ones [Icoz and Stotzky 2008].

Gruber et al. [2012] studied the potential accumulation of Cry1Ab protein as a result of continuous entry of *Bt*-maize MON810 plant residues over more than five vegetation periods on four different sites in South Germany. They found a) complete degradation of the *Bt* maize containing recombinant Cry1Ab protein by the following season and b) no persisting immunoreactive Cry1Ab protein in any soil shortly before the next planting over the three-year period. These results suggested that the recombinant Cry1Ab protein has no extraordinary stability in soil. The results of this field study are consistent with rapid degradation of *Bt*-protein from potted *Bt*-maize MON810 plants in three different soil types found in glasshouse experiments by Badea et al. [2010].

Barriuso et al. [2012] found three prominent phyla in a maize field cultivated continuously for four years with event MON810 expressing Cry1Ab protein: *Proteobacteria*, *Acidobacteria*, and *Actinobacteria*. However, there was no change in their overall distribution between fields grown with non-*Bt* and *Bt*-maize. Besides, fluctuations in climate affected both non-*Bt* and *Bt*-maize almost equally. Earlier studies also found the Cry1Ab protein from event MON810 remaining in the soil even after four years of *Bt*-maize cultivation, while other Cry proteins derived from Cry3Bb1 gene in event MON863 were not [Icoz et al. 2008].

However, a few studies have reported significant differences in microbial community structure in soils grown with *Bt* and non-*Bt* crops using various classical and molecular techniques [Icoz and Stotzky 2008]. These contradictory results may be due to differences in the type of Cry protein, stacked-traits, plant variety, and experimental methods used, as well as soil type and environmental factors [Icoz and Stotzky 2008]. The species and functional diversity of microbial communities in the soil are influenced by numerous direct and indirect environmental factors [Icoz and Stotzky 2008]. Direct effects depend on both the spectrum of activity of the proteins encoded by transgenes [Oger et al. 1997] as well as the quantities of proteins that accumulate in the environment. In contrast, indirect effects are mediated by changes in the chemical composition of plant biomass and root exudates that result from modification in the normal metabolic pathways in plant tissues.

Another question that needs to be addressed is whether cultivation of *Bt* crops affects the yield of subsequent crops, especially non-*Bt* crops grown on the same soils on which *Bt* crops have been grown. If effects on subsequent crops are observed, the duration (e.g., number of seasons) of the effects would be of practical interest [Icoz and Stotzky 2008]. Moreover, the possibility of long-term effects of *Bt* crops cannot be excluded and they must be examined on case-by-case basis.

Issues

The issues concerned with transgenic crops include terminator seeds, intellectual property rights, asynchronous approval of transgenic crops, biopiracy, and coexistence of transgenic crops and food products.

Terminator Seeds

Genetic use restriction technology (GURT), colloquially known as 'terminator technology', produces plants that have sterile seeds. This name was given to methods that restrict the use of genetically engineered transgenic plants by causing second generation seeds to be sterile. Terminator technology uses a terminator gene, also called transcription terminator, a short base sequence at the 3′ end of a gene which causes the RNA polymerase to terminate transcription (vide Chapter 3). This terminator region signals RNA polymerase to release the newly made RNA molecule, which then departs from the gene. If the promoter signals initiation of transcription, terminator signals its end. The end result is that farmers using these seeds are forced to buy fresh seeds for the following season rather than seeds from the current crop. This means that farmers habituated to use seeds saved for the next crop for millennia are now denied to do so. There are two types of terminator technology: V- GURT and T- GURT.

V-GURT and T-GURT

V-GURT. On 3 March 1998, the U.S. Department of Agriculture and Delta & Pine Land Company jointly obtained U.S. patent 5,723,765 for a new genetic technology designed to make sterile seeds and prevent unauthorized seed-saving by farmers. Although originally patented to prevent "seed saving" and as a "technology protection system" for arable crops [United States Patent and Trademark Office 1998], this type of system found its use in ensuring that new seed varieties could be protected against 'unauthorized' use. This varietal genetic use restriction technology (V-GURT) is designed to render seeds sterile at harvest. It affects the reproduction and viability of the whole crop variety.

The complex V-GURT briefly involves insertion of three transgenes into the plant DNA [Gupta 1998]: recombinase repressor gene which contains nucleotide sequence for coding the protein which suppresses the recombinase gene, and which could be controlled by an external stimulus; site-specific recombinase gene (*cre/lox*) which contains the information for a protein which, in turn, cuts the blocking sequence linked to the toxic gene; and toxin gene (called terminator or lethal gene), which codes for a toxic substance and is linked to a blocking sequence preventing the activation of terminator gene. These genes are connected so that: a) the repressor gene prevents the recombinase gene from functioning, b) the recombinase gene, if it functions, allows the toxin gene to activate, and c) the toxin gene produces a toxin that kills the embryo in the seed, so that the seed cannot germinate [Shi 2006].

Once the chemical trigger is released, the repressor gene is switched **off** allowing the recombinase gene to be switched **on** which, in turn, removes the blocking sequence from the terminator gene, resulting in lethal gene expression. Toxins produced by the activated terminator gene destroy the embryo, thereby rendering the seeds sterile. This, however, leaves all the other aspects of plant growth unaffected, because the toxic effects stimulated by chemical treatment only occur during the later stages of embryo development and, therefore, cannot adversely affect final yields [Lehman 1998].

The seed producer can control the system by spraying the first generation seed with the regulator, which then inactivates the repressor gene. Since the repressor gene doesn't function, the recombinase gene is allowed to do its job, as in step (b) above. If the seed producer wishes to protect the intellectual property embedded in the seed, he sprays the seed with the regulator before delivery to the farmer [Shi 2006].

T-GURT. Europe's answer to the American V-GURT is trait-specific genetic use restriction technology (T-GURT). In this, a new chemically activated seed killer, called 'Verminator', kills seeds as claimed by Zeneca (U.K.) in its invention application (WO9403619), by switching on rodent fat genes that have been bio-engineered into crops. This technology [ETC 2003] contains "a gene switch", which is inducible by external application of a chemical inducer and which controls expression of a gene product which, in turn, affects expression of a second gene in the genome. Syngenta received European patent for this technology in 2002, and filed for the U.S. patent in 2005.

T-GURT modifies a crop in such a way that the genetic enhancement engineered into the crop does not function until the crop seed is sprayed with a proprietary chemical, called activator, that is sold by a biotech company. Unlike V-GURT terminator seeds, T-GURT seeds, modified to produce specific traits such as tolerance to salt and drought, do not become sterile after a season's planting. Instead, the farmer can activate the trait only after the seed is treated with the chemical. The seed will still germinate without the chemical treatment, but the plant would not show the modified characteristics.

In T-GURT concept, a trait is switched on and off at will. There are two possible ways in which T-GURTs can be designed [FAO 2001]. In the first method, a chain of genes, similar to the one described in V-GURTs, is constructed. The system can be programmed so that the toxin gene deletes a "trait" gene instead of killing the embryo. Thus, if the seed is sprayed with regulator before delivery to the farmer, the first generation seed will produce the trait embodied in the trait gene, but not the second generation [Jefferson et al. 1999].

The second procedure involves activation of the inserted trait by the farmer by applying an 'activator' compound to the plant or seed. The system can be designed so that subsequent generations of the seed will contain the trait gene, but in an inactive state. Thus, use of the trait in a given year requires the farmer to purchase and apply the activator in that year. The USDA [2001] suggested that a T-GURT can be activated by the farmer's spraying a 'standing crop' with the activator. However, it is not clear if such a technology is feasible, and even if it were feasible, whether the timing of application of the activator would be flexible. If the timing of application were flexible, this would confer an option value upon the T-GURT-protected trait.

Global Reaction to GURT

The world reaction to GURT came very fast and fierce. Ever since this technology came into public attention in 1998, there has been avalanche of public opposition to, and global condemnation of, this set of technologies from many non-governmental and governmental organizations. In 1998–1999, the U.N. Conference on Environment and Development (UNCED), the FAO, and the Rockefeller Foundation, USA have all stated that they were against terminator seeds. The U.N. recommended that products with such technologies should not be approved for field testing until sufficient scientific assessments had been carried out. India, Panama, Ghana, and Uganda said that terminator technology should not be developed. The U.K. government said that it would not allow developing, testing or using breeding material which was designed to prevent seed germination. Although its reaction was against the terminator technology as a whole, V-GURT has been specifically targeted without mentioning T-GURT.

Until the advent of terminator seed technology, farmers have just planted their own seeds and exchanged them among themselves. Millions of farmers around the world have felt that their heritage of saving, exchanging, evolving, selecting, and crossing many traditional varieties to obtain desired characteristics has been forcefully taken away by the terminator technology monopolized by a handful of multinational biotech companies. These practices have become illegal for all plant varieties that are patented or otherwise owned by some entity (often a corporation) [Mechlem and Raney 2007].

In 2000, the noted activist of transgenic crops Vandana Shiva of India said that "A half century after the Bengal Famine (where...during British colonial rule, most of the food grown was exported to U.K. for trade instead of feeding the hungry local people), a new and clever system has been (now) put in place which is once again making the theft of the harvest a right and keeping the harvest a crime. Hidden behind the complex free-trade treaties are innovative ways to steal nature's harvest, the harvest of the seed, and the harvest of nutrition" [Shiva 2000]. In India, this is a cause for some concern as scientists fear for the livelihood of 400 million farmers and for food security in the country. Some marginal Indian farmers have already been driven to suicide. It is feared that this type of technology could be used to make the poorer farmers even more dependent.

In the face of international controversy and massive public opposition to terminator technology, the U.S. Department of Agriculture defended its patent on genetic seed sterilization and negotiated a licensing agreement with Delta & Pine Land. The goal of USDA, according to Willard Phelps, was "to increase the value of proprietary seed owned by US seed companies and to open up new markets in second and third world countries" [Shand and Mooney 1998]. USDA molecular biologist Melvin Oliver, the primary inventor of the technology, explained why the U.S. developed a technology that prohibits farmers from saving seeds: "Our mission is to protect U.S. agriculture and to make us competitive in the face of foreign competition; without this, there is no way of protecting the patented seed technology" [Shand and Mooney 1998].

Consolidation and international partnerships soon began between seed-producing biotech companies around the world. In May 1998, Monsanto announced it would acquire Delta & Pine Land and the merger process was completed in 2007. The soon-to be new owner of V-GURT had pledged in 1999 not to use the technology to which other seed companies also agreed. But Monsanto soon reneged on its promise in 2005 when it announced "Monsanto does not rule out the potential development and use of one of these technologies in the future" [Monsanto 2005]. This controversy led to the introduction of a bill in the Canadian Parliament in May 2007 to "prohibit field testing and commercialization of Terminator seed technology."

Impact of GURTs

There are several potential benefits, costs, and risks associated with genetic use restriction technology.

Potential Benefits. Potential benefits include protection of intellectual property rights, stimulation of private research and development, enhancement of genetic diversity, and containment of transgenes. The degree of potential benefits depends on the groups involved in development, production, and regulation. The groups are not necessarily mutually exclusive.

Private seed companies, the first to benefit from the approval and marketing of GURTs, believe that the absence of intellectual property rights may diminish research and development in plant breeding [Goeschl and Swanson 2003; Pendleton 2004; Lence et al. 2005]. Therefore, GURTs represent a novel mechanism for capturing returns from innovation in the plant breeding industry and potentially increasing private investment into research and development in agriculture and hence higher rate of innovation. In order to protect their investment, seed producing companies felt that GURTs may present a better form of insurance against free use of genetic innovations than patents, rights of plant breeders, and licenses [Visser et al. 2001; Pendelton 2004; Burk 2004]. Thus, plant breeders and seed companies stand to make substantial intellectual and financial gains through implementation of GURTs at the expense of farmers' woes. The companies also believe that the potential for transgene escape may also be reduced through sterile seed technology [Gupta 1998; Visser et al. 2001; Eaton et al. 2002]. This would be beneficial for seed companies because it would decrease the probability of corporate liability for environmental contamination or health risks due to escaped transgenes [Pendleton 2004], but would also reduce the chances of competitors or farmers accessing proprietary genetic material through volunteer or feral crop plants. With enhanced transgene containment through GURTs, seed corporations would save on costs of monitoring farmers' fields for any unauthorized use of copyrighted genetic material or transgene escapes that must be mitigated [Szumigalski 2006].

Seed companies believe that **benefits to farmers** include improved crop yields as a result of increased research and development by them. For example, with application of GURTs, R&D will increase for self-pollinated crops such as wheat, rice, and cotton [Visser et al. 2001]. Thus, farmers may profit in the long-term from these innovations because more productive varieties will become available as breeding efforts increase. For example, the vast majority of improved varieties have been from hybridized crops, with an average annual yield growth of 2.18 percent over the long-term for hybrid crops compared to 1.58 percent for non-hybrid crops in developed countries [Goeschl and Swanson 2003]. Besides, incentives to breed new varieties may enhance genetic diversity in many important crops, thereby providing further long-term benefits associated with biodiversity (e.g., insect resistance) to farmers [Lehmann 1998]. Terminator technology may also eliminate the problem of herbicide-resistant transgenic crop volunteers in farmers' fields because the seeds are sterile [Pilger 2002], reduce potential for cross-pollination with, and increasing fitness of weedy relatives [Visser et al. 2001; Gupta 1998].

The benefits derived by **governments** include reduced funding for agriculture R&D and bio-safety/copyright infringement enforcement programs [Eaton et al. 2002; Pendleton 2004]. When the country, as a whole, increases agricultural production by adopting this technology, the governments may gain politically with policies that support GURTs [Szumigalski 2006].

Society, which in general includes consumers and environment in which they live, may derive some benefits from GURTs as a result of increased productivity [Eaton et al. 2002]. However, as per the forecast model simulations conducted by Goeschl and Swanson [2003], the advanced countries stand to benefit the most in terms of productivity gains from GURTs, while less advanced countries stand to gain the least over the medium to long term (i.e., over 20 years). The general public in developed countries spends around 10 percent, sometimes even less, of its income largely due to higher commodity production in proportion to population and the governmental policies to provide 'cheap food'. Lence et al. [2005] contend that while generally, seed industry might lose from innovations, increased benefits to consumers outweigh producer losses resulting in overall increases with intellectual property protection. This, however, is difficult to comprehend because a seed producer's main objective is pecuniary benefits to be derived from GURTs.

Risks. The risks associated with terminator technology are poorly understood because they are largely buried under rhetoric and non-peer reviewed research on this specific topic. The risks include those associated with **farmer income**, **environment**, and **agro-biodiversity**.

The more devastating impact of terminator technology is on farmers, especially the poor, marginal, and subsistence ones, in both developed and developing nations. The greatest risk is on their input costs because the farmers have to purchase GURT seeds every year. This goes against their traditional practice of saving good seeds. For example, 90 percent of farmers in India save seeds to be replanted [Bhatia 1998] and this practice is also followed worldwide. Besides, increased market control by seed companies threatens farmers' autonomy, food security, and survival. Rich farmers are likely to gain most of the benefits from this technology [Eaton et al. 2002; Goeschl and Swanson 2003] while others could be marginalized. The recurring problem in countries like India and other developing countries is the reliability of germination of seeds supplied by companies which are less than scrupulous. This has been the cause of the growing suicide deaths by the insect-resistant *Bt*-cotton farmers in India. The whole purpose of GURT, as activists contend, is to facilitate monopoly control by agribusiness, the sole beneficiary.

While farmers in developing countries are denied of their age-old privilege of being owners of their own seeds because of TRIPs Agreement as mentioned earlier, the privilege of U.S. farmers, by contrast, is protected by the nation's Plant Variety Protection Act under which they may sell seed up to the amount saved for their own acreage [The Crucible II Group 2001].

The greatest fear is that the sterility trait from first generation seed might spread via pollen to neighboring crops and wild relatives as in the case of transgenic crops discussed earlier. Risks of cross-pollinating a conventional crop through pollen from GURTs are greater than via seed since terminator technology does not inhibit transgenic plants from producing pollen. This type of gene escape of transgenes is unavoidable in crops such as rice, wheat, maize, barley, sorghum, and sugar beet [Giovannetti 2003]. Pollen carrying a terminator trait could make the seeds of next generation sterile, and this would be realized only after these seeds are planted. The reduced seed viability due to terminator pollen contamination may be problematic to farmers who save seed to be planted the next year. The magnitude of these effects depends on the degree of cross-pollination of the crop and distance between the donor and acceptor plants [Visser et al. 2001]. Cross-pollination varies between 1 percent in most self-pollinated crops and 15–20 percent in strongly cross-pollinated crops [Bhatia 1998]. However, cross-pollination in the self-pollinating wheat would be as high as 10 percent for some varieties [Lawrie et al. 2004], and it could occur at a distance of several meters from the source of pollen in field situations [Matus-Cadiz et al. 2004]. In the case of insect-pollinated canola, cross-pollination can occur over several hundred meters [Giovannetti 2003].

As sterile seed technology is designed to eliminate production of viable seeds, some scientists argue that risks associated with transgene escape via seed will be very limited or impossible with GURTs [Lehman 1998; Rakshit 1998; Pendelton 2004]. However, seed treatment with an inducer chemical, possibly tetracycline, is not always 100 percent perfect, and therefore the recombinase gene could remain inactive in some seeds [Giovannetti 2003]. As some seeds could always escape the effects of chemical trigger, their effect could still be potentially large. Such escaped seeds would carry the complete genetic complement of the V-GURT, and could go on to germinate to produce both pollen and more seeds carrying the terminator technology trait [Szumigalski 2006].

Besides the risk of environmental contamination of terminator transgenes, there is a risk of biodiversity, more specifically agro-biodiversity, in farmers' fields due to homogenization of crops. Although reduced crop diversity is already a serious global issue, adoption of GURTs would further aggravate the problem because with sterile seed technology, farmers would not have access to novel genetic traits often utilized to increase agro-biodiversity at the local level [Visser et al. 2001].

The adoption of short-duration, high yielding hybrids during the 'Green Revolution' era has already initiated the erosion of local agro-ecological capital [Visser 1998] and GURTs will exacerbate this situation. Now that farmers have to purchase new seeds every year under GURTs, this technology would be detrimental to local agriculture and economies of emerging nations where farmers habitually save seed. Besides, risks may also arise from grain imports consisting of GURT seeds and these would be detrimental if pollen of plants from these sterile seeds ends up cross-pollinating local crop varieties. Furthermore, since terminator technology, *per se*, results in the destruction of the seed embryo, it could potentially have an adverse effect on the nutritional quality of the harvested grain and consequently health (e.g., type II diabetes) of consumers.

These are the issues that need to be addressed by conducting long-term studies. Given that the terminator technology is still untested on a large scale, farmer and consumer concerns as well as environmental and agro-biodiversity issues continue to remain.

Intellectual Property Rights

Intellectual property (IP) refers to the legal protection to creations of the mind for which the rights of the inventor are recognized. Under IP laws, inventors are granted exclusive ownership rights to a variety of intangible assets, such as literary, music, and artistic works as well as inventions and discoveries in a technical field. The more common types of intellectual property rights, based largely on Western concept, include patents, copyrights, trademarks, etc.

Patents and plant variety protection (PVP) are two different forms of intellectual property rights (IPR). Like IPR, both patents and PVP provide exclusive monopoly rights over a creation for commercial purposes over a period of time. A patent is a right granted to an inventor to prevent all others from making, using, and/or selling the patented invention for 15–20 years [GRAIN 2001]. The criteria for a patent are novelty, inventiveness (non-obviousness), utility, and reproducibility. Although patents were designed for industrial application, biotechnology patent offices grant patents on microorganisms and, in some countries, on all life forms. Over 50 percent of patents delivered by World Intellectual Property Organization are for companies from USA and Japan [Biopiracy Collective 2012].

On the other hand, PVP gives patent-like rights to plant breeders. In this, the genetic makeup of a specific plant variety is protected [GRAIN 2001]. The criteria for protection include novelty, distinctness, uniformity, and stability. PVP laws provide exemptions for breeders who are allowed to use protected varieties for further breeding and crop improvement. At the same time, they allow farmers to save seeds from their harvest. Unlike patenting, PVP is constrained by these exemptions.

The advent of capital-incentive molecular biology and biotechnology for agriculture has heralded the concept of protection of IPRs and the perceived need to protect them for continued research for the benefit of humanity. Initially, application of patents was confined to plant material only to be expanded to gene-based patents which became ubiquitous in biotechnology.

The turning point came on 16 June 1980 when the U.S. Supreme Court recognized the patentability of living organisms in a landmark *Diamondv. Chakrabarty* Case. In 1972, the India-born American microbiologist Ananda Mohan Chakrabarty, working with the General Electric Company, had filed patent application for his invention of a *Pseudomonas* bacterium containing at least two stable energy-generating plasmids, each of them providing a separate hydrocarbon degradative pathway which he believed will have significant commercial value for the treatment of oil spills [U.S. Supreme Court 1980]. When his application was rejected by the lower court, the case was finally referred to the Supreme Court for decision. The Court ruled that a human-made living bacterium was "patentable", provided the "manufacture" was, a) markedly different from those found in nature and b) it possessed potential for significant utility. Long before this case, the U.S. Patent Office had granted a patent (No. 141,072) to Louis Pasteur for a purified yeast cell in 1873 [Robinson and Medlock 2005]. While the Chakrabarty case dealt specifically with a bacterium, its implications covered all living organisms including plants. Subsequent court decisions specifically covered newly developed plant breeds as well. But it was the Supreme Court decision over 34 years ago that gave birth to patenting of genetically modified transgenic seed in agriculture. It was also around the time that *Agrobacterium* became useful for creation of transgenic plants (vide Chapter 4).

There are three different IPR agreements that protect plant varieties and seed producers' rights. These include UPOV Agreement, TRIPs Agreement, and Cartagena Protocol on Biosafety. These treaties, administered by World Trade Organization (WTO), contain a comprehensive set of rules for member countries regarding IPRs over plant varieties, including transgenic crops and GURTs.

UPOV. The treaties of International Union for the Protection of New Varieties of Plants, UPOV, (French: Union internationale pour la protection des obtentions végétales), adopt a *sui generis* (unique or one its kind) system of protection tailored to the needs of plant breeders. The first UPOV Act, drafted in 1961 by industrialized countries led by USA, provides protection for plant breeders in their own and overseas markets. This Act was later revised in 1972, 1978, and 1991, with the last revision coming into force on 24 April 1994. Each of these revisions enhanced the rights for plant breeders while broadening the scope of plant variety. The UPOV Act does not become enforceable in domestic law until the country enacts a national plant variety protection law that conforms to the Act's requirements [Helfer 2004].

The 1991 Act provided uniform provisions on the extent of rights of plant breeders to use protected crop varieties to create new varieties and expanded the IPRs of first generation breeders, but limited the farmers' rights to save seeds and prohibited the sale of seed [Helfer 2004]. UPOV '91 has led to erosion of the free exchange of agricultural genetic material. The countries may provide for simultaneous plant breeder rights (PBRs) and patents on everything from genes to entire crops, while rights of breeders extend to the import and export of protected varieties, and to control of the harvest produced from those varieties without breeder authorization. Thus breeders may now reach all the way into the farmers' fields, and follow production from places without compatible PBR if imported into UPOV countries. UPOV protection is for 25 years for trees and grapevines and 20 years for all other plants in some cases and to new genera [Helfer 2004]. As of 2013, only 71 nations became members of UPOV. These included USA, European countries, Australia, and many US-supported states in Africa and South America, but not India, Pakistan, and Korea among others. A UPOV, which largely supports an agricultural system that is clearly export-oriented [Kent 2002] and favors commercial breeders over farmers and producers, besides private interests over public interests [Dutfield 2011], has been overshadowed by the TRIPS Agreement.

TRIPS. The biotechnology industry and the development of genetically modified crops and foods is a multi-billion dollar industry which is expected to grow in excess of 10 percent annually. With such high stakes, biotech industry advocates for more stringent patent policies that would be more beneficial for them than those provided in UPOV. These policies, as mentioned earlier, are governed by the U.N. Article 28 of the Agreement on Trade-Related Aspects of Intellectual Property Rights (the TRIPS Agreement) adopted in 1994 and became effective on 1 January 1995. It says "planting, harvesting, saving, re-planting, and exchanging seeds of patented plants, or of plants containing patented cells and genes, constitutes use"… "is prohibited by the intellectual property laws of signatory states" [Mechlem and Raney 2007]. This Agreement, aided by the lobbying force of the U.S. and supported by the European Union, Japan, and other developed nations, was the result of the GATT (General Agreement on Tariffs and Trade) that helped establish the World Trade Organization (WTO). Under WTO rules, any country that intends to participate in it must adhere to TRIPS and not infringe, abuse, or misappropriate the patented technology, which also includes seed patents. While the TRIPS agreement strongly favored USA, the world leader in the development of biotechnology and IPRs, other nations, particularly the developing entities, felt it contravened their national, societal, and farmer interests.

Cartagena Protocol on Biosafety. Another international U.N. agreement on biosafety, called "The Cartagena Protocol on Biosafety" came in handy for nations which felt discriminated against by the TRIPs. It seeks to protect biological diversity from the potential risks of genetically modified organisms, termed 'living modified organisms' (LMOs), resulting from modern biotechnology. This legally binding global protocol, adopted at a meeting of members states of the Convention on Biological Diversity (CND) on 29 January 2000 and which became effective on 11 September 2003, employs several mechanisms that give nations control over protecting conventional species from any potential threats posed by LMOs (genetically modified seeds, plants, fish, animals, and microorganisms).

The Protocol [Hyder 2011] allows a) nations to exchange scientific and technical data on LMOs which they can use to make decisions regarding any threats to biodiversity or public health posed by them, b) requires exporters of LMOs to seek permission of the target nation before the first shipment of a new LMO is introduced into the importing nation (if it feels that a particular LMO might pose a threat within its country), and c) mandates specific labeling of shipments of GM commodities such as maize or soya bean intended for direct use as food, feed, or processing. In the last case, shipments composed in whole or part of LMO commodities must state that the shipment "may contain" LMOs and are "not intended for intentional introduction into the environment." As of October 2013, 166 nations excluding Argentina, Australia, Canada, and USA have approved or ratified this U.N.-mediated Protocol.

The Cartagena Protocol, however, did not address issues involving genetically modified foods or food safety. It only applies to LMOs, not to food produced using LMOs.

Despite various international agreements, biotechnologists contend that winning social acceptance of biotechnology and IPR on GMOs and GURTs is extremely important because it is intrinsically connected with providing better food and agriculture in developing and emerging countries, home to a majority of the global population. The message of biotechnologists is: "No IPR, no biotech; No biotech, no progress. But we can cut deals with patent holders to ensure that the poor can access the technology" [Kuyek et al. 2001].

Farmers, the original breeders of crop seeds for millennia, are instrumental in generating the diversity on which the seed producing breeding industry has been building its multi-billion empire since recently. For example, there are nearly 100,000 distinct rice accessions in the gene bank of International Rice Research Institute in Manila, Philippines and the bulk of them were developed by farmers of several countries. Their valuable contributions have been denied by global biotech seed industry which currently holds 90 percent of the current biotech patents on rice genes, transgenic rice plants, or methods to obtain them [Kuyek et al. 2001]. The rights of millions of the original breeders, i.e., farmers, have been ignored. The vital question that confronts the global governments is whether they should continue to deny the intellectual property rights of farmers at the expense of profit-seeking producers of transgenic seeds.

Asynchronous Approval of Transgenic Crops

Commercialization of transgenic crops over the past two decades has become a regulated activity worldwide. However, each nation has a different set of regulatory procedures and standards, with no simultaneous approvals in all countries. Once a transgenic event is approved in the country where it was developed, it may take 2–10 years (or never at all) before other countries grant regulatory approvals. This issue, known as 'asynchronous approval' (AA) of transgenic crops, is a growing issue for its potential economic impact on international trade [Stein and Rodriguez-Cerezo 2009]. In such a situation, traces of new transgenic crops may appear in agricultural commodities exported to countries where these new varieties have not been approved. When such shipments are rejected, it can lead to economic loss to suppliers, more trade disruptions, and ultimately closing access to specific markets.

A problem similar to AA occurs when a developer of a new transgenic crop, designed only for local markets, does not seek approval for commercialization in other countries. In such a situation, there can be an 'isolated foreign approval' by an importing country. However, if there is a "low-level presence" (LLP) in the unauthorized transgenic material, the shipment is liable for rejection. Sometimes, traces of transgenes could end up in a non-transgenic commercial crop due to accidental admixtures. The problem also arises when the unauthorized transgenic crop material, even with LLP, could not be excluded from imports by the importing country because global agriculture trade is an open process. If the importing country operates a 'zero tolerance' policy, imports of crops or food products are liable for rejection if they contain traces of GMOs that are not authorized. In the case of the European Union (EU), the issue of LLP of new GM crops in agricultural imports has already caused trade disruption and economic problems [Stein and Rodriguez-Cerezo 2009].

In August 2006, trace amounts of glufosinate-resistant transgenic rice crop of LLRice601 (LibertyLink rice; Bayer CropScience) of 2005 (not approved for growing and consumption in the U.S.) from Arkansas, have been found in commercial rice samples imported by the European Union from USA. Subsequently, the bilateral rice trade between the two collapsed. In 2005, the EU had imported 32 percent of its rice from the USA, but this came down to just 2.5 percent in 2007, resulting in large economic losses to U.S. exporters as well as EU importers just because of the presence of a miniscule amount of LLRice601 in the overall U.S. supply. A similar case was reported when the EU detected unapproved GM traces in shipments from the U.S. in the case of MON88017 and MIR604 maize events prior to their authorization in the EU [DG Sanco 2010]. The exporters were forced to take back their shipments. Other cases also surfaced in the other parts of the world. Such return of shipments would slow down international trade or even halt it.

With scores of transgenic crop single events and stacked varieties already commercialized and many more in the pipeline, this issue is bound to lead to more serious international trade disruptions and wars. The economic risks of rejections of shipments are inestimable. Part of the problem consists of the 'destination risk', i.e., the official testing for unauthorized GM material only in the port of destination [Stein and Rodriguez-Cerezo 2009]. When compliance with a zero tolerance policy for LLP becomes impossible, the risk of rejection increases, and so will the price [Stein and Rodriguez-Cerezo 2009]. This will affect exporting countries and businesses that are dependent on cheap agricultural imports.

At an international workshop on LLP (low-level presence) held in Vancouver, Canada in March 2012, the 15 participating countries with different geographies and value chains have arrived at two consistent messages: a) global trade of GM products cannot occur in a zero-threshold world and b) LLP should not impede trade [Tranberg 2013]. These reflected more the views of exporting GM-developing countries than the importing non-GM growing nations and GM-growing nations with less stringent or inadequate regulatory standards.

Biopiracy

Biopiracy is defined as the appropriation and commercialization of genetic resources and traditional knowledge of rural and indigenous people of another country. It involves making profit illegally from freely available natural biological materials such as plant parts (plants, seeds, leaves, etc.), plant products, and genetic cell lines by a foreign country or organization without fair compensation to the local people who have been using them for generations or nation in whose territory the materials were originally derived from. Biopiracy is usually done by biotech, pharmaceutical, and seed producing companies with or without the tacit support or knowledge of the countries they belong to. This unauthorized act is considered illegal by the native country. Biopirates draw on biodiversity hotspots in order to create supposedly innovative products and guarantee their monopoly on them through the patent system [Biopiracy Collective 2012].

The issue of biopiracy came to the forefront of genetic engineering in 2006 when Monsanto and its Indian collaborator Mahyco (Maharashtra Hybrid Seeds Company) violated India's National Biological Diversity Act of 2002 by accessing at least 10 local varieties of brinjal (eggplant: *Solanum melongena*) in the southern states of Karnataka and Tamil Nadu to develop their *Bt* versions without prior approval of the National Biodiversity Authority [Sood 2012]. This was considered an act of biopiracy (vide Chapter 8). Brinjal, with its unique 2,500 varieties, is one of the major vegetables of India cultivated for over 5,000 years. The approval of two hybrids of event EE1 of *Bt* brinjal (inserted with a crystal protein gene *Cry1Ac*), the first-ever transgenic crop developed in the country, on 14 October 2009 by the nation's regulator, Genetic Engineering Approval Committee, would have opened the floodgates to a technology that is regarded with suspicion by the overwhelming population [Devraj 2010]. The massive public outcry forced the Indian government to impose a moratorium on 17 February 2010 on transgenic *Bt* brinjal, and it will remain "until the nation arrives at a political, scientific, and societal consensus." Petitions challenging criminal complaints against Monsanto/Mahyco have been dismissed by High Court of the state of Karnataka on 11 October 2013. This verdict paved the way for the first-ever successful criminal case of biopiracy in the country [Prasad 2013]. With this, biotech companies face prosecution if the Supreme Court concurs with the lower court verdict.

Biopiracy is now impacting the global biodiversity and genetic resources currently being used in genetic engineering. The world's most important biodiversity reserves are found in India, the African continent, and South American countries which concentrate great environmental wealth. Although The Convention on Biological Diversity of 1992 and the Nagoya Protocol, ratified in 2010, represent significant progress in the protection of biodiversity in the face of biopiracy, the legal texts do not really indicate the manner in which the nations should control access to resources, nor in which way the populations traditionally using the resources can be consulted [Biopiracy Collective 2012]. In fact, the texts are rather vague and lack legal binding force.

There are numerous examples of biopiracy by technologically advanced countries and organizations. The more prominent of these include the following:

a) Patenting by U.S. Department of Agriculture and W.R. Grace Company in 1995 on a technique to extract an anti-fungal agent from the neem tree (*Azadirachta indica*), native of India and Nepal and whose people have known the traditional medicinal values of plant parts for thousands of years; the European Patent Office which had issued the patent revoked its decision in May 2000 followed by upholding its decision in its entirety on 8 May 2005 [GRAIN 2005].

b) Granting of patent (5,401,504) in May 1995 by U.S. Patent Office to two expatriate Indians at the University of Mississippi Medical Center, Jackson, Mississippi on wound healing properties of turmeric (*Curcurma longa*) which permeates the life of Indians as it is used as spice, in medicines, and to heal wounds since ancient times.

c) Patenting (U.S. Patent No. 5,663,484) by the U.S. corporation of RiceTec of Alvin, Texas on 2 September 1997 of certain hybrids of basmati rice and semi-dwarf rice, native of India, because the U.S. act violated TRIPS Agreement (vide this Chapter); the U.S. Patent Office revoked 13 of the 20 grounds originally cited by the Texas rice producer in 2001, but retained the other claims.

d) Patenting in 2011 by Monsanto of a Closterovirus (Cucurbit-Yellow Stunting Disorder)-resistant melon (*Cucumis melo* var. *agrestis*), a conventionally-bred melon, based on an Indian melon resistant to it; this Monsanto 'invention' was a clear case of biopiracy since the original melon plants came from India and were registered in international seed banks (PI 313970) [Parsai 2012].

e) Patenting (U.S. Patent No. 5,894,079) by an American, Larry Proctor, and his company POD-NRS-LLC, in 1999, of the Enola bean, a landrace of Mexican yellow bean grown by northern Mexican famers for centuries and known to them under the names 'Mayocoba', 'Azufrado', and 'sulfur'; the patent was revoked by a U.S. court on 10 July 2009 [Shashikant and Asghedom 2009].

f) Seeking patent by Cognis, a French company, to use oils and proteins extracted from seeds of sacha inchi [*Plukenetia volubilis* L.), a native Peruvian Amazon plant used by Peruvian Amazonians for over 3,000 years.

Many more medicinal plants native from India have received U.S. patents. These include *Boswellia serrata* (Indian frankincense, referred in Sanskrit as 'shallaki'), *Euphorbia hirta* (Asthma plant, Garden spurge; Hindi: 'dudhi'; Sanskrit: 'Chara'), *Momordica charantia* (Bitter gourd/melon; Hindi: 'karela'; Sanskrit: 'karavella'), *Phyllathus emblica* (Indian gooseberry; Hindi: 'āmla'; Sanskrit 'āmlika'), *Punicia grantum* (Pomegranate; Hindi: 'anār'), *Ricinus communis* (Castor), *Trichosanthes kirilowii* (Snakegourd; Hindi: 'Chamkura'), etc.

With the likelihood of scores of crops in the transgenic engineering pipeline in future, India's brinjal piracy case involving transgenic engineering serves as a wake-up call to biotech companies in their efforts to use biopirated crops and crop varieties from the Old World. The world is not yet flat in regard to transgenic engineering of crops as it is with the Internet. That may take a long time to happen, if ever.

Co-existence of Transgenic Crops and Food Products

Co-existence is the practice of growing crops with different quality characteristics or intended for different markets in the same vicinity without becoming commingled and thereby possibly compromising the economic values of both. It is based on the premise that farmers should be free to cultivate the crops of their choice using the agricultural system they prefer, whether they are transgenic, conventional, or organic. Co-existence is an economic issue, not a safety issue unless the foods derived from transgenic crops pose health risks.

In the context of agricultural production, co-existence is not a new concept because farmers have been practicing it for generations so as to meet demands for different types of products. Many farmers are also accustomed to producing certified seed, hybrid or pure breed, to meet defined purity standards. However, the availability of a vast array of biotech crops for commercial cultivation has made co-existence of transgenic and conventional crops a challenging issue to even farmers with experience. This is because the gene inserted in the transgenic crop is of non-plant origin, and the pollen from this crop could contaminate conventional crops grown nearby.

There is wide variation in the adoption rate of transgenic crops between countries. This is generally due to societal and political opposition as well as disparities in regulatory standards. Society needs regulation whenever the introduction of a new product or technology leads to an externality or market failure [Beckman and Wesseler 2007].

Socio-Economic Consequences

Ensuring co-existence of transgenic and non-transgenic crops and products derived from them will inevitably entail additional costs in several ways. The costs include [Devos et al. 2008a) those required to a) enforce co-existence measures imposed by regulators, both during and after cultivation, b) for testing of crop produce and products, for identifying and quantifying the content of transgenic material in non-transgenic material, and d) for compliance of labeling and traceability requirements. Additionally, farmers may suffer income losses due to restrictions in crop choice and management. Neighboring farmers could impose restrictions if a farmer decides to grow a transgenic crop. If a farmer growing transgenic crops is unable to avoid interference and cannot find mutual agreement with the neighboring conventional crop growers, he would have to renounce growing them on his land. Besides, spatial restrictions, temporal cultivation may occur due to irreversibility. In a field grown where transgenic crop is raised, it could temporarily be difficult to meet the 0.9 percent tolerance threshold if a farmer decides to go back to a non-transgenic cropping system. In this process, a conversion time might be required to deplete transgenic seeds from the seed bank and/or control of volunteers and weedy/wild relatives that may contain the transgene [Devos et al. 2004; Jorgensen et al. 2007; Messéan et al. 2007; D'Hertefeldt et al. 2008].

Co-existence of transgenic and non-transgenic crops in the same region also has social consequences. Farmers who decide to grow transgenic crops need to a) seek approval of neighboring farmers and agreement on their respective cropping situations, b) notify their crop details and seek permission from government regulators, c) consider ethical issues that may arise in connection with the use genes from non-plant sources, d) study the positive and negative effects of transgenic crops in relation to sustainable development, e) assess the risks of the extinction of traditional varieties, f) weigh corporate control of seed, g) bear in mind the legal liability of transgenic crop cultivation, and h) take into account the possibility of litigation by neighboring farmers.

Co-existence of Transgenic and Non-transgenic Crops

Adventious Mixing and Preventive Measures. As agriculture is an open system, certain amount of adventitious mixing is unavoidable. The on-farm sources of such mixing between transgenic and non-transgenic crops include seed impurities, pollen flow between neighboring fields, volunteer plants originating from seeds or vegetative plant parts from previous transgenic crops, and seeds left inside the equipment used at planting, harvesting, processing, storage and transport. Sometimes, cross-fertilization from certain sexually compatible wild/weedy relatives and feral plants may also contribute to adventious mixing.

The existing measures to ensure seed purity in conventional crop production may also be applied within the context of limiting the adventious content of transgenic material in seeds and plant products [Devos et al. 2008a]. These include [Devos et al. 2008a]: a) the use of certified seed, b) spatially isolating fields of the same crop, c) erecting pollen barriers around fields, d) scheduling different sowing and flowering periods, wherever possible, e) limiting carryover of transgenic volunteers into the following crop through the extension of cropping intervals, f) cleaning agricultural machinery and transport vehicles for seed remnants, g) controlling volunteers and wild/weedy relatives, h) applying effective post-harvest tillage operations, i) retaining records of field history, and j) the voluntary clustering of fields. The most drastic preventive co-existence measure is probably banning the cultivation of transgenic crops in a certain region. In addition to stricter on-farm measures, lower thresholds could minimize contamination.

Minimum **isolation distances** can be regulated by individual state and local governments and farmers' cooperatives. These depend on distances pollen travels, distance between donor and recipient fields, flowering synchrony between the donor and recipient plants, rates of cross-fertilization, presence of potential wild relatives, size of pollen grains, local wind conditions, and type of landscape. These distances in the case of transgenic maize vary between a minimum of 15 m to conventional maize as in Sweden and a maximum of 800 m to organic maize as in Hungary [Devos et al. 2009; Devos et al. 2008b]. In Germany, individual states maintain a minimum distance of 800 m to 1,000 m, which is many times more than 30 m suggested for the insect resistant *Bt* maize event TC1507 [EFSA 2012].

Longer isolation distances might be needed for stacked varieties because they contain more than one transgene [De Schrijver et al. 2007]. In the case of stacked maize varieties, cross fertilization results in a higher content of transgenic material being expressed in recipient plants compared with a single trait variety. At the present time, there is no set tolerance threshold for the adventious presence of approved transgenic material in non-transgenic seeds. However, based on a meta-analysis of existing cross-fertilization studies,

Sanvido et al. [2008] had set an isolation distance of 50 m to keep cross-fertilization levels below 0.5 percent at the border of the recipient maize field. More work is needed to determine thresholds for each of the crops containing single-trait and stacked-trait transgenes.

In places where isolation distances cannot be implemented, the first 10–20 m of non-transgenic maize rows facing the transgenic-maize field may be removed.

Pollen barriers, like isolation distances, consisting of the same crop species effectively reduce the extent of cross-fertilization between transgenic and non-transgenic varieties. These "buffer strips or zones" obstruct pollen movement by producing their own competitive transgene-free pollen which may also fertilize the neighboring non-transgenic crop fields. Consequently, crops in the buffer zone, although originally of non-transgenic plants, also catch a lot of transgene pollen and produce transgenic crop. Pollen barriers are considered more efficient than isolation distances in reducing cross-fertilization. A few rows of non-transgenic crop are equivalent to several meters of isolation distances. The width of pollen barriers may be determined by taking into consideration pollen load, wind data (direction and speed), and the extent of cross-fertilization characteristic of a crop. Based on the logarithmic equations of cross-pollination over distances in maize fields, Viljoen and Chetty [2011] determined that a 45 m barrier is sufficient to maintain cross-pollination at the level of < 1.0–0.1 percent, and much wider barriers for more stringent cross-pollination levels. With a barrier of 10–20 m, the recipient non-transgenic maize contains no more than 0.9 percent of transgene material [Devos et al. 2008b].

In the case of insect-resistant *Bt* maize and cotton, farmers are required to use **refuge zones (belts)**. The theory behind the refuge zone strategy is that most of the resistant insects surviving on *Bt*-crops will mate with abundant susceptible insects from refuges, and that the hybrid progeny originating from such matings will be killed by *Bt* crops, if the inheritance of resistance is recessive [Bates et al. 2005]. In *Bt* maize areas larger than 5 ha, a refuge zone of 20 percent of the transgenic area has to be planted with non-transgenic maize in order to delay the potential resistance development in lepidopteran target insects [Devos et al. 2008a]. However, in the case of herbicide-resistant transgenic maize, the cultivation of transgenic and non-transgenic in the same field might create practical challenges since two different weed management strategies need to be applied in a single field [Devos et al. 2008a].

One other problem with refuge zone strategy is farmers' compliance. In India, the refuge zone should be at least five rows of non-transgenic *Bt* cotton or 20 percent of the total area planted, whichever is greater. Farmers are supplied with one unit of non-*Bt* cotton seeds for five units of *Bt* cotton seeds, sufficient for planting in the 'refuge'. Success of this strategy depends on various factors, but farmer compliance, voluntary or otherwise, is vital.

Temporal isolation is another avenue to facilitate co-existence of transgenic and non-transgenic crops. The idea behind temporal isolation is to separate the flowering time of transgenic from non-transgenic crops or totally remove pollen from the transgenic crops. In practice, staggering the sowing times of transgenic and non-transgenic crops may help to reduce gene flow and cross-fertilization because of differences in flowering periods. A time lag in flowering synchrony of at least 8 days between sowings of transgenic and non-transgenic maize crops has significantly reduced the extent of cross-fertilization [Palaudelmàs et al. 2007; Della Porta et al. 2008]. Delayed sowing is not feasible in areas where weather conditions are too short and market prices favor the early-bird crop. Another strategy in temporal isolation is following rotation in which growing of transgenic crops is deferred for a few years/seasons in favor of non-transgenic crops, but this requires close coordination with neighboring farmers.

Altogether a different strategy to prevent transgene flow and cross-fertilization between transgenic and non-transgenic crops is **biological confinement**. In this method, biological tools are employed to prevent transgenic crops from producing pollen at all, as in the case of male-sterile plants, or to develop generating transplastomic plants which produce pollen that does not contain the additional, genetically engineered material.

In **cytoplasmic male sterility** (CMS), specific mutations in mitochondrial DNA induce dysfunctions in the respiratory metabolism occurring in anther tapetum (a layer of nutritive cells) during sporogenesis [Budar et al. 2003], leading to deleterious phenotypes [Chase 2007]. Male sterility is induced by mutations of the mitochondrial DNA and is, therefore, maternally inherited. Therefore, plants either produce no pollen or pollen that is not viable. However, female fertility is not affected by CMS. In fact, CMS has been used since the 1950s in maize seed production to ensure cross-fertilization without the need for mechanical or manual emasculation and to minimize costs. Maize has three types of male-sterile cytoplasm (T, S, and C) and their fertility can be restored by nuclear *rf* (restorer-of-fertility) genes or by interactions with the environment [Weider et al. 2009]. The S and C cytoplasms require a single restorer gene for fertility restoration—*rf3* and *rf4* respectively—whereas two restorer genes, *rf1* and *rf2*, are necessary to restore T cytoplasm CMS (CMS-T) plants to male sterility [Dewey et al. 1987].

Weider et al. [2009] found stable and unstable male sterility occurring in all three CMS types after evaluating 22 CMS versions across 17 environments in Switzerland, France, and Bulgaria. T-cytoplasm hybrids were the most stable, while S-cytoplasm hybrids often showed partial restoration of fertility. C-cytoplasm was similar to T-cytoplasm with regard to maintaining male sterility [Weider et al. 2009]. In the context of co-existence, the cultivation of CMS transgenic maize plants might reduce the release of transgenic pollen by up to 80 percent [Devos et al. 2008b]. To ensure seed set, CMS transgenic maize plants would have to be interplanted with male fertile maize plants, containing either transgenic or non-transgenic characteristics, acting as pollen donors.

In generating **transplastomic plants**, new genes are inserted in the chloroplast DNA, but not in the nuclear DNA. The major advantage of this technology is that in many plant species, plastid DNA is not transmitted through pollen, which prevents gene flow from transgenic plants to other plants. As plastids are inherited maternally in the majority of angiosperm species, they would therefore not be found in pollen grains [Wani et al. 2010]. Plastid transformation is done by biolistic method or by direct DNA uptake by protoplasts. As pollen of transplastomic plants does not contain the transgene, the process increases biosafety and facilitates co-existence of transgenic plants with conventional and organic plants as well in the neighboring weedy or wild relatives. Chloroplast genetic engineering has been found useful in such dicot species as cotton, soya bean, tobacco, tomato, and poplar while its utility in monocot crops like rice and maize remains to be confirmed [Daniell 2007; Verma and Daniell 2007].

Despite the promise plastid transformation holds in ensuring transgene containment, transmission of plastid DNA via pollen [Lu et al. 2003; Wang et al. 2004; Ruf et al. 2004], even very low levels of paternal gene transfer, may be sufficient for the escape and spread of a transgene [Haygood et al. 2004]. Furthermore, the transplastomic transgenic crops would also be fertilized by pollen from other plants resulting in hybridization [Dunwell and Ford 2005; Lu 2003, 2008]. The resulting hybrid seeds could disperse in the environment. Furthermore, chloroplast DNA may be transferred to the nucleus at low frequencies and be integrated into the nuclear genome [Haygood et al. 2004; Medgyesy et al. 1980]. In this case, the transgene is inherited bi-parentally resulting in a failure of the containment strategy [Sheppard et al. 2008; Stegemann and Bock 2006].

Against these possibilities, transplastomic transgenic containment would only be effective if a biotech line is stacked with additional mechanisms such as mitigating genes, genetic use restriction technology (GURT), and/or male sterility [Wang et al. 2004]. The '**mitigator gene**' can be tandemly attached to the gene of choice to be transformed into the chloroplast genome [Gressel 1999]. However, plastid transformation would provide an imperfect bio-containment for transgenic (canola) [Wang 2004].

Another possible molecular strategy suggested for gene containment in certain transgenic crops is **cleistogamy**, a trait to produce seeds by using non-opening and self-pollinating flowers. In this, fertilization occurs before flower opens. This behavior is most widespread in legumes such as groundnut (peanut), bean, and pea. Although cleistogamy reduces gene flow, preliminary results reported in Europe suggested that it is not a consistently reliable tool for biocontainment for some crop species because some flowers may open and release transgenic pollen. The characteristic self-pollination suppresses the creation of genetically superior plants. In rice, which exhibits cleistogamy, genes readily move between cultivated and feral forms of weedy rice, despite predominant self-pollination [Daniell et al. 2002].

The various transgene containment strategies described herein suggest that no single approach can be very effective to confine transgene escape to wild/weedy relatives and landraces as also facilitate co-existence of transgenic and non-transgenic crops. This is because many transgenes that do not provide selective advantage to the host plant in nature may not pose any environmental consequences, and many crops that have extremely low gene frequencies already have a low risk of transgene escape [Lu 2013]. Besides, for some geographical locations where wild relatives or conspecific (belonging to the same species) weeds of the transgenic crops are absent, transgene escape through pollen-mediated gene flow would not be an issue. A strategic combination of transgene confinement from gene flow and mitigation to minimize its impacts in a particular circumstance should provide an effective strategy to manage any environmental consequences caused by transgene escape [Lu 2013].

Co-existence of Transgenic and Non-transgenic Food Products

The development and adoption of transgenic crops are supply-driven. Initially, the target of biotech companies was farmers and sale volumes derived by them. The upstream effect from the demand side is the consumer awareness and increase in the knowledge of genetically modified organisms. In countries where transgenic crop cultivation is largely followed, consumers demand freedom of choice in buying the food they want. As a result, there is a growing demand for more channels in food markets, one for transgenic crop products and others for conventional and organic products. This implies segregation of agricultural products along the vertical food chain and the eventual development of Identity Preservation (IP) and a traceability system together with labeling requirements [Gaisford et al. 2001]. The co-existence of conventional and transgenic products in the food chain introduces new elements in the evaluation of the profitability of transgenic crops and, consequently, in the farmer's adoption decision.

Labeling, a prerequisite for co-existence of transgenic and non-transgenic foods, is an important issue related to biotechnology. Although USA is the birthplace of commercial development of transgenic crops, currently there is no federal or state law that requires food producers to identify whether foods were produced using genetic engineering. About 80 percent of processed and most fast foods consumed by Americans have been derived from transgenic crops. These include foods derived from soya bean, maize, canola, cotton, sugar beet, Hawaiian papaya, and certain varieties of zucchini and yellow squash. Besides, products such as oil, high fructose maize syrup, and sugar created from these crops are added to processed foods. Despite such heavy consumption of transgenic foods by American consumers, the U.S. Food and Drug Administration does not require safety studies of such foods unless they are proved to contain a known allergen. Considering that transgenes have been derived from bacteria and viruses, 9 out 10 people want these foods labeled [Bartolotto 2013]. The biotech companies, however, do not. Several attempts made by many states including California to pass legislations to label transgenic foods have been defeated, largely because of the lobbying of biotech giants such as Monsanto, DuPont Pioneer, Bayer CropScience, Dow Agrosciences, and the Grocery Manufacturers Association [Bartolotto 2013].

Consumers in many parts of the world such as the European Union, Japan, Malaysia and Australia demand labeling so they can exercise choice between foods that have originated from transgenic, conventional, or organic crops. This requires a labeling and traceability system as well as the reliable separation of transgenic and non-transgenic foods at production level and throughout the whole processing chain.

When organic and gluten-free food products coexist in a food market with established labeling and traceability systems and separation of aisles, it is not irrelevant and unreasonable to expect such a system of co-existence with respect to transgenic and non-transgenic foods as well. Basically, consumers have the right to know what is in the foods they consume. It will be a travesty of justice to deny it. It may not be too long before consumers' demands are met by global governments.

References

Albo, A.G., S. Mila, G. Digilio, M. Motto, S. Aime and D. Corpillo. 2007. Proteomic analysis of a genetically modified maize flour carrying Cry1Ab gene and comparison to the corresponding wild-type. Maydica 52: 443–455.

Altieri, M.A. 1998. The Environment Risks of Transgenic Crops: An Agroecological Assessment: Is the failed pesticide paradigm being genetically engineered? Pesticides and You. Spring/Summer 1998.

Andrade, G.J.M. and C.A. Rosolem. 2011. Uptake of manganese in RR soybean under glyphosate application. Rev. Bras. Cienc. Solo. 35: 961–968.

APHIS-USDA. 2007. USDA Provides Update for Farmers on Genetically Engineered Rice. Program Announcement. USDA Safeguaring American Agriculture. Biotechnology Regulatory Services.

Araújo, A.S.F., R.T.R. Monteiro and R.B. Abarkeli. 2003. Effect of glyphosate on the microbial activity of two Brazilian soils. Chemosphere 52: 799–804.

Artemis, D. and I.S. Arvanitoyannis. 2009. Health risks of genetically modified foods. Critical Rev. in Food Sci. and Nutrition. 49: 164–175.

Astwood, J.D., J.N. Leach and R.L. Fuchs. 1996. Stability of food allergens to digestion *in vitro*. Nature Biotechnol. 14: 1269–1273.

Badea, E.M., F. Chelu and A. Lăcătuşu. 2010. Results regarding the levels of Cry1Ab protein in transgenic corn tissue (MON810) and the fate of Bt protein in three soil types. Romanian Biotechnol. Letters 15(1) Suppl.: 55–62.

Bailey, W.A., D.H. Poston, H.P. Wilson and T.E. Hines. 2002. Glyphosate interactions with manganese. Weed Technol. 16: 792–799.

Bakke-McKellep, A.M., E.O. Koppang, G. Gunnes, M. Senden, G.-I. Hemre, T. Landsverk and A. Krogdahl. 2007. Histological, digestive, metabolic, hormonal and some immune factor responses in Atlantic salmon, *Salmo salar* L. fed genetically modified soybeans. J. Fish Dis. 30: 65–79.

Bakshi, A. 2003. Potential adverse health effects of genetically modified crops. J. Toxicol. Environ. Health, Part B. 6: 211–225.

Balsamo, G.M., G.C. Cangahuala-Inocente, J.B. Bertoldo, H. Terenzi and A.C.M. Arisi. 2011. Proteomic analysis of four Brazilian MON810 maize varieties and their four non-genetically-modified isogenic varieties. J. Agric. Food Chem. 59: 11553–11559.

Barbosa, H.S., S.C. Arruda, R.A. Azevedo and M.A. Arruda. 2012. New insights on proteomics of transgenic soybean seeds: evaluation of differential expressions of enzymes and proteins. Anal. Bioanal. Chem. 402: 299–314.

Barriuso, J., J.R. Valverde and R.P. Mellado. 2012. Effect of Cry1Ab protein on rhizobacterial communities of Bt-maize over a four-year cultivation period. PLoS One 7(4): 30 April 2012.

Bartolotto, C. 2013. Why Genetically Modified Foods should be Labeled. Food For Thought. 4 October 2013.

Bartsch, D. and N.C. Ellstrand. 1999. Genetic evidence for the origin of Californian wild beets (genus *Beta*). Theor. Appl. Gene. 99: 1120–1130.

Bates, S.L., J.-Z. Zhao, R.T. Roush and A.M. Shelton. 2005. Insect resistance management in GM crops: past, present, and future. Nat. Biotechnol. 25: 57–62.

Batista, R., I. Martins, P. Jenö, C.P. Ricardo and M.M. Oliveira. 2007. A proteomic study to identify soya allergens—the human response to transgenic versus non-transgenic soya samples. Int. Arch. Allergy Immunol. 144: 29–38.

Baumgarte, S. and C.C. Tebbe. 2005. Field studies on the environmental fate of the Cry1Ab Bt toxin produced by transgenic maize (MON810) and its effect on the bacterial communities in the maize rhizosphere. Mol. Ecol. 14: 2539–2551.

Beckmann, V. and J. Wesseler. 2007. Spatial dimension of externalities and the Coase theorem: implications for co-existence of transgenic crops. pp. 223–242. *In*: W. Heijman (ed.). Regional Externalities. Springer, Berlin Heidelberg, Germany.

Benbrook, C.M. 2012. Impacts of genetically engineered crops on pesticide use in the U.S.—the first sixteen years. Environ. Sci. Europe 24: 24.

Bhalla, P.L., I. Swoboda and M.B. Singh. 2003. Antisense-mediated silencing of a gene encoding a major ryegrass pollen allerge. Pro. Natl. Acad. Sci. USA 96(20): 11678–11680.

Bhatia, C.R. 1998. Terminator transgenics. Current Science 75: 1288–1289.

Biopiracy Collective. 2012. Understanding, resisting, and acting against biopiracy. Collectif Biopiraterie www.biopiraterie.org. (French) 15: 23.

Bohme, H., K. Aulrich, R. Daenicke and G. Flachowsky. 2001. Genetically modified feeds in animal nutrition. 2nd communication: glufosinate tolerant sugar beets (roots and silage) and maize grains for ruminants and pigs. Arch. Anim. Nutrit. 54: 197–207.

Boudry, P., M. Mörchen, P. Saumitou-Laprade, Ph. Vernet and H. Van Dijk. 1993. The origin and evolution of weed beets: consequences for the breeding and release of herbicide-resistant transgenic sugar beets. Theor. Appl. Genet. 87: 471–478.

Bromilow, R.H., K. Chamberlain, A.J. Tench and R.H. Williams. 1993. Phloem translocation of strong acids —glyphosate, substituted phosphonic, and sulfonic acids in *Ricinus communis* L. Pestic. Sci. 37: 39–47.

Brookes, G. and P. Barfoot. 2013. GM crops: global and socio-economic and environmental impacts 1996–2011. PG Economics Ltd, Dorchester, UK. April 2013.

Brookes, G. and P. Barfoot. 2012. GM crops: global socio-economic and environmental impacts 1996–2010. PG Economics Ltd., Dorchester, UK. May 2012.

Brookes, G. and P. Barfoot. 2006. GM Crops: The First Ten Years—Global Socio-Economic and Environmental Impacts. ISAAA Brief No. 36. International Service for the Acquisition of Agri-biotech Applications. Ithaca, NY.

Budar, F., P. Touzet and R. De Paepe. 2003. The nucleo-mitochondrial con-flict in cytoplasmic male sterilities revisited. Genetica 117: 3–16.

Burk, D.L. 2004. DNA rules: legal and conceptual implications of biological "lock-out" systems. Calif. Law Rev. 92: 1553–1587.

Cavalieri, S.D., E.D. Velini, F.M.L. Silva, A.R. São José and G.J.M. Andrade. 2012. Nutrient and shoot dry matter accumulation of two GR soybean cultivars under the effect of glyphosate formulations. Planta Daninha 30: 349–358.

Cellini, F., A. Chesson, I. Colquhoun, A. Constable, H.V. Davies, K.H. Engel et al. 2004. Unintended effects and their detection in genetically modified crops. Food Chem.Toxicol. 42: 1089–1125.

Chase, C.D. 2007. Cytoplasmic male sterility: a window to the world of plant mitochondrial-nuclear interaction. Trends Genet. 23: 81–90.

Cromwell, G.L., B.J. Henry, A.L. Scott, M.F. Gerngross, D.L. Dusek and D.W. Fletcher. 2005. Glufosinate herbicide-tolerant (LibertyLink) rice vs. conventional rice in diets for growing-finishing swine. J. Anim. Sci. 83: 1068–1074.

Daenicke, R., K. Aulrich and G. Flachowsky. 2000. Investigations on the nutritional value of sugar beets and sugar beet leaf silage of isogenic and transgenic plants for muttons. VDLUFASchriftenreihe 55: 84–86.

Dale, P.J., B. Clarke and E.M.G. Fontes. 2002. Potential for the environmental impact of transgenic crops. Nat. Biotechnol. 20: 567–574.

Daniell, H. 2007. Transgene containment by maternal inheritance, effective or elusive? Proc. Natl. Acad. Sci. USA 104: 6879–6880.

Daniell, H., M.S. Khan and L. Allison. 2002. Milestones in chloroplast genetic engineering an environmentally friendly era in biotechnology. Trends Plant Sci. 7: 84–91.

Daniell, H. 2002. Molecular strategies for gene containment in transgenic crops. Nat. Biotechnol. 20(6): 581–586.

Darmency, H., Y. Vigouroux, T. De Garambe, R.-M.M. Gestat and C. Muchembled. 2007. Transgene escape in sugar beet production fields: data from six years farm scale monitoring. Environ. Biosaf. Res. 6: 197–206.

Della Porta, G., D. Ederle, L. Bucchini, M. Prandi, A. Verderio and C. Pozzi. 2008. Maize pollen mediated gene flow in the Po valley (Italy): Source-recipient distance and effect of flowering time. Eur. J. Agron. 28: 255–265.

De Schrijver, A., Y. Devos, M. Van den Bulcke, P. Cadot, M. De Loose, D. Reheul and M. Sneyers. 2007. Risk assessment of GM stacked events obtained from crosses between GM events. Trends Food Sci. Tech. 18: 101–109.

Devos, Y., M. Demont, K. Dillen, D. Reheul, M. Kaiser and O. Sanvido. 2009. Coexistence of genetically modified (GM) and non-GM crops in the European Union. A review. Agron. Sustain. Dev. 29: 11–30.

Devos, Y., M. Cougnon, O. Thas and D. Reheul. 2008a. A method to search for optimal field allocations of transgenic maize in the context of co-existence. Environ. Biosafety Res. 7: 97–104.

Devos, Y., M. Demont, K. Dillen, D. Reheul, M. Kaiser and O. Sanvido. 2008b. Coexistence of genetically modified (GM) and non-GM crops in the European Union: A review. Agron. Sustain. Develop. EDP Sciences. available at http://www.agronomy-journal.org.

Devos, Y., D. Reheul, A. De Schrijver, F. Cors and W. Moens. 2004. Management of herbicide-tolerant oilseed rape in Europe: a case study on minimizing vertical gene flow. Environ. Biosafety Res. 3: 135–148.

Devraj, R. 2010. India bans cultivation of Bt. brinjal. One World South Asia. 10 February 2010.

Dewey, R.E., D.H. Timothy and C.S. Levings, III. 1987. A mitochondrial protein associated with cytoplasmic male sterility in the T-cytoplasm of maize. Proc. Natl. Acad. Sci. USA 84: 5374–5378.

DG SANCO of the EC. 2010. Evaluation of the EU Legislative Framework in the Field of GM Food and Feed FRAMEWORK. Final Report. DG SANCO Evaluation Framework Contract Lot 3 (Food Chain). Food Chain Evaluation Consortium.

D'Hertefeldt, T., R.B. Jørgensen and L.B. Pettersson. 2008. Long-term persistence of GM oilseed rape in the seedbank. Biol. Lett. 4: 314–317.

Diels, J., M. Cunha, C. Manaia, B. Sabugosa-Madeira and M. Silva. 2011. Association of financial or professional conflict of interest to research outcomes on health risks or nutritional assessment studies of genetically modified products. Food Policy 36: 197–203.

Duke, S.O., K.N. Reddy, K. Bu and J.V. Cizdziel. 2012a. Effects of glyphosate on mineral content of glyphosate-resistant soybeans (*Glycine max*). J. Agri. Food Chem. 60: 6764–6771.

Duke, S.O., J. Lydon, W.C. Koskinen, T.B. Moorman, R.L. Chaney and R. Hammerschmidt. 2012b. Glyphosate effects on plant nutrition, crop rhizosphere microbiota, and plant disease in glyphosate-resistant crops. J. Agri. Food Chem. 60(42): 10375–10397.

Duke, S.O. and A.L. Cedeira. 2010. Transgenic crops for herbicide resistance. pp. 133–166. *In*: C. Kole, C. Michler, G. Albert and T.C. Hall (eds.). Transgenic Crop Plants. Springer-Verlag, Berlin, Heidelberg.

Duke, S.O., D.E. Wedge, A.L. Cerdeira and M.B. Matallo. 2007. Interactions of synthetic herbicides with plant disease and microbial herbicides. pp. 277–296. *In*: M. Virruo and J. Gressel (eds.). Novel Biotechnologies for Biocontrol Agent Enhancement and Management. Springer, Dordrecht, The Netherlands.

Duke, S.O., A.M. Rimando, P.F. Pace, K.N. Reddy and R.J. Smeda. 2003. Isoflavone, glyphosate, and aminomethyl phosphonic acid levels in seeds of glyphosate-treated, glyphosate-resistant soybean. J. Agri. Food Chem. 51: 340–344.

Duke, S.O. 1988. Glyphosate. pp. 1–70. *In*: P.C. Kearney and D.D. Kaufman (eds.). Herbicides—Chemistry, Degradation and Mode of Action, vol III. Marcel Dekker, New York.

Dunfield, K.E. and J.J. Germida. 2004. Impact of genetically modified crops on soil- and plant-associated microbial communities. J. Environ Qual. 33: 806–815.

Dunwell, J.C. and C.S. Ford. 2005. Technologies for biological containment of GM- and non-GM-crops. Final Report, Defra Contract CBEC 47.

Dutfield, G. 2011. Food, Biological Diversity and Intellectual Property—The Role of the International Union for the Protection of New Varieties of Plants (UPOV). Global Economic Issue Publications. Intellectual Property Issue Paper No. 9.

Eaton, D., F. Van Tongeren, N. Louwaars, B. Visser and I. Van der Meer. 2002. Economic and policy aspects of 'terminator' technology. Biotechnol. Develop. Monitor 49: 19–22.

EFSA. 2012. Scientific Opinion on the annual Post-Market Environmental Monitoring (PMEM) report from Monsanto Europe S.A. on the cultivation of genetically modified maize MON 810 in 2010. EFSA J. 10: 1–35.

EFSA. 2010. Scientific opinon on the assessment of allergenicity of GM plants and microorganisms and derived food and feed. EFSA J. 8(7): 1700.

EFSA. 2009. Scientific opinion of the GMO and BIOHAZ Panels on the "Use of antibiotic resistance genes as marker genes in genetically modified plants". European Food Safety Authority 1034: 1–82.

Ellstrand, N. 2006. Scientists evaluate potential environmental risks of transgenic crops. http://California Agriculture.ucop.edu. July–Sept.

Ellstrand, N.C., H.C. Prentice and J.F. Hancock. 1999. Gene flow and introgression from domesticated plants into their wild relatives. Ann. Review of Ecol. Systematics 30: 539–563.

ETC Group. 2003. "Terminator Technology—Five Years Later." Communique issue #79, May/June. [available at http://www.etcgroup.org/sites/www.etcgroup.org/files/publication/167/01/termcom03.pdf].

FAO. 2001. Potential Impacts of Genetic Use Restriction Technologies (GURTs) on Agricultural Biodiversity and Agricultural Production Systems. CGRFA/WG-PGR-1/01/7, July 2001.

FAO/WHO. 2001. Evaluation of Allergenicity of Genetically Modified Foods. Report of a Joint FAO/WHO Expert Consultation on Allergenicity of Foods Derived from Biotechnology. Food and Agricultural Organization/World Health Organization. Rome, Italy, 2001.

Flavell, R.B., E. Dart, R.L. Fuchs and R.T. Fraley. 1992. Selectable marker genes: safe for plants? Biotechnology (NY) 10: 141–144.

Food and Drug Administration. 1992. Statement of policy: foods derived from new plant varieties. Fed. Reg. 57: 22984–23002.

Friesen, L., A. Nelson and R.C. Van Acker. 2003. Evidence of contamination of pedigreed canola (*Brassica napus*) Seedlots in Western Canada with genetically engineered herbicide resistant traits. Agronomy J. 95: 1342–1347.

FSANZ. 2004. Final assessment report—Application A509: Food derived from insect protected cotton line COT102.

Fuchs, R.L., J.E. Ream, B.G. Hammond, M.W. Naylor, R.M. Leimgruber and S.A. Berberich. 1993. Safety assessment of the neomycin phosphotransferase II (NPTII) protein. Biotechnology (NY) 11: 1543–1547.

Gaines, T.A., P.F. Byrne, P. Westra, S.J. Nissen, W.B. Henry, D.L. Shaner and P.L. Chapman. 2007. An empirically derived model of field-scale gene flow in winter wheat. Crop Sci. 47: 2308–2316.

Gaisford, J.D., J.E. Hobbs, W.A. Kerr, N. Perdikis and M.D. Plunkett. 2001. The Economics of Biotechnology. Edward Elgar, Cheltenham, UK.

García-Cañas, V., C. Simó, C. León, E. Ibáñez and A. Cifuentes. 2011. MS-based analytical methodologies to characterize genetically modified crops. Mass Spectrom Rev. 30: 396–416.

Gilbert, N. 2013. A hard look at GM crops. Nature: Feature News 497: 24–26.

Giovannetti, M. 2003. The ecological risks of transgenic plants. Rivista di Biologia 96: 207–223.

Goeschl, T. and T. Swanson. 2003. The development impact of genetic use restriction technologies: a forecast based on the hybrid crop experience. Environ. Develop. Economics 8: 149–165.

Gong, C. and Y.T. Wang. 2013. Proteomic evaluation of genetically modified crops: current status and challenges. Frontier Plant Sci. 4: 41.

GRAIN. 2005. EPO upholds decision to revoke neem plant. GRAIN. 9 March 2005.

GRAIN. 2001. Intellectual Property Rights: Ultimate control of agricultural R&D in Asia. 25 March 2001.

Gressel, J. and B.E. Valverde. 2009. A strategy to provide long-term control of weedy rice while mitigating herbicide resistance transgene flow, and its potential use for other crops with related weeds. Pest Manag. Sci. 65(7): 723–731.

Gressel, J. and H. Al-Ahmad. 2006. Mitigating Transgene Flow from Crops. Dept. of Plant Sci., Weizmann Institute of Science, Rehovot 76100, Israel. HYPERLINK: Jonathan.Gressel@ weizmann.ac.il

Gressel, J. 2002. Molecular Biology of Weed Control. Taylor & Francis Group, London, UK, 504 pp.

Gressel, J. 1999. Tandem constructs: Preventing the rise of superweeds. TRENDS in Biotechnology 17: 361–366.

Gruber, H., V. Paul, H.H.D. Meyer and M. Mueller. 2012. Determination of insecticidal Cry1Ab protein in soil collected in the final growing seasons of a nine-year field trial of Bt-maize MON810. Transgenic Res. 21(01): 77–78.

Gupta, P.K. 1998. The terminator technology for seed production and protection: why and how? Current Sci. 75: 1319–1323.

Haas, M., A. Geldreich, M. Bureau, L. Dupuis, V. Leh, G. Vetter et al. 2005. The open reading frame VI product of Cauliflower mosaic virus is a nucleocytoplasmic protein: its N terminus mediates its nuclear export and formation of electron-dense viroplasms. Plant Cell 17: 927–943.

Hall, L., K. Topinka, J. Huffman, L. Davis and A. Good. 2000. Pollen flow between herbicide-resistant *Brassica napus* is the cause of multiple-resistant *B. napus* volunteers. Weed Sci. 48: 688–694.

Haygood, R.A., A.R. Ives and D.A. Andow. 2004. Population genetics of transgene containment. Ecol. Lett. 7: 213–220.

Health Canada. 2000. www.hc-sc.gc.ca. 01 January 2000.

Heap, I. 2014. The International Survey of Herbicide Resistant Weeds. Online. Internet. 10 October. http://www.weedscience.org

Heap, I. 2008. The International Survey of Herbicide Resistant Weeds. Online. Internet. www.weedscience.org

Heinemann, J.A. 2007. A Typology of the Effects of (Trans)gene Flow on the Conservation and Sustainable Use of Genetic Resources. Commission on Genetic Resources for Food and Agriculture, Background Study Paper Number 35 Rev. 1, Food and Agriculture Organizations of United Nations, Rome, Italy. ftp://ftp.org/ag/cgrfa/bsp/bsp35r1e. pdf.

Helfer, L.R. 2004. Intellectual property rights in plant varieties: International legal regimes and policy options for national governments. FAO Legislative Study 85.

Herman, R.A. and W.D. Price. 2013. Unintended compositional changes in genetically modified (GM) crops: 20 years of research. J. Agri. Food Chem. 61(48): 11695–11701.

Herman, E.M. 2003. Genetically modified soybeans and food allergies. J. Exp. Bot. 54(386): 1317–1319.

Herman, E.M., R.M. Helm, R. Jung and A.J. Kinney. 2003. Genetic modification removes an immunodominant allergen from soybean. Plant Physiol. 132(1): 36–43.

Ho, M.W., A. Ryan and J. Cummins. 2000. Hazards of transgenic plants containing the cauliflower mosaic virus promoter. Microb. Ecol. Health Dis. 12(3): 189–198.

Hodgson, J. 2000. Scientists avert new GMO crisis. Nat. Biotechnol. 18: 13.

Hoffmann, T., C. Golz and O. Schieder. 1994. Foreign DNA sequences are received by a wild-type strain of *Aspergillus niger* after co-culture with transgenic higher plants. Current Genetics 27: 70–76.

Hollingworth, R.M., L.F. Bjeldanes, M. Bolger, I. Kimber, B.J. Meade, S.L. Taylor and K.B.Wallace.2003. The safety of genetically modified foods produced through biotechnology. Toxicol Sci. 71(1): 2–8.

Hopkins, D. and E.G. Gregorich. 2003. Detection and decay of the Bt endotoxin in soil from a field trial with genetically modified maize. Eur. J. Soil Sci. 54: 793–800.

Hucl, P. 1996. Out-crossing rates for 10 Canadian spring wheat cultivars. Can. J. Plant Sci. 76: 423–427.

Hyder, J.P. 2011. Cartagena Protocol on Biosafety (2000). pp. 112–114. *In*: B.W. Lerner and K.L. Lerner (eds.). Food: In Context. Vol. 1.

Icoz, I., D. Saxena, D.A. Andow, C. Zwahlen and G. Stotzky. 2008. Microbial populations and enzyme activities in soil *in situ* under transgenic corn expressing cry proteins from *Bacillus thuringiensis*. J. Environ. Qual. 37: 647–662.

Icoz, I. and G. Stotzky. 2008. Fate and effects of insect-resistant Bt crops in soil ecosystems. Soil Biol. Biochem. 40(3): 559–586.

Jefferson, R.A., C. Correa, G. Otero, D. Byth and C. Qualset. 1999. Genetic use restriction technologies: technical assessment of the set of new technologies which sterilize or reduce the agronomic value of second generation seed, as exemplified by U.S. patent 5,723,765, and WO 94/03619. Montreal:

United Nations Convention on Biological Diversity (Technical Report). Available online at: http://www.patentlens. net/ daisy/bios/552/ version/live/ part/4/ data.

Johal, G.S. and D.M. Huber. 2009. Glyphosate effects on diseases of plants. Eur. J Agron. 31: 144–152.

Johal, G.S. and J.E. Rahe. 1988. Glyphosate, hypersensitivity and phytoalexin accumulation in the incompatible bean anthracnose host–parasite interaction. Physiol. Mol. Plant Pathol. 32: 267–281.

Jonas, D.A., I. Elmadfa, K.H. Engel, K.J. Heller, G. Kozianowski, A. König et al. 2001. Safety considerations of DNA in food. Ann. Nutr. Metab. 45(6): 235–254.

Jorgensen, T., T.P. Hauser and R.B. Jørgensen. 2007. Adventitious presence of other varieties in oilseed rape (*Brassica napus*) from seed banks and certified seed, Seed Sci. Res. 17: 115–125.

Kaufman, M. 2000. Biotech critics cite unapproved corn in taco shells. Washington Post. 18 September 2000.

Keese, P. 2008. Risks from GMOs due to horizontal gene transfer. Environ. Biosafety Res. 7(3): 123–149.

Kent, G. 2002. Africa's food security under globalization. African J. Food and Nutr. Sci. 2(1): 22–29.

Kim, Y.T., S.E. Kim, K.D. Park, T.H. Kang, Y.M. Lee, S.H. Lee et al. 2005. Investigation of possible gene transfer from leaf tissue of transgenic potato to soil bacteria. J. Microbiol. Biotechnol. 15: 1130–1134.

Kleter, G.A., A.A.C.M. Peijnenburg and H.J.M. Aarts. 2005. Health considerations regarding horizontal transfer of microbial transgenes present in genetically modified crops. J. Biomed. Biotechnol. 4: 326–352.

Kniss, A. 2013. Can We Please Stop Using the Term "Superweed". Control Freaks. Wyoming Weed Science in (almost Real Time). 1 May 2013.

Kok, E.J. and H.A. Kuiper. 2003. Comparative safety assessment for biotech crops. Trends Biotechnol. 21: 439–444.

Koskella, J. and G. Stotzky. 1997. Microbial utilization of free and claybound insecticidal toxins from *Bacillus thuringiensis* and their retention of insecticidal activity after incubation with microbes. App. Environ. Microbiol. 63: 3561–3568.

Kremer, R.J. and N.E. Means. 2009. Glyphosate and glyphosate-resistant crop interactions with rhizosphere microorganisms. Eur. J. Agron. 31: 153–161.

Kremer, R.J., N.E. Means and S.J. Kim. 2005. Glyphosate affects soybean root exudation and rhizosphere microorganisms. Int. J. Environ. Anal. Chem. 85: 1165–1174.

Kuiper, H.A., E.J. Kok and K.H. Engel. 2003. Exploitation of molecular profiling techniques for GM food safety assessment. Curr. Opin. Biotechnol. 14: 238–243.

Kuiper, H.A., G.A. Kleter, H.P. Noteborn and E.J. Kok. 2002. Substantial equivalence—an appropriate paradigm for the safety assessment of genetically modified foods?. Toxicology 181-182: 427–431.

Kuiper, H.A., G.A. Kleter, H.P.J.M. Noteborn and E.J. Kok. 2001. Assessment of the food safety issues related to genetically modified foods. Plant J. 27: 503–528.

Kuklinsky-Sobral, J., L.A. Welington, R. Mendes, A.A. Pizzirani-Kleiner and J.L. Azevedo. 2005. Isolation and characterization of endophytic bacteria from soybean (*Glycine max*) grown in soil treated with glyphosate herbicide. Plant Soil 273: 91–99.

Kuyek, D., P. Greens, R. Quizano and O.B. Zamora. 2001. Intellectual Property Rights: Ultimate control of agricultural R&D in Asia. GRAIN. 25 March 2001.

Lachman, P. 1999. Health risks of genetically modified foods. Lancet 354: 69.

Lappé, M.A., E.B. Bailey, C. Childress and K.D.R. Setchell. 1999. Alterations in clinically important phytoestrogens in genetically modified, herbicide tolerant soybeans. J. Med. Food 1: 241–245.

Lawrie, R.G., M.A. Matus-Cadiz and P. Hucl. 2006. Estimating out-crossing rates in spring wheat cultivars using the contact method. Crop Sci. 46: 247–249.

Lehmann, V. 1998. Patent on seed sterility threatens seed saving. Biotechnol. Develop. Monitor 35: 6–8.

Lehrer, S.B. 1999a. Safety assessment of foods derived from genetically modified plants: Allergenicity. Presented at a Conference in Beijing, China. 26–27 October 1999.

Lehrer, S.B. 1999b. Potential health risks of genetically modified organisms: How can allergens be assessed and minimized? In Agricultural Biotechnology and the poor. Proc. Intl. Conf. Biotechnol. Washington, DC.

Lence, S.H., D.J. Hayes, A. McCunn, S. Smith and W.S. Niebur. 2005. Welfare impacts of intellectual property protection in the seed industry. Amer. J. Agri. Econ. 87: 951–968.

Levy-Booth, D.J., R.G. Campbell, R.H. Gulden, M. Hart, J.R. Powell, J.R. Klironomos et al. 2008. Real-time polymerase chain reaction monitoring of recombinant DNA entry into soil from decomposing Roundup Ready leaf biomass. J. Agri. Food Chem. 56: 6339–6347.

Loecker, J.L., N.O. Nelson, W.B. Gordon, L.D. Maddux, K.A. Janssen and W.T. Schapaugh. 2010. Manganese response in conventional and glyphosate resistant soybean. Agron. J. 102: 606–611.

Lu, B.-R. 2013. Transgene escape from GM crops and potential biosafety consequences: an environmental perspective. Collection of Biosafety Reviews 4: 66–141.

Lu, B.-R. 2008. Transgene escape from GM crops and potential biosafety consequences: an environmental perspective. International Centre for Genetic Engineering and Biotechnology (ICGEB), Collection of Biosafety Reviews. Trieste: ICGEB Press. 4: 66–141.

Lu, Y., W. Xu, A. Kang, Y. Luo, F. Guo, R. Yang, J. Zhang and K. Huang. 2007. Prokaryotic expression and allergenicity assessment of hygromycin B phosphotransferase protein derived from genetically modified plants. J. Food Sci. 72: M228–M232.

Lu, B.-R. 2003. Transgene containment by molecular means—is it possible and cost effective? Environ. Biosafety Res. 2: 3–8.

Mallory-Smith, C. and E.S. Olguin. 2011. Gene flow from herbicide-resistant crops: it's not just for transgenes. J. Agr. Food Chem. 59: 5813–5818.

Mann, C. 2004. Diversity on the Farm: How Traditional Crops Around the World Help to Feed Us All, and Why We Should Reward the People Who Grow Them. University of Massachusetts, Amherst, MA.

Matus-Cadiz, M.A., P. Hucl, M.J. Horak and L.K. Blomquist. 2004. Gene flow in wheat at the field scale. Crop Sci. 44: 718–727.

Mechlem, K. and T. Raney. 2007. Agricultural technology and the right to food. pp. 131–156. *In*: F. Francioni (ed.). Biotechnologies and International Human Rights. Hart Publishing, Oxford, UK, 438 pp.

Medgyesy, P., L. Menczel and P. Maliga. 1980. The use of cytoplasmic streptomycin resistance: chloroplast transfer from *Nicotiana tabacum* into *Nicotiana sylvestris*, and isolation of their somatic hybrids. Mol. Gen. Genet. 179: 693–698.

Mellon, M. and J. Rissler. 2004. Gone to Seed: Transgenic Contaminants in the Traditional Seed Supply. Union of Concerned Scientists. UCS Publications, Cambridge, MA, USA.

Meridian Institute. 2013. StarLink Resurfaces: GM Corn Banned Decade Ago Found in Saudi Arabia. http://rt.com/ news/banned-gm-resurfaces-saudi-arabia-074/. 28 August 2013.

Messéan, A., C. Sausse, J. Gasquez and H. Darmency. 2007. Occurrence of genetically modified oilseed rape seeds in the harvests of subsequent conventional oilseed rape over time. Eur. J. Agron. 27: 115–122.

Messeguer, J. 2003. Gene flow assessment in transgenic plants. Plant Cell, Tissue and Organ Culture 73: 201–212.

Messeguer, J., C. Fogher, E. Guiderdoni, V. Marfa, M.M. Catala, G. Baldi and E. Mele. 2001. Field assessment of gene flow from transgenic to cultivated rice (*Oryza sativa*) using a herbicide resistant gene as tracer marker. Theor. Appl. Genet. 103: 1151–1159.

Metcalfe, D.D., J.D. Astwood, R. Townsend, A.A. Sampson, S.L. Taylor and R.L. Fuchs. 1996. Assessment of the allergenic potential of foods derived from genetically engineered crop plants. Crit. Rev. Food Sci. Nutr. 36(suppl.): S165–S186.

Miller, H.I. 1999. A rational approach to labeling biotech-derived foods. Science 284: 1471–1472.

Monsanto. 2005. Pledge Report. http://www.monsanto.com/ourcommitments/Documents/ CSR_reports/ Monsanto PledgeReport-2005.pdf.

Moore, J.C., D.E. Walter and H.W. Hunt. 1988. Arthropod regulation of micro- and mesobiota in below-ground detrital food webs. Ann. Rev. Entomol. 33: 419–439.

Munro, I.C. 2002. OECD/FAO Substantial equivalence framework for whole food safety assessment. Toxicologist 66: 381.

Netherwood, T., S.M. Martin Orue, A.G. O'Donnell, S. Gockling, J. Graham, J.C. Mathers and H.J. Gilbert. 2004. Assessing the survival of transgenic plant DNA in the human gastrointestinal tract. Nat. Biotcehnol. 22: 204–209.

Nguyen, H.R. and J.A. Jehle. 2009. Expression of cry3Bb1 in transgenic corn MON88017. J. Agri. Food Chem. 57(21): 9990–9996.

Nordlee, J.A., S.L. Taylor, J.A. Townsend, L.A. Thomas and R.K. Bush. 1996. Identification of a Brazil-nut allergen in transgenic soybeans. N. Engl. J. Med. 334: 726–728.

Novak, W.K. and A.G. Haslberger. 2000. Substantial equivalence of antinutrients and inherent plant toxins in genetically modified novel foods. Food and Chem. Toxicol. 38(6): 473–483.

OECD. 2010. Consensus Document on Molecular Characterisation of Plants Derived from Modern Biotechnology. OECD Environment, Health, and Safety Public. Series on Harmonisation of Regulatory Oversight in Biotechnology. Series on Harmonisation of Regulatory Oversight in Biotechnology No. 51 and Series on the Safety of Novel Foods and Feeds No. 2. ENV/JM/MONO(2010)41. 20 September 2010.

OECD. 1993. Safety evaluation of foods derived by modern technology. Concepts and Principles. OECD. Paris, France.

Oger, P., A. Petit and Y. Dessaux. 1997. Genetically engineered plants producing opines alter their biological environment. Nature Biotechnol. 15: 369–372.

Ortíz-García, S., E. Ezcurra, B. Schoel, F. Acededo, J. Soberón and A.A. Snow. 2005. Absence of detectable transgenes in local landraces of maize in Oaxaca, Mexico. Proc. Natl. Acad. Sci. USA 102: 12338–12343.

Owen. 2008. Weed species shifts in glyphosate-resistant crops. Pest Manage. Sci. 64(4): 377–387.

Padgette, S.R., N.B. Taylor, D.L. Nida, M.R. Bailey, J. MacDonald, L.R. Holden and R.L. Fuchs. 1996. The composition of glyphosate-tolerant soybean seeds is equivalent to that of conventional soybeans. J. Nutr. 126: 702–716.

Pagel-Wieder, S., J. Niemeyer, W.R. Fischer and F. Gessler. 2007. Effects of physical and chemical properties of soils and adsorption of the insecticidal protein (Cry1Ab) from *Bacillus thuringiensis* at Cry1Ab protein concentrations relevant for experimental field sites. Soil Biol. Biochem. 39(12): 3034–3042.

Palaudelmàs, M., J. Messeguer, G. Peñas, J. Serra, J. Salvia, M. Pla, A. Nadal and E. Melé. 2007. Effect of sowing and flowering dates on maize gene flow. pp. 235–236. *In*: A.J. Stein and E. Rodríguez-Cerezo (eds.). Abtracts of the third Internatl. Conf. Coexistence between Genetically Modified (GM) and non-GM-based Agricultural Supply Chains, European Commission.

Paparini, A. and V. Romano-Spica. 2006. Gene transfer and cauliflower mosaic virus promoter 35S activity in mammalian cells. J. Environ. Sci. Health B. 41: 437–449.

Paparini, A. and V. Romano-Spica. 2004. Public health issues related with consumption of the food obtained from genetically modified organisms. Biotechnol. Ann. Rev. 10: 85–122.

Parsai, G. 2012. Opposition to Monsanto patent on Indian melons. 2013. The Hindu Newspaper, 5 February 2012.

Pendleton, C.N. 2004. The peculiar case of "terminator" technology: agricultural biotechnology and intellectual property protection at the crossroads of the third green revolution. Biotechnol. Law Rep. 23: 1–29.

Pessel, D., J. Lecomte et al. 2001. Persistence of oilseed rape (*Brassica napus* L.) outside of cultivated fields. Theor. Applied Gene. 102: 841–846.

Pilger, G. 2002. Terminator could eliminate GM volunteers. Canola Guide May 2002: 16–17.

Piñeyro-Nelson, A., J. Van HeerWaarden, H.P. Perales, J.A. Serratos-Hernández, A. Rangel, M.B. Hufford et al. 2009a. Transgenes in Mexico maize: molecular evidence and methodological considerations for GMO detection in landrace populations. Mol. Ecol. 18(4): 750–761.

Piñeyro-Nelson, A., J. Van Heer Waarden, H.P., Perales, J.A. Serratos-Hernández, A. Rangel, M.B. Hufford et al. 2009b. Resolution of the Mexican transgene detection controversy: error sources and scientific practice in commercial and ecological contexts. Mol. Ecol. 18(20): 4145–4150.

Podevin, N. and P. du jardin. 2012. Possible consequences of the overlap between the CaMV 35S promoter regions in plant transformation vectors used and the viral gene VI in transgenic plants. GM Crops and Food: Biotechnology in Agriculture and the Food Chain 3(4): 296–300.

Powell, J.R., R.G. Campbell, K.E. Dunfield, R.H. Gulden, M.M. Hart, D.J. Levy-Booth et al. 2009. Effect of glyphosate on the tripartite symbiosis formed by *Glomus intraradices*, *Bradyrhizobium japonicum*, and genetically-modified soybean. Appl. Soil Ecol. 41: 128–136.

Powell, J.R. and C.J. Swanton. 2008. A critique of studies evaluating glyphosate effects on diseases associated with *Fusarium* spp. Weed Res. 48: 307–318.

Prasad, S. 2013. Bt Brinjal: HC says Monsanto will have to face bio-piracy case. Bangalore Mirror. 18 October 2013.

Quist, D. 2010. Vertical (Trans)gene Flow: Implications for Crop Diversity and Wild Relatives. TWN Biotechnology & Biosafety Series 11. Third World Network, Penang, Malaysia, 33 pp.

Quist, D. and I. Chapela. 2001. Transgenic DNA introgressed into traditional maize landraces in Oaxaca, Mexico. Nature 414: 541–543.

Rakshit, S. 1998. Terminator technology: science and politics. Current Sci. 75: 747–749.

Robinson, D. and N. Medlock. 2005. Diamond v. Chakrabarty: a retrospective on 25 years of biotech patents. Intellectual Property & Technol. Law J. 17(10): 12–15.

Rong, J., B.-R. Lu, Z.P. Song, J. Su, A.A. Snow, X.S. Zhang et al. 2007. Dramatic reduction of crop-to-crop gene flow within a short distance from transgenic rice fields. New Phytologist 173: 346–353.

Rong, J., Z.P. Song, J. Su, H. Xia, B.-R. Lu and F. Wong. 2005. Low frequency of transgene flow from Bt/CpTi rice to its non-transgenic counterparts planted at close spacing. New Phytologist 168: 559–566.

Rosolem, C.A., G.J.M. Andrade, I.P. Lisboa and S.M. Zoca. 2010. Manganese uptake and redistribution in soybeans as affected by glyphosate. P. Bras. Ci. Solo 34: 1915–1922.

Roseman, T.S., R.D. Graham, H.J. Arnott and D.M. Huber. 1991. The interaction of temperature with virulence and manganese oxidizing potential in the epidemiology of Gaeumannomyces graminis. Phytopathology 81: S1215.

Royal Society of Canada. 2001. Report of the expert panel on food biotechnology. J. Toxicol. Environ. Health A 64: 51–57.

Ruebelt, M.C., M. Lipp, T.L. Reynolds, J.J. Schmuke, J.D. Astwood, D. Dellapenna et al. 2006. Application of two-dimensional gel electrophoresis to interrogate alterations in the proteome of genetically modified crops. 3. Assessing unintended effects. J. Agric. Food Chem. 54: 2169–2177.

Ruf, S., D. Karcher and R. Bock. 2007. Determining the transgene containment level provided by chloroplast transformation. Proc. Natl. Acad. Sci. USA 104: 6998–7000.

Sánchez, G.J., M.M. Goodman and C.W. Stuber. 2000. Isozymatic and morphological diversity in the races of maize of Mexico. Economic Botany 54: 43–59.

Sanvido, O., F. Widmer, M. Winzeler, B. Streit, E. Szerencsits and F. Bigler. 2008. Definition and feasibility of isolation distances for transgenic maize, Transgenic Res. 17: 317–355.

Saxena, D., C.N. Stewart, I. Altosaar, Q. Shu and G. Stotzky. 2004. Larvicidal Cry proteins from *Bacillus thuringiensis* are released in root exudates of transgenic *B. thuringiensis* corn, potato, and rice but not of *B. thuringiensis* canola, cotton, and tobacco. Plant Physiol. Biochem. 42: 383–387.

Saxena, D. and G. Stotzky. 2000. Insecticidal toxin from *Bacillus thuringiensis* is released from roots of transgenic Bt corn *in vitro* and *in situ*. FEMS Microbiol. Ecol. 33(1): 35–39.

Saxena, D., S. Flores and G. Stotzky. 1999. Insecticidal toxin in root exudates from Bt corn. Nature 402: 480.

Sell, P. and G. Murrell. 1996. Flora of Great Britain and Ireland. Vol. 5. Butamaceae—Orchidaceae. Cambridge University Press, Cambridge. UK, 440 pp.

Serra, A.P., M.E. Marchetti, A.C. da Silva Candido, A.C. Ribiero Dias and P. Christoffoleti. 2011. Glyphosate influence on nitrogen, manganese, iron, copper and zinc nutritional efficiency in glyphosate resistant soybean. Ciênc. Rural, Santa Maria 41: 77–84.

Shashikant, S. and A. Asghedom. 2009. Enola bean patent dispute offers lessons for developing countries. Third World Network. Nov/Dec 2009.

Sheppard, A.E., M.A. Ayliffe, L. Blatch, A. Day, S.K. Delaney, N. Khairul-Fahmy et al. 2008. Transfer of plastid DNA to the nucleus is elevated during male gametogenesis in tobacco. Plant Physiol. 148: 328–336.

Shi, G. 2006. Intellectual Property Rights, Genetic Use Restriction Technologies (GURTs), and Stratetic Behavior. Paper No. 156068. Amer. Agril. Econ. Assoc. Ann. Meeting, Long Beach, CA, USA. 23–26 July.

Simard, M., A. Legere, D. Pageau, J. Lajeunesse and S. Warwick. 2002. The frequency and persistence of canola (*Brassica napus*) volunteers in Quebec cropping systems. Weed Technol. 16: 433–439.

Siva, V. 2000. Stolen Harvest: The Hijacking of the Global Food Supply. South End Press, Cambridge, MA, USA.

Sood, J. 2012. Karnataka High Court issues notice to National Biodiversity Authority on charges of paving way for biopiracy. Down to Earth 22 November 2012.

StarLink Corn. 2013. What Happened? University of California, Davis. Retrieved 12 August 2013.

StarLink Corn. 2004. Transgenic Crops: An Introduction and Resource Guide. Colorado State Univ. 11 March 2004.

Stegemann, S. and R. Bock. 2006. Experimental reconstruction of functional gene transfer from the tobacco plastid genome to the nucleus. Plant Cell 18: 2869–2878.

Stein, A.J. and E. Rodriguez-Cerezo. 2009. The global pipeline of new GM crops: Implications of asynchronous approval for international trade. Inst. Prospective Technol. Studies, Joint Res. Centre, European Commission, 110 p.

Stewart, N.C., Jr. 2008. Gene Flow and the Risk of Transgene Spread. 8 March 2008.

Stotzky, G. 2004. Persistence and biological activity in soil of the insecticidal proteins from *Bacillus thuringiensis*, especially from transgenic plants. Plant Soil 266: 77–89.

Stotzky, G. 2002. Clays and humic acids affect the persistence and biological activity of insecticidal proteins from *Bacillus thuringiensis* in soil. Dev. Soil Sci. 28(2): 1–16.

Stotzky, G. 2000. Persistence and biological activity in soil of insecticidal proteins from *Bacillus thuringiensis* and of bacterial DNA bound on clays and humic acids. J. Environ. Qual. 29: 691–705.

Streit, L.G., L.R. Beach, J.C. Register, R. Jung and W.R. Fehr. 2001. Association of the Brazil nut protein gene and Kunitz trypsin inhibitor alleles with soybean protease inhibitor activity and agronomic traits. Crop Sci. 41(6): 1757–1760.

Szumigalski, T. 2006. Literature Review on Genetic Use Restriction Technologies. Canadian Food Grains bank, 21 pp.

Tepfer, M., S. Gaubert, M. Leroux-Coyau, S. Prince and L. Houdebine. 2004. Transient expression in mammalian cells of transgenes transcribed from the Cauliflower mosaic virus 35S promoter. Environ. Biosafety Res. 3: 91–97.

Tepfer, D., R. Garcia-Gonzales, H. Mansouri, M. Seruga, B. Message, F. Leach and M.C. Perica. 2003. Homology-dependent DNA transfer from plants to a soil bacterium under laboratory conditions: implications in evolution and horizontal gene transfer. Transgenic Res. 12(4): 425–437.

The Crucible II Group. 2001. Seeding Solutions. Vol. 2. Options for national laws governing control over genetic resources and biological innovations. International Plant Genetic Resources Institute, Dag Hammarskjold Foundation, Uppsala, Sweden, 243 pp.

Thompson, I.A. and D.M. Huber. 2007. Manganese and plant disease. pp. 139–153. *In*: L.E. Datnoff, W.H. Elmer and D.M. Huber (eds.). Mineral Nutrition and Plant Disease. APS Press, St. Paul, MN, USA.

Tranberg, J. 2013. Developing a policy for low-level presence (LLP): A Canadian case study. AgBio Forum. 16(1): 37–45.

Union of Concerned Scientists. 2000. Risks of genetic engineering. http://www.ucsusa.org/agriculture/ gen.risks.html.

USDA. 2001. Economic issues in agricultural biotechnology. *In*: R. Shoemaker (ed.). Agriculture Information Bulletin No. 762: February 2001. Resource Economics Division, Economic Research Service, USDA.

United States Patent and Trademark Office, 3 March 1998. available at http://patft.uspto.gov/ netacgi/ nph Parser? Sect1=PTO1&Sect2=HITOFF&d=PALL&p=1&u=/netahtml/srchnum. htm&r=1&f=G&l=50 &s1= 5,723,765. WKU. &OS=PN/5,723,765&RS=PN/5,723,765.

United States Supreme Court. 1980. Diamond, Commissioner of Patents and Trademarks v. Chakrabarty. 447 U.S. 303, 206 USPQ 193. 16 June 1980.

Valverde, B.E. 2007. Status and management of grass-weed herbicide resistance in Latin America. Weed Technol. 21: 310–323.

Verma, D. and H. Daniell. 2007. Chloroplast vector systems for biotechnology applications, Plant Physiol. 145: 1129–114.

Viljoen, C. and L. Chetty. 2011. A case study of GM maize gene flow in South Africa. Environ. Sci. Europe 23: 8.

Visser, B., I. van der Meer, N. Louwaars, J. Beekwilder and D. Eaton. 2001. The impact of 'terminator' technology. Biotechnol. Develop. Monitor 48: 9–12.

Visser, B. 1998. Effects of biotechnology on agro-biodiversity. Biotechnol. Develop. Monitor 35: 2–7.

Wang, R., S.B. Chen, Q. Zhuo, L.C. Yang, Z. Zhu and X. Yang. 2005. Expression specificity of hpt gene in sck transgenic rice. J. Hygiene Res. (In Chinese: Wei Sheng Yan Jiu 34(4): 460–462).

Wang, T., Y. Li, Y. Shi, X. Reboud, H. Darmency and J. Gressel. 2004. Low frequency transmission of a plastid-encoded trait in *Setaria italica*. Theor. Appl. Genet. 108: 315–320.

Wani, S.H., N. Haider, H. Kumar and N.B. Singh. 2010. Plant plastid engineering. Curr. Genomics 11(7): 500–512.

Warwick, S.I., A. Legere, M.-J. Simard and T. James. 2008. Do escaped transgenes persist in nature? The case of an herbicide resistant transgene in a weedy *Brassica rapa* population. Mol. Ecol. 17: 1387–1395.

Weider, C., P. Stamp, N. Christov, A. Hüsken, X. Foueillassar, K.H. Camp et al. 2009. Stability of cytoplasmic male sterility in maize under different environmental conditions. Crop Sci. 49: 77–78.

Wilcks, A., A.H. van Hoek, R.G. Joosten, B.B. Jacob and H.J. Aarts. 2004. Persistence of DNA studied in different *ex vivo* and *in vivo* rat models simulating the human gut situation. Food Chem. Toxicol. 42(3): 493–502.

Wellhausen, E., J. Roberts, L.M. Roberts and E. Hernandez. 1952. Races of Maize in Mexico: Their Origin, Characteristics, and Distribution. Harvard University Press, Cambridge, MA.

Winter, C.K. and L.K. Gallegos. 2006. University of California Agricultural and Natural Resource Service. ANR Publication 8180.

World Health Organization. 2000. Safety aspects of genetically modified foods of plant origin. Report of a Joint FAO/WHO Expert Consultation on Foods Derived from Biotechnology.

Yang, L.C., S.X. Zhang, G.H. Pi, Y.H. Li, Z. Zhu and X.Z. Yang. 2005. Preparation of monoclonal antibody against HPT and its application to detecting marker protein in genetically modified Rice. Biomed. Enviro. Sci. 18: 321–325.

Zablotowicz, R.M. and K.N. Reddy. 2004. Impact of glyphosate on the *Bradyrhizobium japonicum* symbiosis with glyphosate-resistant transgenic soybean: a minireview. J. Environ. Qual. 33: 825–831.

Zhuo, Q., J.Q. Piao, Y. Tian, J. Xu and X.G. Yang. 2009. Large-scale purification and acute toxicity of hygromycin B phosphotransferase. Biomed. Environ. Sci. 22: 22–27.

Zobiole, L.H.S., R.J. Kremer, R.S. Oliveira, Jr. and J. Constantin. 2011. Glyphosate affects micro-organisms in rhizospheres of glyphosate-resistant soybeans. J. Appl. Microbiol. 110(1): 118–127.

Zobiole, L.H.S., E.A. Bonini, R.S. Oliveira, Jr., R.J. Kremer and O. Ferrarese-Filho. 2010a. Glyphosate affects lignin content and amino acid production in glyphosate-resistant soybean. Acta Physiol. Plant 32: 831–837.

Zobiole, L.H.S., R.S. Oliveira, Jr., R.J. Kremer, A.S. Muniz and A. Oliviera, Jr. 2010b. Nutrient accumulation and photosynthesis in glyphosate-resistant soybeans is reduced under glyphosate use. J. Plant Nutr. 33: 1860–1873.

Zolla, L.S., S. Rinalducci, P. Antonioli and P.G. Righetti. 2008. Proteomics as a complementary tool for identifying unintended side effects occurring in transgenic maize seeds as a result of genetic modifications. J. Proteome Res. 7: 1850–1861.

Appendix

Table 1. Resistance of weed species (dicots and monocots) to each of the herbicide families with different sites of action [Heap 21 July 2014]. Vide Chapter 2 'Text' for details on site of action of herbicides and 'References' for citations.

No.	HRAC Group	Site of Action (Inhibition)	Chemical Family and Prominent Herbicide/s	Weed Species' Dicots	Monocots	Total
1	A	Acetyl CoA carboxylase (ACCase)	Aryloxyphenoxypropionate; Cyclohexanedione; Phenylpyrazoline (pinoxaden)	0	46	46
2	B	Acetolactate synthase: ALS; (Acetohydroxyacid synthase: AHAS)	Sulfonylurea; Imidazolinone; Triazolopyrimidine; Pyrimidinyl thiobenzoate; Sulfonylaminocarbonyl triazolinone	88	57	145
3	C1	Photosynthesis at PS II	Triazine; Triazinone; Triazolinone (amicarbazone); Uracil; Pyridazinone (pyrazon); Phenylcarbamate	49	23	72
4	C2	Photosynthesis at PS II	Amide; Urea	8	16	24
5	C3	Photosynthesis at PS II	Benzothiadiazinone (bentazon); Nitrile Phenylpyridazinone (pyridate)	3	1	4
6	D	PS I electron diversion	Bipyridilium	22	9	31
7	E	Protoporphyrinogen oxidase (PPO)	Diphenylether; Phenylpyrazole (pyraflufen); N-phenyl-phthalimide; Oxadiazole; Thiadiazole; Triazolinone; Pyrimidinedione	6	0	6
8	F1	Bleaching: Carotenoid biosynthesis at the phytoenedesaturase (PDS)	Pyradazinone (norflurazone); Pyridinecarboxamide; Other (fluridone, fluramone, fluchloridone, beflubutamid)	2	1	3
9	F2	Bleaching: 4-hydroxy phenyl-pyruvate-dioxygenase (4-HPPD)	Triketone; Isoxazole (isoxaflutole); Pyrazole	2	0	2
10	F3	Bleaching: Carotenoid biosynthesis (unknown target)	Triazole; Isoxazolidinone (clomazone); Urea (flumetsulam); Diphenylether (aclonifen)	1	4	5

Table 1. contd....

Table 1. contd.

No.	HRAC Group	Site of Action (Inhibition)	Chemical Family and Prominent Herbicide/s	Weed Species[1] Dicots	Monocots	Total
11	G	EPSP synthase	Glycine (glyphosate)	16	15	31
12	H	Glutamine synthetase	Glufosinate-ammonium, Phosphinic acid	0	2	2
13	I	DHP (dihydropteroate) synthase	Carbamate (asulam)	–	–	–
14	K1	Microtubule assembly	Dinitroaniline; Pyridine; Benzamide; Phosphoroamidate; Benzoic acid (DCPA = chlorthal-methyl)	2	10	12
15	K2	Mitosis/Microtubule polymerization	Carbamate (propham; chloropropham, carbetamide)	0	1	1
16	K3	Cell division (Inhibition of very long chain fatty acids: VLCFS)	Chloroacetamide (acetochlor, alachlor, butachlor, metolachlor, propachlor, pretilachlor); Acetamide (napropamide, diphenamid); Oxyacetamide (flufenacet); Tetra-zolinone (fentrazamide); Anilofos	0	4	4
17	L	Cellulose (cell wall) synthesis	Nitrile (dichlobenil); Benzamide (isoxaben); Quinoline-carboxylic acid (quinclorac, quinmerac)	0	2	2
18	M	Uncoupling (membrane disruption)	Dinitrophenol (DNOC, dinoseb, dinoterb)	–	–	–
19	N	Lipid synthesis—not ACCase inhibition	Thiocarbamate; Benzofuran (ethofume-sate); Phosphoro-dithioate (bensulide): Chlorocarbonicacid (TCA, dalapon)	0	9	9
20	O	Synthetic auxins (action like indoleacetic acid)	Phenoxy-carboxylic acid; Benzoic Acid; Pyridine-carboxylic acid; Quinoline-carboxylic acid (quinclorac)	23	8	31
21	P	Auxin transport (polar transport)	Phthalamate (naptalam); Semicarbazone (diflufenzopyr-Na)	–	–	–
22	Z	Antimicrotubule mitotic disruption	Arylaminopropionic acid (flamprop-M-methyl/isopropyl)	0	3	3
23	Z	Nucleic acid	Organoarsenical	1	0	1
24	Z	Cell elongation	Difenzoquat	0	1	1
25	Z	Unknown	Endothal	0	1	1
-	-	—	[2]**Grand Total**	223	213	436

[1]Weed Species: number of dicot and/or monocots resistant to each herbicide group.

[2]Grand Total: it includes the number of unique cases (species x site of action), and not the number of resistant species. For example, some weed species resistant to group A may also be resistant to other group(s); in other words, these may have evolved resistance to more than one site of action as in the case of *Lolium rigidum* which is resistant to 11 groups and *Echinochloa crus-galli* var. *crus-galli* and *Poaannua* to 9 each, etc.

Table 2. Major gene transfer methods used for transformation of monocotyledonous and dicotyledonous plant species [Review by Barampuram and Zhang 2011; vide Chapter 4 'References' for citations].

Plant Species	Types of Explants	Transfer Method	Gene Transferred	Transformation Efficiency (%)	Reference
Monocots					
Rice	C	*Ag*	*cry1Ac*	2	Kim et al. 2009
	C	BL (MPB)	*shGH*	79.5	Kim et al. 2008
	SA	EP	*npt*II, *ppt*	13.8	De Padua et al. 2001
Hordium vulgare (Barley)	ImE	*Ag*	*hpt*II, *luc*	25	Cho et al. 2003
	EC	Bl (MPB)	*At NDPK2*	0.15	Um et al. 2007
Saccharum sp. (Sugarcane)	AxB	*Ag*	*npt*II, *bar*, *gusA*	50	Manickavasagam et al. 2004
	C	EP	*gusA*	80	Seema et al. 2001
Sorghum bicolor (Sorghum)	C	*Ag*	*gfp*	8.3	Gurel et al. 2009
	ImI	Bl (MPB)	*uidA*, *bar*	3.33	Brando et al. 2012
Triticum sp.	ImE	*Ag*	*bar*, *gusA*	9.7	Wu et al. 2008
Zea mays (Maize/Corn)	ImE	*Ag*	*gusA*, *bar*	12.2	Vega et al. 2008
	ImE	Bl (MPB)	*gusA*, *hpt*II	31	Lowe et al. 2009
Dicots					
Arachis hypogea (Groundnut/ Peanut)	CN	*Ag*	*gusA*	38	Anuradha et al. 2006
	SE	Bl (MPB)	*vp2*, *gusA*	12.3	Athmaram et al. 2006
	EL	EP	*gusA*	3	De Paadua 2000
Brassica oleracea (Cabbage)	ML	Cl-BL (MPB)	*cry1Ab*	11.1	Liu et al. 2008
Cajanus cajan (Pigeon Pea)	CN	*Ag*	*npt*II, *H*	51	Satyavathi et al. 2003
Eucalyptus sp.	ApS	*Ag*	*gusA*	9	Spokevicius et al. 2005
Glycine max (Soya bean)	CN	*Ag*	*bar*, *gusA*	5.5	Zeng et al. 2004
	SE	Bl (MPB)	*Os-mALS*	60	Tougou et al. 2009
	Fl	PTP	*phyA*	13	Gao et al. 2007

Table 2. contd....

Table 2. contd....

Plant Species	Types of Explants	Transfer Method	Gene Transferred	Transformation Efficiency (%)	Reference
Dicots					
Gossypium hirsutum (Cotton)	EC	*Ag*	*cry1Ia5*	83	Leelavathi et al. 2004
	EC	SCW	*AVP1, npt*II	64	Asad et al. 2008
Malus domestica (Apple)	IVS	*Ag*	*Lc*	50	Li et al. 2007
Pinus sp.	EC	*Ag*	*npt II, bar, gusA*	65–98	Charity et al. 2005

Explants: AxB: axillary bud; ApS: apical shoot; C: callus; CN: cotyledonory node; EC: embryo-genic callus; EL: embryonic leaflets; Fl: floral dip; ImE: immature embryo; IMI: immature influorescence; IVS: *in vitro* shoot; L: leaf; ML: mature leaf; SA: shoot apex; SE: somatic embryo. **Transfer Methods**: *Ag*: *Agrobacterium*-mediated transformation; Bl: biolistic method; Cl-MPB: chloroplast-mediated microprojectile particle bombardment; EP: electroporation; MPB: microprojecticle particle bombardment; PTP: pollen tube pathway transformation; SCW: silicon-carbide-whiskers mediated transformation. **Genes Transferred**: *AtNDPK2*: *Arabidopsis* nucleotide diphosphate kinase gene; *AVP1*: vacuolar pyrophosphate; *bar*: bialophos-resistant gene cloned from *Streptomyces hygroscopicus*; *cry1Ab, cry1Ac, cry1Ia5*: *Bacillus thuringiensis*-produced crystalline proteins; *gfp*: green fluorescent protein; *gusA*: β-glucuronidase; *H*: haemoglobin protein; *hpt*II: hygromycin phosphotransferase; *Lc*: maize leaf color regulatory gene; *npt*II neomycin phosphotransferase; *Os-mALS* acetolactate synthase derived from rice; *phyA*: phytase A; ppt: phosphinothricin; *shGH*: synthetic human growth hormone; *uid*A: encoder of β-glucuronidase; *vp2*: parovirus capsid protein sequence 2.

Table 3. Selectable marker genes for transformation in plants (vide Chapter 4).

Selectable Marker Gene	Substrate	Enzyme Encoded	Gene Source
AacC3; aacC4	Gentamycin	Gentamycin-3-N-acyltransferase	*Serratia marcescens; Klebsiella penumoniae*
aadA	Streptomycin, spectinomycin	Aminoglycoside-3-adenyltransferase	*Shigella flexneri*
als	Sulfonylurea herbicides	Acetolactate synthase	*Nicotiana tabacum; Porphyridium* sp.; *Arabidopsis thaliana*
aphIV; hptII	Hygromycin B	Hygromycin phosphotransferase	*E. coli*
ARG7	—	Arginosuccinate lyase locus 7	*Chlamydomonas reinhardtii*
aroA-M1	Glyphosate	5-enolpyruvylshikimate-3-phospahate synthase	*E. coli*
bar	Bialaphos; phosphinothricin (PPT); glufosinate	Phosphinothricin acyltransferase	*Streptomyces hygroscopicus*
BADH	Betaine aldehyde	Betaine aldehyde dehydrogenase	*Spinacea oleracea*
Bla	Penicillin, ampicillin	β-Lactamase	*E. coli*
ble	Bleomycin; phleomycin	Bleomycin resistance	Tn5 & *Streptoalloteichus hindustanus*
bxn	Bromoxynil	Bromoxynil-specific nitrilase	*Klebsiella pneumonia* subsp. *Ozaenae*
cat	Chloramphenicol	Chloramphenicol acetyltransferase	Bacteriophage p1Cm
csrt-1	Sulfonylurea herbicides	Acetolactate synthase	*Arabidopsis thaliana*
dhfr	Methotextrate	Dihydrofolate reductase	Plasmid R67
dhps	S-Aminoethyl; L-cystein	Dihydrodipicolinate synthase	*E. coli*

Table 3. contd....

Table 3. contd.

Selectable Marker Gene	Substrate	Enzyme Encoded	Gene Source
ept	Streptomycin	Streptomycin transferase	Tn5
EPSPS; cp4epsps; goxv247; gat	Glyphosate	5-enolpyruvylshikimate-3-phosphate synthase	*Petunia hydrida*
gox	Glyphosate	Glyphosate oxidoreductase	*Achromobacter* LBAA
hpp	HPPD herbicides (diketonitrile)	Hydroxyphenylpyruvate dioxygenase	Tobacco
manA	D-Mannose	Phosphomannose isomerase	*E. coli*
nptII/neo	Kanamycin, paranomycin, neomycin, geneticin (G418)	Neomycin phosphotransferase II	*E. coli* Tn5 (transposon)
nptIII	Amikacin, kanamycin, neomycin, geneticin (G418), paromomycin	Neomycin phosphotransferase III	*Streptococcus faecalis* Rplasmid
Pat	Glufosinate, L-phosphinothricin, bialaphos	Phosphinothricin acyltransferase	*Streptomyces viridochromogenes*
psbA	Atrazine	Q_B protein	*Amaranthus hydridus*
psbA	Metribuzin, DCMU	Q_B protein	*Chlamydomonas*
sul	Sulfonamide	Dihydropteroate synthase	Plasmid R46
Tdc	4-Methyl tryptophan (4-mT)	Tryptophan decarboxylase	*Catharanthus roseus*
tfdA	2,4-D	2,4-D monooxygenase (2,4-dichlorophenol)	*Alcaligenes eutrophus*
uidA/GUS	β-Glucuronidase	Cytokinin glucuronides	*E. coli*
xylA	D-Xylose	Xylulose isomerase	*Streptomyces rubignosus*

Table 4. Transgenic plants for enhanced phytoremediation of xenobiotics including herbicides in various plant species by various genes [based on Review by Abhilash et al. 2009; vide Chapter 7 'References' for citations].

Target Plant	Gene(s)	Enzymes	Source	Transgenic Effects	Reference
Rice	*CYP1A1*	Cytochrome P450 monooxygenase	Human	Enhanced metabolism of chlortoluron, norflurazon	Kawahigashi et al. 2007, 2008
Rice	*CYP1A1*	Cytochrome P450 monooxygenase	Human	Remediation of atrazine, simazine	Kawahigashi et al. 2005b
Rice	*CYP1A1, CYP2B6, CYP2C19*	Cytochrome P450 monooxygenase	Human	Phytoremediation of atrazine, metolachlor	Kawahigashi et al. 2006
Rice	*CYP2C9*	Cytochrome P450 Monooxygenase	Human	Tolerance to sulfonylurea herbicides	Hirose et al. 2005
Rice	*CYP2B6*	Cytochrome P450 monooxygenase	Human	Metabolism of metolachlor, ethofumesate, benfuresate	Kawahigashi et al. 2005a, 2005c
Rice	*CYP2B6*	Cytochrome P450 monooxygenase	Human	Remediation of metolachlor	Kawahigashi et al. 2005a
Rice	*CYP2B22 CYP2C49*	Cytochrome P450 monooxygenase	*Sus crofa* (wild pig)	Tolerance to several herbicides	Kawahigashi et al. 2005c
Rice	*Protox* genes	Protoporphyrinogen IX oxidase	Bacteria	Tolerance to diphenyl ether herbicide oxyfluorfen	Jung et al. 2008
Potato; Rice	*CYP1A1, CYP2B6, CYP2C19*	Cytochrome P450 monooxygenase	Human	Resistance to sulfonylurea and other herbicides	Inui and Ohkawa 2005
Tobacco	*onr*	Pentaerythritol tetranitrate reductase (PETN)	*Enterobacter cloacae* PB2	Enhanced denitration of glycerol trinitrate (GTN) and TNT	French et al. 1998

Table 4. contd....

Table 4. contd.

Target Plant	Gene(s)	Enzymes	Source	Transgenic Effects	Reference
Tobacco	*CYP105A1*	Cytochrome P450 monooxygenase	*Streptomyces griseolus*	Tolerance to sulfonylurea herbicides	O'keefe et al. 1994
Tobacco	*CYP450 2E1*	Cytochrome P450 monooxygenase	Human	Oxidation of TCE and ethylene dibromide	Doty et al. 2000
Tobacco; *A. thaliana*	*CYP71A10*	Cytochrome P450 monooxygenase	*Glycine max*	Tolerance to phenyl urea herbicides	Siminszky et al. 1999
Tobacco	*tpx1, tpx2*	Peroxidases (Px)	*Lycopersicum esculentum*	Hairy cultures of transgenic tobacco: phenol removal	Alderete et al. 2009
Tobacco	*CYP76B1*	Cytochrome P450 monooxygenase	*Helianthus tuberosus*	Tolerance to herbicide	Didierjean et al. 2002
Tobacco	*Gst1-6His*	Glutathione S-transferases	Maize	Higher tolerance to alachlor	Karavangeli et al. 2005
Tobacco	*bphC*	2,3, dihydroxybiphenyl-1,2-dioxygenase	*Pseudomonas testosteroni* B-356	Enhanced degradation of polychlorinated biphenyls (PCBs)	Chrastilova et al. 2007
A. thaliana, Tobacco	*Mn peroxidase*	Mn-peroxidase (MnP)	*Coriolus versicolor* (mushroom)	Enhanced pentachlorophenol (PCP) removal	Iimura et al. 2002
Tobacco	*ophc2*	Organophosphorus hydrolase (OPH)	*Pseudomonas pseudoalcaligenes*	Enhanced degradation of organophosphorus (methyl parathion)	Wang et al. 2008
Tobacco	*CYP450E1*	Cytochrome P450 monooxygenase	Human	Enhanced degradation of anthracene, chlorpyriphos	Dixit et al. 2008
Tobacco	*GST*	Glutathione s-transferase	*Trichoderma virens*	Enhanced accumulation of cadmium in *Nicotiana tabacum*	Dixit et al. 2011

Appendix **415**

Tobacco	*LAC*	Fungal laccase	*Coriolus versicolor* (mushroom)	Laccase secretion into the rhizosphere; removal of bisphenol A and PCP	Sonoki et al. 2005
Tobacco	Biphenyl di-oxygenase	Biphenyl dioxygenase	*Burkholderia xenovorans*	Catalyse the oxygenation of 4-chlorobiphenyls	Mohammadi et al. 2007
Tobacco	*DhaA*	Aromatic-cleaving extradiol dioxygenase	*Terrabacter* sp.	Enhanced detoxification of 1-chlorobutane in rhizosphere	Uchida et al. 2005
Tobacco	*bphC*	Biphenyl catabolic enzymes	*Pandoraea pnomenusa*	Enhanced degradation of PCBs	Frankova et al. 2003; Noakova et al. 2009
Tobacco	*NfsI*	Nitroreductase	*E. cloaceae*	Transgenic plants remove TNT; reduces it to 4-hydroxylamino-2,6-dinitrotoluene	Hannink et al. 2002, 2007
A. thaliana	*DbfB*	Aromatic-cleaving extradiol dioxygenase	*Terrabacter* sp.	Enhanced detoxification of 2,3-dihydroxybiphenyl (2,3-DHB)	Uchida et al. 2005
A. thaliana	*XplA; XplB*	Cytochrome P450 monooxygenase	*Rhodococcus rhodochorus*	Enhanced RDX degradation	Jackson et al. 2007
A. thaliana	*NfsA*	Nitroreductase	*E. coli*	Greater nitroreductase activity; 7–8 times higher uptake than wild plants	Kurumata et al. 2005
A. thaliana	*743B4, 73C1*	Glycosyltransferases (UGTs)	*A. thaliana*	Overexpression of UGTs: enhanced detoxification of TNT and root growth	Gandia-Herrero et al. 2008

Table 4. contd....

Table 4. contd.

Target Plant	Gene(s)	Enzymes	Source	Transgenic Effects	Reference
A. thaliana	*LAC1*	Root-specific laccase	Cotton	Laccase secreted to rhizosphere; enhanced resistance to phenolic alleleochemicals and 2, 4, 6-trichlorophenol	Wang et al. 2004
Alfalfa; Tobacco	*atzA*	Atrazine chlorohydrolase	Bacteria (*Pseudaminobacter* sp.)	Enhanced metabolism of atrazine	Wang et al. 2005
Brassica juncea	*γ-ECS, GS*	γ-Glutamycysteine synthetase; Glutathione synthetase	*Brassica juncea*	Overexpression of ECS and GS resulted in enhanced tolerance to atrazine, 1-chloro-2,4-dinitroben zene, phenan-threne, metolachlor	Flocco et al. 2004
Lycopersicon esculentum	*pxl*	Peroxidases (Px)	Roots of *L. esculentum*	Overexpression of *tpxI* gene in transgenic tomato hairy roots: enhanced phenol removal	Oller et al. 2005
Populus trichocarpa	*γ-ECS*	γ-Glutamycysteine synthetase	Poplar	Overexpression of Γ-ECS: increased tolerance to chloro-acetanilide herbicides	Gullner et al. 2001
Hybrid aspen *P. tremula* x *P. remuloides*	*pnrA*	Nitroreductase	*Pseudomonas putida*	Transgenic aspen (hybrid) tolerates and takes up greater amounts of TNT from contaminated water and soil	Van Dillewijin et al. 2008
Hybrid poplar *Populus tremula* x *Populous alba*)	*CYP450 2E1*	Cytochrome P450 Monooxygenase	Rabbit	Increased removal of TCE, vinyl chloride, carbon tetrachloride, benzene, and chloroform from hydroponic solution and air	Doty et al. 2007

Glossary

Terms Useful in Transgenic Engineering

Abiotic stress
The effect of nonliving factors (drought, cold, flooding, ozone, pollutants, intense light, nutrient imbalance, etc.) that can harm living organisms.

Agrobacterium
A genus of bacteria that includes several plant pathogenic species, causing tumor-like symptoms: crown gall, hairy root culture, etc.

Agrobacterium tumefaciens
A rod-shaped flagellated bacterium that causes crown galls in some plants. It infects a wound and injects a short stretch of DNA into some cells around the wound. The DNA is from a large plasmid. This Ti (tumor induction) plasmid is transferred to the plant cell, where it causes the cell to grow into a tumor-like structure which synthesizes specific opines that only the pathogen can metabolize.

Allele
One of a pair, or series, of variant forms of a gene that occur at a given locus in a chromosome. In a diploid cell there are two alleles of every gene (one inherited from each parent, although they could be identical). Within a population, there may be many alleles of a gene.

Allergen
A substance, usually a protein that can cause or provoke an immune response causing an allergy or allergic reaction in the body.

Allergy
A reaction by the body's immune system after exposure to a particular substance, often a protein.

Allopolyploid
A polyploid having multiple sets of chromosomes derived from different species. Hybrids are sterile as they do not have sets of homologous chromosomes and so pairing cannot take place. If doubling of the chromosome number occurs in a hybrid derived from two diploid (2n) species, the resulting tetraploid (4n) is a fertile plant, because it contains two sets of homologous chromosomes and pairing may occur. This tetraploid is an allotetraploid.

Amplified Fragment Length Polymorphism (AFLP)
A type of DNA marker, generated by digestion of genomic DNA with two restriction enzymes to create many DNA fragments, ligation of specific sequences of DNA (called adaptors) to the ends of these fragments, and PCR amplification of restriction endonuclease treated DNA. A small proportion of restriction fragments

	is amplified in any one reaction, so that AFLP profiles can be visualized by gel electrophoresis.
Annealing	The process of heating (de-naturing) and slowly cooling (re-naturing) double-stranded DNA to allow the formation of hybrid DNA or complementary strands of DNA or of DNA and RNA via hydrogen bonding.
Antibiotic resistance	Genes (usually of bacterial origin) used as selection markers in transgenesis **marker genes.** They code for antibiotic resistance used in genetic modification. They allow the survival of cells in the presence of normally toxic antibiotic agents.
Antisense DNA	One of the two strands of double-stranded DNA, usually that which is complementary ('anti') to all or part of the mRNA, i.e., the non-transcribed strand.
Antisense gene	A gene that produces a transcript (mRNA) that is complementary to the pre-mRNA or mRNA of a normal gene (usually constructed by inverting the coding region relative to the promoter).
Antisense RNA (asRNA)	A single-stranded RNA that is complementary to an mRNA strand transcribed within a cell.
Arabidopsis	A genus of flowering plants in the Cruciferae. *A. thaliana* is used as a model plant because it has a small, fully sequenced genome (5 pairs of chromosomes: $2n = 10$), can be cultured and transformed easily. It has a rapid generation time of two months.
Associative genomics	The branch of genomics that searches for mutations in populations by linkage disequilibrium analysis as also by direct assessment of association between alleles and phenotypes.
Autopolyploid	A polyploid that has multiple and identical or nearly identical sets of chromosomes (genomes), all derived from the same or nearly from the same progenitor.
***Bacillus thuringiensis* (*Bt.*)**	A bacterium that produces a toxin against certain insects, particularly Coleoptera and Lepidoptera; some of the toxin genes are important for transgenic approaches to crop protection.
Backcrossing (BC)	Crossing of a hybrid with one of its parents or an individual genetically similar to its parent. An F_1 hybrid crossed with one of its parents produces a BC_1 hybrid. In the next generation, BC_1 hybrid is crossed back to the same parent to produce a BC_2 hybrid. It is used to incorporate specific traits into elite lines and transfer of transgenes from transformed lines to parental lines.
Barnase	A bacterial ribonuclease which, when transformed into plants and expressed in the anthers, generates a male sterile phenotype. It is used for F_1 hybrid seed production; relies on the ability to genetically sterilize genotypes to ensure that all seed borne on the plant are the result of outcrossing. The sterility phenotype is suppressed by the **barstar** protein, which is used to reverse the sterility.

Base pair (bp) The arrangement of two nucleotides binding together across the double helix. The two strands that constitute DNA are held together by specific hydrogen bonding between a purine and a pyrimidine, one from each strand. The length of a nucleic acid molecule is expressed as the number of base pairs it contains: thousand or kilo base pairs: Kb; million base pairs: Mb.

B-chromosome Supernumerary, also called, accessory chromosome; a major source of intra-specific variation in nuclear DNA amounts in numerous plant species.

β-lactamase (beta-lactamase) An enzyme that detoxifies penicillin group antibiotics, such as ampicillin. The β-lactamase gene is commonly used as a selectable marker for transformation, where only transformed cells are able to tolerate the presence of ampicillin.

β-galactosidase An enzyme that catalyzes the formation of glucose and galactose from lactose.

Binary vector A two-plasmid system in *A. tumefaciens* for transferring into plant cells a segment of T-DNA that carries cloned genes. One plasmid contains the virulence gene (responsible for transfer of the T-DNA), and another contains the T-DNA borders, the selectable marker, and the DNA to be transferred.

Biolistic method A technique to generate transgenic cells, in which DNA-coated small metal (tungsten or gold) particles (of a fraction of a μm) are propelled across by various means fast enough to puncture target cells. It is used to transform animal, plant and fungal cells, and even mitochondria inside cells. *Synonym*: microprojectile bombardment.

Biopharming The use of genetically transformed crop plants and livestock animals to produce compounds, especially pharmaceuticals. *Synonym*: molecular pharming.

Biopiracy The appropriation, patenting, and commercialization of genetic resources, collections, and traditional knowledge of rural and indigenous people of another country without their consent.

Bioremediation A process that uses living organisms to remove contaminants, pollutants or unwanted substances from soil or water.

Biotic stress Stress resulting from living organisms which can harm plants, such as herbicides, weeds, insects, and pathogens.

Biotransformation The conversion of one chemical or material into another, by using a biocatalyst, usually an enzyme, or a fixed whole, dead microorganism, which contains an enzyme or several enzymes.

Blot **As a verb:** Transfer of DNA, RNA or protein to an immobilizing matrix. **As a noun:** Usually refers to the autoradiograph produced during the Southern, Northern, and Western blotting procedures.

Bt crop Crop that is genetically engineered to carry a gene from the soil bacterium *Bacillus thuringiensis* (*Bt*).

The bacterium is toxic to some insects and non-toxic to humans and other mammals.

Carotenoid
A group of chemically red to yellow pigments responsible for the characteristic color of many plant organs or fruits, such as tomato, carrot, etc. Oxygen-containing carotenoids are called **xanthophylls**. Carotenoids serve as light-harvesting molecules in photosynthetic assemblies and also play a role in protecting prokaryotes from the deleterious effects of light.

Cauliflower mosaic virus (CaMV)
A DNA virus affecting cauliflower and many other dicot species. Its importance is due to the promoter of its 35S ribosomal DNA, which is constitutively active in most plant tissues; the widely used promoter for the expression of transgenes.

cDNA (complementary DNA)
A DNA strand synthesized *in vitro* from a mature RNA template using reverse transcriptase. DNA polymerase is used to create a double-stranded molecule. It differs from genomic DNA by the absence of introns.

cpDNA (Chloroplast DNA)
The DNA present in the chloroplast. Although the chloroplast has a small genome, the large number of chloroplasts per cell ensures that chloroplast DNA is a significant proportion of the total DNA in a plant cell.

Chloroplast transit peptide (CTP)
A transit peptide that, when fused to a protein, acts to transport that protein into plant chloroplasts. Once inside, the transit peptide is cleaved off the protein; used to target transgene expression to the chloroplast.

Cisgenic transformation
A variant of native-gene transfer method; the inserted gene is unchanged with its own introns and regulatory sequences in the normal sense orientation. In this, the recipient plant is modified with a natural gene from a crossable-sexually compatible-plant.

Co-integrate vector
A plant transformation vector containing both the T-DNA and the virulence in a singular circular molecule. In this, the sequences to be transferred to the plant genome reside on the same plasmid as the *Vir* genes. One plasmid is engineered to carry a T-DNA segment incorporating the gene(s) to be introduced into *A. tumefaciens*.

Comparative genomics
The study of the relationship of genome structure and function across different biological species or strains.

Conjugation
1. Attachment of sugar and other polar molecules to less polar compounds, thus making them water-soluble. 2. The unidirectional transfer of plasmid DNA from one bacterium cell to another, involving cell-to-cell contact. The plasmid usually encodes the majority of the functions for its own transfer. 3. Direct transfer of gene material between two bacterial cells.

Constitutive promoter
An unregulated promoter that allows for continuous transcription of its associated gene. It is required to ensure that a specific gene transferred into a plant will be functional in all plant tissues.

Construct (DNA construct)
An engineered chimeric DNA designed to be transferred into a cell or tissue. Typically, the construct comprises

the gene or genes of interest, a marker gene, a promoter, and appropriate control sequences as a single package. A repeatedly-used construct may be called a **cassette**.

Co-transformation
The transformation process: multiple genes are introduced simultaneously followed by integration in the plant genome. The genes are either present on the same plasmid or separate plasmids. When two plasmids are used, one carries a selectable marker, and the other the transgene. The transformed cells will have incorporated both plasmids, possibly at different genomic loci.

Cross resistance (herbicide)
The phenomenon of a plant population developing simultaneous resistance to more than one class of herbicides with similar mechanisms and sites of action.

Crown gall
A tumorous growth at the base of certain plants characteristic of infection by *A. tumefaciens*. The gall is induced by the transformation of the plant cell by portions of the Ti plasmid.

Comparative genomics
Comparative genomics is the study of the relationship of genome structure and function across different biological species or strains. It uses information from different species and assists in understanding gene organization and expression and evolutionary differences.

Cry proteins
A class of crystalline proteins produced by strains of *B. thuringiensis* and engineered into crop plants to give resistance against insect pests. These proteins are toxic to certain categories of insects; harmless to mammals and most beneficial insects (**Synonym: δ-endotoxins**).

Cytochrome P450
A highly diversified set (more than 11,500) of heme-containing proteins that exist in mammalian, bacterial, fungal, and plant systems. They are involved in different types of monooxygenation and hydroxylation reactions in the primary and secondary metabolism of many natural and xenobiotic compounds.

Cytoplasmic male sterility
Total or partial male sterility in plants caused by specific nuclear and mitochondrial interactions. Genetic defect is due to faulty functioning of mitochondria in pollen development, preventing the formation of viable pollen. Commonly found or inducible in many plant species and exploited for some F_1 hybrid seed programs.

Cytosol
The fluid portion of the cytoplasm, i.e., the cytoplasm minus its organelles.

Dalton (Da)
Unit of mass roughly equivalent to the mass of a hydrogen atom 1.67×10^{-24} g. Used in shorthand expressions of mol. wt. such as kilo-daltons (kDa) or mega-daltons (MD), which are equal to respectively to 1×10^3 and 1×10^6 daltons.

Denature
To disrupt the normal *in vivo* conformation of a nucleic acid (more usually) a protein by physical or chemical means, accompanied by the loss of activity.

Denatured DNA
Double-stranded DNA converted to single strands by breaking the hydrogen bonds linking complementary nucleotide pairs; achieved by heating; often reversible.

De-repression
The process of "turning on" the expression of a gene or set of genes whose expression has been repressed (turned off), usually by the displacement of a repressor protein from a promoter region of DNA. When attached to the DNA, the repressor protein prevents RNA polymerase from initiating transcription ("turning on" of a gene).

Digest
To treat DNA molecules with one or more restriction endonucleases in order to cleave them into smaller fragments.

Diploid
1. The status of having two complete sets of chromosomes, most commonly one set of paternal origin and the other of maternal origin; 2. An organism or cell with a double set of chromosomes (one each of paternal and maternal origin) or referring to an individual containing a double set of chromosomes per cell.

DNA Barcode
Refers to a sequence-based identification system that may be constructed of one locus or a combination of two or more loci used together as a complementary unit. It provides quick taxonomic identification of a plant species without involving the morphological cues; based on a standard short genomic region.

DNA fingerprinting
The unique pattern of DNA fragments identified originally by Southern hybridization (using a probe that binds to a polymorphic region of DNA) or by polymerase chain reaction (PCR).

DNA ligase
An enzyme that catalyzes a reaction to link two separate DNA molecules via the formation of a phosphodiester bond between the 3'-hydroxyl end of one and the 5'-phosphate of the other. Its natural role lies in DNA repair and replication. An essential tool in recombinant DNA technology, as it enables the incorporation of foreign DNA into vectors.

DNA microarray
A large set of cloned DNA molecules immobilized as a compact and orderly pattern of sub-microliter spots onto a solid matrix (typically a glass slide). It is used to analyze patterns of gene expression, presence of markers, or nucleotide sequence. The major advantage is that it enables large numbers of individuals to be simultaneously genotyped at many loci.

DNA sequencing
Procedure to determine the nucleotide sequence of a DNA fragment. It is done by two common methods: Maxam-Gilbert and Sanger Sequencing. In both, the DNA fragments are separated according to length by polyacrylamide gel electrophoresis, enabling the sequence to be read directly from the gel.

DNA probe
A labeled segment of DNA that is able, after a DNA hybridization reaction, to detect a specific DNA sequence in a mixture of sequences. If the tagged

	sequence is complementary to anyone in the mixture, the two sequences will form a double helix. This will be identified by its label (either radioactive or fluorescent).
Double-stranded DNA (dsDNA) **dscDNA**	Two complementary strands of DNA annealed in the form of a double helix. A double-stranded complementary DNA: A double-strand DNA molecule created from a cDNA template.
Ecosystem	The complex of a living community and its environment, functioning as an ecological unit in nature.
Electrophoresis	A technique that separates charged molecules, such as DNA, RNA or protein, on the basis of relative migration in an appropriate matrix (such as agarose gel or polyacrylamide gel) subjected to an electric field. Examples: agarose gel electrophoresis; polyacrylamide gel electrophoresis (PAGE); pulsed-field gel electrophoresis (PFGE).
Electroporation	1. An electrical treatment of cells that induces transient pores, through which DNA can enter the cell. 2. The introduction of DNA or RNA into protoplasts or other cells by the momentary disruption of the cell membrane through exposure to an intense electric field.
ELISA	A test used to detect the presence of the protein encoded by the transgene in a sample of plant tissue. It is ideal for quantitative and qualitative detection of many types of proteins by using antigen-antibody reaction.
Event (Transgenic)	The original transformed plant and its progeny that contain a heterologous DNA construct as also the regenerated population resulting from the insertion of the transgene into the genome of the plant. Also refers to the progeny produced by sexual backcrosses between a donor inbred line (the original transformed line and the progeny) comprising the insert DNA and the adjacent flanking genomic sequences and a recipient inbred line (or recurrent line) that does not.
Expression vector	Also known as expression construct, it is used to introduce a specific gene into a target cell. The vector is usually a plasmid or virus designed for protein expression in cells. The plasmid is engineered to contain regulatory sequences that act as enhancer and promoter regions and lead to efficient transcription of the gene carried on the expression vector.
Fitness	It is the ability of the organism to survive and produce offspring in a given environment. It is the central idea in evolutionary theory.
Functional genomics	The study of genes, function, and regulation, and assigning of functions to each and every gene identified through structural genomics.
Gene	The molecular unit of heredity of a living organism transmitted from one generation to another during sexual or asexual reproduction. Genes hold information to build and maintain the cells of an organism and pass genetic traits to offspring.

Gene expression	The process by which a gene produces mRNA and protein, and hence exerts its effects on the phenotype of an organism.
Gene flow	Also known as gene migration, it is the transfer of genetic material or alleles from one plant to another and from one site to another.
Gene promoter (Promoter)	A region of DNA that initiates transcription of a particular gene by serving as a sort of "On" switch. It is a DNA sequence just upstream from a gene that acts as a binding site for transcription factors and RNA polymerase II. It also serves as an "On/Off" switch to control when and where in the plant the gene will be expressed.
Gene/Trait Stacking	The process of combining two or more genes of interest into a single plant. Also called 'Gene pyramiding'. A transgenic line or hybrid plant expressing both herbicide resistance and insecticide resistance genes/traits derived from two parent plants is termed 'pyramided stack' or 'stack'.
Genetic code	The set of rules by which information is encoded within genetic material (DNA or mRNA sequences) is translated into proteins by living cells. Also called triplet code, it contains 64 nucleotide triplets (codons) specifying the 20 amino acids and termination/stop codons (UAA, UAG, UGA). It specifies the correspondence during protein translation between codons and amino acids.
Genetic engineering	The technique of removing, modifying, or adding genes to a DNA molecule to change the information it contains. By changing this information, genetic engineering changes the type or amount of proteins an organism is capable of producing, enabling it to make new substances or perform new functions.
Genetically modified organism (GMO)	An organism that has been modified by the application of recombinant DNA technology.
Genetic marker (Marker gene)	A DNA sequence used to "mark" or track a particular location (locus) on a particular chromosome.
Genetic recombination	The process of producing transgenic plants by genetic engineering; the breaking and rejoining of DNA strands to form new molecules of DNA encoding a novel set of genetic information.
Genetic transformation	The transfer of extracellular DNA among and between species by using bacterial or viral vectors.
Genome	1. The entire complement of genetic material (genes plus non-coding sequences of the DNA/RNA) present in each cell of an organism, or in a virus or organelle to build and maintain a living example of that organism. 2. A complete set of chromosomes (hence of genes) inherited as a (haploid) unit from one parent. Genome size is generally given as its total number of base pairs (bp).
Genomics	The study of the organization, evolution, and function of the genes and non-coding regions of the genome. Helps in identifying all the genes in a plant as well as genetic

	properties and networks that contribute to the development of a superior plant.
Genome sequencing	Determination of the precise order of nucleotide bases (A, G, C, and T) in a DNA strand and illustration of how the genes are encoded within the genome. Also called DNA sequencing.
Genotype	The genetic identity of an individual, often evident by outward characteristics. It also indicates: 1. The genetic constitution (gene makeup) of an organism; 2.The pair of alleles at a particular locus, e.g., *Aa* or *aa*; 3. The sum total of all pairs of alleles at all loci that contribute to the expression of a quantitative trait. Also known as DNA fingerprinting of plants.
Genotyping (plant)	Also known as DNA fingerprinting of plants, it is used to identify genetic diversity within a breeding population. Plant genotype analysis is used for the identification of plants in commerce, plant breeding, and research.
Glutathione S-transferases (GSTs)	The dimeric, multifunctional plant enzymes involved in the cellular detoxification and excretion of many physiological and endogenous substances. Their major function is detoxification of a wide variety of hydrophobic, electrophilic, and usually cytotoxic compounds by catalyzing their conjugation with the tripeptide glutathione commonly found in the reduced form.
Gluten	A mixture of two proteins, gliadin and glutenin, occurring in the endosperm of cereal (particularly wheat) grain. Their amino acid composition varies, but glutamic acid (33 percent) and proline (12 percent) predominate.
Glycosylation	The covalent addition of sugar or sugar-related molecules to proteins or polynucleotides.
Glycoprotein	A protein molecule modified by the addition of one or several oligosaccharide groups.
Haploid	A cell or organism containing only one representative from each of the pairs of homologous chromosomes found in the normal diploid cell.
Heat shock protein (HSP)	A class of protein chaperones which are typically over-expressed as a response to heat stress. Two such proteins, HSP 90 and HSP 70, have a role in ensuring that crucial proteins are folded into the correct conformation.
Heterosis (Hybrid vigor)	The phenomenon of increased physiological vigor of crosses between species, or between distantly related and dissimilar variants within a species, compared with parents. It is achieved by crossing highly inbred lines of crop plants.
Histones	A group of water-soluble chromatin proteins rich in basic amino acids, closely associated with DNA. They are involved in the coiling of DNA in chromosomes and in the regulation of gene activity. They guide the interactions between DNA and other proteins, and help control which parts of the DNA are transcribed.

Homologous recombination	Rearrangement of related DNA sequences on a different molecule by crossing over in a region of identical sequence.
Horizontal gene transfer	Transmission of DNA between species, involving close contact between the donor's DNA and the recipient, uptake of DNA by the recipient, and stable incorporation of the DNA into the genome of the recipient.
Hybrid	Offspring resulting from the mating of two genetically distinct individuals. It carries two different alleles of the same gene.
Hybridization	1. The process of forming a hybrid by cross pollination of plants (or by mating animals) of different types. 2. The production of offspring of genetically different parents, normally from sexual reproduction, but also asexually by the fusion of protoplasts or by transformation. 3. The pairing of two DNA strands, often from different sources, by hydrogen bonding between complementary nucleotides.
Hygromycin	An antibiotic used as selective agent in bacterial and transgenic plant cell gives a purple-black color when viewed under a light microscope.
Inbred line	The product of inbreeding, i.e., the mating of individuals that have ancestors in common; it refers to populations resulting from at least six generations of selfing or 20 generations of brother-sister mating, so that they have become, for all practical purposes, completely homozygous.
Inducer	A low-molecular-weight compound or a physical agent that is bound by a repressor so as to produce a complex that can no longer bind to the operator; thus the presence of the inducer turns on the expression of the gene(s) controlled by the operator.
Inducible promoter	The activation of a promoter in response to either the presence of a particular compound, i.e., the inducer, or to a defined external condition, e.g., elevated temperature. A promoter activated by exogenous (external) factors such as biotic (herbicide, insect, pathogen) and abiotic (heat, light, water, salinity, chemicals, etc.) stresses. It can also be tissue or development stage specific.
Integrating vector	A vector that is designed to integrate cloned DNA into the host's chromosomal DNA.
Intellectual property (IP)	Refers to the legal protection to creations of the mind, wherein the rights of the inventor are recognized.
Inter-simple sequence repeat	A PCR-based genotyping technique that uses variation found in the genome region **(ISSR)** between microsatellite loci. It involves amplification of DNA segments present at an amplifiable distance in between two identical microsatellite repeat regions oriented in opposite direction.
Intragenic transformation	The transformation process in which the inserted DNA can be a new combination of DNA fragments from the

same species or from a cross-compatible species in a sense or antisense orientation. It enables using plant-derived transfer (P-) DNAs that consist of only native genetic elements, without affecting the overall structure of the plant's genome.

Introgression **(Introgressive hybridization)**	The introduction/movement of new alleles or gene(s) into a population from an exotic source, usually another species. This is achieved by repeated backcrossing of an initial interspecific hybrid with one of its parent species in order to eliminate all genetic changes except for the desired new gene(s).
Intron (Intervening sequence)	A nucleotide sequence within a gene that is removed by RNA splicing in a process called spliceosome (except for self-splicing introns) while the final mature RNA product of a gene is being generated. It refers to both the DNA sequence within a gene and the corresponding sequence in RNA transcripts.
Isozyme	A variant of a particular enzyme. In general, all the isozymes of a particular enzyme have the same function and sometimes the same activity, but they differ in amino-acid sequence.
Kanamycin	An antibiotic of the aminoglycoside family that inhibits translation by binding to the ribosomes. Important as a substrate for selection of plant transformants.
Kilobase (kb)	A length unit equal to 1000 base pairs of a double-stranded nucleic acid molecule. One Kb of double-stranded DNA has a mass of about 660 kilodalton (kDa).
Landrace	In plant genetic resources, an early, cultivated form of a crop species, evolved from a wild population, and generally composed of a heterogeneous mixture of genotypes.
Ligase	The process of joining two or more DNA fragments.
Lyase	Any of a class of enzymes that catalyze either the cleavage of a double bond and the addition of new groups to a substrate, or the formation of a double bond.
Marker gene (Genetic marker)	A gene required to identify and 'mark' the introduced genes, and finally to enable the selective growth of transformed cells. It helps to determine if a nucleic acid sequence has been successfully inserted into an organism's DNA.
Marker-assisted selection (MAS)	The use of DNA markers to improve response to selection in a population. The markers will be closely linked to one or more target loci, which may often be quantitative trait loci.
Maxam Gilbert sequencing	Also called chemical sequencing, it uses chemicals to break the DNA into fragments at specific bases; in both, the DNA fragments are separated according to length by polyacrylamide gel electrophoresis, enabling the sequence to be read directly from the gel.
Megabase (Mb)	A length of DNA consisting of 10^6 base pairs (if double-stranded) or 10^6 bases (if single-stranded). 1 Mb = 10^3 kb = 10^6 bp.

Metabolomics	The large-scale study of the full complement of secondary metabolites produced by a given species in all its tissues and growth stages.
Michaelis constant (K_m)	A measure of how efficiently an enzyme converts a substrate into product: has a theoretical upper limit of 10^8–10^{10}/M.s; enzymes working close to this, such as fumarases, are termed super-efficient.
Microarray	Hybridization between two single stranded DNA strands. **See DNA microarray**.
Microinjection	The injection of DNA into the nucleus of an oocyte, embryo, or other cell by injection through a very fine needle.
Microprojectile bombardment	A procedure for modifying cells by shooting DNA-coated metal (tungsten or gold) particles into them. **See 'Biolistics'**.
MicroRNA (miRNA)	It encodes noncoding RNAs that regulate gene expression by post-transcriptional or translational repression during plant development.
Microsatellite	A segment of DNA with a variable number of copies (5–50) of a sequence of around 5 or fewer bases (called a repeat unit). At any one locus (genomic site), there are usually several different "alleles" in a population, each allele identifiable according to the number of repeat units.
Minisatellite	Repeated segments of the same sequence of multiple triplet codons, each segment varying between 14 and 100 base pairs (14-100 bp); used for DNA fingerprinting following Southern hybridization. Generally concentrated at the ends of chromosomes and in regions with a high frequency of recombination.
Molecular assisted breeding	It involves applying tools of molecular biology, often molecular markers, sometimes called DNA or genetic markers, to track the makeup of plants during variety development process.
Molecular assisted selection	The first step in marker-assisted breeding. In this, a trait of interest is selected, not based on the trait itself, but on a marker linked to it.
Molecular genetics	The area of knowledge concerned with the genetic aspects of molecular biology, especially with DNA, RNA, and protein molecules.
Molecular marker	A fragment of DNA associated with a certain location within the genome. It may arise due to mutation or alteration in the genome loci. The fragment is either a short sequence, with a single base-pair change (single nucleotide polymorphism), or a long one, like mini-satellites.
Multiple resistance	Evolution of resistance in a weed or crop biotype to two or more herbicides with different mechanisms of action and resistance.
Mutagen	An agent or process capable of inducing mutations (e.g., irradiation, alkylating agents).

Mutagenesis	A process by which an organism is genetically changed, resulting in a mutation, which is a change in the DNA sequence of a gene. It may occur naturally, for example, due to natural exposure to a mutagen, i.e., a physical or chemical agent that raises the frequency above the spontaneous rate.
Mutation	A sudden, heritable change appearing in an individual as the result of a change in the structure of a gene (gene mutation); changes in the structure of chromosomes (chromosome mutation); or in the number of chromosomes (genome mutation).
Mutation breeding	Commonly used practice in plant breeding and other areas in which a chemical or radiation is applied to whole plant or cell to elicit changes in the organism's DNA. These will confer resistance to biotic and abiotic stresses, higher yield, improvement in quality, etc.
Negative cross resistance	The phenomenon of a plant resistant to one herbicide may develop increased sensitivity to other herbicide families due to changes in other physiological processes.
Nitroreductases	Enzymes which catalyze the reduction of nitro-substituted compounds using FMN or FAD as prosthetic groups and NADH or NADPH as reducing agents. They break down complex nitrogen-containing molecules to provide plants with a useable form of nitrogen.
Northern blot	A biological technique to study gene expression by detection of RNA (or isolated mRNA) in a sample. It involves the use of electrophoresis to separate RNA samples by size and detection with a hybridization probe complementary to part of or the entire target sequence.
Oligomer	A molecule formed by the covalent joining of a small (undefined) number of monomers.
Oligonucleotide	A nucleotide oligomer. A short molecule (usually 6 to 100 nucleotides) of single-stranded DNA.
Oligonucleotide ligation assay (OLA)	A gynotyping technique to determine the presence or absence of a specific nucleotide pair within a target gene, often indicating whether the gene is wild type (normal) or mutant (defective). In this, the target sequences immediately surrounding the variable position are simultaneously hybridized (annealed) with two adjacent oligonucleotide probes so that their junction occurs at the site of variation.
Oligosaccharide	Carbohydrate consisting of several linked monosaccharide units.
Operon	A functionally integrated genetic unit for the control of gene expression in bacteria. It consists of one or more genes that encode one or more polypeptide(s) and the adjacent site (promoter and operator) that controls their expression by regulating the transcription of the structural genes.

Open Reading Frame (ORF)	A sequence of nucleotides in a DNA molecule that has the potential to encode a peptide or protein: it starts with a start triplet (ATG), followed by a string of triplets, each of which encodes an amino acid, and ends with a stop triplet (TAA, TAG or TGA). This term is often used when, after the sequence of a DNA fragment has been determined, the function of the encoded protein is not known. The existence of open reading frames is usually inferred from the DNA (rather than the RNA) sequence.
Opine	The condensation product of an amino acid with either a keto-acid or a sugar produced by the plant host as a result of *Agrobacterium* infection, and used exclusively by the *Agrobacterium* as a carbon source for growth and reproduction within the plant. Opine synthesis is a unique characteristic of tumor cells.
Phenotype	The visible appearance of an individual or a plant (with respect to one or more traits), which reflects the reaction of a given genotype with a given environment.
Phytodegradation	Breakdown of contaminants taken up by plants through metabolic processes within the plant, or the breakdown of those surrounding the plant through the effect of compounds such as enzymes produced by the plants.
Phytoextraction (Phytoaccumulation)	The process of using plants or algae to remove contaminants from soils, sediments, or water into harvestable plant biomass.
Phytoremediation	The process by which green plants detoxify soils, sediments, and aquatic sites contaminated with organic and inorganic pollutants.
Phytostabilization	The process using certain plant species to immobilize the contaminant in the soil through absorption and accumulation by roots, adsorption onto roots, or by way of precipitation, complexation and metal valence reduction within the root zone.
Phytovolatilization	The uptake and transpiration of the contaminant by growing trees and plants which release the contaminant or its modified form into the atmosphere.
Plant-incorporated protectants (PIPs)	Formerly referred to as plant-pesticides, they are substances that act like insecticides (pesticides) produced and used by a plant to protect it from pests such as insects, viruses, and fungi.
Plasmid	A small extrachromosomal, autonomous circular DNA molecule found in certain bacteria, capable of independent replication. Plasmids can transfer genes between bacteria and are important tools of transformation in genetic engineering.
Plastid	A general term for a number of plant cell organelles which carry non-nuclear DNA; includes the pigment-carrying bodies: a) chloroplasts in leaves, b) chromoplasts in flowers, and c) the starch-synthesizing amyloplasts in seeds.

Plastoquinone	One of a group of compounds involved in the transport of electrons as part of the process of photosynthesis.
Pleiotropy	The phenomenon whereby a particular gene affects multiple phenotypic traits (Adj. pleiotropic).
Ploidy	The number of sets of chromosomes per cell, e.g., haploid, diploid, polyploidy.
Point mutation	A change in DNA at a specific site in a chromosome. Includes nucleotide substitutions and the insertion or deletion of one or a few nucleotide pairs.
Polyacrylamide gel	Often referred to incorrectly as acrylamide gel. These gels are made by cross-linking acrylamide N,N'-methylene-*bis*-acrylamide. They are used for the electrophoretic separation of proteins, DNA and RNA molecules.
Polymerase chain reaction (PCR)	A procedure that amplifies a particular DNA sequence. It involves multiple cycles of denaturation, annealing to oligonucleotide primers, and extension (polynucleotide synthesis), using a thermostable DNA polymerase, deoxyribo-nucleotides, and primer sequences in multiple cycles of denaturation-renaturation-DNA synthesis.
Polymerization	Chemical union of two or more molecules of the same kind such as glucose or nucleotides to form a new compound (starch or nucleic acid) having the same elements in the same proportions but a higher molecular weight and different physical properties.
Polymorphism	The occurrence of two or more alleles at a locus in a population; also known as genetic polymorphism.
Polyploid	Tissue or cells with more than two complete sets of chromosomes, resulting from chromosome replication without nuclear division or from union of gametes with different number of chromosome sets: triploid (3x), tetraploid (4x), pentaploid (5x), hexaploid (6x), heptaploid (7x), octoploid (8x).
Polysaccharide	Long-chain molecules, such as starch and cellulose, composed of multiple units of a monosaccharide.
Probe	1. For diagnostic tests, the agent that is used to detect the presence of a molecule in a sample. 2. A DNA or RNA sequence labeled or marked with a radioactive isotope or that is used to detect the presence of a complementary sequence by hybridization with a nucleic acid sample.
Probe DNA	A labeled DNA molecule used to detect complementary-sequence nucleic acid molecules by molecular hybridization. To localize the probe DNA sequence and reveal the complementary sequence, autoradiography or fluorescence is used.
Protoplast	A plant or bacterial cell for which the relatively rigid wall has been removed either chemically or enzymatically, leaving its cytoplasm enveloped by only a delicate peripheral membrane. Protoplasts are spherical and smaller than the elongate, angular shaped and often vacuolated cells from which they have been released. They are utilized for selection or hybridization at the cellular for a variety of other purposes.

Protoplast culture	Plant protoplasts isolated by mechanical means or by enzymatic digestion of plant tissues or organs, or cultures derived from these.
Random Amplified Polymorphic DNA (RAPD)	A technique that detects DNA segments that are amplified at random. It is done by creating several arbitrary, short primers (8–12 nucleotides), and then uses the PCR using a large template of genomic DNA, hoping that fragments will amplify. By resolving the resulting patterns, a semi-unique profile can be gleaned from a RAPD reaction.
Rapid Fragment Length Polymorphism (RFLP)	Technique involving fragmentation of a sample of DNA by a restriction enzyme which can recognize and cut DNA wherever a specific short sequence occurs, in a process known as a restriction digest.
Recombinant DNA (rDNA) technology	A laboratory gene-splicing procedure in which the DNA of the donor organism is cut into pieces using restriction enzymes followed by insertion of one of these fragments into the DNA of the host plant.
Recombination	The process by which progeny derive a combination of genes different from that of either parent.
Recombinant protein	A protein that results from the expression of recombinant DNA within living cells. Its amino acid sequence is encoded by a cloned gene.
Recombinant RNA	A term used to describe RNA molecules joined *in vitro* by T4 RNA ligase.
Reporter gene (Screenable marker gene)	A gene that is attached to a regulatory sequence of the gene of interest. It is used to confirm transformation, determine transformation efficiency, and monitor gene or protein activity. Examples: *GUS, gfp, luc,* and *lacZ* genes. **See Text and Acronyms**.
Repression	Inhibition of transcription by preventing RNA polymerase from binding to the transcription initiation site: a repressed gene is "turned off."
Repressor	A protein which binds to a specific DNA sequence (the operator) upstream from the transcription initiation site of a gene or operon and prevents RNA polymerase from commencing mRNA synthesis.
Resistance (Crop/weed to Herbicide)	The inherited ability of a plant species/biotype to survive and reproduce after exposure to a dose of herbicide normally lethal to the wild type. In both crop plants and weeds, herbicide resistance may be naturally occurring or induced by genetic engineering or selection of variants produced by tissue culture or mutagenesis.
Restriction endonuclease	A class of endonucleases that cleaves DNA after recognizing a specific sequence, e.g., *Bam*H1 (5′GGATCC3′), *Eco*RI (5′GAATTC3′), and *Hin*dIII (5′AAGCTT3′).
Restriction Fragment Length Polymorphism (RFLP)	The occurrence of variation in the length of DNA fragments that are produced after cleavage with a type II restriction endonuclease. The differences in DNA lengths are due to the presence or absence of recognition site(s)

for that particular restriction enzyme (type I, II, III, or IV) which can recognize and cut DNA wherever a specific short sequence occurs, in a process called restriction digest.

Rhizodegradation
The breakdown of contaminants in the rhizosphere through microbial activity that is enhanced by the presence of plant roots. Also called "Rhizoremediation."

Rhizofiltration
The process of adsorption or precipitation onto plant roots (or absorption into the roots) of contaminants in solution surrounding the root zone. Similar to phytoextraction, but plants are used to clean up contaminated groundwater rather than soil.

Ribulose biphosphate (RuBP)
A five-carbon (keto-pentose) sugar that is combined with carbon dioxide to form a six-carbon intermediate in the first stage of the dark reaction of photosynthesis.

Ri plasmid
A class of large conjugative plasmids found in the soil bacterium *Agrobacterium rhizogenes*. Ri plasmids are responsible for hairy root disease of certain plants. A segment of the Ri plasmid is found in the genome of tumor tissue from plants with hairy root disease.

RNA Interference (RNAi)
A double-stranded RNA (dsRNA) that inhibits gene expression, typically by causing the destruction of specific mRNA molecules.

Sanger Sequencing
Called the di-deoxy or chain-terminating method using DNA polymerase to make new DNA chains, with di-deoxy nucleotides (chain terminators) to stop the chain randomly as it grows.

Selectable marker gene
A gene that protects the organism from a *selective agent* that would normally kill or prevents its growth. It usually encodes resistance to an antibiotic added to a gene construct to allow easy selection of cells that contain the construct from the large majority of cells that do not. The more widely used marker genes: *npt*II, *hpt*II, and *bar*. **See Text and Acronyms.**

Selection pressure
An interaction between natural variation in a species and factors in its environment that cause a certain plant to have an advantage over the others. Its effectiveness is measured in terms of differential survival and reproduction, and consequently in change in the frequency of alleles in a population.

Sense RNA
A primary transcript (RNA) containing a coding region (contiguous sequence of codons) that is translated to produce a polypeptide.

Sequencing
The determination of the order of nucleotides in a DNA or RNA molecule, or that of amino acids in a polypeptide chain.

Single-nucleotide Polymorphism (SNP)
A DNA sequence variation occurring when a single nucleotide (A, T, C, or G) in the genome (or other shared sequence) differs between members of a biological species. SNP represents a single nucleotide difference between two individuals at a defined location. For

example, two sequenced DNA fragments from different species, AAGCCTA to AAGCTTA, contain a difference in a single nucleotide.

Single resistance
Confinement of resistance of a weed species to only one herbicide or one with single site of action.

Site-specific mutagenesis
A technique to change one or more specific nucleotides within a cloned gene in order to create an altered form of a protein with one or more specific amino acid changes. Also known as oligonucleotide-directed mutagenesis and oligonucleotide-directed site-specific mutagenesis.

Southern blot
A method used in detection of a specific DNA sequence in DNA samples. It combines transfer of electrophoresis-separated DNA fragments to a filter membrane and subsequent fragment detection by probe hybridization.

Statistical genomics
The branch of genomics which determines the size of the genome and the number of genes present in the entire genome by sequencing individual genes, gene segments, or entire genomes.

Substantial equivalence
The concept in which a transgenic food is substantially equivalent in composition and nutritional characteristics to an existing food.

***Taq* polymerase**
A heat-stable DNA polymerase isolated from the thermophilic bacterium *Thermus aquaticus,* and used in PCR. *See* polymerase.

Telomerase
A specialized reverse transcription enzyme. It is a ribonucleoprotein that maintains telomere ends by the addition of the telomere repeat TTTAGGG, as in *Arabidopsis thaliana* (5′ to 3′ toward the end). It carries an RNA template from which it synthesizes DNA repeating sequence, or "junk" DNA.

Terminator (of transcription)
1. A DNA sequence just downstream of the coding segment of a gene, which is recognized by RNA polymerase as a signal to stop synthesizing mRNA. 2. A name given to antisense DNA inserted in plants to make impossible the use of a second generation of seed by a farmer.

Terminator region
A DNA sequence that signals the end of transcription.

Tetraploid
An organism whose cells contain four haploid ($4x$) sets of chromosomes.

T-GURT
The technology that contains "a gene switch" which is inducible by external application of a chemical inducer and which controls expression of a gene product which, in turn, affects expression of a second gene in the genome. It regulates of the expression of a specific transgenic trait in plants while enabling plants to remain fertile and set viable seeds.

Ti plasmid
Tumor-inducing plasmid. A giant plasmid of *Agrobacterium tumefaciens* that is responsible for the induction of tumors in infected plants. Ti plasmids are used as vectors to introduce foreign DNA into plant cells. **See vector**.

Tolerance **(Crop/weed to Herbicide)**	The inherent ability of a species to survive and reproduce after herbicide treatment. It implies that there is no selection or genetic manipulation to make the plant tolerant; it is naturally tolerant.
Transcription factor (TF)	A master regulator of cellular processes, it coordinates the functions of groups of genes and performs these functions alone or with other proteins in a complex, by promoting (as an activator), or blocking (as a repressor) the recruitment of RNA polymerase.
Transduction	The process of injecting foreign DNA by a bacteriophage virus into the host bacterium. In this, the desired genetic material is first inserted into a suitable virus, and this modified virus is allowed to infect the plant. Also called viral transformation.
Transfection	The process of deliberately introducing DNA by non-viral, direct methods into eukaryotic cells. In this, a small portion of which becomes covalently associated with the host cell DNA. It can be two types: transient and stable.
Transformation	The process of moving genes from one species to another. It is also the process used to describe the insertion of new genetic material into nonbacterial cells, including animal and plant cells.
Transformation efficiency	The number of cells that take up foreign DNA as a function of the amount of added DNA during a transformation process; expressed as transformants per μg of added DNA.
Transgene	A gene from one genome that has been incorporated into the genome of another organism; often refers to a gene introduced into a multicellular organism.
Transgenesis	The process of introducing an exogenous gene (transgene) into a living organism so that it will exhibit a new property and transmit that gene and property to its offspring in successive generations.
Transgenic	An organism in which a foreign gene (a transgene) is incorporated into its genome. This organism contains genes from an unrelated organism. Taking genes from one species and inserting them into another species to get that trait expressed in the offspring.
Transgenic mitigation (TM)	A concept in which mitigator genes are tandemly linked to the desired primary transgene, which would reduce the fitness of hybrids and their rare progeny, considerably reducing risk.
Transposon	A transposable or movable genetic element. A relatively small DNA segment that has the ability to move (mobile genetic element) from one chromosomal position to another. An example is 'Tn 5', a bacterial transposon that carries the genes for resistance to the antibiotics neomycin and kanamycin and the genetic information for insertion and excision of the transposon.
Vector	A plasmid containing DNA from the donor. It is a vehicle to transfer the genetic material such as DNA sequences

from the donor organism to the target cell of the recipient organism.

Vertical transmission   Inheritance of a gene from parent to offspring.

V-GURT   The varietal genetic use restriction technology which uses, in its design, complex inducible gene expression systems to make plants produce seeds sterile at harvest under specific conditions.

vir **genes**   A set of genes on a Ti plasmid that prepare the T-DNA segment for transfer into a plant cell.

Western blot   Also called protein immunoblot, the technique that uses gel electrophoresis to separate native proteins by 3-D structure or denatured proteins by the length, molecular weight, and thus by type, of the polypeptide. The proteins are then transferred to a membrane (nitrocellulose or PVDF), where they are stained with antibodies specific to the target protein.

Whole-genome sequencing   Also known as full-genome, complete-genome, or entire-genome sequencing is a laboratory process that determines the complete DNA sequence of a plant's genome at a single time.

Index

Color Plate Section

Chapter 2

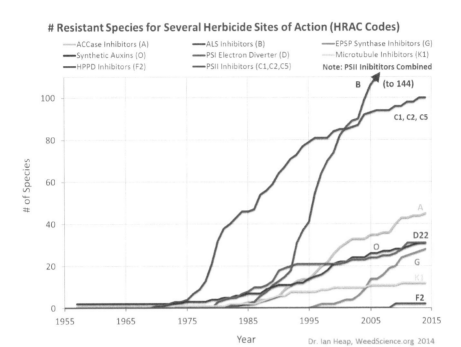

Fig. 2.3. Increase in weed species to herbicides of several sites of action beginning 1957 [Heap 2014].

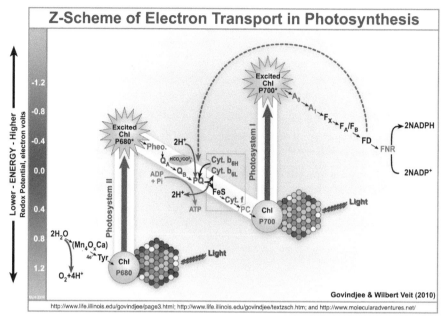

Fig. 2.4. Z-Scheme, an energy diagram for electron transfer in the 'light reactions' of plant photosynthesis. It shows the pathway of electron transfer from water to $NADP^+$. The process involves the oxygen-evolving complex, reduction potentials, and electron flow during photosynthesis, reduction of CO_2, and NADP phosphorylation. Light energy absorbed by special pair pigments, P680 of photosystem II (PS II) and P700 of photosystem I (PS I), drives electron flow uphill. Acronyms (from left to right of the diagram): Mn: a manganese complex containing 4 Mn atoms, bound to PS II reaction center; Tyr: a particular tyrosine in PS II; O_2: oxygen; H^+: protons; P680: the reaction center chlorophyll (Chl) in PS I, the primary electron donor of PS II; P680*: excited (Chl) P680 that has the energy of the photon of light; Pheo: pheophytin molecule (the primary electron acceptor of PSII {it is like a chlorophyll *a* molecule where magnesium (in its center) has been replaced by two "H"s}; Q_A (also called PQ_A): a plastoquinone molecule tightly bound to PS II; Q_B (PQ_B): another plastoquinone molecule that is loosely bound to PS II; FeS: Rieske Iron Sulfur protein; Cyt. f: Cytochrome f; $Cytb_6$ (Cyt. b_6L and Cyt. b_6 H): Cytochrome b_6 (of Low and High Energy); PC: copper protein plastocyanin; ChlP700: the reaction center chlorophyll (Chl: actually a dimer, i.e., two molecules together) of PSI—it is the primary electron donor of PSI; Excited ChlP700*: P700 that has the energy of the photon of light; Ao: a special chlorophyll a molecule (primary electron acceptor of PSI); A_1: a phylloquinone (Vitamin K) molecule; F_X, F_A, and F_B: three separate Iron Sulfur Centers; FD: ferredoxin; and FNR: Ferredoxin NADP oxido Reductase (FNR). Three major protein complexes are involved in running the "Z" scheme: (1) PS II; (2) Cytochrome bf complex (containing $Cytb_6$; FeS; and Cytf); and (3) PS I. The diagram does not show where and how ATP is made [Govindjee and Veit 2010].

Chapter 3

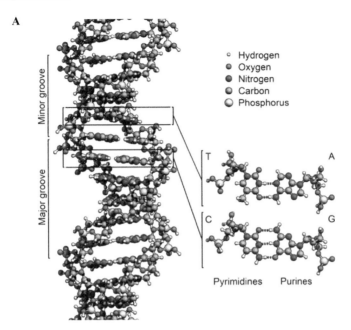

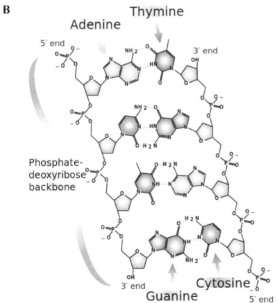

Fig. 3.1. A. The structure of the DNA double helix showing the four bases (adenine, cytosine, guanine, and thymine), and the location of the major and minor groove. The atoms of the structure are color-coded by element and the detailed structure of two base pairs is shown in the bottom right [Wikipedia Commons 2012]. B. The chemical structure of the DNA with colored label identifying the four bases and the phosphate and deoxyribose components of the backbone [Zephyris 2011].

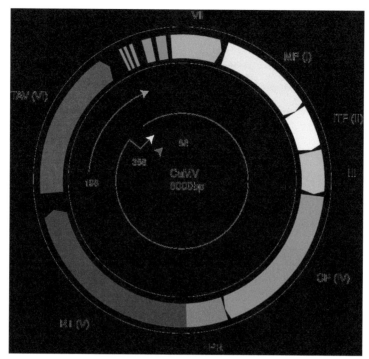

Fig. 3.2. Genomic CaMV with the 35S RNA, containing a highly structured 600 nucleotide long leader sequence with six to eight short open reading frames (ORFs) [Fütterer et al. 1988; Pooggin et al. 1998; Wikipedia Commons 2005; free use under The GNU Free Documentation License].

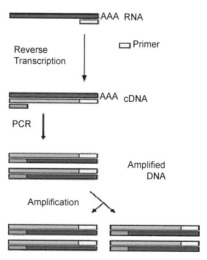

Fig. 3.3. Reverse transcription polymerase chain reaction [JPark623; Wikipedia Commons 2012].

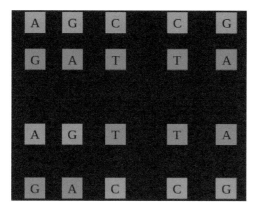

Fig. 3.4. Genetic recombination involving breakage and rejoining of parental DNA molecules (M, F), leading to new combinations of genes (C1, C2) on the chromosome that share DNA from both parents [Eccles 2006; Wikipedia Commons].

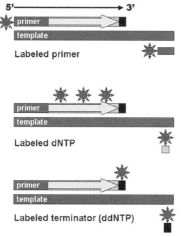

Fig. 3.5. DNA fragments are labeled with a radioactive or fluorescent tag on the primer (1), in the new DNA strand with a labeled dNTP, or with a labeled ddNTP [Lakdawalla 2007; Wikipedia Commons 2007].

Chapter 4

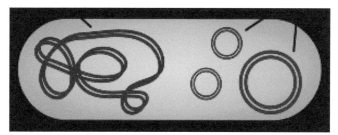

Fig. 4.1. The image showing a line drawing of a bacterium with its chromosomal DNA and several plasmids within it. The bacterium is drawn as a large oval. Within the bacterium, small to medium size circles illustrate the plasmids, and one long thin closed line that intersects itself repeatedly illustrates the chromosomal DNA [Spully 2007].

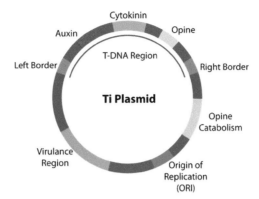

Fig. 4.2. Simplified two-vector system located in the same *Agrobacterium* strain having a T-region in one vector (A) and a *vir* region in another vector (B) [Patent Lens 2007].

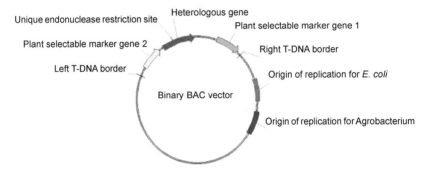

Fig. 4.3. Typical structure of a binary vector with key components [Patent Lens 2007].

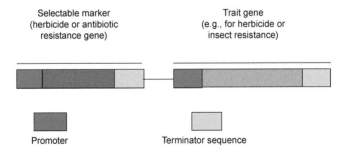

Fig. 4.4. Simplified presentation of a transgene construct that contains a gene (transgene), a promoter, and a terminator inserted into a plant's genome to perform the regulatory functions that lead to the production of a protein. The DNA sequence between the promoter and terminator determines the type of protein that is produced [Byrne 2014].

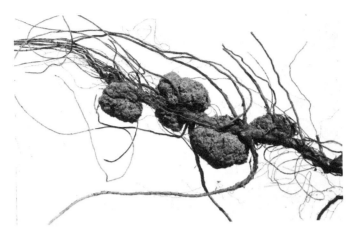

Fig. 4.5. *Agrobacterium tumefaciens*-induced galls at the root of *Carya illinoinensis* (pecan) [Clemson University-USDA Cooperative Extension Slide Series 2002].

Chapter 7

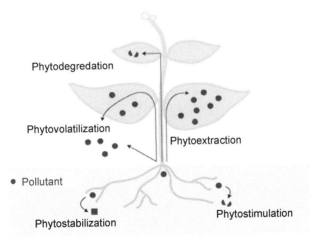

Fig. 7.1. Different processes of phytoremediation. The pollutant or contaminant (*red circles*) can be sequestered or degraded inside the plant tissue, extracted by the foliage, degraded or stabilized in the rhizosphere, or volatilized [Pilon Smits 2005].

Chapter 8

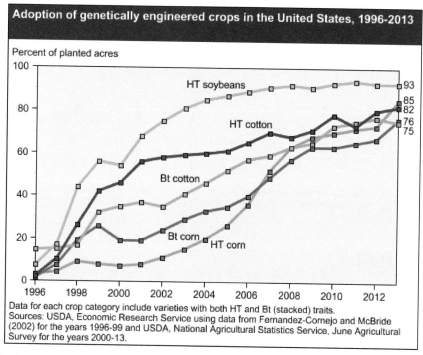

Fig. 8.1. Area under different herbicide-tolerant (HT) and *Bt* insect-tolerant transgenic crops and their adoption rates in USA during 1996–2013 [Fernandez-Cornejo 2013].